INTERNATIONAL UN
MONOGRAPHS ( ᷄RAPHY

INTERNATIONAL UNION OF CRYSTALLOGRAPHY
BOOK SERIES

This volume forms part of a series of books sponsored by the International Union of Crystallography (IUCr) and published by Oxford University Press. There are three IUCr series: IUCr Monographs on Crystallography, which are in-depth expositions of specialized topics in crystallography; IUCr Texts on Crystallography, which are more general works intended to make crystallographic insights available to a wider audience than the community of crystallographers themselves; and IUCr Crystallographic Symposia, which are essentially the edited proceedings of workshops or similar meetings supported by the IUCr.

IUCr Monographs on Crystallography

1 *Accurate molecular structures: Their determination and importance*
  A. Domenicano and I. Hargittai, *editors*
2 *P. P. Ewald and his dynamical theory of X-ray diffraction*
  D. W. J. Cruickshank, H. J. Juretschke, and N. Kato, *editors*
3 *Electron diffraction techniques, Volume 1*
  J. M. Cowley, *editor*
4 *Electron diffraction techniques, Volume 2*
  J. M. Cowley, *editor*
5 *The Rietveld method*
  R. A. Young, *editor*
6 *Introduction to crystallographic statistics*
  U. Shmueli and G. H. Weiss
7 *Crystallographic instrumentation*
  L. A. Aslanov, G. V. Fetisov and J. A. K. Howard
8 *Direct phasing in crystallography: Fundamentals and applications*
  C. Giacovazzo
9 *The weak hydrogen bond in structural chemistry and biology*
  G. R. Desiraju and T. Steiner
10 *Defect and microstructure analysis by diffraction*
  R. L. Snyder, J. Fiala and H. J. Bunge

IUCr Texts on Crystallography

1 *The solid state: From superconductors to superalloys*
  A. Guinier and R. Jullien, *translated by* W. J. Duffin
2 *Fundamentals of crystallography*
  C. Giacovazzo, *editor*
3 *The basics of crystallography and diffraction*
  C. Hammond

The Weak Hydrogen Bond

In Structural Chemistry and Biology

GAUTAM R. DESIRAJU

School of Chemistry
University of Hyderabad
desiraju@uohyd.ernet.in

and

THOMAS STEINER

Institut für Kristallographie
Freie Universität Berlin
steiner@chemie.fu-berlin.de

OXFORD
UNIVERSITY PRESS

OXFORD

UNIVERSITY PRESS

Great Clarendon Street, Oxford OX2 6DP

Oxford University Press is a department of the University of Oxford.
It furthers the University's objective of excellence in research, scholarship,
and education by publishing worldwide in

Oxford New York

Athens Auckland Bangkok Bogotá Buenos Aires Cape Town
Chennai Dar es Salaam Delhi Florence Hong Kong Istanbul Karachi
Kolkata Kuala Lumpur Madrid Melbourne Mexico City Mumbai Nairobi
Paris São Paulo Shanghai Singapore Taipei Tokyo Toronto Warsaw

with associated companies in Berlin Ibadan

Published in the United States
by Oxford University Press Inc., New York

First published 1999
First published in paperback 2001

A catalogue record for this book is available from the British Library

Library of Congress Cataloging in Publication Data

Desiraju, G. R. (Gautam R.)
The weak hydrogen bond in structural chemistry and
biology / Gautam R. Desiraju and Thomas Steiner.
(International Union of Crystallography monographs on crystallography; 9)
Includes bibliographical references and indexes.
1. Hydrogen bonding. I. Steiner, Thomas. II. Title. III. Series.
QD461.D44 1999 541.2′24 – dc21 99-20408

ISBN 0 19 850970 7 [pbk.]
Typeset by Best-set Typesetter Ltd., Hong Kong
Printed in Great Britain
on acid-free paper by
Bookcraft (Bath) Ltd,
Midsomer Norton, Somerset

Preface

Research in the area of hydrogen bonding is an evergreen endeavour and hydrogen bonds continue to manifest themselves in myriad ways in structural chemistry and biology. The weak hydrogen bond is a part of this extended domain and the facts that have accumulated concerning this particular interaction type have just about acquired a critical enough mass, that the writing of this book appeared to be both timely and useful. This project was commissioned in 1996 and even during the course of its execution, we were only too aware of the intense and rapidly growing interest in this newest vista of hydrogen bond research. Fully a quarter of all the citations here represent publications that appeared after we commenced our efforts.

As mentioned by Jeffrey and Saenger in their 1991 preface, the undertaking of any book in an area as oceanic as hydrogen bonding requires that the subject matter be contained. Yet, this is not easy in the present context because there is no general consensus as to what constitutes a *weak hydrogen bond*. Chemists and biologists differ a little in their perception as to what is weak and what not, and the use of the term *weak* also presupposes that one knows what is meant by the term *strong*. It is appropriate to state here that the title originally planned for this work was *The non-conventional hydrogen bond*. Indeed there are several merits to using this latter term and we have not abandoned it in this work, in the realization that the ambits of the *weak hydrogen bond* and the *non-conventional hydrogen bond* sometimes intersect and at others not. This dichotomy is revealing. In particular, the reader should appreciate that we are discussing an interaction type that defies rigid compartmentalization.

This book is divided into three sections. The first (Chapter 1) provides an introduction to the weak hydrogen bond in relation to hydrogen bonds in general and defines the scope of the work. The second and largest section (Chapters 2 and 3) deals with the development of the concept of the weak hydrogen bond. This has been done in two different ways. In Chapter 2, we have selected the C—H$\cdots$O bond as the prototype of the entire interaction type, and have analysed it in detail in order to justify its inclusion in the larger hydrogen bond family. In Chapter 3, we have then extended these arguments to a very wide range of hydrogen bond donors and acceptors. This includes weak acceptors such as π-systems and weak donors in which the H atom is covalently bonded to phosphorus, chalcogen and transition metal atoms. The third and final section is concerned with the ways in which weak hydrogen bonds may be employed in supramolecular chemistry and crystal engineering (Chapter 4) and how they influence biological structure and function (Chapter 5). A brief recapitulation (Chapter 6) concludes the work.

It is a pleasure to acknowledge the helpful and extended discussions we have had with several collaborators over the years. In particular we would like to mention Frank Allen, Roland Boese, Dario Braga, Jack Dunitz, Jenny Glusker, Fabrizia Grepioni, Judith Howard, Jan Kanters, Gertraud Koellner, Jan Kroon, Bert Lutz, Ashwini Nangia, Wolfram Saenger, K. Subramanian, Joel Sussman, Matthias Tamm and the late Nora Veldman.

We are grateful to a number of individuals for their contributions to the preparation of this manuscript. These include Srinivasan Kuduva and Venkat Thalladi who helped with some of the illustrations and Madhav Desiraju, Ram Jetti and P. O. Koshy who assisted with the references. Indeed, we wish to record our appreciation to the entire Hyderabad research group for their interest and cooperation. Some on-line searches of Chemical Abstracts were carried out at the National Chemical Laboratory, Pune and we thank the authorities there for making available this facility to us.

One of us (T. S.) worked on this book at the Freie Universität Berlin to which he is affiliated and for a more extended period at the Weizmann Institute of Science, Rehovot. A year long stay at the Department of Structural Biology of the latter institution, hosted graciously by Professor Joel L. Sussman, was made possible by a fellowship awarded by the Minerva Society, Munich. This stay provided the opportunity for writing substantial parts of the present manuscript. The library and other facilities offered there are gratefully acknowledged.

Many of our own papers in the area of weak hydrogen bonding have appeared in journals published by the Royal Society of Chemistry, and it is a particular pleasure for us to mention that one of us (T. S.) was a recipient of a travel grant from the RSC Journals Grants for International Authors. This grant enabled us to work together in Hyderabad when we were finalizing this manuscript.

We would like to thank the International Union of Crystallography, in particular the Book Series Committee and its Chairman, Professor Philip Coppens for their encouragement. This is a structural text and we are pleased therefore that it appears in a series sponsored by the IUCr.

This book could not have become a reality without the patience and support extended to us by our wives, Krishna and Traude.

The topic of weak hydrogen bonding is capable of being interpreted in varied ways and as the interaction under consideration becomes weaker, these interpretations become more diffuse and subjective. Separating experiment from instinct, fact from opinion, and logic from impulse is a constant and continuing challenge in this area. It is our hope that the reader will find enough in this work to further sieve and sift through these matters.

Hyderabad   G. R. D.
October 1998   T. S.

Contents

Acknowledgements

The authors would like to thank the following for permission to reproduce published material:

Academic Press
Journal of Molecular Biology: Figs 3.5, 5.3, 5.8, 5.9, 5.12, 5.17, 5.28, 5.38, 5.50, 5.68, 5.69
Biochemical and Biophysical Research Communications: Fig. 5.37
Academy of Sciences of the USA
Proceedings of the National Academy of Sciences of the USA: Fig. 5.51(a)
American Association for the Advancement of Science
Science: Fig. 5.24(b)
American Chemical Society
Accounts of Chemical Research: Fig. 2.38
Biochemistry: Fig. 5.14
Chemical Reviews: Figs 3.124(a), 3.138
Inorganic Chemistry: Figs 3.65, 3.72
Journal of the American Chemical Society: Figs 2.15, 2.22, 2.25, 2.31, 3.11(a), 3.60, 3.61, 3.62, 3.63, 3.68, 3.96, 3.133, 3.136, 4.12, 4.15, 4.16, 4.17, 5.23, 5.40, 5.49, 5.51(b), 5.52, 5.55, 5.58, 5.59, 5.60, 5.61(a), 5.62, 5.71
Journal of Organic Chemistry: Fig. 4.7
Journal of Physical Chemistry: Figs 2.26, 2.27, 2.28, 2.29, 3.6, 5.19, 5.20
Organometallics: Figs 3.9, 3.10, 3.70(a), 3.70(b), 3.75, 3.123, 3.127, 3.128, 3.129, 3.131
Johann Ambrosius Barth
Zeitschrift für Anorganische und Allgemeine Chemie: Fig. 3.115
Biophysical Society
Biophysical Journal: Fig. 5.41
Cambridge University Press
Protein Science: Fig. 5.70
Elsevier Science Publishers
Biochimica et Biophysica Acta: Fig. 2.4
FEBS Letters: Fig. 5.15
Journal of Molecular Structure: Figs 3.40, 3.42, 3.44, 3.45, 3.49, 3.53(a), 3.88, 3.102, 4.31
Journal of Organometallic Chemistry: Fig. 3.34
Tetrahedron: Fig. 3.58

Trends in Biological Sciences: Fig. 5.48
Gordon and Breach
Crystallography Reviews: Figs 2.1, 2.10, 2.11, 2.14, 2.24, 4.5, 4.10, 5.53, 5.61
International Union of Crystallography
Acta Crystallographica, Section B: Figs 3.12(b), 3.25(b), 3.50, 3.51, 3.52, 3.66, 3.82, 3.83, 3.84, 3.99, 3.124(b), 3.125, 4.6, 4.9
Acta Crystallographica, Section C: Figs 3.27, 3.47, 3.104, 3.112
Acta Crystallographica, Section D: Figs 5.5, 5.6, 5.10, 5.21, 5.22, 5.43, 5.44, 5.45, 5.63, 5.64, 5.66
JAI Press
Advances in Molecular Structure Research: Fig. 3.31
Macmillan
Nature: Figs 3.17, 5.24(a)
Munksgaard
International Journal of Protein and Peptide Research: Fig. 5.7(b)
National Research Council of Canada
Canadian Journal of Chemistry: Fig. 3.15
Oldenbourg Verlag
Zeitschrift für Kristallographie: Fig. 3.111
Oxford University Press
G. A. Jeffrey (1997). *An introduction to hydrogen bonding*: Fig. 5.57
Royal Society of Chemistry
Chemical Communications: Figs 2.12, 2.13, 2.16, 2.17, 2.18, 2.19, 2.34, 3.26, 3.35, 3.71, 3.72, 3.97, 3.98, 3.107, 3.126, 3.137, 3.140, 4.13, 4.14, 4.19, 4.26, 4.27
Faraday Discussions: Figs 5.77, 5.78, 5.79
Journal of the Chemical Society, Faraday Transactions: Fig. 3.7
Journal of the Chemical Society, Perkin Transactions 2: Figs 2.21, 2.23, 3.12(a), 3.36
New Journal of Chemistry: Fig. 3.139
VCH-Wiley
Chemistry, a European Journal: Figs 4.8, 5.42
Angewandte Chemie, International Edition in English: Figs 3.77, 3.122, 4.11, 5.46, 5.47, 5.54
Verlag der Zeitschrift für Naturforschung
Zeitschrift für Naturforschung B: Figs 3.11(a), 3.101

1

Introduction

1.1 The hydrogen bond

The hydrogen bond is a unique phenomenon in structural chemistry and biology. Its fundamental importance lies in its role in molecular association. Its functional importance stems from both thermodynamic and kinetic reasons. In supramolecular chemistry, the hydrogen bond is able to control and direct the structures of molecular assemblies because it is sufficiently strong and sufficiently directional. This control is both reliable and reproducible and extends to the most delicate of architectures. In mechanistic biology, it is of vital importance because it lies in an energy range intermediate between van der Waals interactions and covalent bonds. This energy range is one that permits hydrogen bonds to both associate and dissociate quickly at ambient temperatures. This twin ability renders the interaction well suited to achieving specificity of recognition within short time spans, a necessary condition for biological reactions that must take place around room temperature. For these reasons, the subject of hydrogen bonding is of major interest and remains relevant with each new phase in the kaleidoscope of chemical and biological research.

1.1.1 Historical background

The earliest references to concepts that would be termed *hydrogen bonds* in modern parlance, occur in the German literature. Werner (1902) and Hantzsch (1910) employed the term *Nebenvalenz* (secondary valence) to describe the binding situation in ammonia salts, **1**. A paper by Pfeiffer (1913) entitled 'Zur Theorie der Farblacke, II' gives the structural formula **2** to explain the reduced reactivities of compounds with C=O and OH groups placed adjacently, with amines and hydroxides. The phenomenon was termed *Innere Komplexsalzbildung*. This may be one of the first reports of a hydrogen bond in organic chemistry. Moore and Winmill (1912) used the term *weak union* to describe the weaker basic properties of trimethylammonium hydroxide relative to tetramethylammonium hydroxide. Latimer and Rodebush (1920) in discussing structure **3**, suggested that 'a free pair of electrons on one water molecule might be able to exert sufficient force on a hydrogen held by a pair of electrons on another water molecule to bind the two molecules together' and that 'the hydrogen nucleus held between two octets constitutes a weak *bond*'.

CO

H₃N--------HCl

H:N:H:O:H

1

C

O···'H'O

3

2

However, with the growing interest in the much stronger covalent, ionic and metallic bonds during the following years, interest in these matters declined. Possibly, the understanding of covalency had to mature before chemists could turn their attention to violations of the octet rule.

The mid-thirties witnessed a qualitative change. Four articles that appeared during 1935–6 have been identified as 'definitive' by Jeffrey (1997) in his recent monograph. The most important of these is by Pauling (1935) who used the term 'hydrogen bond' for the first time and freely, to account for the residual entropy of ice. Other papers on diketopiperazine (Corey 1938) and glycine (Albrecht and Corey 1939) mention 'hydrogen bonds' while the paper by Senti and Harker (1940) on acetamide speaks of 'N–H–O bridges'. The term *bridge* derives from the work of Huggins (1936) and its German equivalents *Wasserstoffbrückenbindung* and *Wasserstoffbrücke* continue to be used today. These terms are of some interest. They may even be of some utility given the complex nature of the hydrogen bond interaction and the several misunderstandings that have arisen from the use of the word *bond* for its description.

It was, however, the chapter on hydrogen bonding in *The nature of the chemical bond* (Pauling 1939) that drew the subject of hydrogen bonding into the chemical mainstream. Pauling was clear and unambiguous in the use of the word *bond* when he stated that 'under certain conditions an atom of hydrogen is attracted by rather strong forces to two atoms, instead of only one, so that it may be considered to be acting as a bond between them'. The reader will note that the word *bond* is used here mostly in a linguistic sense though the chemical overtones are clear enough. In a configuration such as X−H···A, it is the H atom that is considered to be the seat of bonding and not the entity H···A. Given such an interpretation, the use of the word *bridge* is hardly objectionable, and the sometimes heated discussions as to whether or not an interaction of a particular geometry is a hydrogen *bond* are perhaps unnecessary. After all, if the H atom is accepted as a bridging or bonding agent between the elements X and A, then this should suffice for an operational definition of the hydrogen bond.

The second core idea to emerge from Pauling's work is that the hydrogen bond is an electrostatic interaction. He states thus: 'It is now recognized that

Table 1.1 Electronegativities of the elements (Pauling 1939)

H 2.1					
Li 1.0	B 2.0	C 2.5	N 3.0	O 3.5	F 4.0
Na 0.9	Al 1.5	Si 1.8	P 2.1	S 2.5	Cl 3.0
K 0.8	Ga 1.6	Ge 1.8	As 2.0	Se 2.4	Br 2.8
Rb 0.8	In 1.7	Sn 1.8	Sb 1.9	Te 2.1	I 2.5

the hydrogen atom, with only one stable orbital (the $1s$ orbital), can form only one covalent bond, that the hydrogen bond is largely ionic in character, and that it is formed only between the most electronegative atoms'. These attributes of a hydrogen bond have been the subject of intensive study and discussion. The electrostatic nature of the hydrogen bond and indeed the unique ability of the H atom to form these associations, arises from the fact that the solitary electron on the H atom is on time-average situated between H and X, and that with increasing electronegativity of X, the H atom is increasingly deshielded in the forward direction. Pauling assumed that only if X and A are very electronegative, would the deshielding of H and in turn the electrostatic attraction between H and A be sufficiently high to term the interaction a hydrogen bond. In practical terms, this means that the hydrogen bond phenomenon would be restricted to interactions $X-H\cdots A$, where X and A can be any of the following elements: F, O, Cl, N, Br and I (Table 1.1).

1.1.1.1 *Definition of a hydrogen bond*

Both these ideas were developed and refined further culminating in the definition of a hydrogen bond by Pimentel and McClellan (1960). This is the first of the modern definitions of the phenomenon: *A hydrogen bond is said to exist when (1) there is evidence of a bond, and (2) there is evidence that this bond sterically involves a hydrogen atom already bonded to another atom.* It is important to realize that the Pimentel and McClellan definition makes no assumptions about the nature of X and A and that it enables an evaluation of the hydrogen bonding potential of groups like $C-H$, $P-H$ and $As-H$ among others, and of π-acceptors. Because its single electron is involved in the covalent bond $X-H$, the H atom is always deshielded in the forward direction. This deshielding occurs irrespective of the nature of the X atom. Does this mean then that an $X-H$ group is always a potential hydrogen bond donor, even if there is no accumulation of electron density on the X-atom? Jeffrey and Saenger (1991) pose the question: 'Should the $C-H\cdots O=C$ interaction be referred to as a hydrogen bond, even though there is every reason to suspect that the carbon atom is not electronegative and may even carry a positive charge? By Pauling's definition, the answer is no. By Pimentel and McClellan's definition, the answer is yes'. Refinement of the latter definition led to a quantification by Steiner and Saenger (1993a) who consider a hydrogen bond as '*any* cohesive interaction $X-H\cdots A$ where H carries a positive and A a negative (partial or full) charge and the charge on X is more negative than on H'. Note now that a positive charge on the atom X is not precluded. This definition is incomplete in that it highlights only the electrostatic character of hydrogen bonds and is restrictive with respect to borderline cases, but it is still a useful working definition for many of the kinds of hydrogen bonds being studied today.

1.1.1.2   *The weak hydrogen bond*

This book is concerned with the weak hydrogen bond, which may be defined as an interaction $X-H\cdots A$ wherein a hydrogen atom forms a bond between two structural moieties X and A, of which one or even both are only of moderate to low electronegativity. Of course, the phrase '*weak* hydrogen *bond*' appears to be an oxymoron. The two central ideas of Pauling on hydrogen bonds, namely that they are bonds and that they are electrostatic, are related through the concept of strength. Bonding would seem to imply strength and unless the electrostatic nature of the association $X-H\cdots A$ was pronounced, it would not seem to be particularly strong. However, we shall show that while the most familiar properties of hydrogen bonds depend on their electrostatic character, it is not necessary for a hydrogen bond to be strong to retain many of these characteristics. A hydrogen bond, in keeping with Pimentel and McClellan, is defined then on phenomenological rather than energetic grounds.

1.1.1.3   *Other books*

Several recent monographs deal with the subject of hydrogen bonding. Chief among them is the authoritative *Hydrogen bonding in biological structures* by Jeffrey and Saenger (1991). The most general and widely applicable parts of this book have been condensed and extended by Jeffrey (1997) in his useful text entitled *An introduction to hydrogen bonding*. The book by Scheiner (1997) entitled *Hydrogen bonding. A theoretical perspective*, is timely given that computational results date much faster than experimental work. As for weak hydrogen bonds, the recent work on *The CH/π interaction* by Nishio, Hirota and Umezawa (1998) deals with just one specific interaction but studied with a wide range of techniques. The much older work by Green (1974) on *Hydrogen bonding by $C-H$ groups* is mainly spectroscopic in emphasis and is mentioned here if only because it was the lone book specifically devoted to the subject of weak hydrogen bonds for many years.

1.1.2 *Geometrical parameters and definitions*

Modern concepts of the hydrogen bond lead to both geometrical and energetic implications. In this section, we take up the geometrical characterization of hydrogen bonds. Part of the difficulty in studying weak hydrogen bonds lies in the inappropriateness of applying geometrical criteria which are suited only for strong bonds for the identification of weak bonds. In that this is a work dealing mainly with structural aspects of weak hydrogen bonding, these difficulties will be of major concern to us, and will be commented upon later.

1.1.2.1 *Distances and angles. Bifurcation*

The general hydrogen bond is constituted with a donor $X-H$ and an acceptor A, and is referred to in this work as $X-H\cdots A$. The bond may be described in terms of the d, D, θ and r as shown in Fig. 1.1. Clearly, only three of these parameters are independent. In the older literature, the focus was on the heavy atom distance D because the H atom position often could not be determined. Today, it is common practice to use the three parameters involving the H atom, d, θ and r, as the independent set, and to consider D as an auxiliary parameter. If the hydrogen bond is extended on the acceptor side as $X-H\cdots A-Y$, an acceptor angle ϕ, $H\cdots A-Y$ may also be defined. A stringent description of hydrogen bond geometries requires the use of even more independent parameters, the number and nature of which depend on the particular system. For a fuller description of the pair of diatomic molecules shown in Fig. 1.1, one might need to consider also the torsion angle around $H\cdots A$, making the number of parameters five. For multi-atom acceptors, some convention is needed to define the position of A. In a triple bond, distances are usually measured to the centre of the bond (M), and in phenyl rings, the centroid is taken as the point of reference.

Because hydrogen bonds are long-range interactions, a group $X-H$ can be bonded to more than one acceptor A at the same time. If there are two acceptors A_1 and A_2, this is called a *bifurcated hydrogen bond* $X-H\cdots(A_1, A_2)$, Fig. 1.2. Hydrogen bonds with three acceptors are called *trifurcated*, accordingly.

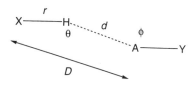

Fig. 1.1. Definition of the geometrical parameters d, D, r, θ and ϕ for a hydrogen bond.

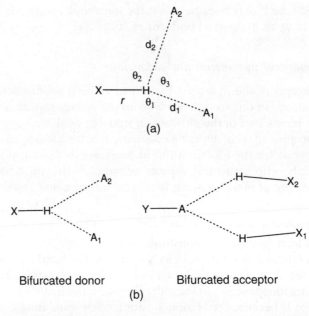

Fig. 1.2. The bifurcated hydrogen bond. (a) Geometrical parameters. (b) Definition of a bifurcated donor (*left*) and bifurcated acceptor (*right*).

Bifurcated hydrogen bonds are characterized by the distances r, d_1, d_2 and the angles θ_1, θ_2, θ_3. The elevation of the H atom from the plane formed by the three heavy atoms, as measured by the sum of angles $\theta_1 + \theta_2 + \theta_3$, is an inverse indicator of the efficacy of a bifurcated bond.

The term 'bifurcated' was commented upon unfavourably by Jeffrey and Saenger (1991) who preferred the term 'three-centre' indicating that the H atom is bonded to three other atoms. However, we feel that the term 'bifurcated' is of some utility because it permits a distinction between the two geometries shown in Fig. 1.2(b) as *bifurcated donor* and *bifurcated acceptor*. The latter is of relevance for weak hydrogen bonds because many organic and organometallic systems are donor rich when weak donors are taken into account, and accordingly bifurcated acceptors do occur frequently.

1.1.2.2 *Location of the H atom*

Some comments on the X-ray diffraction method are pertinent here. With the tremendous advances in the construction and capabilities of diffractometers and low temperature facilities for data collection, highly precise structural information is now available, but the location and refinement of H atom positions often still remains at the limits of the technique (Glusker *et al.* 1994). A more fundamental concern lies in the fact that in X-ray structure determinations, the distances of the H atoms to the bonded heavier atoms (C—H,

Table 1.2 X—H bond lengths in an X-ray and a neutron crystal structure of α-glycine, distances between H atom positions in these structures (Δ) and angles between the X—H directions (δ) (Olovsson and Jönsson 1976, see also Koetzle and Lehmann 1976)

Bond	$X-H_x$ (Å)	$X-H_n$ (Å)	Δ (Å)	δ (°)
N—H(1)	0.996(19)	1.054(2)	0.066	1.8
N—H(2)	0.982(18)	1.037(2)	0.067	2.2
N—H(3)	0.959(16)	1.025(2)	0.070	1.3
C(2)—H(4)	0.963(16)	1.090(2)	0.130	2.1
C(2)—H(5)	0.966(18)	1.089(2)	0.133	2.8

N—H, O—H) are on the average 0.1–0.2 Å shorter than the internuclear distances. This happens because X-rays are scattered by electrons and the atomic position derived for an H atom from an X-ray analysis approximates the centroid of the electron density. The latter is not centred around the H nucleus, but is displaced towards the atom X. The use of neutron diffraction analysis avoids this problem since the scattering centres are the atomic nuclei themselves. The distances derived from neutron analysis therefore correspond nearly to the interatomic distances and, accordingly, neutron diffraction is a most important technique in the determination of accurate hydrogen bond parameters. As an example the X—H bond lengths in glycine as determined by X-ray and neutron diffraction are given in Table 1.2. A comprehensive analysis of the matter has been performed by Allen (1986). It has been argued that while the neutron-derived distances are more accurate, this does not necessarily mean that they are the most chemically meaningful. This is because one cannot simply identify an atom with its nucleus, but rather consider it as being composed of nucleus and electrons (Cotton and Luck 1989, Aakeröy and Seddon 1993a). In any event, neutron distances have established themselves as benchmarks in hydrogen bond research (Hamilton and Ibers 1968; Jeffrey 1992).

1.1.2.3 *Normalization of X—H bonds*
All this leads to the technique of distance normalization, used in this book and in many recent papers. In this procedure, the distances obtained in an X-ray analysis are corrected by extending the X—H bond vector to the average neutron derived distance of X—H. If no neutron-determined value is available, values from gas phase spectroscopy can be used. A list of standard X—H distances is given in Table 1.3. The hydrogen bond distances d are typically shorter in normalized than in non-normalized geometries, and the θ values are slightly lower. Normalization procedures are standard in many modern computer programs used in the analysis of crystal structures.

Normalization is unproblematic for H atoms attached to $C(sp^2)$ and $C(sp)$

Table 1.3   Standard X—H bond lengths (Å)

B—H	C—H	N—H	O—H	F—H
1.19[a]	1.083[b]	1.009[b]	0.983[b]	0.917[c]
Al—H	Si—H	P—H	S—H	Cl—H
1.59[c]	1.50[a]	1.42[c]	1.338[a]	1.27[c]
Ga—H	Ge—H	As—H	Se—H	Br—H
1.62[d]	1.51[c]	1.52[c]	1.46[c]	1.42[c]
	Sn—H	Sb—H	Te—H	I—H
	1.71[c]	1.70[c]	1.69[c]	1.61[c]

[a] Mean neutron diffraction value. [b] CSD default, neutron based. [c] Gas phase value for XH_n. [d] Gas phase mean bond length in Ga_2H_6.

atoms where it may be routinely performed. The procedure may be somewhat unreliable for sp^3-hybridized C atoms because the conformational positions of the C—H bonds are unclear and also for —NHR and —NH_2 groups because the pyramidal character of the N atom is unknown. In routine normalization, the so-called 'neutron' value of X—H is assumed to be constant, and this disregards modification of X—H by the hydrogen bond itself. For weak hydrogen bonds, this modification is so small (<0.01 Å, Section 2.2.4) that normalization to a constant value is justified. For strong hydrogen bonds, X—H is elongated by several hundredths of an Å, making standard normalization a more approximate but still reasonable procedure. In very strong hydrogen bonds, X—H is lengthened by more than 0.05 Å and standard normalization becomes questionable. For strong hydrogen bonds, the elongation $r = f(d)$ has been parametrized using neutron diffraction data and can be used in a refined normalization procedure where this effect is taken into account (Steiner 1998a).

1.1.2.4   *The hydrogen bond as a group–pair interaction*

It is most important to note that when one refers to *the hydrogen bond*, one is actually referring to the whole group of three atoms, X, H and A. In most cases of hydrogen bonding, one of the two bonds formed by the hydrogen atom, namely X—H, is much stronger than the other, H···A. Coloquially, it is just the latter that is termed 'hydrogen bond', but this informal terminology has led to many misconceptions as to what constitutes a hydrogen bond and what not. For weak hydrogen bonds, X—H dominates energetically over H···A in an overwhelming manner. However, the entities X—H and H···A are not really independent of each other; hydrogen bonds are not atom–pair but group–pair interactions. Reducing the interaction $R_1 — X — H···A — R_2$ to the central bond H···A leads to simplistic views that overlook the inherent modifications of the donor by the acceptor group, and vice versa. The distance

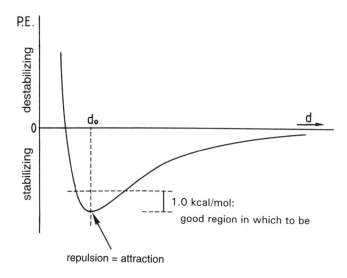

Fig. 1.3. Schematic representation of a hydrogen bond potential as a function of distance.

$X-H$ is not a constant but is affected by $H \cdots A$. The $H \cdots A$ distance is greatly dependent on the nature of R_1, X and also R_2, and likewise $A-R_2$ experiences an influence of the donor group. Analogously, in bifurcated hydrogen bonds, the term *hydrogen bond* would, strictly speaking, refer to the group of four atoms, X, H, A_1 and A_2.

1.1.3 *Energetic parameters and definitions*

Let us first examine Fig. 1.3, which shows schematically the distance profile of potential energy of a typical intermolecular interaction, say a hydrogen bond. The energy is lowest at the equilibrium distance d_0, is negative for all distances $d > d_0$ and also for distances somewhat shorter than d_0, and is positive only for very short distances. The zero-energy line separates what one may call *stabilizing* ($E < 0$) and *destabilizing* ($E > 0$) regions (Dunitz 1996a). Any deviation from the equilibrium distance costs an enthalpic penalty, but this penalty is large only for large deviations in d.

Let us now move from energies to forces. At the equilibrium distance, the force is zero. For distances $d \neq d_0$, a force arises that tries to establish optimal geometry for the system. For all distances $d > d_0$, this force is *attractive*, and for all distances $d < d_0$, it is *repulsive*. The strongest attractive force occurs at the inflection point of the curve, which therefore represents quite an unstable geometry. The repulsive force becomes very large as d becomes short. The curvature at the minimum is the force constant; the sharper the minimum, the larger the force constant and the larger the forces that arise from distortions.

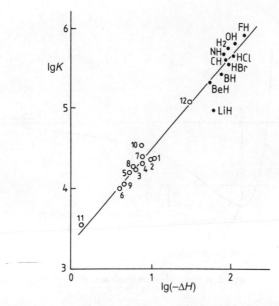

Fig. 1.4. Linear logarithmic relation between force constant k (in dyn/cm) and hydrogen bond energy $-\Delta H$ (in kcal/mol). Black dots: diatomic hydrides. Open circles: hydrogen bonded complexes as (1) phenol–triethylamine; (2) phenol–pyridine; (3) HCl–dimethyl ether; (4) HCl–diethyl ether; (5) HBr–dimethyl ether; (6) N-methylacetamide dimer; (7) formic acid catemer; (8) water; (9) polymeric alcohols; (10) formic acid dimer; (11) HCl–HCl; and (12) potassium bifluoride (after Iogansen and Rozenberg 1971).

In hydrogen bonds, every donor–acceptor combination has its own potential energy curve. For the stronger combinations, the minimum is deeper and shifted to shorter distances. There is a linear relationship between the depth of the potential energy curve and the force constant, so that the stronger bonds are more difficult to distort than the weaker ones. With spectroscopic methods, this relationship can be experimentally verified and found to neatly extrapolate from the H$\cdots$A component of hydrogen bonds all the way to the X—H covalent bond of hydrides, Fig. 1.4 (Iogansen and Rozenberg 1971). All this is strictly valid only for gas phase dimers. In condensed media, the potential energy curve itself is a function of the surroundings and environment, and a phenomenon like cooperativity can exert a pronounced influence.

In crystal structures, only few hydrogen bonds can adopt distances very close to d_0, and most are distorted to some degree. Severe distortions, however, are unlikely because of the enthalpic penalty that would then have to be paid. In a combined statistical and theoretical study on O—H$\cdots$O hydrogen bonds in carbohydrates, Kroon *et al.* (1975) have shown that the distribution of distances d in crystals represents roughly a Boltzmann population of the potential energy curve. Most of the hydrogen bonds have energies not more than 1.0 kcal/mol above the minimum.

In the potential energy curve shown in Fig. 1.3, only the distance d is varied, whereas all other parameters are kept constant. If the angle θ were to be varied also, a two-dimensional potential energy surface would result. Then, a bending of θ would lead to a restoring force that tries to straighten the bond. Notably, such force vectors have components perpendicular to the hydrogen bond and they do not fall into the repulsive/attractive categories. In real hydrogen bonds in the condensed state, *all* the variable parameters namely d, θ, φ and others deviate from the optimal values, corresponding to a general point on a multi-dimensional potential energy surface.

Progressing from individual hydrogen bonds to the situation in a molecular crystal, one notes that there are many attractive and repulsive forces acting on any given molecule and it is trivial to note that in an equilibrium situation, the attractive and repulsive forces on each atom balance out exactly. One may also note that because attractive forces between uncharged molecules are long range while repulsive forces come into play only at short distances, the attractions between molecules occur mainly between relatively distant atoms, whereas the repulsions occur between the atoms that form the shorter contacts (Fig. 1.3).

In this book, we shall consistently use the terms *attractive* and *repulsive* to represent forces, and *stabilizing* and *destabilizing* to represent energies. Weak hydrogen bonds are characterized by shallow potential energy curves, large equilibrium distances d_0 and, in the extreme, an easy passage into the destabilizing region upon compression. Energies do not change significantly over large distance ranges and this means that pronounced geometrical distortions are possible. Like geometrical criteria, it is also difficult to employ energetic criteria for hydrogen bonding in this domain.

1.2   The weak or non-conventional hydrogen bond — scope of this work

Hydrogen bonds cover a wide and continuous energy scale from around −0.5 to nearly −40 kcal/mol. As an illustration, calculated energies are given in Table 1.4 for a number of hydrogen bonds covering the whole energy range of the phenomenon. The very weakest of hydrogen bonds are barely distinguishable from van der Waals interactions while the strongest ones are stronger than weak covalent bonds. Some element of subjectivity is therefore unavoidable when a hydrogen bond is qualified with the epithet *weak*, for it presupposes what is meant by *strong*. Any energy cut-off between strong and weak bonds is arbitrary and therefore disputable. In principle, one could categorize hydrogen bonds as 'weak' and 'strong' according to an energetic criterion (an energy cut-off value), phenomenological criteria (distances or IR wave number shifts), or operational criteria (what they can do). The results of such alternative classifications are not necessarily consistent, because there are always cases where a hydrogen bond is, say 'strong' in terms of energy and 'weak' in

Table 1.4 Calculated energies and equilibrium distances D for different kinds of hydrogen bonds

Hydrogen bond	Energy (−kcal/mol)	D (Å)	Reference
$[F-H\cdots F]^-$	39	2.30	Gronert 1993
$[OH_3\cdots OH_2]^+$	33	2.48	Del Bene et al. 1985
$[NH_4\cdots NH_3]^+$	24	2.85	Del Bene et al. 1985
$[OH_2\cdots OH]^-$	23	2.44	Gronert 1993
$NH_4^+\cdots OH_2$	18.9	2.77	Del Bene 1988
$OH_2\cdots Cl^-$	13.5	3.27	Del Bene 1988
$[NH_3\cdots NH_2]^-$	10.2	2.91	Gronert 1993
$O=C-O-H\cdots O=C-O-H^{(a)}$	7.4		Neuheuser et al. 1994
$Cl-H\cdots OH_2$	5.4		Hinchliffe 1984
$H_2O\cdots H_2O$	5.0		Feyereisen et al. 1996
$N\equiv C-H\cdots OH_2$	3.8	3.12	Turi and Dannenberg 1993
$Me-OH\cdots Ph$	2.8		Malone et al. 1997
$OH_2\cdots F-CH_3$	2.4		Howard et al. 1996
$H-C\equiv C-H\cdots OH_2$	2.2	3.26	Turi and Dannenberg 1993
$Cl-H\cdots SeH_2$	2.0		Hinchliffe 1984
$H-C\equiv C-H\cdots C\equiv C-H$	1.4		Philp and Robinson 1998
$H_2S\cdots H_2S$	1.1	4.16	Woodbridge et al. 1986
$CH_4\cdots OH_2$	0.6		Novoa et al. 1991
$CH_4\cdots SH_2$	0.4		Rovira and Novoa 1998
$CH_4\cdots FCH_3$	0.2		Howard et al. 1996

[a] Cyclic dimer.

terms of geometry, or the other way round. Hydrogen bonds may also be classified as *conventional* and *non-conventional*. Because the stronger hydrogen bond types were studied first and more intensively, the categories 'strong' and 'conventional' have a very large intersection. Nevertheless, there do exist strong hydrogen bonds that are novel and in this respect non-conventional, and there are completely conventional hydrogen bond types that are very weak. In this section, we discuss these categories more fully and outline the scope of the present work.

1.2.1 Classification of hydrogen bonds

Table 1.5 lists properties of hydrogen bonds that we classify as very strong, strong and weak. These properties are geometrical, energetic, thermodynamic and functional in nature. The table is meant to be used as a guide only and is not intended to divide hydrogen bonds into watertight compartments. This would be misleading because the energies and indeed all the mentioned prop-

Table 1.5 Some properties of very strong, strong and weak hydrogen bonds

	Very strong	Strong	Weak
Bond energy (−kcal/mol)	15–40	4–15	<4
Examples	$[F \cdots H \cdots F]^-$	$O-H \cdots O=C$	$C-H \cdots O$
	$[N \cdots H \cdots N]^+$	$N-H \cdots O=C$	$O-H \cdots \pi$
	$P-OH \cdots O=P$	$O-H \cdots O-H$	$Os-H \cdots O$
IR v_s relative shift	>25%	5–25%	<5%
Bond lengths	$H-A \approx X-H$	$H \cdots A > X-H$	$H \cdots A \gg X-H$
Lengthening of $X-H$ (Å)	0.05–0.2	0.01–0.05	≤0.01
$D(X \cdots A)$ range (Å)	2.2–2.5	2.5–3.2	3.0–4.0
$d(H \cdots A)$ range (Å)	1.2–1.5	1.5–2.2	2.0–3.0
Bonds shorter than vdW	100%	Almost 100%	30–80%
$\theta(X-H \cdots A)$ range (°)	175–180	130–180	90–180
kT (at room temperature)	>25	7–25	<7
Effect on crystal packing	Strong	Distinctive	Variable
Utility in crystal engineering	Unknown	Useful	Partly useful
Covalency	Pronounced	Weak	Vanishing
Electrostatics	Significant	Dominant	Moderate

erties of hydrogen bonds lie in continuous ranges. The row entitled 'examples' is of particular importance because it provides a broad chemical basis for our classification. The reader will easily recognize that the examples in the three categories are different from one another. To assign a hydrogen bond in the borderline regions, chemical considerations are more advisable than numerical criteria and cut-off definitions.

Different terminologies have been employed by others. Jeffrey and Saenger (1991) have classified hydrogen bonds as 'strong' and 'weak' while Jeffrey (1997) has elaborated this further to 'strong', 'moderate' and 'weak'. The reader will note that there is a near correspondence between the category we term 'strong' and Jeffrey calls 'moderate'. Jeffrey's terminology is in keeping with the biological literature where bonds such as $O-H \cdots O-H$, and especially $O_w-H \cdots O_w-H$, are not taken to be particularly 'strong', and $N-H \cdots O$ definitely veers towards what is understood as 'weak'. The reason we refer to the middle category as 'strong' originates from supramolecular considerations. By 'strong' we mean hydrogen bonds that are able to control crystal and supramolecular structure effectively. This certainly includes $O-H \cdots O=C$, $N-H \cdots O=C$ and $O-H \cdots O-H$. By 'weak' we mean hydrogen bonds whose influence on crystal structure and packing is variable. In this sense, a strong hydrogen bond is one which is much stronger than a van der Waals interaction while a weak hydrogen bond is one which is not. In the end though, these discussions are semantic in nature. Still, terminologies are necessary even if they are subjective and it is necessary for the reader to know at an early stage what we mean by the terms *strong* and *weak*.

Very strong hydrogen bonds are formed by unusually activated donors and acceptors, often in an intramolecular situation. Frequently, they are formed between an acid and its conjugate base, $X-H\cdots X^-$, or between a base and its conjugate acid, $X^+-H\cdots X$. Very strong hydrogen bonds are of great importance in the context of chemical reactivity, and have therefore been studied for many years (Hamilton and Ibers 1968; Hibbert and Emsley 1990). The distinctive characteristic of these bonds, and one which is responsible for many of their properties listed in Table 1.5, is their substantial covalent character (Gilli *et al.* 1994). For instance, the $X-H$ and $H\cdots A$ distances are comparable. These bonds may therefore be studied by many of the methods that are used to study covalent bonds proper. We have assigned to this category an energy range of 15–40 kcal/mol. The transition from very strong to *strong hydrogen bonds* (4–15 kcal/mol) represents a transition from quasi-covalent to electrostatic character. All hydrogen bonds are electrostatic, but this particular characteristic is dominant in this large and most familiar category of hydrogen bonds. To many structural chemists and biologists, the properties associated with this category exemplify hydrogen bonding as a whole. Molecules that contain functional groups capable of forming strong hydrogen bonds always do so, unless there are adverse steric factors.

Weak hydrogen bonds (<4 kcal/mol) are the subject matter of this book and form the final category. Though numerous, they have not always been identified in a general sense till recently. Partly, this is because molecules which contain groups capable of forming weak hydrogen bonds do not necessarily need to associate in this particular manner. These bonds are electrostatic but this characteristic is modified by variable dispersive and charge-transfer components that depend substantially on the nature of the donor and acceptor group. The strongest of these, say bonds such as $O-H\cdots Ph$ and $C\equiv C-H\cdots O$, are quite electrostatic and comparable to a bond like $O-H\cdots O-H$. They lie in the energy range -2 to -4 kcal/mol. The weakest of these are formed by unactivated methyl groups and are barely stronger than van der Waals interactions (about -0.5 kcal/mol). All kinds of hydrogen bonds are different and, likewise, all kinds of weak hydrogen bonds are also different. For a further classification of weak hydrogen bonds, it may be convenient to subdivide them on the basis of the donor and acceptor groups. This has been attempted in Table 1.6.

Bonds formed with weak donors and strong acceptors have been studied the longest, especially by crystallographers. This set includes $C-H\cdots O/N$ (Chapter 2) and more recently $M-H\cdots O$ where M = Tr (Section 3.5). The second set is constituted with strong donors like $O-H$ and $N-H$ and weak acceptors like $C\equiv C$, Ph and M (Chapter 3). Bonds formed with weak donors and weak acceptors are at the end of the hydrogen bonding regime. These include $C-H\cdots\pi$, $S-H\cdots\pi$, $C-H\cdots M$ and $C-H\cdots F-C$ (Chapter 3). Other varieties which may not, strictly speaking, be even considered hydrogen bonds are the agostic interaction (Section 3.5), the formyl hydrogen bond

Table 1.6 Some categories of weak hydrogen bonds

1.	*Weak donor—strong acceptor*		
	C—H$\cdots$O=C	C—H$\cdots$N	C—H$\cdots$Cl$^-$
	P—H$\cdots$O	Mo—H$\cdots$O≡C	Ir—H$\cdots$Cl—Ir
2.	*Strong donor—weak acceptor*		
	N—H$\cdots$Ph	Cl—H$\cdots$C≡C	O—H$\cdots$C=C
	O—H$\cdots$F—C	O—H$\cdots$Se	N—H$\cdots$Co
3.	*Weak donor—weak acceptor*		
	C≡C—H$\cdots$C≡C	C≡C—H$\cdots$Ph	C—H$\cdots$F—C
	C—H$\cdots$Pt	C—H$\cdots$Cl—C	C—H$\cdots$H—Re
4.	*Other varieties*		
	Agostic	N—H$\cdots$H—B	Formyl hydrogen bond
5.	*Conventional but weak (not discussed)*		
	$^-$O—H$\cdots$O—H		

(Section 2.2.11) and debatably, the inverse hydrogen bond (Section 3.6). Some hydrogen bonds formed with conventional donors and acceptors are actually quite weak but are not discussed by us. This set includes, for example, O—H$\cdots$O—H, and X—H$\cdots$N with N being a poorly pyramidalized N(sp^2) atom. We generally do not discuss hydrogen bonds that are formed by strong donors and acceptors, but are still weak; typical examples would be minor components of bifurcated arrangements, very elongated N—H$\cdots$O bonds such as are seen frequently in biomolecules, and strongly bent hydrogen bonds which occur in sterically adverse situations.

1.2.1.1 *Conventional and non-conventional bonds*

Another way of classifying hydrogen bonding interactions is based on the 'conventionality' of the donor and acceptor groups. Conventional, that is traditional, donors include N—H, O—H, halogen—H and perhaps S—H. Conventional acceptors include these same atoms in different hybridizations as appropriate. Generally but not always, these conventional donors and acceptors are also fairly acidic or basic. Table 1.7 lists hydrogen bonds according to the strong/weak and conventional/non-conventional categories. Most weak hydrogen bonds are also non-conventional and most strong bonds are conventional; however, there are exceptions, and the table delimits the bond types we wish to discuss. Not all weak bonds have been covered by us with equal emphasis, and also not all the non-conventional ones. This is admittedly arbitrary in part and to some extent also reflects our own research interests, but it will be noted that the very large majority of bonds that may be termed weak and/or non-conventional find a mention in this work.

Table 1.7 Two modes of classifying hydrogen bonds. The bond types discussed in this work are highlighted

	Very strong	Strong	Weak
Conventional	$F-H\cdots F^-$ Proton sponges $X^+-H\cdots A^-$	$N-H\cdots O=C$ $O-H\cdots O-H$ $Hal-H\cdots O$ $Water\cdots Water$ $O-H\cdots O=C$ $O-H\cdots Hal^-$	$^-O-H\cdots O$ Bifurcated $O-H\cdots S$ $S-H\cdots S$ $O-H\cdots F-C$ $O-H\cdots Cl-C$
Non-conventional	**Agostics**	$N^+-H\cdots\pi$ $O-H\cdots\pi$ $N-H\cdots H-B$ $X-H\cdots|C$	$C-H\cdots O$ $C-H\cdots N$ $O/N-H\cdots\pi$ $C-H\cdots\pi$ $O-H\cdots M$ $M-H\cdots O$ $P-H\cdots O$ $O/N-H\cdots P$ $O/N-H\cdots Se$ $C-H\cdots F-C$ $Si-H\cdots O$

The reader should note that terms like *conventional* and *non-conventional* are even more subjective than the terms *weak* and *strong* and in using both these criteria subjectively, we hope that we have been able to assemble a collection of hydrogen bond types that are chemically related. A few inconsistencies between Tables 1.6 and 1.7 may also be noted. In the former, the $O-H\cdots\pi$ has been classified as 'weak' while in the latter it is 'strong'; $Re-H\cdots H-C$ is weak but $Ga-H\cdots H-N$ is strong. The agostic interaction is formally included in Table 1.6 with the other 'weak' bonds but it is listed in Table 1.7 as 'very strong' because it has much covalent character. Such inconsistencies are unavoidable in a subject as murky as the classification of borderline interactions.

The emphasis of this book is primarily directed towards organic structures. Biological structures are necessarily based on organic molecules and we have discussed them separately in Chapter 5. Organometallic structures are distinct and have been included because of the novelty of hydrogen bond types in this varied group of compounds (Section 3.5). With a few exceptions, we do not consider purely inorganic structures, although weak hydrogen bonds of various types also occur in these systems.

1.2.2 *The nature of the hydrogen bond interaction and its limits*

The preceding sections show that when one speaks of *a hydrogen bond*, one refers to a structural entity composed of at least three atoms. These atoms constitute a group and hydrogen bonding is then a group property. Accordingly, one should also anticipate that *the hydrogen bond* is not a simple interaction but a complex conglomerate of at least four component interaction types. In this regard, we share the view expressed by Jeffrey and Saenger (1991) that the term *hydrogen bridge* is perhaps a more appropriate descriptor. This latter terminology does not carry with it the implication that the hydrogen bond is like a covalent bond but only much weaker.

These component interactions of hydrogen bonding have been well studied in the past (Kollman and Allen 1970; Morokuma 1977; Price and Stone 1987) and it is not our intention to elaborate on them in great detail here. Suffice it to say there are different ways of theoretically partitioning the total hydrogen bond interaction, with the currently best established one being that of Morokuma (Morokuma 1971; Umeyama and Morokuma 1977). In this concept, the hydrogen bond is decomposed into terms that represent electrostatics, polarization, exchange repulsion, charge transfer and dispersion. Of these terms, only the exchange repulsion is repulsive, whereas the others are attractive at all distances. One can group these contributions globally into directional and non-directional (isotropic) terms. The isotropic terms are exchange repulsion and dispersion, the sum of which is often called 'van der Waals interaction'.

The exchange repulsion roughly follows an r^{-12} law that means it is strong at short distances but diminishes very rapidly with increasing distance. This is the interaction that keeps molecules apart from one another. The omnipresent attractive dispersion interaction diminishes with r as $-r^{-6}$, and can be considered as 'the universal glue that leads to the formation of condensed phases' (Price 1997). Isotropic van der Waals contributions are present in *all* intermolecular interactions. The electrostatic, polarization and charge transfer contributions are highly variable between different kinds of hydrogen bonds, and define their specific and directional identities. Of course and as stated earlier, the electrostatic term is dominant in most hydrogen bonds (Dykstra 1988).

Charge transfer involves transfer of electrons from an occupied orbital of one molecule to the unoccupied orbitals of the other, and is therefore conceptually similar to covalency. Very strong hydrogen bonds have a quasi-covalent nature (Gilli *et al.* 1994) with a large charge transfer contribution. Electrostatics is dominant in strong hydrogen bonds, where it contributes 60–80 per cent of the attractive terms. In weak hydrogen bonds, the relative contribution of electrostatics is smaller, and in the weakest $C-H\cdots O$ bonds, such as those formed by methyl groups, the electrostatic term can be of the same magnitude or even smaller than the dispersion term (van Mourik and van Duijneveldt 1995). Indeed, one can easily imagine that for $X-H$ groups of gradually falling polarity, the electrostatic contribution to an intermolecular interaction falls gradually whereas the dispersion term remains unchanged. In the end, as electrostatics become even weaker than dispersion, the total interaction becomes less and less directional and the hydrogen bonds fade into the van der Waals continuum.

Continuing along this line of thought, the weakest hydrogen bonds may be compressed by an external force into the destabilizing region without any real change in their chemical nature. The consequences are intriguing and are discussed further in Sections 1.2.3 and 2.2.11. Bonding is associated with stabilization and accordingly one might argue that the dividing line between stabilization and destabilization (zero potential energy in Fig. 1.3) constitutes the limit of hydrogen bonding. However, if one takes the view that the interaction remains essentially chemically unaltered in the destabilizing region, though it is now very far from optimal geometry, then there is little basis for introducing such an energetic limit to hydrogen bonding. Can one define the limits of the hydrogen bonding domain in any other way?

Limits are important, however, because otherwise any approach of any X—H group towards any species A will have to be considered a hydrogen bond, and the term itself will lose its meaning and utility. We feel that such a limit could be chosen on the basis of chemical considerations, such as distance fall-off character. Distance fall-off is important in the context of supramolecular chemistry, molecular recognition and crystal engineering. Hydrogen bonds, even weak ones, are so important in organizing molecules into predictable arrays because they are of long-range character (Desiraju 1997a). Therefore, their orienting effect is felt when molecules approach one another prior to nucleation, and before the effects are felt of the dispersive interactions that determine close-packing and stabilization energies. From a simple electrostatic viewpoint, one can break up the electrostatic contribution to intermolecular interactions into multipole terms, i.e. interactions between monopoles, dipoles, quadrupoles etc. with different characteristic fall-offs. A monopole–monopole interaction, that is the interaction between point ions, falls off as $-r^{-1}$. Formal monopole–dipole and dipole–dipole interactions fall off as $-r^{-2}$ and $-r^{-3}$, respectively. As the charges get more diffuse, one progresses into the dipole–quadrupole interaction with an $-r^{-4}$ fall-off. The interaction of an O—H group with a π-base would be an example of this. A quadrupole–quadrupole interaction (such as Ph···Ph) falls off even faster, as $-r^{-5}$, and finally we reach dispersive interactions with their approximately $-r^{-6}$ energy fall-off. This is a formal description. Multipole expansions are only valid if the distances between the interacting species are much greater than the dimensions of the dipole or quadrupole and this criterion is clearly not fulfilled for hydrogen bonds. However, this simple model helps one to understand the directional and deformable nature of hydrogen bonds.

1.2.2.1 *Limits of the hydrogen bond interaction*

With this model in hand, we would like to suggest an operational limit of weak hydrogen bonding between an $-r^{-4}$ and an $-r^{-5}$ fall-off. Our limit would then include almost all C—H···O/N interactions as hydrogen bonds. Methyl donors would be borderline cases because of their large dispersion contribution. The π-acceptors also lie in the borderline area. The $C-H\cdots\pi$ $(-r^{-4})$ of highly acidic donors behaves almost like a $C-H\cdots O$ interaction and could be called a hydrogen bond. With less activated $C-H$ groups, there is a problem and one is definitely in the borderline region. The $Ph\cdots Ph$ or herringbone interaction is not isotropic, in other words not of the van der Waals type, but it still falls off too quickly $(-r^{-5})$ to be termed a hydrogen bond. The reader should appreciate that we are well and truly in the grey area here, and that these opinions are, to some extent, subjective. Nishio *et al.* (1998) for example would term the $CH_3\cdots Ph$ interaction as the 'weakest kind of hydrogen bond'. Others would not label any $C-H\cdots\pi$ interaction as a hydrogen bond. Our limit is an approximate one too, and there are cases where an activated or cooperativity assisted methyl group forms $C-H\cdots\pi$ interactions with the character of hydrogen bonds, Section 3.1.3.5 (Madhavi *et al.* 1997). Still, it is important to note that the grey area between hydrogen bonds and van der Waals interactions occurs in the region of the $C-H\cdots\pi$ interaction.

We should also mention, finally, that the simple electrostatic approach is not only approximate but sometimes even erroneous. This is true even in domains of strong hydrogen bonding where electrostatics is dominant. An example is the directionality behaviour of the water molecule. It is well established experimentally that in the water dimer, the preferred approach is not along the water dipole but inclined to it in the direction of an electron lone pair (Legon and Millen 1987*a*). This can be explained only with molecular orbital considerations.

1.2.3 *Differences between strong and weak hydrogen bonds*

The strong or conventional hydrogen bond is very well known, indeed so well known that there is sometimes a tendency to judge weak hydrogen bonds from that more familiar yardstick. The reader should therefore appreciate that strong and weak bonds are quite different in certain respects. Despite a continuous energy scale, we have classified them separately in Section 1.2.1, and have justified this demarcation in general terms. Here we detail some specific ways in which they are different. Some of these ideas have been mentioned earlier in this chapter and also occur elsewhere in the book. However, it is useful to collect them together in this section.

1.   For strong bonds, most of the inter-atomic distances cluster in a narrow range which is 0.1–0.2 Å around the energy minimum (d_0). Cases with calculated energies of more than 1.0 kcal/mol above the optimum are very rare. A compression of the bond makes it repulsive but can never move it into the destabilizing region. For weak bonds, however, such a transit into the positive energy region is possible because the energy of the bond itself is of this order. For the weakest of hydrogen bonds, such transits may actually be rather

Table 1.8   Van der Waals radii of the elements (Å), taken from Bondi (1964)

H			
1.20			

C	N	O	F
1.70	1.55	1.52	1.47
Si	P	S	Cl
2.10	1.80	1.80	1.75
	As	Se	Br
	1.85	1.90	1.85
		Te	I
		2.06	1.98

frequent. The concept that experimentally observable interactions can be attractive at longer separations and repulsive at shorter separations is quite unfamiliar to many crystallographers and spectroscopists who study strong hydrogen bonds.

2. The van der Waals cut-off criterion in the H⋯A distance for the assignment of hydrogen bond character is inappropriate for weak hydrogen bonds. This has been stated repeatedly (Jeffrey and Saenger 1991; Desiraju 1996a; Jeffrey 1997; Steiner 1997a) but it is so important that the matter is emphasized again. The van der Waals cut-off criterion requires that in a hydrogen bond, d must be smaller than the sum of the H and A van der Waals radii (Table 1.8). This criterion does not stand on experimental or theoretical ground, but has only been established for reasons of apparent convenience. The criterion works reasonably well for strong hydrogen bonds which are almost always short enough to fulfil it. Even for these, however, minor components of bifurcated hydrogen bonds are often excluded and in sterically adverse situations, bonds like N—H⋯O can also be elongated beyond the van Waals separation. Weak hydrogen bonds are longer and application of the van der Waals cut-off criterion can have catastrophic consequences. For C—H⋯O interactions with less activated donors, d_0 is around 2.5 Å and longer, and a cut-off at such a distance would roughly bisect a d-distance distribution. X—H⋯π hydrogen bonds are even longer and certain types are omitted altogether by a van der Waals cut-off. We shall refer to this matter again at several appropriate points.

The van der Waals cut-off also carries with it the peculiar implication that the hydrogen bond as such is 'switched off' at a particular critical distance and that it becomes a van der Waals interaction at longer separations. Such a view is severely misleading in particular when looking at molecular recognition phenomena. In reality, if there is *any* interaction at very long separations, it is more likely than not electrostatic rather than dispersive and hardly the other way round.

3. The results of crystallographic and spectroscopic investigations do not need to match one another for weak bonds as well as they do for strong bonds. We have stated earlier that all hydrogen bonds may be characterized in both geometrical and energetic terms, using, respectively, the classical techniques of crystallography and spectroscopy. For strong hydrogen bonds, there are good correlations between crystallographic and spectroscopic properties. The geometry of weak bonds is, however, deformed very easily, and with their shallow potential energy surfaces, large distortions are possible with very little expenditure in terms of energy. The correlation between crystallographic and spectroscopic properties is therefore quite variable and the weaker the bond, the more the scatter between such measurements.

4. The hydrogen bond in general can be regarded as the incipient state of a proton transfer process (Bürgi and Dunitz 1983), and this is true for strong as well as for weak bonds. However, only for strong hydrogen bonds do such proton transfer processes occur with significant rates.

1.3 Methods of studying weak hydrogen bonds

We shall be necessarily brief here since much of what is relevant and valid for hydrogen bonds in general also holds for weak hydrogen bonds. The reader is referred to more general texts for detailed descriptions of the various experimental and computational techniques (Jeffrey and Saenger 1991; Jeffrey 1997; Scheiner 1997).

1.3.1 *Crystal structure analysis and statistical treatment of these results*

1.3.1.1 *Diffraction methods*

This is a structural text and we shall rely most heavily on the results of X-ray and neutron diffraction analysis. Diffraction methods are especially important to the researcher of weak hydrogen bonds because these interactions, in contrast to strong hydrogen bonds, are not generally detected with ease by spectroscopic methods. The reasons for the rapid growth and development of X-ray and neutron crystallography have been described in detail elsewhere (Desiraju 1989*a*; Glusker *et al.* 1994). Crystallography has become an overwhelmingly popular technique for three broad reasons:

(1) a near automation in the 'direct methods' for crystal structure solution and the subsequent refinement has been achieved;

(2)   extremely powerful yet inexpensive computing facilities are now acces-
      sible nearly everywhere; and

(3)   diffractometers continue to improve significantly.

Let us see how these reasons have brought about a revolution of sorts.

Diffraction data for crystal structures with less than a hundred non-H atoms
in the asymmetric unit may now be phased routinely, even as 'direct methods'
enthusiasts grapple with macromolecular structures. Workstations and PCs are
no longer specialist items and the results of crystal structure determinations
may be analysed and displayed in a number of aesthetically pleasing ways.
Access to Internet facilities means that a large amount of public domain software may be gainfully employed for the understanding and dissemination of
crystallographic knowledge. Programs that are used to solve and refine crystal
structures produce outputs in standard formats (such as the CIF) that will
interface with all kinds of other software. Even the publication process has
come under the sway of these computational advances and *Acta Crystallographica Section C* was, quite appropriately, the first journal to demand sub-
missions exclusively in an electronic format for refereed contributions. As for
diffractometers, one notes that the first cycle of fully computer-controlled four-
circle instruments, which has lasted for around 25 years, is now giving way to
the new generation of charge coupled device (CCD) diffractometers. These
instruments are based on image-plate detectors and can record diffraction
intensities of many reflections simultaneously, and the time required for a data
collection has shrunk by nearly an order of magnitude. All this means that we
may reasonably expect a very large number of accurate crystal structure determinations in the near future. The number of organic and organometallic crystal
structures has grown to around 200 000 during the last 70 or so years. It is
expected that this number will touch nearly 500 000 in just over a decade
from now. Crystal structures that are obtained in such huge numbers will not
constitute just archiving statistics. They will compel structural chemists and
biologists to think in new and different ways about research problems that
can and should be undertaken. No aspect of structural science will remain
untouched by this deluge.

Neutron diffraction is closely associated with the field of hydrogen bonding,
because only with this technique is the crystallographer able to establish
accurately the position of the most crucial atom in the interaction, namely
hydrogen (Hamilton and Ibers 1968; Jeffrey 1992; see also Section 1.1.2).
This technique is of unusual importance in the study of hydrogen bonding
to multi-atom π-bases (Chapter 3), where it is not possible to infer the H-
atom positions on the basis of geometrical considerations. Other sophisticated
and specialized techniques continue to push the crystallographic frontiers
further. The use of X-ray diffraction to measure experimental charge distributions in crystals has depended on advances in diffractometry, computation and low temperature techniques (Coppens 1997). This method

is invaluable for the evaluation of theoretical results and improves our fundamental knowledge of intermolecular interactions. Low temperature methods in general require much patience, skill and experience but the rewards are great. These cryocrystallographic techniques can be used to determine the crystal structures of substances with freezing points far below room temperature.

1.3.1.2 *Crystallographic databases*

Another consequence of the ever-growing number of small-molecule crystal structure determinations has been the development of methods for the storage and retrieval of crystallographic data. This has in turn led to the widespread use of crystallographic databases. Four databases are of relevance to the subject matter of this book. These are the Cambridge Structural Database (CSD), the Protein Data Bank (PDB), the Nucleic Acids Data Bank (NDB) and the Inorganic Crystal Structure Database (ICSD). These have been discussed elsewhere in general terms (Allen *et al.* 1987) and some salient facts are presented in Table 1.9.

The CSD (Allen *et al.* 1991; Allen 1998) is of critical importance to the study of hydrogen bonding as such. This is best illustrated with reference to Fig. 1.5 which contains d–θ scatterplots for strong $(O-H\cdots O)$ and weak $(C-H\cdots O)$ hydrogen bonds in patterns of the types **4** and **5**, respectively. These parameters have been obtained from around 500 and 150 accurately determined crystal structures, respectively. The geometrical parameters of the strong $O-H\cdots O$ bonds cluster within a narrow d–θ region, Fig. 1.5(a). Therefore, one can obtain reliable information on the metrics of this interaction type from any of a large number of structures. Of course, the crystal structure selected would need to be accurately determined but beyond

Table 1.9 Important crystallographic databases

1. Cambridge Structural Database (Allen *et al.* 1991)
   CSD, 183 000 organics and metallo-organics
   Cambridge Crystallographic Data Centre, Cambridge, UK

2. Protein Data Bank (Abola *et al.* 1997)
   PDB, 6500 entries
   Brookhaven National Laboratory, Long Island, NY, USA

3. Nucleic Acids Data Bank (Berman *et al.* 1992)
   NDB, 731 entries
   Rutgers University, Piscataway, NJ, USA

4. Inorganic Crystal Structure Database (Bergerhoff *et al.* 1983)
   ICSD, 55 000 inorganics and minerals
   Fachinformationszentrum, Karlsruhe, Germany

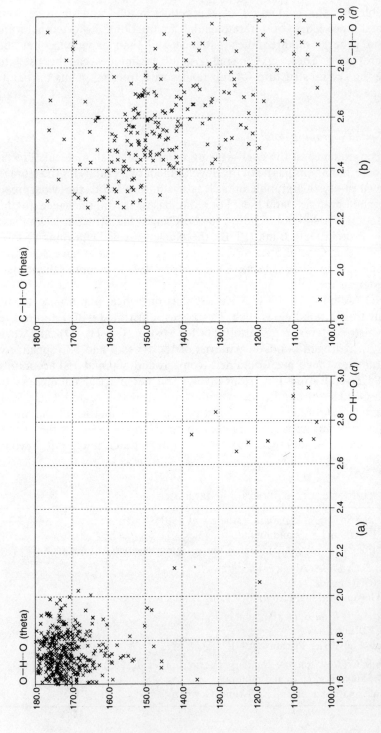

Fig. 1.5. Difference between strong and weak hydrogen bonds. θ–d scatterplots for (a) O−H⋯O and (b) C−H⋯O hydrogen bonds in synthons **4** and **5**, respectively (CSD data). The weak hydrogen bonds span much wider ranges of angles and distances than the tightly grouped sample of strong hydrogen bonds. Note, however, the roughly inverse θ–d correlation in the right hand side plot, that is characteristic of a hydrogen bonding interaction.

that, it would matter little as to which particular structure was chosen. For the $C-H \cdots O$ bonds, however, this is certainly not the case. Figure 1.5(b) shows that these interactions have lengths and angles that vary over relatively wide ranges. This happens because they are easily deformed by other interactions. Accordingly, estimates of the 'normal' length and angle of a $C-H \cdots O$ hydrogen bond would vary widely and in fact, no single structure determination would enable such quantitities to be reasonably determined. However, it is significant that the $d-\theta$ scatterplot for the $C-H \cdots O$ bonds shows the overall directionality behaviour that is characteristic of all hydrogen bonds. So, the $C-H \cdots O$ interaction is a hydrogen bond, but it is 'soft'.

Statistical methods are of vital importance in the study of these easily deformable interactions. In Section 2.2.1 and elsewhere, we shall show that the mean values of the geometrical parameters in a large number of structures are representative of the interaction type. Unless one samples a large number of crystallographic observations, one will not generally obtain reliable quantitative information on these weak interactions. The weaker the interaction, the larger the number of structures that need to be sampled.

Crystallographic databases also permit the analysis of more extended supramolecular patterns and networks in crystals. This was realized early in the development of databases (Allen *et al.* 1983), and the more personalized approaches to pattern recognition characteristic of those times (Murray-Rust and Glusker 1984; Leiserowitz and Hagler 1983; Sarma and Desiraju 1986) gradually evolved into more automated methods that make use of sophisticated graphics capabilities and other accessories (Lommerse *et al.* 1996; Allen *et al.* 1997*a*). Some of these extended networks are of unusual importance in determining stable crystal packing arrangements and have been designated as *supramolecular synthons* (Desiraju 1995*a*). A graph set notation for their description has been proposed (Etter 1990; Bernstein *et al.* 1995). The identification of synthons is an attempt to reduce and simplify a crystal structure down to its structural and functional kernel (Nangia and Desiraju 1998*a,b*) and such an exercise is greatly facilitated by the use of the CSD. From the viewpoint of the present work, the reader should note that the synthon description of a crystal structure takes into account mediation by strong and weak hydrogen bonding in the network structure. These matters will be discussed further in Chapter 4.

1.3.2 *Vibrational spectroscopy*

Vibrational spectroscopy is the classical method for the study of hydrogen bonding in condensed phases. Almost all kinds of hydrogen bonds were discovered with this technique. Its applicability ranges from the strongest to the weakest hydrogen bond types, and in both solution and the solid state. The probe here is the vibrational frequency(ies) of the atomic groups involved in hydrogen bonding, and because these frequencies can be measured very accurately, very subtle effects can be detected. In the context of weak hydrogen bonds, the method can in many instances give definitive answers as to whether or not a given $X-H\cdots A$ contact in a crystal is a hydrogen bond. Relatively small shifts, say around $20\,cm^{-1}$, in the $X-H$ stretching frequency with respect to the 'free' molecule are considered indicative of weak hydrogen bonding. For strong hydrogen bonds, it is even possible to observe the vibrational behaviour of the $H\cdots A$ interaction.

Correlations of structural and vibrational parameters have been established for various strong hydrogen bond types (Jeffrey 1997). For weak hydrogen bonds, this proves to be much more difficult, presumably because hydrogen bond geometries are so deformable. Examples where such correlations could be established will be shown in Sections 2.2.1 and 3.1.4. Correlations between spectral parameters and hydrogen bond energies date back to Badger and Bauer (1937) and have continually been refined by several groups of authors. We do not comment much on this matter, and mention only the rules of Iogansen that are applicable for $O-H\cdots O$ hydrogen bonds in solution: $\Delta H = -0.31\,(\Delta v_{OH})^{1/2}$ and $\Delta H = -2.9\,(\Delta A_{OH})^{1/2}$ where v is given in cm^{-1}, A is the relative intensity of the v_{OH} absorption band and ΔH is given in kcal/mol (Iogansen 1981).

Despite the many benefits, the spectroscopic method is not free from drawbacks. The effects of weak hydrogen bonds on vibrational spectra are not always as clear as for strong bonds, and can be quite dissimilar for different kinds of weak hydrogen bond. Even for relatively simple systems, spectral complexity can prevent proper interpretation; this is the case in particular for systems exhibiting vibrational coupling. In consequence, the weakly activated (yet important) $C-H$ donor types $-CH_3$, $>CH_2$ and Ph are notoriously difficult to study with IR spectroscopic methods. Furthermore, it is not always easy to distinguish between intra- and inter-molecular effects based on vibrational spectra. For example in monoalcohols, the 'free' $O-H$ stretching frequency varies between different conformers by over $10\,cm^{-1}$ because of the varying intramolecular surrounding (Lutz and van der Maas 1998).

Recent improvements in instrument technology and the parallel growth of interest in weak hydrogen bonds has led to a renaissance of IR techniques in the hydrogen bonding field, in particular for solid samples.

1.3.3 *Gas-phase rotational spectroscopy*

In this relatively new experimental method, geometrical parameters of gas phase adducts are determined. The adducts can be observed in the ground vibrational state, free of interference from other molecules. This allows one to gain information on hydrogen bonds in the 'pure' or undisturbed form. Apart from the geometries, dissociation energies, force constants and other parameters may also be derived (Legon and Millen 1987*a*). This kind of information is of fundamental importance because it is not obtainable from experiments in condensed phases where the equilibrium geometry is not directly observable. Information on the vibrational behaviour of hydrogen bonds, and on interconversion between alternative interaction geometries, can also be derived from rotational spectra. A disadvantage of this method is that the H atom makes the smallest contribution to the molecular mass in a hydrogen bonded aggregate, and so the positions of these atoms are determined with the least accuracy.

1.3.4 *Computation*

Quantum chemistry complements crystallography and spectroscopy in the study of weak intermolecular interactions. Theoretical methods can provide benchmark values for the energies of intermolecular interactions without the complicating effects of the solid state or solution environment. These effects, as has been mentioned, can be quite significant for weak hydrogen bonds. Crystallography and spectroscopy provide information on equilibrium geometries, as indeed most experimental techniques do. Computational methods on the other hand can be used to study domains of the potential energy surface which are far from the equilibrium structure (Scheiner 1997). Such investigations have their own merits, in terms of mapping out interconversions between various geometries, studying the electronic redistributions that accompany the formation of hydrogen bonds or even in an examination of a phenomenon such as polymorphism or crystallization. Because of the widespread availability of powerful and yet low-cost computers, theoretical methods are now accessible to a large number of structural chemists, including experimentalists. Theory coupled with database research has emerged as an effective way of studying weak intermolecular interactions (Nobeli *et al.* 1997).

  An obvious limitation of computational chemistry is that it is severely limited by the computing power that is available at the time. The rapid recent advances in the 'number-crunching' ability of computers has meant therefore that computational results age much faster than experimental results. As an example, let us consider the fact that around a decade ago most computational work on intermolecular interactions was carried out using semi-empirical methods. These methods were 'state-of-the-art' at that time but were just not designed or parametrized to treat intermolecular interactions properly. They were used nevertheless because there was really no alternative. Today, it is doubtful if one would want to rigorously consider any of this older work, though general chemical trends might be correctly predicted. Sometimes, even these trends are at variance with what is derived from present day *ab initio* methods. The latter are preferred today because they give reliable interaction energies in most cases to various degrees of approximation. Even for *ab initio* methods, however, the results are very dependent on the choice of basis set, the use of density functional theory (DFT) and the various stratagems by which electron correlation effects are handled. All these aspects and limitations must be considered carefully when computational work is assessed.

The entire area of computational supramolecular chemistry, and especially the study of weak hydrogen bonding, constitutes a new challenge. Theory enables one to compartmentalize the various contributions to hydrogen bond energy and also to dissect and study the very important phenomenon of co-operativity (Section 2.2.7). Of more recent interest is the computational treatment of small molecular clusters, or synthons (Section 4.3.4) and to examine their role in crystal growth and stabilization. In the biological area, molecular dynamics (MD) may be quite useful because it enables a study of the dynamics of the breaking and making of weak hydrogen bonds (Section 5.5.3.2). In this way not only the energetics but also the lifetimes of weak hydrogen bonds may be studied.

1.4   Summary

The foregoing discussion shows that the *hydrogen bond* is a composite interaction that spans wide ranges of geometry and energy. This spread results from the great chemical variation that exists among the hydrogen bond donor and acceptor groups. Despite this, the multifarious interactions that may be termed hydrogen bonds of one sort or another have several features in common. These features may pertain to form or function. The aim of the present work is to assess one particular type of hydrogen bond, namely the *weak hydrogen bond*, in the overall context of the hydrogen bonding phenomenon. We wish to compare and contrast the weak hydrogen bond with other hydrogen bond types and to view it in the broader perspective of supramolecular chemistry and structural biology. We begin this exercise therefore with the oldest and most frequently studied example of the weak hydrogen bond, the $C-H\cdots O$ or just as appropriately the $C-H\cdots N$ interaction.

2

Archetypes of the weak hydrogen bond – C—H···O and C—H···N interactions in organic and organometallic systems

Many, if not all, attributes of the classical hydrogen bond, X—H···A, would seem to derive from the fact that X and A are electronegative elements and from the unique electronic configuration of the H atom that permits close approaches of the X and A atoms. Indeed, it is only this combination of circumstances that makes this type of association possible, thus rendering the term '*hydrogen* bond' so appropriate. The earlier literature makes no real distinction between X and A, and both are assumed to be quite electronegative. The difference between strong and weak hydrogen bonding lies in the fact that in the latter situation, one or both of the atoms X and A are of moderate electronegativity only. This leads to several questions. What are the implications when X and A are very different? When the electronegativities of X and A differ widely, which is the more significant element in terms of influencing the properties of the interaction? Finally, does one actually leave the domain of hydrogen bonding if the electronegativity of X or A falls below some threshold?

2.1 Historical developments

Possibly, the earliest indication that an H atom attached to carbon can form hydrogen bonds is found in the paper of Kumler (1935) who studied the relationship between the dielectric constants and the dipole moments of several organic liquids. It was noted that HCN behaves similarly to other simple hydrogen bonded substances like HF, glycol, aliphatic carboxylic acids and their amides. It was specifically suggested that HCN molecules are associated with hydrogen bonds (thus designated in the original publication) as H—C≡N···H—C≡N···H—C≡N, an arrangement that is identical to that found later in the crystal (Fig. 2.1(a); Dulmage and Lipscomb 1951). Of course, HCN is an extremely strong carbon acid and is in many respects atypical. Even so, this paper is noteworthy because it was published more than 60 years ago. A few years later, Pauling (1939) noted that the boiling point of acetyl chloride is substantially higher (51°C) than that of trifluoroacetyl chloride. He proposed that this difference could be due to hydrogen bond formation. The formal introduction of the weak hydrogen bond into the chemical literature is, however, usually attributed to Glasstone (1937). It was long known that

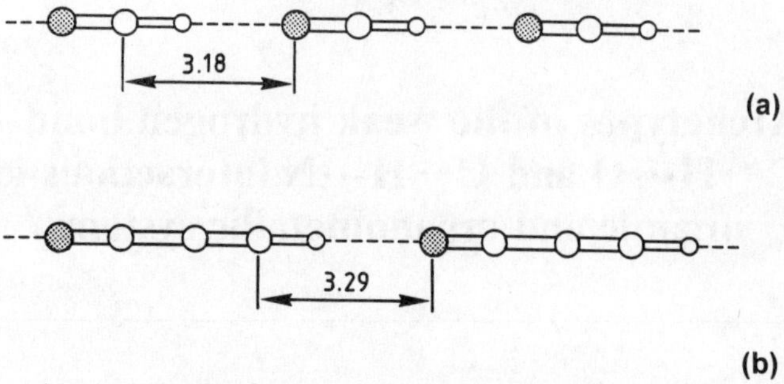

(a)

(b)

Fig. 2.1. (a) Crystalline HCN (Dulmage and Lipscomb 1951). (b) The related structure of cyano-acetylene (Shallcross and Carpenter 1958). From Steiner (1996a).

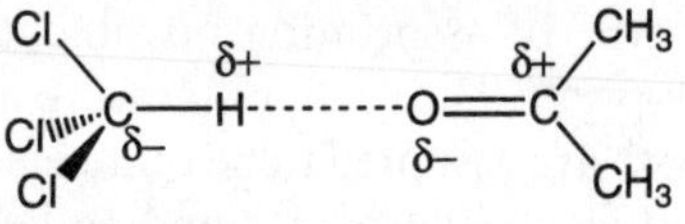

6

mixtures of chloroform with acetone or ether have abnormal physical properties, such as vapour pressures, viscosities and dielectric constants. Glasstone investigated such systems by polarization measurements on liquid complexes of haloforms with acetone, ethers and quinoline. He found that the molar polarizations of the mixtures are larger than those of the pure components, and this was explained by the association of molecules by directional electrostatic interactions of the type shown in **6**.

This idea was rapidly accepted by spectroscopists and by 1939, Gordy had already termed this interaction a *hydrogen bond* on the basis of IR evidence. Owing to the importance of IR spectroscopy in the study of strong hydrogen bonds, numerous related studies were subsequently performed. These focused on the diminution of the $C-H$ stretching frequency in the IR spectrum, v_{CH}, in the presence of electronegative atoms. The largest bathochromic shifts, more than $100\,cm^{-1}$, which approach the v_{XH} shifts in $O-H\cdots A$ or $N-H\cdots A$ bonds, are observed for activated $C-H$ groups such as are seen in alkynes, $C\equiv C-H$, or in $C-H$ groups adjacent to electronegative groups. Much of this work was reviewed and nicely interpreted by Allerhand and Schleyer (1963). Haloform donors and acceptors such as dimethyl sulfoxide and pyridine were highlighted. It was noted by these authors that v_{CH} of 1,3,5-trichlorobenzene is depressed by $35\,cm^{-1}$ in the presence of pyridine and this could well be the first reported evidence of hydrogen bond formation by aromatic H atoms. Among the important conclusions in this review, many portions of which are valid even today, is that 'the ability of a $C-H$ group to act as a proton donor depends on the carbon hybridization, $C(sp)-H\,>$

$C(sp^2)-H > C(sp^3)-H$, and increases with the number of adjacent electron withdrawing groups.' Interestingly, these authors also concluded that while a significant spectral shift in the IR is definite evidence of $C-H\cdots O$ hydrogen bonding, lack of shift does not necessarily indicate the absence of such bonding. This becomes particularly relevant in the modern context when the donor abilities of poorly activated $C-H$ groups such as—CH_3 and non-activated groups such as $-CHO$ are being scrutinized.

With regard to activated systems, the spectroscopic evidence was always more convincing. The relatively high acidity of the $C(sp)-H$ group results in marked and consistent spectral shifts upon hydrogen bonding. The monograph by Green (1974) provides a comprehensive review of all spectroscopic and crystallographic work, and mentions many alkynes. Satisfyingly, the formally forbidden $C\equiv C$ stretching vibration in symmetrical alkynes appears in the IR spectrum in the presence of electron donor solvents like acetone. Hydrogen bonding by acetylenes remains an active field of investigation (Steiner 1998b). The acetylene–water, acetylene–dimethyl ether and acetylene–acetone systems have all been investigated spectroscopically in matrix isolation and the uniform conclusion is that hydrogen bonding mediates in the stabilization of these complexes (DeLaat and Ault 1987).

Given, however, that the hydrogen bonds formed by $C-H$ groups are, on the whole, weaker than the classical hydrogen bonds formed by $O-H$ and $N-H$ groups, and given also that spectroscopic measurements are concerned with energy levels, it causes little surprise that there was a shift in emphasis in studies of $C-H\cdots O$ and $C-H\cdots N$ hydrogen bonding, from spectroscopy to crystallography. The paper of Dulmage and Lipscomb (1951) on HCN has been mentioned already. Another early crystal structure for which $C-H\cdots N$ hydrogen bonding is discussed is that of cyanoacetylene (Fig. 2.1(b); Shallcross and Carpenter 1958). Both structures are composed of infinite linear chains and the authors readily interpreted the short $H\cdots N$ distances as hydrogen bonds. This was also well supported by IR data; in solid HCN, v_{CH} is 180 cm^{-1} lower than in the gaseous state, a shift almost half the corresponding one in ice.

Another relevant early crystal structure is that of dimethyl oxalate, reported by Dougill and Jeffrey (1953). The authors noted that the carbonyl O atoms coordinate tightly around the methyl groups ($D = 3.35, 3.54, 3.57$ Å), in the expected directions of the $C-H$ bonds (the H atoms were not seen), and invoked a significant bonding interaction that they termed 'polarization bonding'. Further, the authors suggested that these interactions are the reason for the anomalously high melting point of the substance (54°C), which is about 100°C higher than that of related carboxylic acid esters. The structure was redetermined much later and a dense network of $C-H\cdots O$ contacts may now actually be demonstrated (Fig. 2.2; Jones et al. 1989). The d distances are long (2.5–2.8 Å) but one may suppose that due to their large number, they are in fact responsible for the unusually stable crystalline association of dimethyl oxalate. The study of Dougill and Jeffrey can therefore be taken as the first evidence of a hydrogen bonded $-CH_3$ group. To be sure, the melting point of

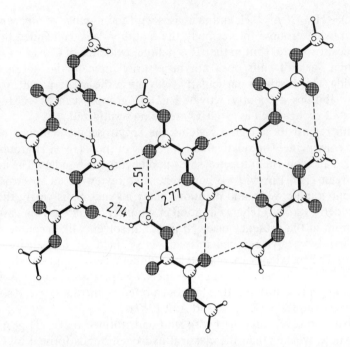

Fig. 2.2. Dimethyl oxalate. The methyl C—H group perpendicular to the plane of paper forms a contact to an O atom in the next layer with $d = 2.62$ Å. Drawn using coordinates from Jones *et al.* (1989). Distances d are normalized.

an organic compound depends not only on enthalpic but also on entropic factors, but given the observed crystal structure of dimethyl oxalate, the latter possibility seems not so important.

2.1.1 *Sutor's study*

The first systematic approach to C—H$\cdots$O hydrogen bonds in crystals was undertaken by Sutor in the early sixties (1962, 1963). Her field of study was the structure analysis of purines and pyrimidines, a group of compounds where short contacts of this kind occur frequently and attract attention not only by their geometry but also by their *repetitivity*. The latter kind of observation, that is the occurrence of similar contacts in many different crystal structures, would later become one of the strongest general arguments for the functional importance of C—H$\cdots$O interactions. Assuming a van der Waals distance of 2.6 Å for the H$\cdots$O contact, Sutor held in a paper in *Nature* in 1962 that C—H$\cdots$O hydrogen bonds exist in crystalline theophylline ($d = 2.25$ Å), caffeine ($d = 2.12$ Å), uracil ($d = 2.20, 2.27$ Å) and a few other compounds that contain C(sp^2)—H groups. A full paper in the *Journal of the Chemical Society* the following year extended this argument to other related compounds. Caffeine

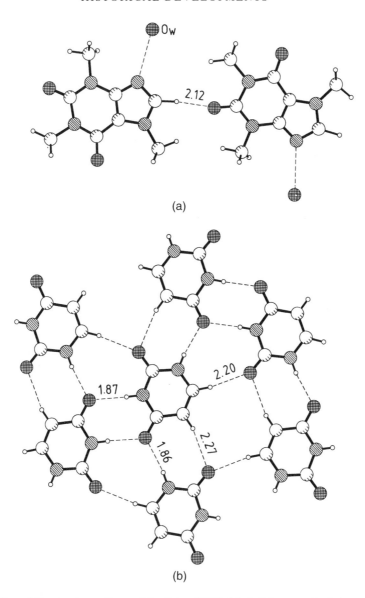

Fig. 2.3. Two of the structures discussed by Sutor (1963). (a) Caffeine monohydrate, water H atoms were not located (Sutor 1958). (b) Uracil (Stewart and Jensen 1967). Sutor used the earlier structure of Parry (1954). Distances d are normalized.

monohydrate and uracil are particularly interesting. The former is one of Sutor's own structures and there is a short $C-H \cdots O$ contact ($D = 3.18$, $d = 2.12$ Å) that is around 0.5 Å shorter than the van der Waals separation (Fig. 2.3(a)). This is a good example of a $C-H$ donor which is activated by two adjacent electronegative atoms. In uracil also, activated $C(sp^2)-H$ donors bond to $C=O$ acceptors and all the H atoms in the molecule are engaged in hydrogen bonding (Fig. 2.3(b)). While one $C=O$ group accepts two $N-H\cdots O$ bonds of similar geometry, the other accepts two $C-H\cdots O$ bonds. If the $C-H\cdots O$ interactions are neglected, this would represent a highly unlikely situation in that one O atom is a double acceptor while the other is completely unsatisfied. This circumstance was later taken by Leiserowitz (1976) as a strong argument in favour of the structural importance of $C-H\cdots O$ hydrogen bonding. An intramolecular $C-H\cdots O$ hydrogen bond was proposed by Sutor also in muscarine iodide, a suggestion that was confirmed many years later by Kroon *et al.* (1990) for a whole group of muscarinic agonists.

In 1968, Sutor's interpretation was criticized forcefully, and as it turned out later, incorrectly, by Donohue (1968) who called into question the value of 2.6 Å as the 'normal' $O\cdots H$ contacting distance. Using values suggested by Ramachandran *et al.* (1963) for 'normally allowed' (2.4 Å) and 'outer limit' (2.20 Å) $O\cdots H$ distances, he claimed that no special bonding effects were needed to explain the $C-H\cdots O$ geometries of Sutor and that in effect the 2.6 Å value itself was inappropriate. Ironically, Ramachandran had always believed the $C-H\cdots O$ interaction to be a stabilizing one and used the term '$C-H\cdots O$ bond' in a straightforward way and without any qualification in papers on collagen and polyglycine (Ramachandran and Sasisekharan 1965; Ramachandran *et al.* 1966). Figure 2.4 is a reproduction of the polyglycine II structure taken from the 1966 publication and the reader will note that the $N-H\cdots O$ and $C-H\cdots O$ bonds are given equal emphasis. Apparently, Ramachandran had deemed the phenomenon of $C-H\cdots O$ hydrogen bonding to be so unexceptional that he did not elaborate further on this issue! Indeed, according to Ramakrishnan (1988), Ramachandran had suggested the quoted distance ranges merely to alert crystallographers to the possibility of occurrence of unusually short $H\cdots O$ separations with carbon donors. In retrospect, though, while Donohue was incorrect and perhaps even harsh in his criticism of Sutor's analysis, her choice of structural examples was in some ways unfortunate. One of the structure determinations has an incorrect space group (acetylcholine bromide), one is a projection structure in which two of the three projections are very poor (muscarine iodide) and in one (glycyl-L-tyrosine hydrochloride), the relevant O atom belongs to a water molecule and shows abnormal thermal motion. So while Sutor's hypothesis has been amply confirmed in subsequent years, there was some scope for criticism on technical grounds at the time she published her work.

The entire episode is fascinating in terms of the sociology of scientific research in that it illustrates what can happen when a very prominent scientist is dramatically in error. Donohue's criticism of Sutor sensitized an entire generation of crystallographers against the idea of the $C-H\cdots O$ hydrogen bond. Some opposed the idea directly while others were equivocal at best. In a neutron study of the dipeptide glycylglycine, Koetzle *et al.* (1972) noted a short $C-H\cdots O$ contact of 2.43 Å. In a cautious formulation, these authors

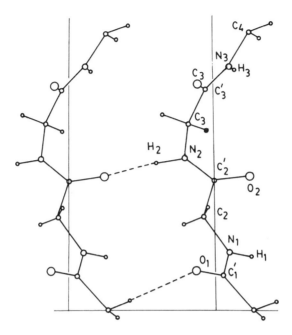

Fig. 2.4. Fibre diffraction model of polyglycine II proposed by Ramachandran *et al.* (1966). There are two intermolecular hydrogen bonds, $N-H\cdots O$ with $D = 2.73\,Å$ and $C-H\cdots O$ with $D = 3.20\,Å$ (redrawn from the publication).

stated that though one might not wish to call these contacts $C-H\cdots O$ hydrogen bonds, they probably help in stabilizing crystal structures. Of course none of this need be particularly surprising if one considers the then prevailing opinion that the $H\cdots A$ separation should be considerably less than the van der Waals value in any hydrogen bond (Sundaralingam 1998). Many years later though, Donohue himself (1983) was compelled to state that 'there are many features which deserve further study, among them are the stringency of linearity, the less frequently encountered bonds such as $C-H\cdots O$'. As it turned out, Donohue had also been equally strict about linearity, deeming it as being an essential attribute of hydrogen bonds and accordingly terming bifurcated hydrogen bonds, 'rare but not extinct'.

2.1.2 The dark ages

In the hiatus that followed Donohue's paper, the dark ages of the weak hydrogen bond as it were, only a few papers on this subject even appeared. Ferguson and co-workers reported several $C-H\cdots O$ hydrogen bonds in terminal alkynes, one of which is shown in Fig. 2.5, and the structural studies were supported by spectroscopic data (Ferguson and Tyrrell 1965; Ferguson and Islam 1966; Cameron *et al.* 1969). Sim and co-workers carried out studies on the crystallographic and spectroscopic properties of acetylenic esters, wherein the

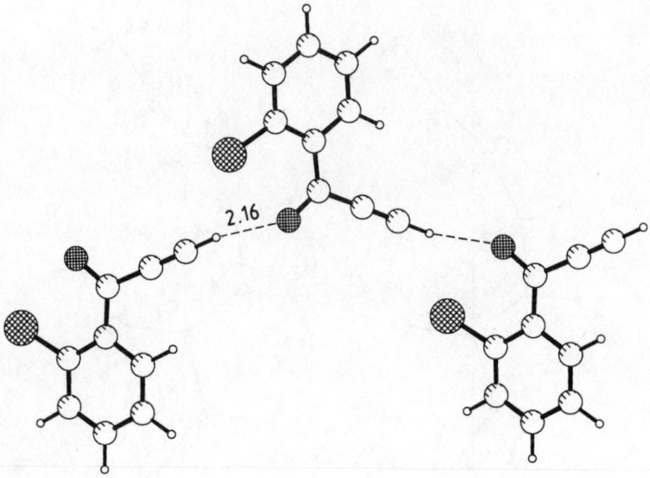

Fig. 2.5. *o*-Chlorobenzoacetylene with a short $C\equiv C-H\cdots O$ hydrogen bond (Ferguson and Islam 1966). Distance *d* is normalized.

ethynyl group forms $C-H\cdots O$ interactions (Calabrese *et al.* 1966, 1970). In propargyl 2-bromo-3-nitrobenzoate, for example, there is an $H\cdots O$ separation of 2.39 Å and a spectral shift of 47 cm^{-1} between solution and the solid state. However, clean correlations between crystallographic and spectroscopic properties were not seen. A review by Bernstein, Cohen and Leiserowitz (1974) on the structural chemistry of quinones is more interesting. Two ideas were elaborated by these authors:

(1)   a correlation of the strength of $C-H\cdots O$ interactions in terms of $H\cdots O$ distances may be misleading – a better indicator of the existence of hydrogen bonding is the occurrence of a constant geometrical pattern of molecules;

(2) $C-H\cdots O$ interactions are 'lateral' in the crystal structures of planar molecules.

These conclusions have been thoroughly substantiated during the last two decades and it is impressive that these authors were able to arrive at them, surveying but a very small number of crystal structures. The review is also important in that it highlighted the important crystal structure of 1,4-benzoquinone (Fig. 2.6). This structure is to the entire subject of weak hydrogen bonding, what naphthalene is to close-packing and rocksalt is to ionic interactions. The formation of $C-H\cdots O$ hydrogen bonds in benzoquinone is facilitated by the acidity of the $C(sp^2)-H$ group. Translationally related molecules are linked by $C-H\cdots O$ interactions to form a linear ribbon. Adjacent ribbons, related by 2_1 screw axes, are also linked by $C-H\cdots O$ interactions.

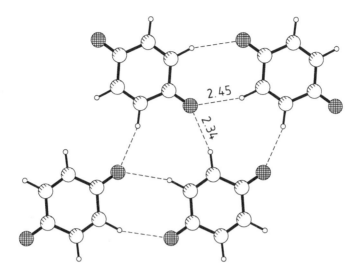

Fig. 2.6. 1,4-Benzoquinone (Trotter 1960). Notice the centrosymmetric dimer arrangement. Distances d are normalized.

Such an arrangement results in a characteristic triangular pattern or supramolecular synthon (Chapter 4) consisting of two molecules in the reference ribbon and one in the adjacent ribbon. This synthon is discussed again in Chapter 3 in the context of $C-H\cdots F$ interactions.

Another important paper during this period is a review by Leiserowitz (1976) of carboxylic acid crystal structures. Here, the involvement of the $C-H\cdots O$ interaction in crystal packing is revealed through changes in the conformational preferences of α,β-unsaturated acids. Noting that the conformation of the carboxyl group is almost exclusively *syn*planar in α,β-saturated acids while there is a roughly equal possibility of *syn* and *anti*planar conformations in α,β-unsaturated acids, the author was able to correlate the appearance of the unexpected *anti* conformation in the unsaturated acids with stabilization from $C-H\cdots O$ interactions. A typical example is fumaric acid which has dimorphs with similar layered crystal packing (Fig. 2.7). Since the *syn* conformation is slightly more stable in the isolated molecule, the *anti* conformation must be stabilized by other interactions which are identified as intralayer lateral $C-H\cdots O$ bonds. The behaviour of the higher analogue, *trans, trans*-muconic acid was also discussed at length. In α-*trans*-cinnamic acid too, the *anti* conformation is stabilized by lateral $C-H\cdots O$ interactions.

Other papers published during this period need to be mentioned simply because there were so few of them. Crystallographic and spectroscopic evidence (NMR, IR) were used by Sammes *et al.* (1976) to postulate an intramolecular $C-H\cdots N$ hydrogen bond in 3-cyanomethylsulfonyl-2-morpholinocyclohexane. In the light of the breakthrough Taylor–Kennard

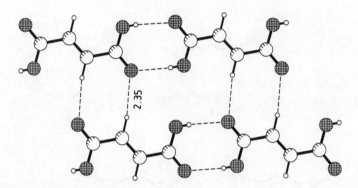

Fig. 2.7. Triclinic form of fumaric acid (Bednowitz and Post 1966). Distances *d* are normalized. The monoclinic form is very similar (Berkovitch-Yellin and Leiserowitz 1982).

paper (see below), these authors were able to substantiate their suggestions later (Li and Sammes 1983*a,b*; Harlow *et al.* 1984*a,b*) with related molecules. An early *ab initio* study (Vishveshwara 1978) analysed the proton donor ability of the C−H group in terms of its hybridization and substitution pattern, while Batail *et al.* (1981) confidently stated that a C−H$\cdots$O hydrogen bond exists in the tetrathiafulvalene (TTF)–chloranil molecular complex. In their most interesting study, these authors found that this complex undergoes a neutral to ionic phase transition at 84 K accompanied by a structural contraction within a plane particularly rich in C−H$\cdots$O hydrogen bonds. It is noteworthy that in a paper written so long ago the authors have stated that when the TTF molecules become ionic, the C−H groups of TTF$^+$ become activated and that this should lead to an increase in the strength of the C−H$\cdots$O interactions, leading in turn to the structural contraction. Unusually strong C(sp^2)−H$\cdots$O bonds, $d = 2.12$ Å, were found by Goldberg and Kosower (1982) in 9,10-dioxa-*syn*-(hydro-chloro)bimane. Turning to the bio-organic field, the occurrence of short C−H$\cdots$O hydrogen bonds in nucleosides and nucleotides was convincingly demonstrated by Shefter and Trueblood (1965), Sundaralingam (1966), Sussman *et al.* (1972) and Takusagawa *et al.* (1979). C−H$\cdots$O hydrogen bonding in complex formation of crown ethers was shown by Goldberg (1975). However, papers such as these were few and far between during the entire period. In general, a survey of the literature between 1968–82 conveys a scattered and sporadic impression that simply seems to lack a focal point, a point of departure into systematic research activity.

2.1.3   *The Taylor–Kennard paper*

Such a focus was provided, and in ample measure, by the very important paper of Taylor and Kennard in 1982. This sophisticated study employed the Cambridge Structural Database (CSD), which was at the time a new technique, and was based on 113 high quality organic crystal structures determined by neutron diffraction. The exclusive use of neutron data eliminated many vexatious and controversial issues concerning the exact location of the H atoms in these structures. The study provides conclusive evidence of the existence of C—H···O hydrogen bonds in crystals. The authors analysed the nearest-neighbour contacts of each H atom bonded to C and made the following observations and conclusions:

1. The frequency of contacts from H(C) to O, C, H, etc. is compared with the stoichiometry of the crystal structures. It is revealed that H(C) atoms have a statistical preference for contacts to O rather than to C or to other H atoms.

2. C—H···O contacts frequently have d distances less than 2.4 Å. Since the van der Waals distance is 2.6 Å, this was regarded as a certain indication of hydrogen bonding.

3. The geometrical characteristics of short C—H···O contacts are similar to those of O—H···O and N—H···O hydrogen bonds. This includes the property of directionality on the donor as well as on the acceptor side of the contact.

4. From these directionality properties, one may conclude that the C—H···O bond is a cohesive and stabilizing electrostatic interaction.

5. The tendency of H(C) to form a short contact with O increases, if C is bonded to an N atom. This tendency is most pronounced when this N atom bears a formal positive charge.

6. There is also evidence that C—H···N, C—H···Cl and maybe also C—H···S contacts represent hydrogen bonds. The shortest C—H···Cl contacts have d distances around 2.6 Å.

These conclusions are fully valid today. However, it should be noted that van der Waals cut-off criteria were applied rigidly in the study. This was probably inevitable, considering the earlier background, but it gives the impression that the arguments are valid only for the shortest contacts. In the years that followed, researchers have somewhat uncritically adopted these restrictions and have considered C—H···O geometries as hydrogen bonds *only* if $d < 2.4$ Å, just because this is the upper limit in the tabulations of Taylor and Kennard. This matter is of some import and is discussed in greater detail in Section 2.2.2. One notes further that Taylor and Kennard did not address the question of the importance of C—H···O hydrogen bonds in determining organic crystal packing, merely stating that 'the frequency with which they occur suggests that they play a significant role' in this regard. Yet, it was mostly through a consideration of the effects of C—H···O and C—H···N hydrogen bonds on crystal packing that a relaxation of the original distance criteria became necessary (Desiraju 1991a). This having been said, there is little doubt that it was the Taylor and Kennard paper that put the subject of weak hydrogen bonds in a proper, scientific and non-emotional perspective. We have no problem therefore in terming this study as the end of the historical phase of the subject.

2.2 General properties

The period following the Taylor–Kennard article saw a rapid increase in the number of papers on weak hydrogen bonds and it gradually became possible to discern more clearly the nature and manifestations of these interactions. The following sections describe the most important properties of $C-H\cdots O$ and $C-H\cdots N$ bonds. Vibrational spectroscopy has always been regarded as the ultimate criterion for assigning hydrogen bond character to an interaction. Length and angle properties can be readily obtained from crystallographic databases but what has been more difficult to establish is the chemical meaning and possible validity of various geometrical cut-off limits. $X-H$ bond lengthening and reduction of thermal vibrations are classical manifestations of hydrogen bonding that are also observed for $C-H$ groups. Efforts to study $C-H\cdots O$ hydrogen bonding with high level computations are still evolving, as are descriptions of a $C-H\cdots O$ interaction as a soft hydrogen bond. Of a somewhat less contentious nature is the importance of cooperative effects in establishing patterns of weak hydrogen bonds. The repetitive character of $C-H\cdots O$ and $C-H\cdots N$ patterns in molecular crystals has since early times been taken as evidence for their importance as stabilizing interactions. The understanding of intramolecular bonds is much less satisfactory and their properties have yet to be properly evaluated. An intriguing possibility of $C-H\cdots O$ geometries that are actually destabilizing rather than stabilizing is provided by some very weak $C-H$ donor groups. Finally, there is a small though growing number of reports wherein these weak hydrogen bonds are implicated in solution processes. In the following subsections, these ideas will be elaborated further.

2.2.1 *Vibrational spectroscopy*

When comparing the IR or Raman vibrational spectra of a 'free' $R-X-H$ group and an $R-X-H\cdots A$ hydrogen bond, the hydrogen bond formation may have several manifestations:

(1)   the $X-H$ stretching frequency v_s is reduced;

(2)   the band width of v_s increases;

(3)   the intensity of v_s increases; and

(4)   the frequency of the $R-X-H$ bending vibration v_b increases.

Further, for strong hydrogen bonds, a new band may appear in the far infrared region for the H···A stretching vibration. Bathochromic or red shifts of v_s are usually accepted as a criterion of hydrogen bond formation. Whereas such shifts are observed routinely in conventional hydrogen bonds, this is true only in activated systems for C—H···O hydrogen bonds. As mentioned in Chapter 1, spectroscopic measurements are a measure of bond energies while diffraction techniques yield information on bond metrics. For weak bonds therefore, spectroscopic shifts are considered to be more reliable indicators of hydrogen bond formation than changes in H···A separations. However, the lack of a spectral shift does not in itself constitute evidence for the absence of hydrogen bonding.

Of all the C—H donors, the C≡C—H group is by far the best suited for IR studies in the solid state. There are several reasons for this assessment. Because of the high acidity of this group, the hydrogen bonds formed and the associated spectral effects are relatively strong. The C(sp)—H stretching band is intense and displaced from the bands of all the other C—H groups, and this means that it can be unambiguously identified in the absorption spectrum. Furthermore, the C(sp)—H stretching vibration is robust with respect to interference from other C—H oscillators, so that vibrational coupling effects do not play a major role. Because of all these reasons, C≡C—H···O hydrogen bonds have been studied for many years with a combination of crystallography and vibrational spectroscopy. However, the early results are not unambiguous in quantitative terms (Calabrese et al. 1966). It was only just over a decade back that Desiraju and Murty (1987) showed that v_s clearly correlates with D (black dots in Fig. 2.8) in nine terminal alkynes. In the years since, a number of further combined crystallographic and spectroscopic studies have appeared (for instance see, Brodkin and Foxman 1996), the results of which are included as square symbols in the figure. These subsequent studies considerably enlarge the range of the correlation, in the directions of both short and long hydrogen bonds. The short end is now occupied by the C≡C—H···OH$_2$ hydrogen bond in (diethynylbenzene)·(H$_2$O)$_2$·(triphenylphosphine oxide)$_4$ [D = 3.06 Å; v_s, 3106 cm^{-1}]. This hydrogen bond possesses both geometric and spectroscopic characteristics of a moderately strong hydrogen bond and is not dissimilar to an O—H···O—H hydrogen bond (Kariuki et al. 1997). The data point at the long end of the correlation (D = 3.71, d = 2.92 Å, θ = 130°; v_s = 3293 cm^{-1}; Δv_s = −22 cm^{-1}) originates from the crystal structure of the tricarbonyl compound [Cr(CO)$_3$\{η^6-[7-exo-(C≡CH)C$_7$H$_7$]\}], **7**. This C≡C—H···O contact to a CO ligand is possibly the longest for which effects on the vibrational spectrum have been experimentally demonstrated (Steiner et al. 1998a).

Figure 2.8 is essentially a geometry–energy plot and shows a good correlation that is valid over a very wide D-range from 3.0 to 3.7 Å. While long bonds with D > 3.7 Å are very much weaker than short bonds with D ∼ 3.0 Å there

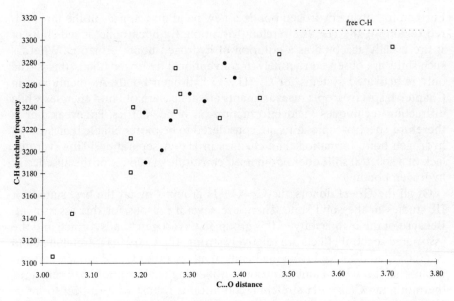

Fig. 2.8. Correlation of the infrared C—H stretching frequency (in cm^{-1}) with the C$\cdots$O distance (in Å) in some C$\equiv$C—H$\cdots$O hydrogen bonds. The black dots are from the paper of Desiraju and Murty (1987) and the squares show data taken from later publications. The data points at the extremities are: $D = 3.02$ Å, $\nu_{CH} = 3106$ cm^{-1} (Kariuki *et al.* 1997); $D = 3.06$ Å, $\nu_{CH} = 3144$ cm^{-1} (Steiner *et al.* 1997b); $D = 3.71$ Å, $\nu_{CH} = 3293$ cm^{-1} (Steiner *et al.* 1998a). Free R—C$\equiv$C—H groups have ν_{CH} values typically around 3310 cm^{-1} with some variation depending on R.

is no indication that the hydrogen bond is 'switched off' at a critical distance. Rather, the figure shows that it gradually becomes weaker over a broad distance range. Concerning distance cut-off criteria for hydrogen bonds, Fig. 2.8 clearly demonstrates the debilitating effect of a short cut-off value like $D = 3.2$ Å used in former times – the larger part of the correlation would be simply lost and only the very strongest C—H$\cdots$O bonds would remain.

The data points in Fig. 2.8 contain a large scatter. The reader should note

8

that this scatter is not due to experimental uncertainties but that it is intrinsic. Hydrogen bonds with very similar v_s values can have significantly different geometries and, conversely, bonds with similar geometries can have widely differing v_s values. This lack of correspondence between D and v_s is to a large degree due to the influences of the molecular surroundings and environment. An example of the influence of cooperativity is discussed in detail in Section 2.2.7; in crystalline mestranol there are two $C{\equiv}C-H\cdots O$ bonds with almost the same D values but they are in different cooperative situations (Steiner *et al.* 1997*a*). One is fortified by an accepted $O-H\cdots C{\equiv}C$ interaction and exhibits a lower v_s ($D = 3.44\,\text{Å}$; $v_s = 3248\,\text{cm}^{-1}$) while the other is in an anticooperative situation and has a much higher v_s ($D = 3.48\,\text{Å}$; $v_s = 3287\,\text{cm}^{-1}$).

Studies of $C-H\cdots O$ bonding in crystalline alkynes have been further elaborated by the Utrecht group who have observed several subtle effects. For the strong $C{\equiv}C-H\cdots O$ bond in 2-ethynyladamantan-2-ol, Steinwender *et al.* (1993) have studied the effects on absorption bands other than v_s. They reported wavenumber shifts upon hydrogen bonding of around $-130\,\text{cm}^{-1}$ for v_s, $+98\,\text{cm}^{-1}$ for the in-plane bending and $+114\,\text{cm}^{-1}$ for the out-of-plane bending modes. In a study of eight 17-hydroxy-17-ethynyl steroids, Lutz *et al.* (1994) observed that v_s of the strongest $C(sp)-H\cdots O$ hydrogen bonds are sensitive to temperature variation as is seen in $O-H\cdots O$ hydrogen bonded systems. Typically, the hydrogen bonds become shorter upon cooling and Δv_s increases. However, spectroscopic shifts between dilute solution and the solid state were seen to be more quantitative indicators of hydrogen bonding than shifts observed upon temperature variations in crystals. IR and Raman spectra were studied by Lutz *et al.* (1996) for the drug *N*-(2,6-dimethylphenyl)-5-methylisoxazole-3-carboxamide, **8**, which shows a short $C(sp^2)-H\cdots O$ interaction ($D = 3.29$, $d = 2.29\,\text{Å}$, $\theta = 159°$). Very interestingly, the frequency v_s is in this case almost unaffected by hydrogen bonding, whereas the band intensity increases dramatically. Similar behaviour is long known for chloroform for which the v_s band broadens and intensifies, but does not shift.

Some NMR work has been done in connection with the effects of $C-H\cdots O$ hydrogen bonding. Scheffer *et al.* (1985) have noted that for $C-H\cdots O{=}C$ hydrogen bonds formed in crystalline quinones, the ^{13}C solid state NMR chemical shifts of the carbonyl C atoms move downfield by

1–3 ppm relative to the corresponding solution values. In comparison, the analogous shift in a conventional $O-H\cdots O=C$ hydrogen bond is around 8 ppm. A correlation between ^{17}O NMR chemical shifts in solution and $C-H\cdots O$ bond forming ability is detailed in Section 2.2.12.

2.2.2 Length properties

The length, d (or D), of a hydrogen bond $X-H\cdots A$ is one of its most characteristic and distinctive attributes. In conventional or strong hydrogen bonds, the $H\cdots A$ and $X\cdots A$ distances (respectively, d and D) are considerably shortened when compared to the sums of the respective van der Waals radii. For example, in most kinds of $O-H\cdots O$ and $N-H\cdots O$ hydrogen bonds, the $O\cdots O$ and $N\cdots O$ distances are centred around distinct values (such as 2.60 Å for carboxylic acid dimers and 2.82 Å for hydrogen bonds between water molecules). Typically, the entire distance range falls shorter than the sum of the $H\cdots A$ or $X\cdots A$ van der Waals radii and clusters within a narrow range, typically 0.2 Å wide. For the weaker $C-H\cdots O$ and $C-H\cdots N$ hydrogen bonds, the situation is far less clear-cut. By their very nature, such bonds can be distorted easily by other crystal forces and $C\cdots O$ (or $C\cdots N$) distances with $3.0 < D < 4.0$ Å or $2.0 < d < 3.0$ Å are usual. At this stage, van der Waals radii are no longer helpful because the van der Waals sum occurs roughly around the mid-point of the distance range. Since it is customary to assess the strengths of hydrogen bonds by their lengths, a not necessarily rigorous or even correct procedure, much effort has been expended in establishing the distance ranges within which it may be reasonably expected to find hydrogen bond type behaviour.

2.2.2.1 Very short $C-H\cdots O$ hydrogen bonds

The shortest $C-H\cdots O$ bonds known today have d and D separations of around 2.0 and 3.0 Å, respectively, and are formed by highly activated H atoms. Such activation arises because of acidity or cooperativity reasons. Highly acidic groups include those in $CHCl_3$, $CH(NO_2)_3$, HCN, $C\equiv CH$ and N^+-CH. A number of very short $C-H\cdots O$ hydrogen bonds formed by such highly activated donors is shown in Fig. 2.9. The strongest donor in the figure is $CH(NO_2)_3$, which forms a crystalline adduct with dioxane with $D = 2.94$ Å (Bock *et al.* 1993). Particularly short bonds are formed if a strong donor is combined with a strong acceptor like $P=O$. A deliberately designed example is shown in Fig. 2.9(c) with the complex triphenylsilylacetylene–triphenylphosphine oxide (Steiner *et al.* 1997b). Activation by cooperativity effects can be very pronounced too, and will be described in more detail in Section 2.2.7.

$C-H\cdots N$ interactions are somewhat longer than $C-H\cdots O$ bonds. Even for the very strong carbon acid HCN, D is clearly longer than in the $C-H\cdots O$ examples shown in Fig. 2.9. An early paper by Jaskólski (1984) reports a short $C-H\cdots N$ bond ($D = 3.20$, $d = 2.30$ Å, $\theta = 156°$) that is rationalized on the basis that the $C-H$ group is located between two N atoms one of which bears a partial positive charge.

It should be noted, however, that very short D separations need not invariably be associated with very short d separations. In fact the converse is often true with say D being less than $3.0\,\text{Å}$, but θ far from $180°$ and d greater than $2.5\,\text{Å}$. A reliably determined example for an intermolecular contact of this kind is shown in Fig. 2.10 ($D = 2.97$, $d = 2.52\,\text{Å}$, $\theta = 104°$). This occurs in the neutron crystal structure of 3-amino-1,6-anhydro-3-deoxy-β-D-glucopyranose (Noordik and Jeffrey 1977). A less extreme case, this time for an intramolecular contact ($D = 2.95$, $d = 2.22\,\text{Å}$, $\theta = 123°$) has been provided by Kapteijn *et al.* (1993). Noting that hydrogen bond character decreases with decreasing θ values essentially vanishing around $\theta = 90°$, it may be concluded that these highly bent and short $C\cdots O$ geometries do not really correspond to significant hydrogen bonding interactions. Such geometries will often have close to zero interaction energies (see Section 2.2.6), be repulsive or in the extreme case, even destabilizing (see also Fig. 1.3 and Section 2.2.11), and intramolecular cases (Section 2.2.9) must always be viewed with some circumspection. Finally it may be noted that while short $C\cdots O$ separations are often accompanied by small θ angles as described above, the chemically unreasonable combination of a short $H\cdots O$ distance and a small θ angle is seldom observed.

2.2.2.2   *Distribution of distances*

The separations d in $C-H\cdots O$ hydrogen bonds cover a wide range. If large numbers of crystal structures are analysed, one observes statistical distribution functions for the d values and these functions have different shapes for different donor types. Since the distribution of d values is not yet well documented for most types of $C-H$ groups, the situation is somewhat loosely sketched in Fig. 2.11.

For activated $C-H$ donors, the d distributions have a well-defined maximum, probably corresponding to an optimal value, and fall to lower frequencies at longer separations. This sort of behaviour is reminiscent of conventional hydrogen bonds. For weak $C-H$ donors, no maximum is observed and the frequency of $H\cdots O$ contacts increases with increasing distance. This is found for example in carbohydrates and in phenylalanine residues in proteins. For moderately activated $C-H$ donors, the situation lies between these extremes. For the weaker donors, the long distance region of the distribution cannot be interpreted satisfactorily and there is a gradual and ill-defined transition from hydrogen bonds to non-directional contacts. This region is even more blurred for the weakest of hydrogen bonds, namely $C-H\cdots\pi$ and $C-H\cdots F-C$, and very difficult to understand. The problems that arise in the assignment of hydrogen bonding character based upon distance arguments have been explicitly identified (Steiner and Saenger 1992*a*; Braden and Gard 1996), and incidentally these problems also occur with long $O-H\cdots O$

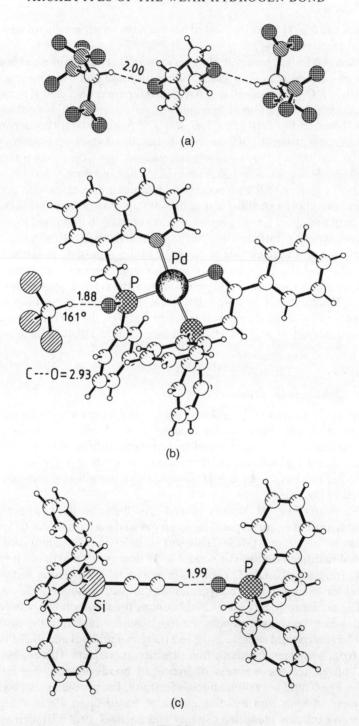

Fig. 2.9.

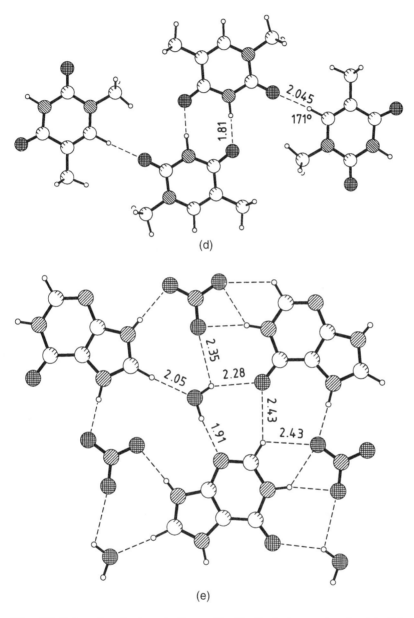

(d)

(e)

Fig. 2.9. Extremely short and linear C—H···O hydrogen bonds. (a) 2:1 Adduct trinitromethane–dioxane, d = 2.00, D = 2.94 Å, θ = 143° (Bock *et al.* 1993). (b) Chloroform bonding to a P=O acceptor in a Pd complex (Balegroune *et al.* 1988). (c) 1:1 Complex triphenylsilylacetylene–triphenylphosphinoxide, θ = 171° (Steiner *et al.* 1997*b*). (d) Neutron structure of 1-methylthymine (Kvick *et al.* 1974). (e) Hypoxanthinium nitrate monohydrate, θ = 173° (Rosenstein *et al.* 1982). Geometries are normalized.

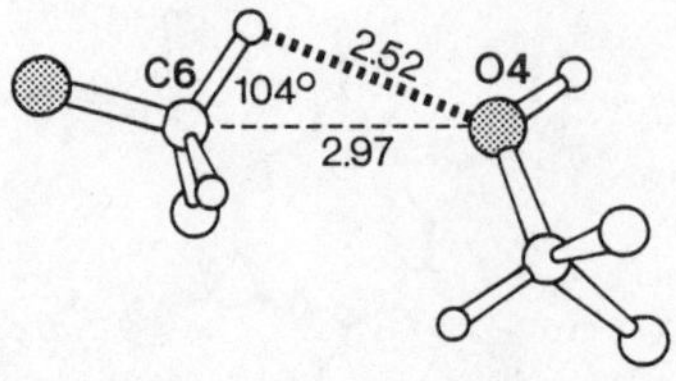

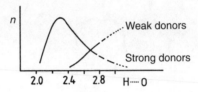

Fig. 2.10. Very short but non-linear intermolecular C—H···O contact in the neutron crystal structure of 1,6-anhydro-3-deoxy-β-D-glucopyranose (Noordik and Jeffrey 1977). From Steiner and Saenger (1992a).

Fig. 2.11. Typical shapes of d-distance distributions in crystals for C—H···O interactions formed by strong and weak C—H donors, respectively. Moderately activated donors have an intermediate behaviour (from Steiner 1996a).

hydrogen bonds (Steiner and Saenger 1992b). None of this should come as a surprise. It must be emphasized that the entire phenomenon of hydrogen bonding ranges from the very strong to the very weak. All attributes of hydrogen bonding, be they geometrical, crystallographic or spectroscopic reflect this continuum of character.

2.2.2.3   *Statistical studies and distance cut-offs*

The basic problem in evaluating the length attributes of a weak interaction such as the C—H···O hydrogen bond is the necessarily high noise-to-information ratio. Since the interaction is easily deformed, the variation in bond geometry is considerable and if the number of crystal structures sampled is small, even a careful crystallographic study may not lead to sensible conclusions. If large numbers of structures are examined, however, the occurrence of trends in the data may lead to meaningful chemical conclusions. This then is the essence of the database philosophy for the study of weak intermolecular interactions (Desiraju, 1989a). The existence of crystallographic databases such as the Cambridge Structural Database (CSD) that contain large amounts of high-quality crystallographic data (Allen *et al.* 1991), means therefore that increasingly reliable studies of weak intermolecular interactions are possible (see also Section 1.3.1). Contrary to belief held in some quarters, the use of databases is not merely archival, bibliographic or review-like in character. The use of statistical techniques on data that has been retrieved from crystallographic databases (Allen *et al.* 1983) makes it possible to draw *chemical* conclusions that are diverse and yet certain. In general, such conclusions cannot be obtained by examining individual structures or even a handful of structures. The larger the data sample, the weaker the interaction that may be probed. This way, through the mediation of statistics, chemistry emerges from crystallography.

The geometrical criteria which are selected to characterize a particular $C-H\cdots O$ geometry as a significant intermolecular interaction or as a hydrogen bond have also been the subject of discussion. There appear to be two distinct viewpoints:

(1) Use a conservative $C\cdots O$ threshold such as 3.25 or 3.3 Å and refer to longer separations as van der Waals interactions;

(2) use a more liberal cut-off threshold.

Historically, the Sutor–Donohue controversy in the 1960s arose because of the tendency to accept the former viewpoint. More recent analysis, however, has shown that many longer $C-H\cdots O$ contacts ($3.5 < D < 4.0$ Å) and contacts formed by the weaker $C-H$ acids have angular characteristics and effects on crystal structures that resemble the shorter contacts ($3.1 < D < 3.5$ Å) and contacts formed by strong carbon acids. We maintain therefore that the $C-H\cdots O$ bond is not really a van der Waals interaction but is primarily electrostatic, falling off much more slowly with distance and hence is viable at distances which are equal to or longer than the conventional van der Waals limit. Therefore the outer cut-off $C\cdots O$ distance must be liberal. In this respect, we regard the van der Waals cut-off employed by Taylor and Kennard (1982) as too limiting. This is perhaps the one criticism that could be raised against an otherwise impeccable study. However, these authors might be excused this stringent cut-off in that they were attempting to prove the very existence of the $C-H\cdots O$ hydrogen bond.

Others have expressed a more extreme opinion. Recently, Cotton *et al.* (1997), almost harking back to Donohue (1968), have stated that most $C-H\cdots O/N$ geometries are nothing more than classical van der Waals contacts and in this context *it must be clearly and unambiguously restated that the use of the van der Waals cut-off in assessing a $C-H\cdots O/N$ contact as a possible hydrogen bond is wrong.* Only for the most acidic of $C-H$ groups can this lead to results that are other than misleading. Mascal (1998) while rebutting the contention of Cotton *et al.* has shown that there are many $C-H\cdots N$ contacts shorter than the van der Waals limit and that these are mostly genuine hydrogen bonds. Mascal is correct but his interpretation is, in a sense, incomplete. Weak hydrogen bonds are just that and they do not become van der Waals interactions simply because they are long (Section 1.2.3). For the $O-H\cdots O$ distance distribution, the van der Waals limit is much greater than the maximum $O\cdots O$ distance. Therefore, the use of a van der Waals cut-off, though wrong, is harmless. For $C(sp)-H\cdots O$ geometries, the situation is largely similar. For the general $C-H\cdots O/N$ distribution, however, where distances d are generally longer, the van der Waals cut-off roughly bisects the distribution and the use of such a cut-off is now definitely ambiguous. For the even weaker $C-H\cdots\pi$ hydrogen bond, the van der Waals cut-off is nearly at the beginning of the distance distribution. Here, the use of a van der Waals cut-off would be catastrophic. And yet, the $O-H\cdots O$, $C-H\cdots O/N$ and $C-H\cdots\pi$ interactions have many similar attributes, properties and consequences.

In practical terms, the question is not whether this or that cut-off is correct or incorrect but rather that one uses a consistent and high value throughout. In the context of database studies, for this is where the issue of a cut-off arises most frequently, no cut-off indeed need be specified. For operational simplicity many workers have used a d cut-off of 2.8 Å for H-normalized data but even longer values have been considered as being justified and necessary, on occasion. More complex and accurate prescriptions have been suggested in studies of small, homogeneous groups of crystal structures but the above cut-off provides generality with the attendant simplicity. Of course mean geometries will inevitably depend on the particular d range chosen but this is not a problem if comparisons are being made, say within the overall group of structures being studied. Needless to say, while evaluating individual $C-H\cdots O$ hydrogen bonds, the greatest weight should be given to those contacts wherein short $H\cdots O$ separations ($2.0 < d < 2.3$ Å) are accompanied by large hydrogen bond angles ($150 < \theta < 180°$).

2.2.2.4 *Effects of donor acidity and acceptor basicity*

One of the strongest types of evidence for the hydrogen bond nature of the $C-H\cdots O/N$ interaction is that $H\cdots O$ distances decrease systematically with increasing acidity of the $C-H$ group, and also with increasing basicity of the O/N atom. As mentioned earlier, different types of $C-H$ groups have different abilities to act as hydrogen bond donors. This ability, or donor strength, generally depends on the carbon hybridization $C(sp) > C(sp^2) > C(sp^3)$, and within a particular hybridization state, it increases if the C atom is bonded to electron-withdrawing atoms or groups. This effect has been referred to as activation; unfortunately this term may lead to the faulty conclusion that non-activated $C-H$ groups may not act as donors at all. Indeed, relatively non-acidic groups can also form $C-H\cdots O$ hydrogen bonds, as in say acetic acid.

The earliest systematic study of the effects of donor acidity on $C-H\cdots O$ bond lengths deals with CSD studies for chloroalkyl compounds and for alkenes and alkynes (Desiraju, 1989b). Figure 2.12 is a histogram of $C\cdots O$ distances in chloroalkanes $(Cl_{3-n}R_n)C-H\cdots O$ with $0 \le n \le 3$ and where R is not a strongly electron-withdrawing substituent. Using a distance cut-off of $D < 3.90$ Å, mean $C\cdots O$ distances are 3.32(2) Å for $CHCl_3$, and 3.40(3), 3.46(2) and 3.59(1) Å for $RCHCl_2$, R_2CHCl and R_3CH, respectively. The histograms

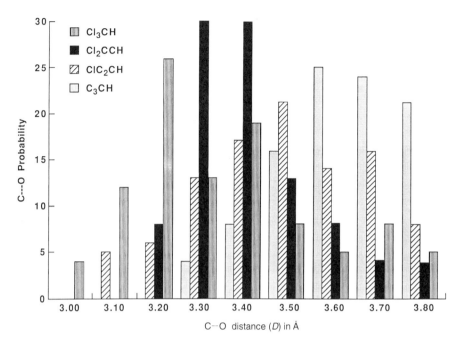

Fig. 2.12. Distributions of distances D in $C-H\cdots O$ hydrogen bonds formed by the donors Cl_3CH, Cl_2CCH, ClC_2CH and C_3CH (from Desiraju 1989b).

Table 2.1 Mean $d(H\cdots A)$ distances in hydrogen bonds of different $C-H$ solvent donors to the acceptors O, N and Cl^- (Steiner 1998c)

Donor	O acceptors (Å)	N acceptors (Å)	Cl^- acceptors (Å)
CHCl$_3$	2.31(1) [222]	2.37(3) [32]	2.38(3) [16]
C≡C−H	2.40(2) [145]	2.40(6) [12]	2.56(4) [8]
CH$_2$Cl$_2$	2.492(8) [356]	2.53(2) [58]	2.59(3) [34]
N≡C−Me	2.567(9) [323]	2.59(1) [114]	2.72(3) [20]
DMSO	2.56(1) [203]	2.65(2) [18]	2.94 [3]
Acetone	2.60(1) [92]	2.64(3) [11]	2.98 [1]

Distances d normalized, numbers of contacts given in brackets.

for these four groups of compounds are offset regularly showing that shorter $C-H\cdots O$ bonds are obtained for more acidic $C-H$ groups. A similar analysis was performed for $C-H\cdots O$ bonds involving alkyne ($-C\equiv C-H\cdots O$) and alkene ($>C=C-R)H\cdots O$ groups (Desiraju 1990). The mean $C\cdots O$ distance for 105 alkyne geometries was found to be 3.46(2) Å while that for 622 alkene geometries is 3.64(1) Å. This large difference can only be rationalized on the basis of carbon acidity differences between alkyne and alkene $C-H$ groups. It is very important that the sequence of donor strengths is independent of the acceptor type A. This is shown in Table 2.1 where mean d values

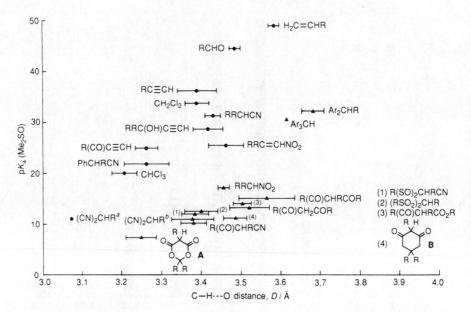

Fig. 2.13. Plot of distances $D(C\cdots O)$ against acidities pK_a (in Me$_2$SO) for C—H$\cdots$O hydrogen bonds donated by different kinds of C—H groups. Sterically unhindered carbon acids are shown as •, while for those marked as ▲, the approach of the C—H and O groups is more difficult. For (CN)$_2$CHR compounds, [a] represents a single unclathrated structure while [b] represents the average of all structures (from Pedireddi and Desiraju 1992).

are listed for hydrogen bonds of six C—H donor types to the acceptors O, N and Cl⁻.

More importantly, it became possible to generalize these results to a large variety of carbon acids (Pedireddi and Desiraju 1992). The CSD analysis was extended to around twenty categories of acids that include chloro- and cyano-alkanes, alkynes and α-substituted alkynes, active methylene compounds, nitroaliphatics and polyarylmethanes. A broad range of D values in the interval 3.0–4.0 Å was found (Fig. 2.13). However, these lengths are not distributed randomly, and mean D values for the various chemically distinct types of C—H group correlate well with mean pK_a values in DMSO for representative compounds within each of the functional groups considered. It is interesting to note that this correlation extends to C$\cdots$O distances as long as 4.0 Å. This follows from the dominantly electrostatic nature of the C—H$\cdots$O interaction. Further it may be noted that almost any kind of C—H group, ranging from those in weakly acidic alkanes to strongly acidic alkynes and cyclopropanes, forms C—H$\cdots$O hydrogen bonds.

Another feature to emerge from Fig. 2.13 is that acidity effects in the solid state depend on both electronic and steric factors. There are two broad categories of carbon acids in the figure. In the first category the C—H groups have single atoms as α-substituents (CHCl$_3$, CH$_2$Cl$_2$), are part of an sp^2 system

(RR'C=CH—NO$_2$, H$_2$C=CHR) or are otherwise sterically unhindered (RC≡C—H, RCHR'CN, RCHO; • in Fig. 2.13). Here, the mean D values are well correlated with pK_a values for the respective substituent group with the correlation being as good as the internal correlation between DMSO and gas phase acidities. In the second category, the C—H group is part of a bulkier moiety and the mean D values are significantly longer (3.38 to 3.52 Å) than those for the sterically unhindered C—H groups of the same acidity [(RSO$_2$)$_2$CHR, RR'CH—NO$_2$, (RCO)$_2$CHR; ▲ in Fig. 2.13]. These observations suggest that mean C···O distances in crystals are a good measure of carbon acidity when there is little or no steric hindrance to the approach of an O atom to a C—H group. When such steric hindrance exists, the mean approach of O does not fully reflect the group acidity. This steric factor is valid for other kinds of weak hydrogen bonds too. For example, hydrogen ligands bound to the metal core in crystalline organometallics can be completely embedded within the ligand sheath and, as such, are unable to form M—H···O hydrogen bonds, in contrast to sterically more accessible M—H groups that do form such bonds (Section 3.5; Braga *et al.* 1996a).

These results constitute in effect a crystallographic scale of carbon acidity, a property of fundamental importance in physical organic chemistry. This novel scale is of general applicability and can, in principle, be used for compounds which cannot be measured by the traditional gas phase and solution methods – compounds containing unstable anions, reactive substances, organometallics and cluster compounds.

Basicity effects. For conventional hydrogen bonds it has been shown that the basicity of the acceptor also affects the hydrogen bond length much like the acidity of the donor. As an example, mean distances d and D in hydrogen bonds of water molecules and C—NH$_3^+$ groups to a number of O atom acceptors are shown in Table 2.2. Note that H$_2$O, C=O, S=O and C(sp^3)OH have very similar acceptor strengths, whereas P=O is significantly stronger and C—O—C, NO$_2$ and M—CO are much weaker. That such dependence on acceptor basicity must also be expected for C—H···O hydrogen bonds has been explicitly stated by Taylor and Kennard (1982). In chemically heterogeneous structure samples, however, C—H donor acidities and acceptor basicities vary simultaneously. To reveal the anticipated basicity effects it is therefore best to hold acidity effects constant. Accordingly, a CSD analysis was performed on invariant C—H donors in varying crystal environments (Steiner 1994a). As an example, Fig. 2.14 shows histograms of H···O separations in C—H···O hydrogen bonds donated by CH$_2$Cl$_2$ and accepted by C=O and C—O—C, respectively. The contacts accepted by the ketonic acceptors are on average shorter than those accepted by ethereal acceptors, reflecting the higher acceptor strength of C=O. Considering further the bases P=O and M—CO, a series of observations with six different donors and eight different

Table 2.2 Mean distances $d(H\cdots O)$ and $D(X\cdots O)$ in hydrogen bonds of H_2O and NH_3^+ donors with various O atom acceptors (Steiner 1998c)

Acceptor	H_2O donor (Å)		$C-NH_3^+$ donor (Å)	
Mean d(H$\cdots$O) distance				
P=O	1.871(7)	[374]	1.856(7)	[336]
H_2O	1.903(3)	[3002]	1.918(8)	[325]
>C=O	1.903(3)	[3280]	1.918(6)	[1035]
>S=O	1.906(7)	[650]	1.936(12)	[162]
$C(sp^3)-OH$	1.921(5)	[795]	2.00(2)	[130]
C—O—C	2.018(11)	[307]	2.07(2)	[72]
$C-NO_2$	2.158(12)	[175]	2.19(2)	[37]
M—CO	2.29	[2]	2.35	[3]
Mean D(X$\cdots$O) distance				
P=O	2.793(5)	[374]	2.817(4)	[336]
H_2O	2.808(2)	[3002]	2.851(5)	[325]
>C=O	2.823(2)	[3280]	2.839(3)	[1035]
>S=O	2.824(5)	[650]	2.859(7)	[162]
$C(sp^3)-OH$	2.834(4)	[795]	2.90(1)	[130]
C—O—C	2.894(8)	[307]	2.94(1)	[72]
$C-NO_2$	2.988(9)	[175]	3.05(2)	[37]
M—CO	3.04	[2]	3.24	[3]

Distances d normalized, numbers of contacts given in brackets.

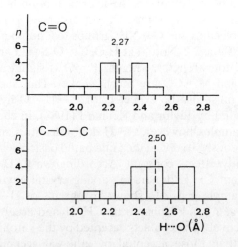

Fig. 2.14. Effect of acceptor basicity on mean $C-H\cdots O$ hydrogen bond lengths: histograms of distances d with CH_2Cl_2 donors and C=O and C—O—C acceptors, respectively (from Steiner 1994a).

Table 2.3 Mean distances $d(H\cdots O)$ in hydrogen bonds of different $C-H$ donors with various O atom acceptors (Steiner 1998c)

	CHCl₃ (Å)		C≡C—H (Å)		CH₂Cl₂ (Å)		N≡C—Me (Å)		DMSO (Å)		Acetone (Å)	
P=O	1.97	[2]	2.02	[5]	2.27(7)	[6]	2.40(4)	[4]	—	[0]	2.49	[2]
H₂O	2.11	[1]	2.11	[3]	2.16	[2]	2.65	[3]	2.73	[3]	2.61(9)	[4]
>C=O	2.22(3)	[52]	2.29(3)	[40]	2.41(2)	[123]	2.53(3)	[42]	2.52(2)	[73]	2.60(2)	[67]
>S=O	2.25(6)	[21]	2.50(6)	[9]	2.40(4)	[31]	2.57(3)	[14]	2.57(2)	[58]	2.53(8)	[5]
C(sp^3)—OH	2.31(8)	[5]	2.40(3)	[31]	2.47(11)	[4]	2.52(6)	[11]	2.52(2)	[73]	2.53(19)	[4]
C—O—C	2.33(5)	[16]	2.45(4)	[19]	2.55(2)	[57]	2.56(2)	[56]	2.65(4)	[4]	2.64(3)	[14]
C—NO₂	2.52(6)	[4]	2.63(3)	[6]	2.48(4)	[15]	2.59(3)	[17]	2.60(3)	[22]	2.57(5)	[5]
M—CO	2.49(3)	[20]	2.48(12)	[4]	2.57(1)	[176]	2.63(3)	[21]	—	[0]	2.61(2)	[29]

Distances d normalized, number of contacts given in brackets.

Table 2.4 Difference of mean distances d to P=O and M—CO acceptors (using data in Table 2.3)

Donor	$d(M-CO) - d(P=O)$ (Å)
CHCl₃	0.52
C≡C—H	0.46
CH₂Cl₂	0.30
N≡C—Me	0.22
Acetone	0.12

acceptors are listed in Table 2.3 (Steiner 1998c). The effect of donor acidity shows up in the columns of the table, while the effect of acceptor basicity is seen in the rows. Based on this data, the ranking of O atom acceptor strengths in $C-H\cdots O$ interactions is identical to that found in conventional $O-H\cdots O$ hydrogen bonds. The range of variations of d caused by the basicity effect is about as large as that caused by the acidity effect (both over 0.5 Å). However, the extremes of the scale (P=O and M—CO) do not occur frequently in purely organic structures and for the most common O atom acceptors studied, the variation in d is actually quite small.

The basicity effect is qualitatively the same for all kinds of donors, but quantitatively there are large differences. If the acceptor strength is reduced from P=O to M—CO, the d values increase by 0.52 Å for the strong donor CHCl₃, but only by 0.12 Å for the weak donor acetone (Table 2.4). This shows that hydrogen bonds from strong donors are markedly influenced by variations in acceptor strength, whereas hydrogen bonds from weak donors are affected only slightly. Still, it is important to note that even very weak donors are not completely insensitive to variations in acceptor basicity; complete insensitivity would be an indication that the contacts do not represent hydrogen bonds but mere van der Waals interactions. Analogous observations have also been

Table 2.5 Mean distances d in hydrogen bonds of different C—H donors with two kinds of N atom acceptors. The $C(sp^3)OH$ donor is included for comparison (Steiner 1998c)

Donor	C—N=C (Å)	N≡C—C (Å)
CHCl$_3$	2.42(5) [16]	2.53 [2]
CH$_2$Cl$_2$	2.49(5) [13]	2.59(6) [8]
N≡C—Me	2.60(5) [10]	2.61(1) [112]
DMSO	2.61(2) [8]	2.67(9) [9]
$C(sp^3)OH$	1.888(6) [418]	2.02(2) [35]

Distances d normalized, numbers of contacts given in brackets.

Table 2.6 Mean distances $D(C \cdots O)$ in intermolecular C—H$\cdots$O bonds formed by terminal and bridged CO ligands (M—C≡O, t-CO and $M_2 >$ C=O, b-CO; Braga *et al.* 1995)

	t-CO (Å)	b-CO (Å)
Ti	3.53(2) [29]	[0]
V	3.49(1) [88]	3.5(1) [4]
Cr	3.510(3) [2415]	3.56(5) [9]
Mn	3.510(4) [1510]	3.48(2) [51]
Fe	3.510(2) [4205]	3.450(9) [285]
Co	3.520(5) [916]	3.47(1) [257]
Ni	3.52(2) [30]	3.50(2) [47]

Numbers of contacts given in brackets.

made for N atom acceptors, but because the number of C—H$\cdots$N hydrogen bonds in the CSD is about a factor of ten smaller than the number of C—H$\cdots$O bonds, the results are more qualitative. Nevertheless, it is shown that C—H$\cdots$N contacts to amines are about 0.1 Å shorter than those to cyano groups (Table 2.5; Steiner 1998c).

Organometallics. Another means of revealing basicity effects, given that they tend to be convoluted with acidity effects, is to study a very large number of compounds in the expectation that acidity effects will cancel out in a similar way that basicity effects cancel out in the analysis of donor acidities shown in Figs 2.12 and 2.13. This corresponds to the row averages in Table 2.3. This approach was taken in a study of C—H$\cdots$O bonds to metal-bound CO ligands (Braga *et al.* 1995). Table 2.6 gives mean intermolecular C$\cdots$O distances for C—H$\cdots$O bonds formed by terminal (M—C≡O, t-CO) and bridged

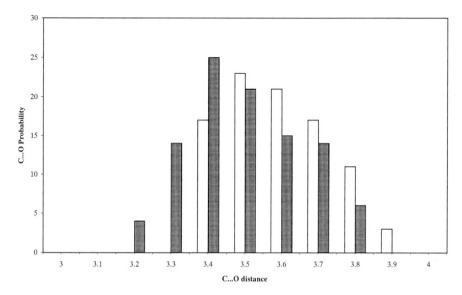

Fig. 2.15. Histograms of distances $D(C\cdots O)$ for intermolecular $C-H\cdots O$ contacts to carbonyl ligands bonded to Fe. *Open bars* represent terminal CO ligands, $-C\equiv O$, and *shaded bars* represent bridging ligands, $>C=O$. Note that the histogram for terminal ligands is offset systematically towards longer distances (from Braga *et al.* 1995).

$(M_2 > C=O, b\text{-}CO)$ CO ligands. For most of the metal atoms considered, the mean distance D for t-CO is longer than the corresponding distance for the b-CO. For example, while the mean D value for 4205 $C-H\cdots O$ bonds formed by t-CO in Fe atom organometallics is 3.510(2) Å, the corresponding value for 285 bonds formed by b-CO is 3.450(9) Å. The shortening of the mean D value of 0.06 Å is just detectable but real, and the chemical conclusion one may draw is that t-CO is less basic that b-CO. Such a conclusion is in accord with the more efficient π-accepting capability of CO in the bridging as opposed to the terminal bonding mode.

The above arguments are based on mean $C\cdots O$ distances. It is also instructive to examine Fig. 2.15 which shows histograms of $C\cdots O$ distances for intermolecular t-CO$\cdots$H$-$C and b-CO$\cdots$H$-$C geometries in the case of ligands bound to cobalt, chosen as a test case. These histograms represent 916 and 257 hits, respectively, and so are statistically significant samples. Here, the full range of $C\cdots O$ distances is considered rather than just the mean values, and it may be observed that the entire histogram for t-CO is offset towards longer $C\cdots O$ distances relative to the histogram for b-CO. These histograms strongly reinforce the conclusions that have been drawn from the data in Table 2.6 and show that:

(1) a consideration of the mean $C\cdots O$ distance alone is representative of the entire sample;

(2)   C—H···O hydrogen bonds to b-CO ligands are shorter than those to t-CO ligands.

In summary, one may conclude with some confidence that the basicity of CO ligands has a definite effect on C—H···O hydrogen bond lengths formed by them.

Other examples of short hydrogen bonds.   To conclude this section, a number of related examples are mentioned. These may be easily rationalized using the principles that have been laid out above. Numerous highly activated systems are known which form very short C—H···O/N hydrogen bonds. Freyhardt and Wiebcke (1994) have noted that among the several C—H···O bonds formed in a tetraethylammonium borate salt, the shortest are the ones accepted by the deprotonated O atoms of the $[BO(OH)_2]^-$ anions, while Kiplinger *et al.* (1995) have noted that a C—H···$F^{\delta-}$—$W^{\delta+}$ interaction in a W(II) metallocyclo-propene complex has a d separation of 2.15 Å, a value much less than the van der Waals separation. CSD analysis has shown that the C—H groups in cyclo-propanes, aziridines and oxiranes form short C—H···O hydrogen bonds, indicative of the C(ring)—H group having an acidity intermediate between $C(sp^1)$—H and $C(sp^2)$—H groups (Allen *et al.* 1996a). The high carbon acidity of three-membered rings is well known. In N-heterocycles, there is usually an enhancement of C—H acidity, and Anderson and Muchmore (1995) report a short C—H···O bond ($D = 3.01, d = 2.30$ Å, $\theta = 156°$) in a pyrazole *N*-oxide. A nice example of an intramolecular—CF_2H···O=C hydrogen bond was demonstrated theoretically in a substituted pyrazole by Erickson and McLoughlin (1995) and also corroborated spectroscopically.

2.2.3   *Angular properties*

The angular attributes of a hydrogen bond X—H···A—C are defined in terms of the hydrogen bond angle θ (X—H···A) and the bending angle at the accep-tor atom ϕ (H···A—C). Hydrogen bonds are *per se* directional interactions with a preference for linearity ($\theta = 180°$). The degree to which this preference is adhered to or not provides valuable information on the nature of the inter-action itself. Linearity is preferred because this geometry would optimize the electrostatic interaction (dipole–monopole, dipole–dipole). For strong hydro-gen bonds, the angular range is typically $150 < \theta < 180°$. For weak hydrogen bonds, the range is broader. In any analysis of the angles θ and ϕ from X-ray studies, the H atom positions must be corrected to neutron normalized values. Additionally, in statistical studies, the distribution of angles should be cor-rected for the solid angle factor ($\sin \theta$ or $\sin \phi$). This procedure, known as the cone correction (Kroon and Kanters 1974) is necessary because the solid angle covered by an angular interval $\Delta\theta$ or $\Delta\phi$ is smaller for linear than for bent angles. In effect, any distribution of θ or ϕ angles obtained from the CSD is a

consequence of a convolution of chemical and geometrical factors. By employ-
ing the cone correction, the geometrical bias is accounted for and the resulting corrected distribution is characteristic of chemical factors alone. The
reader should note here that effects of steric conflict are included in the 'chemical' category.

2.2.3.1 *Donor directionality and the angle* θ

For $C-H\cdots O$ hydrogen bonds, the tendency towards linearity was first
demonstrated by Taylor and Kennard (1982). They determined the θ distribution for short $(d < 2.4\text{Å})$ contacts in their neutron data sample. They found
that θ most frequently lies between 150–160° with a mean value of 152°. Upon
performing the cone correction, this distribution is found to peak at a linear
angle, clearly showing a statistical preference for linearity and hence showing
the directional nature of the $C-H\cdots O$ interaction (Fig. 2.16). Following a

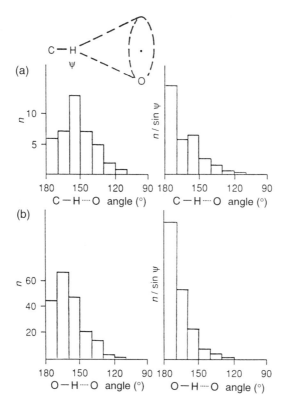

Fig. 2.16. Distribution of angles ψ in crystal structures. (a) 41 $C-H\cdots O$ hydrogen bonds in
neutron crystal structures, $d < 2.4\text{Å}$ (Taylor and Kennard 1982). (b) 196 $O-H\cdots O$ hydrogen
bonds in carbohydrate X-ray structures (Kroon and Kanters 1974). The inset illustrates the cone
correction (weighting by $1/\sin\psi$) that relates the left and the right histograms (from Steiner
1997a). The angle ψ in these figures corresponds to θ in the rest of the discussion.

suggestion of Kroon (1995), related histograms for $O-H\cdots O$ hydrogen bonds in carbohydrates are drawn in the same figure and immediately show the close similarity of the angular characteristics of $O-H\cdots O$ and short $C-H\cdots O$ hydrogen bonds (Fig. 2.16). A very similar histogram was obtained subsequently for $C-H\cdots O$ bonds donated by $C_\alpha-H$ of amino acids (Steiner 1995a). For interactions longer than 2.4 Å, the directionality rapidly becomes weaker. This could be due to the fact that some of these longer contacts are part of a bifurcated scheme, but these aspects have not been evaluated rigorously. The efficacy of a $C-H\cdots O/N$ interaction diminishes with falling θ and in practice, we recommend a lower limit of 110° when accepting a $C-H\cdots O$ geometry as a hydrogen bond. As for all cut-off criteria in this subject, this or any other angular delimiter should not be used indiscriminately. However, since the electrostatic interaction is zero at $\theta = 90°$, this perpendicular geometry represents the formal end of hydrogen bonding.

While $C-H\cdots O$ hydrogen bonds are, in principle, linear they are also easily bent because of their weakness and so directionality is often blurred by steric hindrance and competition with other hydrogen bonding groups. This is clearly observed for $C-H\cdots O$ geometries in carbohydrates. In this category of compounds, interconnected $O-H\cdots O$ hydrogen bonds are dominant and form a frame within which the weaker $C-H\cdots O$ interactions must adjust. Figure 2.17 is a scatterplot of θ versus d for intermolecular $C-H\cdots O$ contacts in carbohydrate neutron crystal structures (Steiner and Saenger 1992a). Most of the donors shown in the figure have activated $C(sp^3)-H$ groups. Despite the

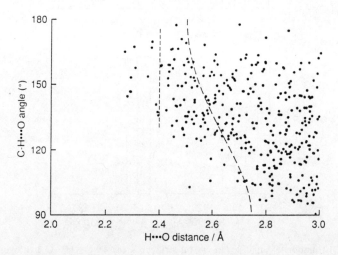

Fig. 2.17. Scatterplot of angles θ against distances d in intermolecular $C-H\cdots O$ contacts in carbohydrate neutron structures. The straight line at $d = 2.4$ Å shows the cut-off used by Taylor and Kennard (1982), the curved line shows the $H\cdots O$ van der Waals separation based on a non-spherical H atom (from Steiner 1997a).

seemingly unfavourable competitive situation, numerous $C-H\cdots O$ interactions with d separations down to around 2.25 Å are observed. The shortest of these contacts show the same preference for approximate linearity as those in Fig. 2.16 but at increasing distances, the directionality rapidly becomes softer, with a nearly isotropic angular distribution beyond 2.70 Å. This quasi-isotropic behaviour at longer distances should not be regarded as a sign of lack of directionality; it is merely a consequence of competition with stronger forces. Furthermore, and as mentioned above, other obscuring effects such as bifurcation are very difficult to account for unless one were to actually carry out a manual inspection of all the crystal structures.

We believe directionality to be a common property of all types of hydrogen bonds, strong and weak, and in view of the debate on the distinction between the weak hydrogen bond and the van der Waals interaction, have extracted angular distributions of θ for a variety of $X-H\cdots A$ geometries (Steiner and Desiraju 1998). Structural data were retrieved for $C-H\cdots O$ contacts involving the prototypes of the $C(sp)-H$, $C(sp^2)-H$ and $C(sp^3)-H$ groups, that is ethynyl, vinyl and ethyl groups. For comparison, data for conventional hydrogen bonds from hydroxyl donors are also presented. To reduce chemical inhomogeneity, only organic carbonyl acceptors were considered. For the $H\cdots O$ distance cut-off, a long value of 3.0 Å was selected, which is the van der Waals sum plus 0.3 Å. Data for $C-H\cdots H-C$ van der Waals contacts were also retrieved to the distance limit of the van der Waals sum plus 0.3 Å ($= 2.7$ Å). To examine the degree to which linear contact geometries are preferred, cone corrected histograms of angular $C-H\cdots O$ distributions were generated (Fig. 2.18).

The histogram for hydroxyl donors shows the well-known directional behaviour of conventional hydrogen bonds [mean $\theta = 154.0(4)°$]. For the acidic ethynyl donors $C\equiv C-H$, the mean $C-H\cdots O$ angle θ is only slightly smaller, $152(2)°$, and the angular distribution is only slightly broader. For vinyl donors, the mean angle θ falls to $143(1)°$ and the angular distribution widens considerably. For the very weakly polarized methyl donor of the ethyl group, the mean angle θ falls further to $137.1(7)°$ and the angular distribution is correspondingly softened, but it still shows directional behaviour with linear contact geometries being favoured. This means, in other words, that the approach of an O atom to a methyl group is not indifferent as regards the contacting angle but preferably takes place along the direction of the $C-H$ vectors. Finally, the mean $C-H\cdots H$ angle for $C-H\cdots H-C$ contacts of methyl groups is $128.6(3)°$. Here, however, the angular distribution is almost ideally isotropic in the range 120 to 180° and this is exactly the picture that is expected for the non-directional van der Waals interaction.

The sequence of histograms in Fig. 2.18 clearly shows a decrease of directionality for $C-H\cdots O$ interactions with decreasing $C-H$ acidity. For alkyne donors, the directionality behaviour is that of conventional hydrogen bonds such as those formed by hydroxyl groups. For vinyl donors, the directionality

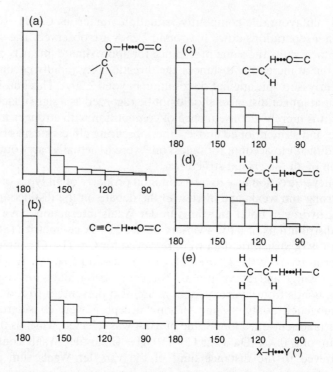

Fig. 2.18. Directionality of C—H···O hydrogen bonds formed by C—H groups of different acidity. Shown are cone-corrected θ angle distributions for the donors C≡C—H, C=CH$_2$ and CH$_2$—CH$_3$ in combination with C=O acceptors ($d < 3.0$ Å, no θ angle cut-off). For comparison, histograms are also shown for C(sp^3)—O—H···O=C hydrogen bonds and for van der Waals contacts CH$_2$—CH$_3$···H—C ($d < 2.7$ Å) (from Steiner and Desiraju 1998).

is weaker, but is still clearly pronounced. For methyl groups, the directionality is the weakest, but is definitely different from the perfectly isotropic behaviour for C—H···H—C contacts. Since C—H···O interactions of alkyl groups are *not* isotropic (even at the long distance cut-off of 3.0 Å), they should definitely not be classified as mere van der Waals contacts. The observed differences in directionality behaviour between any kind of C—H···O hydrogen bond and the van der Waals interaction is a consequence of the fundamentally different distance and angle fall-off characteristics of these interactions. In general, the tendency towards linearity for C—H···O/N hydrogen bonds is a good confirmation of the electrostatic nature of these interactions.

One possible way to view the directionality of dipole–dipole interactions is to look at the non-bonded repulsions between the negatively charged C atom and O or N atoms. These repulsions would obviously tend to straighten out the bonds, leading to a preference for linearity. The extent of straightening would be more pronounced for the shorter bonds because the undesirable

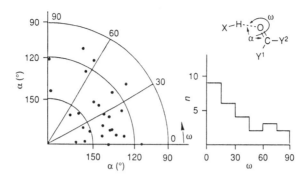

Fig. 2.19. Approach of sterically unhindered C—H donors to C=O acceptors, geometrical definitions are shown in the inset. The data show weak but significant acceptor directionality (from Steiner 1997a).

Coulombic repulsions would be larger in a bent situation. In summary, the shorter the bond, the more likely it is to be linear. Conversely, the less acidic the C—H group, the easier it would be for deviations from linearity.

2.2.3.2 Acceptor directionality and the angle φ

O and N atom acceptors exhibit complex and weak directionality behaviour that depends on their chemical nature and state of hybridization. Significant acceptor directionalities were noted for N—H···O and O—H···O hydrogen bonds by Murray-Rust and Glusker (1984) who carried out a comprehensive study of these aspects. Similar directionalities may be expected for C—H···O bonds, and while no systematic study has been published in this context, there is a consensus that for carbonyl acceptors the H···O=C angle φ is distributed around 120°. Often the C—H group tends to lie near the plane of the O atom lone pairs. For ethereal acceptors, C—O—C, the X—H bond could lie near the plane of the O atom lone pairs but without any particular directional preference within this plane towards the conventional ('rabbit ears') lone pair directions.

The angular properties of C—H···O=C interactions have been examined for acidic and sterically unhindered C—H donors which are not in competitive situations (CHCl$_3$, CH$_2$Cl$_2$, C≡CH) (Steiner et al. 1996a). A polar scatterplot of the angle φ versus the torsion angle ω (H···O=C—Y) and a histogram of ω angles are shown in Fig. 2.19. The distribution of ω angles indicates a preference for in-plane contacts (ω = 0°). The acceptor directionality is very soft but surely discernible. Because the acceptor directionality is soft even for the strongest C—H donors, it must be expected to be even softer for the weaker donor types.

Acceptor directionality for C—H···O hydrogen bonds formed by CO ligands in organometallic complexes provides valuable information on the nature of the CO ligand itself (Braga *et al.* 1995). Figure 2.20 is a d–ϕ scatter-plot of intermolecular $C-H\cdots O$ bonds formed by terminal ligands, t-CO (see Section 2.2.2.4 for a further definition of this term). The most significant feature here is that data on as many as 4205 hits from 1137 Fe atom containing crystal structures are included. The distribution of intermolecular $CO\cdots H$ angles is unexpected. There is a large angular distribution ($90 < \phi < 180°$) at long $O\cdots H$ separations ($d > 2.60\,\text{Å}$) which narrows as the $O\cdots H$ separation decreases, sharpening around 125–135° for the shortest $C-H\cdots O$ hydrogen bonds. Notice that here too, the hydrogen bond acceptor directionality is more pronounced for the shorter and generally stronger bonds. Cone correction of the ϕ distribution for the bonds formed at t-CO results in a maximum around 150°. All this is suggestive of ketonic character for the carbonyl group, and this is surprising because in a simple valence bond picture, this should be true only for the bridging b-CO group where there are presumably two lone pairs on the O atom available to accept H bonds, at around 120°. According to such a depiction there should be a potential for the formation of only one hydrogen bond in a t-CO group, that is towards the single lone pair along the $M-C=O$

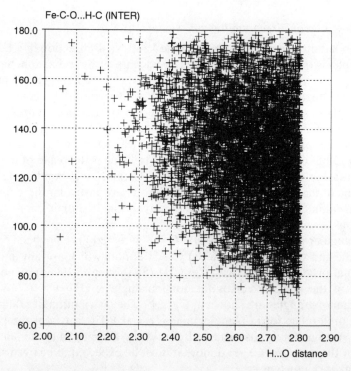

Fig. 2.20. Scatterplot of angles $\phi(H\cdots OC)$ versus distances $d(H\cdots O)$ for 4205 intermolecular $C-H\cdots O$ contacts in terminal $CO-Fe$ carbonyl complexes. Notice that the shortest distances are more likely to have ϕ values around 140° (adapted from Braga *et al.* 1995).

vector ($\phi \sim 180°$). Such results are interesting because they may be used to adduce evidence for the shapes of lone pairs on the O atoms of terminal (t-CO) and bridged (b-CO) CO ligands in organometallic complexes.

Donor and acceptor directionalities of hydrogen bonds are important in the context of crystal engineering requirements. For weak hydrogen bonds, directionalities are so soft that they are negotiable in the crystal design exercise but it is helpful to keep them in mind (see Section 4.3). Acceptor directionality at the C≡N group, though of interest in the above context, is yet to be investigated systematically.

2.2.3.3 *Joint consideration of angles and lengths*

There is little disagreement regarding very linear and very bent C—H⋯O geometries. However, the majority of situations lie somewhere between and this leads to the question of evaluation of the general C—H⋯O geometry. This ambiguous situation is very characteristic of weak hydrogen bonds and a cautious approach is required. It is helpful to jointly consider bond lengths and angles graphically. Figure 2.21 is a scatterplot of θ angle versus d (H atom positions normalized) for all C—H⋯O bonds in a series of nine carboxylic acid complexes formed from dimethylamino, methoxy and nitro-substituted cinnamic and benzoic acids. In eight cases, hetero-dimers such as **9** are formed by O—H⋯O hydrogen bonding between dissimilar acid molecules. The ninth, **10**, is a 1:1 molecular complex of 4-(*N,N*-dimethylamino)benzoic acid and 3,5-

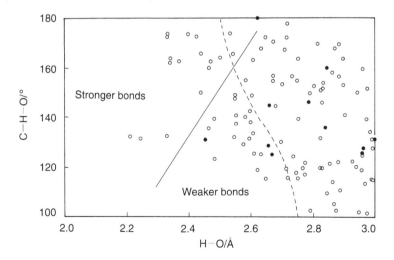

Fig. 2.21. C—H⋯O hydrogen bonds in nine carboxylic acid complexes of N(CH$_3$)$_2$, OCH$_3$ and NO$_2$ substituted cinnamic and benzoic acids. Scatterplot of angles θ against normalized distances d. The black dots show C—H⋯O hydrogen bonds in homodimer **10** and the rest show heterodimers of the form **9**. The dashed line shows the H⋯O van der Waals distance based on a non-spherical H atom. The solid line is eyeballed (adapted from Sharma *et al.* 1993).

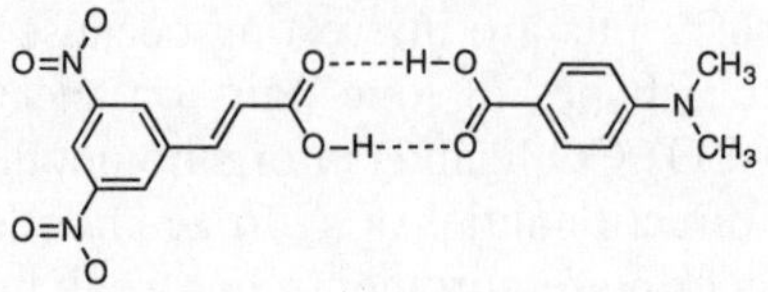

9

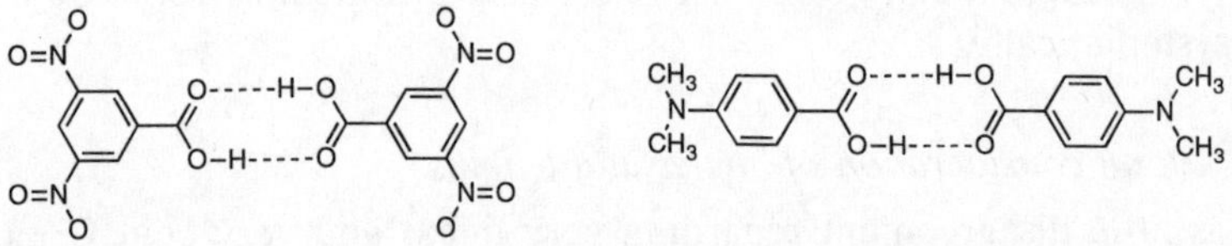

10

dinitrobenzoic acid and contains hydrogen bonded homodimers. A key step which leads to rationalization of the formation of the unexpected homodimer is the awareness that $C-H\cdots O$ hydrogen bonding in **10** is much poorer than in the eight hetero-dimer complexes. Figure 2.21 contains $C-H\cdots O$ hydrogen bonds of differing strengths as indicated by their length and linearity attributes. $C-H\cdots O$ bonds in complex **10** are shown as filled circles (•) from which it is clear that these bonds are neither the shortest nor the most linear in the group. In contrast, all the eight hetero-dimer complexes contain at least a few $C-H\cdots O$ bonds in the strong bonds region.

2.2.3.4 *Bifurcation and variability of coordination*

Bifurcated or three-centre hydrogen bonds, $X-H\cdots(A_1, A_2)$ are very common in $O-H\cdots O$ and $N-H\cdots O$ hydrogen bonded structures (Jeffrey and Saenger 1991; see Section 1.1.2). They are also observed in $C-H\cdots O/N$ patterns. Three-centre bonds are longer than the 'normal' or two-centre bonds and the H atom usually lies close to the plane formed by the atoms X, A_1 and A_2. This then is the basis of the commonly used criterion for bifurcation (Parthasarathy 1969), that measures the deviation from 360° of bond angles around the bifurcated donor H atom. The reader will note that this tendency of a bifurcated arrangement towards planarity follows directly from hydrogen bond linearity. One might say that the tendency towards linearity for a two-centre hydrogen bond and that towards planarity for a three-centre hydrogen bond stem from the same reason, the electrostatic nature of all hydrogen bonds. All the four atoms H, X, A_1 and A_2 constitute the hydrogen bond, and so the deviation of the H atom from the plane of X, A_1 and A_2 is as relevant an angular parameter for the estimation of the strength of the interaction as are the individual angles $X-H\cdots A_1$ or $X-H\cdots A_2$.

In $N-H\cdots O$ and $O-H\cdots O$ systems, bifurcation usually means that the

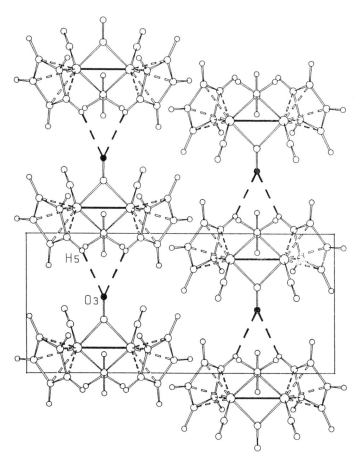

Fig. 2.22. Example of a bifurcated acceptor. $C-H\cdots O$ hydrogen bonds accepted by the bridging CO ligand in $(\eta^5\text{-}C_5H_5)_2Fe_2(CO)_2(\mu\text{-}CO)(\mu\text{-}CHCH_3)$ (from Braga *et al.* 1995).

donor is bifurcated because in the typical hydrogen bond forming organic molecule there is an excess of acceptor atoms over donor atoms. For $CH\cdots O/N$ and other weakly hydrogen bonded systems, however, it is better to specify whether one is referring to a bifurcated donor or to a bifurcated acceptor, $A\cdots(H_1-X, H_2-X)$; the latter is common because most organic and organometallic molecules become donor-rich if weak donors are counted. Figure 2.22 for instance shows the bifurcated acceptor interaction established by the bridging CO in crystalline $(\eta^5\text{-}C_5H_5)_2Fe_2(CO)_2(\mu\text{-}CO)(\mu\text{-}CHCH_3)$. Because of the crystallographic symmetry, the unique bridging CO is at an equal distance from both Cp ligands of a neighbouring molecule $(d = 2.48\,\text{Å}, \theta = 123.6°)$ and in such a way the crystal can be pictorially described as being formed by chains of molecules linked via $C-H\cdots O$ interactions.

In extreme cases of acceptor multifurcation, many C—H groups approach a single basic atom (Desiraju *et al.* 1993). Sometimes, as many as four or five C—H groups form close contacts with a C=O group. Naturally, ϕ values will deviate from 120° in such cases. The formation of such multiple C—H$\cdots$O contacts is not unexpected in donor-rich systems. Such variable effects may be more likely for weak hydrogen bonds than for strong ones, in that close-packing effects also need to be considered.

2.2.4   C—H bond lengthening

It is an inherent property of hydrogen bonds that the covalent bonds of the groups involved therein are modified. So, in an interaction of the type R_1—X—H$\cdots$A—R_2, the X—H and A—R_2 covalent bonds are weakened, resulting in their lengthening. A smaller and concomitant shortening of the R_1—X bond is also observed. Corresponding spectroscopic shifts in the IR arising from a reduction of the force constants of the X—H and A—R_2 bonds are also easily seen. These effects are very pronounced for strong hydrogen bonds, where X and A are highly electronegative atoms. The greatest effect is observed for the X—H covalent bond. Based on neutron diffraction data, lengthening of the X—H bond has been characterized for the hydrogen bond types O—H$\cdots$O (Olovsson and Jönsson 1976; Bürgi and Dunitz 1983; Steiner and Saenger 1994; Bertolasi *et al.* 1996), N—H$\cdots$N (Steiner 1995*b*) and a number of heteronuclear hydrogen bonds such as O/N—H$\cdots$O/N, O/N—H$\cdots$S, O/N—H$\cdots$Cl⁻ (Steiner 1998*a*). In hydrogen bonds such as those found in alcohols, the X—H bond is elongated by several hundredths of an Å unit. In very strong hydrogen bonds, the lengthening can be as much as 0.2 Å. The modifications of the R_1—X and A—R_2 bonds are much smaller and can be reliably quantified only for the strongest of hydrogen bond types (Bertolasi *et al.* 1996).

For the weak C—H$\cdots$O hydrogen bonds, analogous effects should, in principle, be expected. Spectroscopic evidence of C—H bond weakening is found for activated C—H donors and is discussed in Section 2.1. The magnitude of C—H lengthening in C—H$\cdots$O interactions can be estimated using the valence bond model of the hydrogen bond, which is based on a strict relation between bond distances and valences (Brown 1992; Steiner 1998*a*). The results of such an exercise are given in Table 2.7. In the distance range of normal C—H$\cdots$O hydrogen bonds, the C—H lengthening is expected to be ≤ 0.01 Å, and this would be difficult to detect even with neutron diffraction. In consequence, covalent bond lengthening effects have not been investigated extensively. In the single crystal correlation study to date (Steiner 1995*a*), the lengthening of the C—H bond upon C—H$\cdots$O hydrogen bond formation is just discernible but even so is satisfying. A slight lengthening of the C—H bond in C—H$\cdots$O hydrogen bonds formed by the activated C_α—H bonds in α—amino acids was detected in an analysis of

Table 2.7 Bond orders s of $H\cdots O$ and lengthening of $C-H$ in $C-H\cdots O$ hydrogen bonds. Estimation using the valence model of the hydrogen bond (Steiner 1998a)

$H\cdots O$ (Å)	Bond order (valence units)	Bond lengthening (Å)
1.9	0.076	0.029
2.0	0.058	0.022
2.1	0.044	0.017
2.2	0.034	0.013
2.3	0.026	0.010
2.4	0.020	0.007
2.5	0.015	0.005
2.6	0.011	0.004
2.8	0.007	0.002
3.0	0.004	0.001

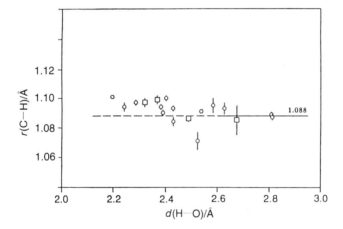

Fig. 2.23. Correlation of the covalent bond length $r(C-H)$ with the distance $d(H\cdots O)$ in $C_\alpha-H\cdots O$ hydrogen bonds in amino acid neutron crystal structures. The horizontal line shows the average value r for distances $d > 2.6$ Å; notice that all values r for distances $d < 2.4$ Å are scattered above this line (from Steiner 1995a).

neutron diffraction data on such compounds (Fig. 2.23). The lengthening is around 0.01 Å for contacts with $d \sim 2.30$ Å and was shown to be statistically significant at the 99 per cent reliability level. The experimental data are in accord with the valence bond model which associates a d separation of 2.3 Å with a bond order of 0.026 valence units and a $C-H$ lengthening of 0.01 Å.

The strongest $C-H\cdots O$ hydrogen bonds known today have distances around $d \sim 2.0$ Å and the expected lengthening Δd of $C-H$ is ~ 0.02 Å or more (Table 2.7). A relevant neutron crystal structure is also available, that of 2-ethynyladamantan-2-ol (Allen *et al.* 1996*b*). In this compound, an alkynic $C-H$ bond forms a very short contact to a hydroxylic O atom ($d = 2.07$Å; see also Section 2.2.2). Happily, there are two symmetry-independent molecules in the crystal and the second ethynyl group is 'free'. By comparing these two groups, it is possible to state that the hydrogen bonded $C-H$ bond is lengthened by 0.025(14) Å corresponding to its weakening by roughly 0.05 valence units. More studies are required on this aspect of weak hydrogen bonding but even the limited data available at present point to the bonding nature of the $C-H\cdots O$ interaction.

2.2.5 *Reduction of thermal vibrations*

Hydrogen bonding reduces the thermal vibrations of the engaged residues. In an exemplary study, this has been demonstrated with neutron diffraction data for $O-H\cdots O$ hydrogen bonds formed by water molecules (Eriksson and Hermansson 1983). For $C-H\cdots O$ bonds, the effect was first shown for terminal alkynes (Steiner 1994*b*); these compounds are good probes in this context because they are among the most acidic of $C-H$ donors while simultaneously enjoying pronounced vibrational freedom. The situation is sketched in Fig. 2.24 from which it may be appreciated that a free $C(1)\equiv C(2)-H$ moiety can vibrate strongly in a direction normal to the group and the vibration amplitude of the terminal atom C2 is much larger than that of the internal atom C1. If the group donates a

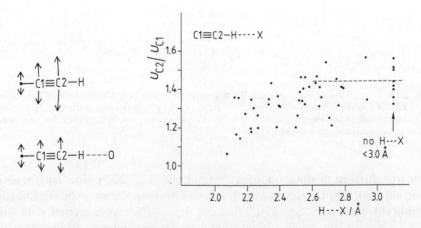

Fig. 2.24. Correlation of the *U*-ratio defined as $U_{eq}(C2)/U_{eq}(C1)$ with the distance d in 51 $C\equiv C-H\cdots X$ hydrogen bonds (X = O, N, π). The horizontal line shows the mean value for $C\equiv C-H$ groups that donate no hydrogen bond with $d < 2.8$ Å (adapted from Steiner 1994*b*).

hydrogen bond, the alkynyl vibration is more restricted. Though the vibration amplitude is still greater for C2 than for C1, the relative difference between the two amplitudes is smaller than in a freely vibrating group.

A rigorous analysis of the above situation from diffraction data would require that the mean-square displacements of C1 and C2 perpendicular to the $C\equiv C$ bond are known, and furthermore the libration of the entire molecule would have to be taken into account. This is not feasible if many structures are analysed because the needed anisotropic displacement parameters U_{ik} are normally not published. Accordingly, a simpler method has been used, based on the more readily available equivalent isotropic displacement parameters U_{eq}. It is apparent that the ratio $U(C2)/U(C1)$ reflects, at least qualitatively, the vibration behaviour of the $C\equiv C-H$ group. This ratio should be significantly larger than unity for a free $C\equiv C-H$ group, reduce with increasing hydrogen bond strength, and approach 1.0 for the strongest hydrogen bonds. This U-ratio was compared with distances $d(A)$, $(A = O, N, \pi)$ for 51 $C\equiv C-H$ groups in 42 crystal structures (Fig. 2.24). The correlation plot clearly demonstrates the anticipated effect. On the whole, the $U(C2)/U(C1)$ values reduce continuously with a shortening of the $d(A)$ distance, indicating reduction of thermal vibrations. Significantly, the effect is observed not only for the shortest $C-H\cdots A$ bonds but also for those with $2.60 < d < 2.80\,Å$, showing again the long range nature of the interaction and justifying long distance cut-offs in crystal correlation studies.

The complementary effect on the acceptor side has been shown by Braga *et al.* (1995) from an analysis of the neutron determined crystal structure of an organometallic complex, $(\mu_3\text{-}H)FeCo_3(CO)_9(P(OMe)_3)_3$. Here, eight distinct $C-H\cdots O$ bonds are found and they are of interest because they include intramolecular and intermolecular bonds to both terminal and bridged CO ligands with both bifurcated and non-bifurcated acceptors. Each of the bonds is formed by a different methyl group while the ninth and final CO ligand in the molecule is 'free' in that it does not accept a $C-H\cdots O$ bond. This structure is ideal for the purpose of ascertaining thermal vibrations of the $M-CO\cdots H-C$ hydrogen bond because a wide variety of basicities are found under constant acidity conditions and all in a single crystal structure whose geometry has been precisely determined by low temperature neutron diffraction analysis. Figure 2.25 is a plot of the $U(O)/U(C)$ ratio for the nine CO ligands versus the 14 $H\cdots O$ distances formed by eight of these ligands. The plot shows a separation between terminal-CO and bridged-CO ligands. The b-CO ratios are lower, in the range 1.42–1.53, while the t-CO ratios are higher, in the range 1.59–1.78. This means that there is lower relative motion of O with respect to C for b-CO as compared to t-CO, as might be expected, since t-COs protrude more from the surface and have a larger bending motion. More interestingly, for the

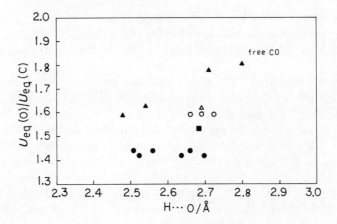

Fig. 2.25. Plot of the U-ratio $U_{eq}(O)/U_{eq}(C)$ versus the distance $d(H \cdots O)$ for the $C-H \cdots O$ hydrogen bonds in the neutron structure of $(\mu_3\text{-H})FeCo_3(CO)_9(P(OMe)_3)_3$. (▲) CO-t, intermolecular, non-bifurcated; (△) CO-t, intramolecular; (○) CO-t single acceptor trifurcated to three CH groups; (■) CO-b, intramolecular; (•) CO-b, inter- and intra-molecular, bifurcated (adapted from Braga *et al.* 1995).

single acceptor t-CO bonds, there is a very good correlation between the U-ratio and the $H \cdots O$ distance with the non-hydrogen bonded ligand (U-ratio 1.81) also following the same trend. All this shows that as the $C-H \cdots O$ bond gets stronger, the O atom vibrates less. The entire exercise shows that an analysis of thermal vibrations at the basic centre is as reliable as examining the acidic centre in an assessment of the bonding character of a $C-H \cdots O$ interaction. In this respect, subtle differences between t-CO and b-CO bonds and between bifurcated and non-bifurcated bonds are also revealed and, significantly, these shades of difference are apparent even for $C-H \cdots O$ hydrogen bonds formed by $C(sp^3)-H$ atoms. It is of interest to note that the above analysis was not successful when ADPs from low temperature X-Ray measurements were used, showing the necessity of using only very accurate data in analyses of thermal vibrations.

The demonstration of reduction of H atom vibrations in $C-H \cdots O$ bonds is more difficult. The first analysis along these lines was from carbohydrate crystal structures and it failed because of excessive scatter of the experimental data (Steiner and Saenger 1992*a*). In a subsequent and more recent neutron diffraction study, Wozniak *et al.* (1996) reported such an effect for a small set of $C-H \cdots O$ bonds (range 2.3–2.8 Å) in the 1:1 molecular complex of 1,8-*bis*(dimethylamino)naphthalene and 1,2-dichloromaleic acid. Here, it was seen that the d values correlate with the $U(H)/U(C)$ ratios though poorly. In summary, both statistical analyses of X-ray data and individual analyses of neutron data appear to provide good information on the reduced vibrations of $C-H \cdots O$ hydrogen bonds and hence on their tethering ability in crystal structures.

2.2.6 *Computational studies and hydrogen bond energies*

Computational studies on C—H···O hydrogen bonds date back to the early 1970s, but most of these calculations are only of historical interest today. Of late, a number of theoretical studies have been seen to appear with frequency. We do not attempt to cover this area completely nor even to evaluate it fully. What is presented here is a brief survey of the literature. This, hopefully will serve as an entry into the field for the interested reader.

An early study that is still worthy of mention is the attempt of Berkovitch-Yellin and Leiserowitz (1982) to find reasons for the failure of close-packing models to describe adequately the crystal structures of carboxylic acids. Using atom–atom potential calculations, these authors calculated the energy of the C—H···O=C hydrogen bonded dimer fragment as occurring in fumaric acid (Fig. 2.7) and arrived at a value of –3.2 kcal/mol for the dimer. This energy is composed of 80 per cent electrostatic and 20 per cent dispersion contributions. Along similar lines, Cox *et al.* (1981) estimated the energy of the C—H···O hydrogen bond in acetic acid, **11**. Because the energy of this mixed dimer was calculated as –4.25 kcal/mol and that of the centrosymmetric dimer as –6.5 kcal/mol, the energy of the C—H···O interaction was estimated to be around –1.0 kcal/mol.

However, these sorts of calculations led to less realistic conclusions in other cases. The semi-empirical calculations too that followed were not really able to treat intermolecular interactions properly. For example, Kumpf and Damewood (1988) concluded on the basis of semi-empirical calculations that while malononitrile can form C—H···O hydrogen bonds with crown ethers, nitromethane cannot. This is in conflict with more recent results which attribute weak donor capability to the methyl group (Novoa *et al.* 1991), and is a good example to show how quickly theoretical papers can become dated.

Today, fully corrected *ab initio* molecular orbital methods are the computational standard. A number of high level *ab initio* studies have been carried out on simple C—H···O and C—H···N bonded systems, and a few examples will be given. Turi and Dannenberg (1993, 1994, 1995) carried out *ab initio* calculations at the Hartree–Fock and second order Møller–Plesset (MP2) levels on complexes of acetylene and HCN with H_2O, HCHO and O_3. The

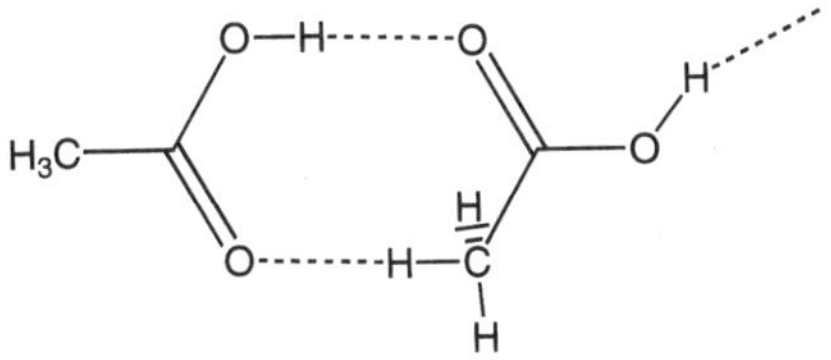

11

importance of correction for basis set superposition errors (BSSE) was noted. As a typical example, the $C-H\cdots O$ hydrogen bond in the complex $H-C\equiv C-H\cdots OH_2$ was calculated to have an energy of $-2.19\,kcal/mol$. $C-H\cdots O$ hydrogen bonds were seen to provide secondary stabilization to $O--H\cdots O$ hydrogen bonds in the crystal structure of the mono enol form of 1,3-cyclohexanedione. Detailed calculations using various basis sets up to D95++(d,p) were performed on several geometries of the $1:1$ $CH_3NO_2-NH_3$ complex. In the most favoured geometry **12**, there is one $C-H\cdots N$ and two $N-H\cdots O$ interactions. The total stabilization was estimated to be somewhat greater than $3.0\,kcal/mol$. The roughly linear $C-H\cdots N$ interaction was termed a hydrogen bond, while the $N-H\cdots O$ interactions were character- ized as electrostatic and together allotted only around $-1.0\,kcal/mol$. Semi- empirical calculations were also performed for comparison and it was noted that while the AM1 results agree with these *ab initio* calculations, the PM3 and SAM1 results did not.

The prototypical system of a weakly activated donor and a strong acceptor, CH_4-OH_2, has been studied by several groups: Latajka and Scheiner (1987), Novoa *et al.* (1991), Szczesniak *et al.* (1993), and most recently van Mourik and van Duijneveldt (1995). The latter authors obtained a bond energy of $-0.53\,kcal/mol$ for a linear hydrogen bonded adduct **13**, and only -0.07 kcal/mol for the alternative geometry **14**. The results on geometry **14** are of interest because this interaction mode allows close approach of an O atom to a methyl group almost without a net interaction energy. For the approach to $R-CH_3$ in the direction of the $R-C$ vector, the potential energy curve devi- ates very little from zero, until exchange repulsion becomes dominant (compare the related situation in Fig. 2.10). Bond energies were also calcu- lated for methyl–water interactions $R-CH_3\cdots OH_2$ in geometry **14**. For the weakly activated donor with $R = NH_2$, the energy was $-0.61\,kcal/mol$, while for the strongly activated donor with $R = NH_3^+$, it was found to be $-9.3\,kcal/mol$.

Substitution effects have been studied in detail for interactions of fluoromethanes with H_2O. In the series of donors CH_3F, CH_2F_2 and CHF_3 combined with H_2O, Alkorta and Maluendes (1995) calculated the hydrogen bond to form along the $H-O-H$ bisector and the inclusion of each H

−0.53 kcal/mol   −0.07 kcal/mol

12   **13**   **14**

atom to strengthen the hydrogen bond by about 1 kcal/mol. This is shown in Table 2.8. The hydrogen bond distance shortens concomitantly by about 0.1 Å per F atom. Novoa and Mota (1997) also commented in detail on this system, and pointed out that their energy value for $F_3C-H\cdots OH_2$, −2.41 kcal/mol, is eight times as high as for $H_3C-H\cdots OH_2$. They concluded that substituent effects indeed play an important role in $C-H\cdots O$ hydrogen bonding.

In a study of the α-2-hydro nitronyl nitroxide radical, Deumal et al. (1997) concluded that their best BSSE corrected interaction energy of a given $C(sp^3)-H\cdots O-N$ contact is −0.40 kcal/mol, a value that is about one tenth of the interaction energy of an isolated $C(sp^2)-H\cdots O-N$ contact. For the ferromagnetic implications of this result, the reader is directed to Section 4.3. Again the $C-H\cdots S$ energy in the CH_4-H_2S complex was estimated to be around −0.42 kcal/mol, comparing well with a value of −0.84 kcal/mol, for the $S-H\cdots C$ interaction (Rovira and Novoa 1998). Ornstein and Zheng (1997) have calculated $C-H\cdots OH_2$ energies for a variety of donors and their results will be discussed in greater detail in Section 5.5. All these computations, the quality and reliablity of which will undoubtedly increase with increasing computing power, give an estimate of $C-H\cdots O$ and $C-H\cdots N$ energies in representative chemical systems.

A combined *ab initio* and statistical investigation on the model $C-H\cdots O=C$ bonded dimer **15** as occurring in 1,4-benzoquinone (Fig. 2.6) has been carried out recently by van de Bovenkamp et al. (1999). From a CSD

Table 2.8 Energy computations on $C-H\cdots O$ bonded complexes $H_2O-CH_mF_n$ (MP2/6-31G** level; Alkorta and Maluendes 1995)

Complex	ΔE (kcal/mol)	d (H$\cdots$O) (Å)
$CH_4\cdots OH_2{}^a$	−0.53	
$CFH_3\cdots OH_2$	−1.41	2.51
$CF_2H_2\cdots OH_2$	−2.13	2.39
$CF_3H\cdots OH_2$	−3.16	2.28

[a] From van Mourik and van Duijneveldt (1995).

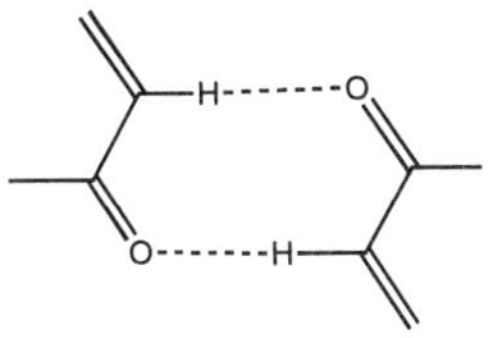

15

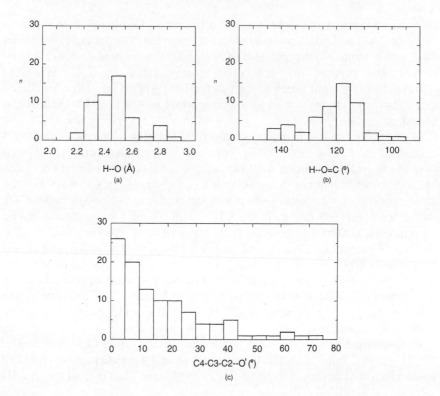

Fig. 2.26. Geometry of the $C-H\cdots O=C$ bonded synthon **15**. CSD analysis considering only centrosymmetric geometries with $d < 3.0\,\text{Å}$. (a) Distribution of normalized distances d. (b) Distribution of angles ϕ at the acceptor. (c) Distribution of a torsion angle ρ that describes the deviation from coplanarity; the atom labelling within the donor group is $C1=C2(H)-C3(=O)-C4$. Only relatively planar arrangements with $\rho < 20°$ are considered in (a) and (b) (using data from van de Bovenkamp *et al.* 1999).

survey, it was found that this commonly occurring supramolecular synthon exists with a wide range of geometries. These distortions are shown pictorially in Fig. 2.26, with the distributions of the parameters d and θ, and also of a torsion angle that defines the degree of non-planarity of the synthon. Notably, the $C-H\cdots O=C$ interactions in **15** are, on average, shorter and more linear than those in non-cyclic fragments (Table 2.9). Such behaviour is general. Hydrogen bond geometries in a specific pattern are typically better than those taken from random structures (see also Section 4.3).

In *ab initio* calculations on this system, a pronounced basis-set dependence of the energy was found, while the optimal geometry is relatively basis-set independent (Fig. 2.27). With the assumption of planarity, a two-dimensional potential energy surface was calculated as shown in Fig. 2.28(a). The global minimum was found at $d = 2.43\,\text{Å}$, $\theta = 176.8°$ with an energy of $-17.9\,\text{kcal/mol}$ for the dimer. The optimal geometry is non-linear because this way the

Table 2.9 Mean C—H···O geometries in the cyclic pattern **15** and in non-cyclic hydrogen bonds of the fragment C=CH—CO—C (van de Bovenkamp *et al.* 1999)

	15 (cyclic)	Non-cyclic
n	53	69
$d(\text{H}\cdots\text{O})$ (Å)	2.47(2)	2.58(2)
$D(\text{C}\cdots\text{O})$ (Å)	3.53(2)	3.52(2)
$\theta(\text{C}-\text{H}\cdots\text{O})$ (°)	165.7(9)	148(2)
$\phi(\text{H}\cdots\text{O}=\text{C})$ (°)	120(3)	130(3)

Distance d normalized.

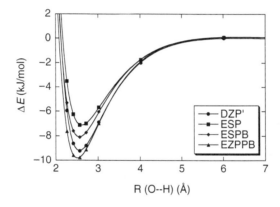

Fig. 2.27. Basis-set dependence of *ab initio* calculations, shown for the example of synthon **15** with $\theta = 140°$. The minimum energy is strongly basis set dependent, whereas the posititon of the minimum varies only slightly (from van de Bovenkamp *et al.* 1999).

intradimer H···H repulsion is reduced, and the electrostatic interaction between the antiparallel carbonyl groups is optimized compared to a dimer with linear C—H···O bonds. Whilst studying the schemes in Fig. 2.29, the reader should note that the optimal geometry is of the type (a), whereas geometries of the type (c) are disfavoured. The total energy is composed of an approximately −12 kcal/mol contribution from the two hydrogen bonds, the rest coming from electrostatic interactions between the carbonyl groups. This shows that the geometry and the energetics of synthon **15** are not determined exclusively by hydrogen bonding but also to a meaningful extent by other interactions. The description of **15** as the sum of two C—H···O hydrogen bonds is, therefore, inadequate. The broad shape of the potential energy surface of **15** allows for large distortions from optimal geometry, and the experimental points in Fig. 2.28(a) show that large distortions actually do occur

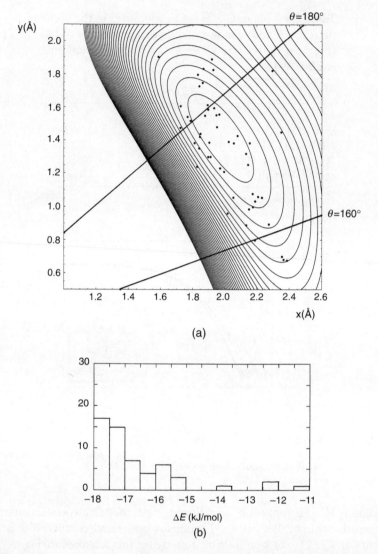

(a)

(b)

Fig. 2.28. Combined *ab initio* and statistical study on the C—H···O=C bonded synthon **15**. (a) Potential-energy surface for the dimer benzoquinone–propenal. The spacing between contour levels is 0.5 kJ/mol, the lowest contour level corresponds to −17.5 kJ/mol. The straight lines show geometries with hydrogen bond angles of $\theta = 180°$ and $160°$, respectively. Notice that the global minimum occurs for a slightly bent angle θ. The dots represent geometries occurring in crystal structures. (b) Calculated energies of the experimental points in (a); the distribution shows the density of experimental data points between given contour lines (from van de Bovenkamp *et al.* 1999).

in crystals. Even geometries of the type (c) of Fig. 2.29 are observed frequently. Satisfactorily, all experimental geometries are found in the lower region of the potential energy surface, and only four out of 53 dimers have calculated energies greater than 1 kcal/mol above the global minimum (Fig. 2.28b). The geo-

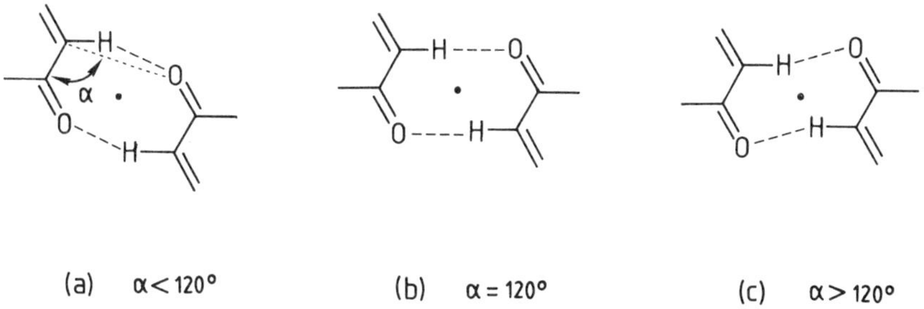

(a) α < 120°   (b) α = 120°   (c) α > 120°

Fig. 2.29. Geometries of the C—H···O=C bonded synthon **15**. (a) Sheared geometry with α < 120°. This geometry is favoured because it optimizes the stabilizing interactions between the C=O groups. (b) Geometry with α = 120°. If the synthon were governed by the hydrogen bonds alone, this would be the ideal geometry. (c) Sheared geometry with α > 120°. This is unfavourable because of the short H···H contact (after van de Bovenkamp *et al.* 1999).

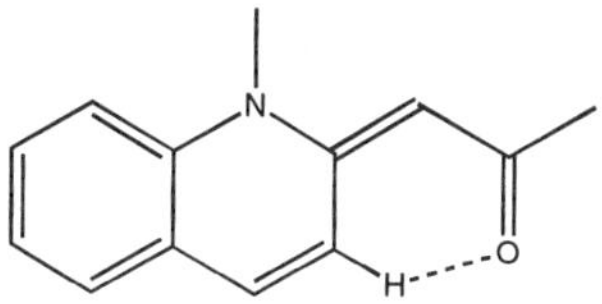

16

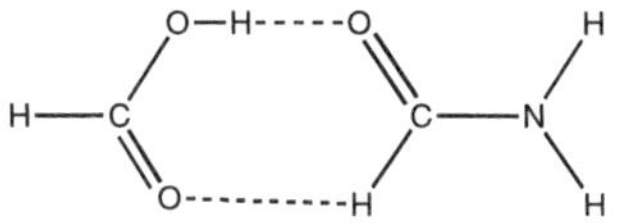

17

metrical variability **15** is of great importance for its function as a supramolecular synthon – to be effective in molecular recognition, such a pattern must allow geometrical flexibility so that it can adjust to the multitude of extraneous influences that always occur in the solid state. In a related context, it is also of interest to mention the recent study of Allen *et al.* (1998*a*) where the general importance of stabilizing C=O···C=O interactions in crystals was demonstrated.

Extension of computational work to the understanding of specific chemical phenomena involving weak hydrogen bonding is also an active area of research. Gough and Millington (1995) have characterized a weak intramolecular C—H···O bond in the crystal structure of 2-acetonylidene-1,2-dihydro-1-methylquinoline, **16**, with *ab initio* methods, and identified a bond critical point between the H and O atoms. Bond critical points in C—H···O contacts occurring in a proton sponge complex have been identified by Mallinson *et al.* (1997*a*) in a charge density analysis. *Ab initio* calculations at the MP2/6-31G*//RHF/6-31G* level provide data on the intermediacy of a C—H···O interaction in the hetero Diels–Alder reaction of SO_2 to 1,3-butadiene and isoprene, and account for the observed stereoselectivity and regioselectivity (Suárez *et al.* 1994). Computations show that in the cyclic complex of formic acid with formamide, **17**, the C—H···O hydrogen bond provides around 2.0–3.5 kcal/mol to the complex interaction energy (Neuheuser *et al.*

1994). The authors have further noted that this is a clear indication that the formyl $C-H$ group may yet participate in a $C-H\cdots O$ hydrogen bond (see also Sections 2.2.12 and 2.2.13). Evidence for $C-H\cdots N$ hydrogen bonding has been found in the Hofmann-type clathrates by Ruiz and Alvarez (1995).

The theory of atoms in molecules has also been invoked to show the bonding nature of the $C-H\cdots O$ interaction. Popelier and Bader (1992) have commented on the $C-H\cdots O$ hydrogen bond between a methyl group and a negatively charged O atom in the biomolecule creatine. Koch and Popelier (1995) have proposed a set of criteria to establish hydrogen bonding in $C-H\cdots O$ situations, including bifurcated interactions. Molecular dynamics (MD) studies have also been used to demonstrate the viability and stability of $C-H\cdots O$ interactions. According to MD computations of Politzer and co-workers (Seminario $et\ al.$ 1995), these interactions exist in liquid nitromethane, and according to Monte Carlo simulations of Cordeiro (1997) also in liquid formamide, N-methylformamide and N,N-dimethylformamide. In extensive MD computations, Auffinger and Westhof (1996) have shown that $C-H\cdots O$ hydrogen bonds between specific sites stabilize the structure of the tRNA^{Asp} anticodon hairpin. This work is discussed in greater detail in Chapter 5. A number of other computational studies that are not mentioned here also indicate the stabilizing nature of the $C-H\cdots O$ interaction.

All this gainsaid, the calculated values of $C-H\cdots O/N$ bond energies may not be reliable except those from the most elaborate of calculations, and these are only possible for very small, relatively uninteresting systems. Even for the $CH_4\cdots OH_2$ system, published values from different groups vary significantly. In general, the main problem with all computations of weak interactions is that to describe the diffuse electron density far from the nuclei, one needs a very large basis set, while to estimate the dispersion energy accurately, extensive electron correlation calculations are necessary. Scheiner (1997) provides a detailed chronology of events that illustrate the pitfalls encountered when applying low level or incomplete theoretical treatments to weak hydrogen bonding interactions. Scheiner repeatedly refers to these as not being 'true' hydrogen bonds (without defining the term) and does not apply the epithet $hydrogen\ bond$ in an unqualified way to any of the $C-H\cdots O$ interactions discussed in his monograph. He, however, also states that 'one person's H-bond is another person's van der Waals complex, or another's Coulombic interaction'. To summarize then, the various computational studies of weak hydrogen bonding are still in a stage of evolution and refinement, and much progress is required in this area.

2.2.7 Cooperativity

Hydrogen bonds possess the important property of cooperativity. Accordingly, the energy of an array of say n interlinked hydrogen bonds is larger than the sum of n isolated hydrogen bonds. This non-additive property arises because the ability of donor and acceptor groups to form hydrogen bonds is further increased by an increase in polarity when the hydrogen bonds form part of a collective ensemble. There are two different mechanisms that can produce this effect:

1. Functional groups acting simultaneously as hydrogen bond donors and acceptors form extended chains or rings in which the individual hydrogen bonds enhance each other's strengths by mutual polarization. This is illustrated in say alcohols which form patterns of the type $O^{\delta-}-H^{\delta+}\cdots O^{\delta-}-H^{\delta+}\cdots O^{\delta-}-H^{\delta+}\cdots$

2. Charge flow in suitably polarizable π-bond systems increases donor and acceptor strengths. This is sometimes referred to as resonance assisted hydrogen bonding (Gilli *et al.* 1989).

Given the importance of polarization in both cases, there is every reason to believe that cooperativity is of importance for weak hydrogen bonds.

The simplest case of cooperative hydrogen bonding is a finite chain of the type $X-H\cdots Y-H\cdots Z$ in which the mutual polarization occurs through the σ-bonds. Since most $C-H$ groups can act only as donors and not as acceptors, they can only assume the role of the promoter group $X-H$. An example with a finite chain $C\equiv C-H\cdots O-H\cdots O=C$ has been discussed by Lakshmi *et al.* (1995*a*) with regard to the crystal structure of cycloheptane **18** (Fig. 2.30). In a chain $C-H\cdots O-H\cdots A$, the strengths of *both* hydrogen bonds are enhanced. On the one hand, the $O-H\cdots A$ hydrogen bond benefits from the accepted $C-H\cdots O$ bond. This benefit increases with increasing $C-H$ donor strength. On the other, the $C-H\cdots O$ bond is itself enhanced by the polarization which the $O-H$ group achieves in its hydrogen bond to A. This latter effect increases with the acceptor strength of A.

An example of cooperativity enhancement of a $C-H\cdots O$ hydrogen bond in a finite arrangement $C\equiv C-H\cdots O-H\cdots O=P$ has been discussed for the 4:2:1 complex of Ph_3PO, H_2O and 1,4-diethynylbenzene (Kariuki *et al.* 1997). Figure 2.31 shows that the water molecule donates two hydrogen bonds to very strong $O=P$ acceptors, making it in turn a much stronger acceptor than it normally is. In consequence, the $C\equiv C-H\cdots O_w$ hydrogen bond is exceptionally short ($D = 3.02$, $d = 1.96\,\text{Å}$) and strong ($\Delta\nu_{CH} = 202\,\text{cm}^{-1}$).

18

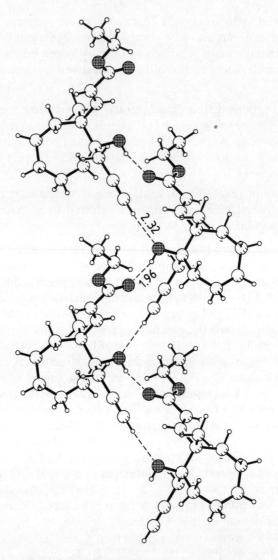

Fig. 2.30. Cooperative hydrogen bond chain C≡C—H···O—H···O=C in 1β-hydroxy-1α-propargyl-2β-methyl-2(2-ethoxycarbonylvinyl)-cycloheptane, **18** (Lakshmi *et al.* 1995*a*). Distances *d* are normalized.

Polarization via a combination of σ and π bonds is also common. A pattern with C—H···N bonding that is immediately suggestive of cooperativity is provided by the infinite chain structure in crystalline HCN (Section 2.1). In an *ab initio* quantum chemical study of (HCN)$_n$ chains, it was calculated that the H···N distances and the C—H stretching frequencies reduce with increasing system size, while the C—H bond lengths increase. The bond energies increase

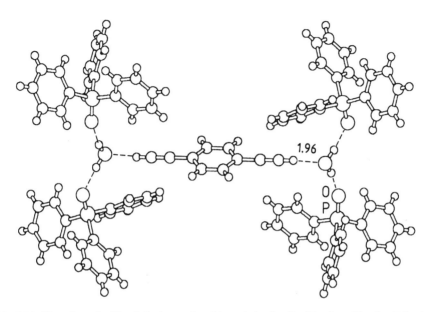

Fig. 2.31. Very short C—H···O hydrogen bond in a chain C≡C—H···O$_W$—H···O=P in the ternary complex (1:2:4) 1,4-diethynylbenzene–H$_2$O–triphenylphosphine oxide (adapted from Kariuki *et al.* 1997). Distance *d* is normalized.

from −4.7 kcal/mol for the dimer, −5.4 kcal/mol for the trimer, −5.7 kcal/mol for the tetramer to −6.0 kcal/mol for the pentamer (Kofranek *et al.* 1987).

Quantum chemical calculations have also been performed on the infinite molecular ribbons in crystalline *N,N*-dimethylnitroamine (Sharma and Desiraju 1994). The reader is directed initially to the hydrogen bonded dimer **19** formed by a pair of adjacent molecules. The pair of hydrogen bonds in **19** render the free nitro group more basic and the free methyl groups more acidic and the hydrogen bonds formed in the trimer, **20**, the tetramer and the pentamer are successively stronger. According to semi-empirical molecular orbital calculations the average C—H···O bond energy (per dimer unit) is −3.2, −3.6, −3.8 and −3.9 kcal/mol for **19, 20**, tetramer and pentamer, at which stage the cooperativity effect begins to level off. The energy enhancement due to cooperativity is thus greater than 20 per cent in the pentamer, a value similar to that found for O—H···O—H···cooperativity. The very short C—H···O bond in 2-ethynyladamantan-2-ol (Allen *et al.* 1996*b*) has already been noted in Section 2.2.1. Neutron analysis shows that the C—H···O and O—H···O bonds form a cooperative network, **21** and it is this O—H···O and O—H···π cooperativity that affords the great strengthening of the C—H···O bond (Fig. 2.32).

The ethynyl group is of particular relevance to the phenomenon of cooperativity because it can simultaneously donate a C—H···A and accept an

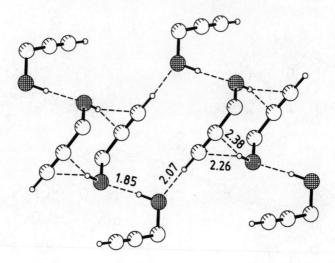

Fig. 2.32. Cooperative hydrogen bonds in 2-ethynyladamantan-2-ol, **21**, as determined by neutron diffraction at 100 K (Allen *et al.* 1996*b*). The adamantyl groups are omitted for clarity. There are two symmetry-independent molecules; one ethynyl group forms a very short C—H···O hydrogen bond (d = 2.07 Å, θ = 173.3°), whereas the other one forms no hydrogen bond at all. The O—H···C≡C hydrogen bond has a geometry with respect to the bond midpoint of d(M) = 2.26, D(M) = 3.22 Å, θ(M) = 179.0°.

19

20

21

X—H···π hydrogen bond. This dual donor–acceptor character is reminiscent of the hydroxy group and indeed this similarity has been explicitly commented upon (Steiner *et al.* 1995*a*). Three patterns, **22**, **23** and **24**, that are topologically equivalent may be considered and one might expect that these patterns have analogous cooperativity properties, even if of different strengths. In this

O—H·····O—H·····O   **22**

C—H·····O—H·····O   **23**

X—H·····C≡C—H·····O   **24**

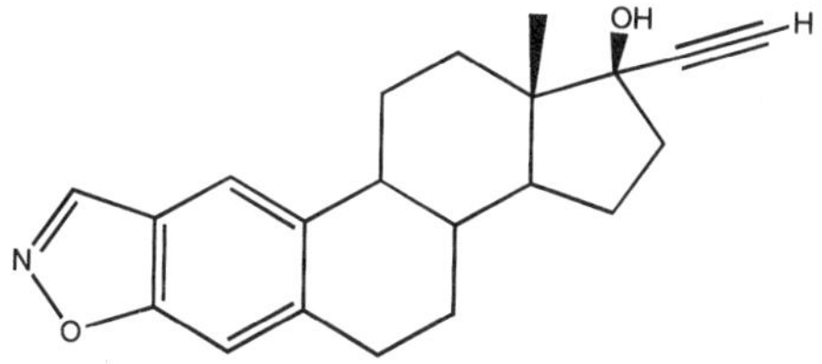

25

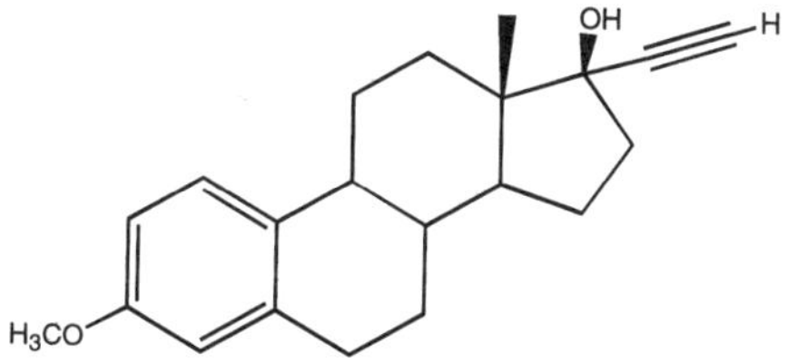

26

regard, compounds with a geminal attachment of ethynyl and hydroxy groups have a particularly rich structural chemistry. Around 100 crystal structures in the CSD contain this molecular grouping, with a significant proportion of 17-substituted steroids, and a very large number of different cooperatively assisted hydrogen bonded patterns are observed. Indeed, the number of distinct patterns might even be termed bewilderingly large and the reason for this structural variability still remains unexplained. With reference to cooperativity though, two steroids, danazole, **25**, and mestranol, **26**, are noteworthy.

In danazole, an unusual example of a steroid with an isoxazole moiety fused to the A-ring, infinite chains of C—H···O ($d = 2.27$ Å) and O—H···π ($d = 2.40$ Å) hydrogen bonds form a cooperative pattern, C≡C—H···O—H···C ≡C—H···O—H, that corresponds topologically to an infinite hydrogen bond chain of the O—H···O—H···O—H···type (Fig. 2.33, Viswamitra *et al.* 1993). Very unusually, for the hydroxy group is sterically unhindered, the structure does not contain any O—H···O hydrogen bonds. In mestranol there are

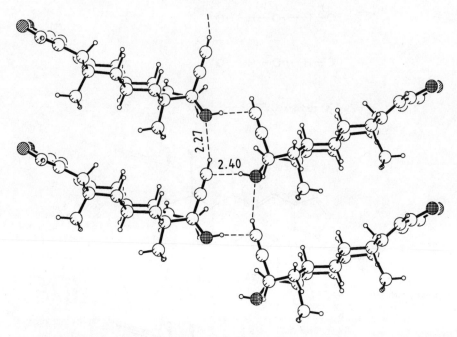

Fig. 2.33. Cooperative hydrogen bond chain C≡C—H···O—H···C≡C—H···O—H in dana-zole, **25** (Viswamitra *et al.* 1993).

two symmetry-independent molecules, A and B, with virtually identical con-formations but having different intermolecular contacts (Steiner *et al.* 1997*a*). A section of the packing is shown in Fig. 2.34. The two symmetry-independent ethynyl groups donate C—H···O hydrogen bonds of very similar geometries (A: $D = 3.443$, $d = 2.44$ Å, $\theta = 153°$. B: $D = 3.478$, $d = 2.48$ Å, $\theta = 151°$). However, while A is part of an infinite cooperative array of C—H···O, O—H···O and O—H···π hydrogen bonds, the C—H···O bond formed by B is isolated. Interestingly, the IR spectrum of the compound shows different solution–solid shifts for the two ≡O—H groups, 21.3 and 59.8 cm^{-1}. It is straightforward to suggest that the C—H···O interaction in the infinite chain is the stronger one (Section 2.2.1). The strongly shifted band is thus assigned to molecule A and the weakly shifted one to molecule B. The nearly three times larger bathochromic shift in A relative to B is interpreted as a direct consequence of the cooperative effect. Other ethynyl hydroxy compounds and their coopera-tivity attributes are further discussed in Chapter 3.

2.2.8 *Hardness and softness*

The efficacy of formation of hydrogen bonds depends substantially on donor acidity and acceptor basicity (Section 2.2.2), but it may also depend on the

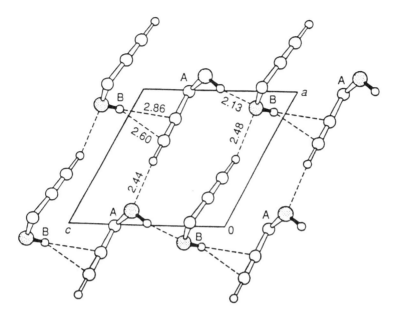

Fig. 2.34. Hydrogen bond pattern in mestranol, **26** (from Steiner *et al.* 1997*a*). Geometries are normalized.

ability of the hydrogen bond donor to polarize and the acceptor to become polarized. This is called the *hardness* or *softness* of the acid or base (HSAB). Hardness and softness are attributes which cannot be precisely measured, but only qualitatively described (Pearson 1963, 1968). However, this HSAB concept has been useful in mechanistic organic chemistry to describe the relative reactivities of various electrophiles and nucleophiles because it has been found that hard acids prefer to bond with hard bases and soft acids with soft bases. This is called the HSAB principle. The rule has nothing to do with acid and base strength but merely says that an acid–base adduct will have extra stability if both the acid and base are hard or if both the acid and base are soft. This rule is not a theory but a generalization based on experimental observations. Taking the viewpoint that all intermolecular interactions are incipient reactions, a hydrogen bond becomes an incipient acid–base reaction. Indeed, the concept of a hydrogen bond as an arrested protonation process is well known (Bürgi and Dunitz 1983). Thus it should be possible to use HSAB ideas to explain the formation or non-formation of some hydrogen bonds. To clarify these ideas further, it is instructive to enumerate hard and soft acids and bases in the hydrogen bond context and to provide some examples. For a general treatment of the HSAB principle, the reader is advised to consult the literature (Ho 1975, 1977; Parr and Pearson 1983).

Hard hydrogen bond donors A — H are constituted with small atoms A and

Table 2.10   Hard and soft hydrogen bonds. The hard–hard and soft–soft combinations are highlighted

		Very hard F—C	Hard O, N	Soft S, Se	Very soft π
Very hard	F—H		F—H···N		F—H···π
Hard	O—H	**O—H···F—C**	**O—H···O**	O—H···S	O—H···π
Hard	N—H	N—H···F—C	**N—H···O**	N—H···S	N—H···π
Soft	S—H		S—H···O	**S—H···S**	S—H···π
Very soft	C—H	C—H···F—C	C—H···O	C—H···S	**C—H···π**

have a high positive charge on H. They do not contain many unshared electron pairs and if they do, these electron pairs are not easily donated. They are of low polarizability and are electropositive. The naked proton would be the ultimate example of hardness. Groups like OH, NH_2, NHR, FH, ClH are hard donors. Soft donors are made up of large atoms and/or atoms with moderate electronegativities. They could contain electron pairs that may be easily donated. They are polarizable to varying degrees. Typical examples are CH, PH, SH, TrH. Hard hydrogen bond acceptors are constituted with small, electronegative atoms. They are of low polarizability and are hard to oxidize. Typical examples are hydroxylic or phenolic O, amino N, C=O, C—O—C, C—F, C—Cl. Finally, soft acceptors contain large, easily polarizable atoms or π-clouds. They are relatively easy to oxidize. Examples are C—Br, C—I, thiol S, Tr, C≡C, phenyl. The reader will be able to extrapolate easily within and between these categories. Accordingly $C(sp)$—H will be harder than $C(sp^2)$—H, with $C(sp^3)$—H being the softest donor in this group. According to such a scheme of things, one could arrange hydrogen bonds in a matrix, with donor hardness varied in the rows and acceptor hardness in the columns. This is shown in Table 2.10, which contains examples that are actually observed in crystal structures.

Some illustrative facts begin to emerge. For example, the HSAB principle predicts straightaway that O—H···π, N—H···π and N—H···S and C—H···F hydrogen bonds with their hard–soft or soft–hard combinations are less likely than the hard–hard and soft–soft combinations. This is indeed borne out in practice, given the relative commonness of some these donors and acceptors. Specifically, the weakness of the N—H···S hydrogen bond has been commented upon (Allen *et al.* 1997*a,b*), though not in terms of violations of the HSAB rule. The X—H···O and X—H···N (X = C, N, O) bonds are common enough but the X—H···F bond is very rare (Shimoni and Glusker 1994; Dunitz and Taylor 1997). This is counterintuitive to what would be expected from electronegativity arguments but the HSAB principle provides a plausible rationalization for this fact. Even OH may not be hard enough for C—F. In their discussion of mestranol, Philp and Robinson (1998) comment that the O—H···π hydrogen bond (–1.5 kcal/mol) is not much stronger than the $C-H\cdots\pi$ bond (-1.0 kcal/mol). They ascribe this to the weak basicity of the ethynyl group. Considering, however, that donor acidity is more important than acceptor basicity and also that the hardness and softness are mismatched and matched, respectively, in these two bonds, a nicer rationalization is provided with the HSAB rule.

Hydrogen bonding in organometallic crystals (Section 3.5) provides other examples of the HSAB principle. It was found, for example (Braga *et al.* 1995), that while CO ligands in organometallic clusters do not act as good acceptors of strong hydrogen bonds, they readily form $C-H\cdots O$ hydrogen bonds in a competitive situation. The basicity of the $C\equiv O$ ligand is not the issue here. The soft acceptor CO prefers to hydrogen bond with the soft donor $C-H$, or alternatively the hard donor $O-H$ prefers to form bifurcated or solvated hydrogen bonds with hard acceptors rather than waste itself on the soft acceptor CO, or in other words, the $C-H$ group matches the CO ligand in softness. A large number of organometallic hydrogen bonds, such as $M-H\cdots O\equiv C$, $C-H\cdots M$, $M-H\cdots Cp$ and above all $H\cdots H$, are good examples of an interaction between a soft donor and a soft acceptor.

These polarization effects have been quantified with charge density studies. Carroll and Bader (1988) reported a link between the softness of an acid and the increased penetration of its H atom by the basic atom. Koch and Popelier (1995) calculated that for $C-H\cdots O$ bonds, the non-bonded charge density on the H atom (soft acid) is penetrated more than that on the O atom (hard base). As in mechanistic chemistry, the observation of more anomalous or unexpected observations is necessary before HSAB arguments are placed on a surer footing. However, there is no chemical reason to presuppose that they will be of any less relevance and importance in the supramolecular context.

2.2.9 *Intramolecular phenomena*

Intramolecular contacts are conveniently considered according to two distinct phenomenological situations. The first pertains to intramolecular $C-H\cdots O$ and $C-H\cdots N$ hydrogen bonds and the second to effects of *inter*molecular $C-H\cdots O/N$ hydrogen bonds on *intra*molecular conformations. Both situations belong to the less well-understood aspects of weak hydrogen bonding.

With regard to intramolecular phenomena there are notable differences between weak and strong hydrogen bonds. For strong and very strong hydrogen bonds, there is little distinction between between inter- and intra-molecular cases in terms of effects on physical and spectroscopic properties, and on structural significance. Even though geometries of intramolecular bonds are affected by steric constraints, hydrogen bonds may be identified relatively easily from just the contact geometries. Problems arise only if the intramolecular $O/N-H\cdots O/N$ contact has low θ angles, say less than $120°$, and this generally occurs only in bifurcated arrangements. The energies of intramolecular strong and very strong hydrogen bonds amount normally to several kcal/mol, that is more than typical rotation barriers and energy differences between molecular conformations. Therefore, strong intramolecular $O/N-H\cdots O/N$ hydrogen bonding is often a determinant of molecular conformation.

For weak hydrogen bonds, the situation is very different. Because of their weakness, intramolecular $C-H\cdots O$ interactions can hardly compete with conformational and other intramolecular effects. Justification for their existence from geometries alone is normally not possible, and their influence on molecular conformation is variable. Even if such an influence exists, it is very difficult to demonstrate. Still there is good evidence that intramolecular $C-H\cdots O/N$ bonds are operative in a number of cases.

2.2.9.1 *Intramolecular interactions*

Whilst considering intramolecular hydrogen bonds, it is useful to distinguish two basic categories. In the first, donor and acceptor are separated by a long and flexible chain of covalent bonds that allows the hydrogen bond to more or less freely adopt its optimal geometry. In this case, the intramolecular hydrogen bond does not experience strong conformational restraints, and can exhibit structural and energetic properties that are very similar to those possessed by intermolecular hydrogen bonds. These bonds can therefore be termed 'quasi-intermolecular'. In the second category, the covalent linkage between donor and acceptor is short and relatively rigid. For $C-H\cdots O/N$ contacts of this kind, it is normally very difficult to decide whether they are really hydrogen bonds or forced consequences of the molecular structure. Of course, no definite borderline exists between these categories and numerous intermediate cases are observed.

It is convenient to discuss first the simpler of these two categories, namely the quasi-intermolecular $C-H\cdots O$ bonds. Interactions of this kind can occur only in relatively large molecules and are very frequently found in macrocycles and in biological macromolecules. Two typical examples are shown in Fig. 2.35. In uncomplexed 18-crown-6, an elliptical molecular conformation is stabilized by some intramolecular $C-H\cdots O$ interactions, Fig. 2.35(a) (Maverick *et al.* 1980; Goldberg 1984). In the neutron crystal structure of a modified nucleoside shown in Fig. 2.35(b) (Takusagawa *et al.* 1979), a very short intramolecular $C-H\cdots O$ bond is formed. In both cases, the $C-H\cdots O$ contacts, were they destabilizing, could easily be avoided or be made less repulsive by conformational changes. Numerous examples for quasi-intermolecular $C-H\cdots O$ hydrogen bonds will be described for proteins and nucleic acids in Chapter 5. Particularly in proteins, it is not unusual that the donor and acceptor of an intramolecular hydrogen bond are separated by hundreds of covalent bonds down the main chain. If the covalent link between donor and acceptor is short and rigid, however, structural interpretation becomes much more difficult and geometry alone does not provide a reliable yardstick. A seemingly favourable hydrogen bonding geometry could be the fortuitous

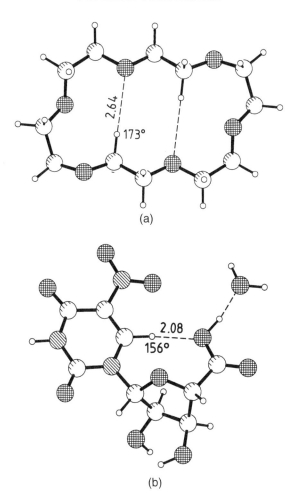

(a)

(b)

Fig. 2.35. Intramolecular C—H···O hydrogen bonds with quasi-intermolecular geometries. These interactions cannot be considered as 'forced' contacts because the flexible molecules can easily avoid them. (a) Uncomplexed 18-crown-6 at 100 K, distance d normalized (Maverick *et al.* 1980, discussed by Goldberg 1984). (b) Neutron structure of a modified nucleoside (Takusagawa *et al.* 1979).

result of other stronger intramolecular effects, while a true intramolecular hydrogen bond could be distorted because of other factors. Some independent evidence, be it spectroscopic, theoretical or geometrical, would therefore definitely increase the confidence level in characterizing an intramolecular C—H···O/N hydrogen bond. We will discuss the most typical problems with relevant literature examples.

An intramolecular C—H···O hydrogen bond was postulated to exist in 1,3-bis(*m*-nitrophenyl)urea, **27**, and related compounds by Etter and Panunto

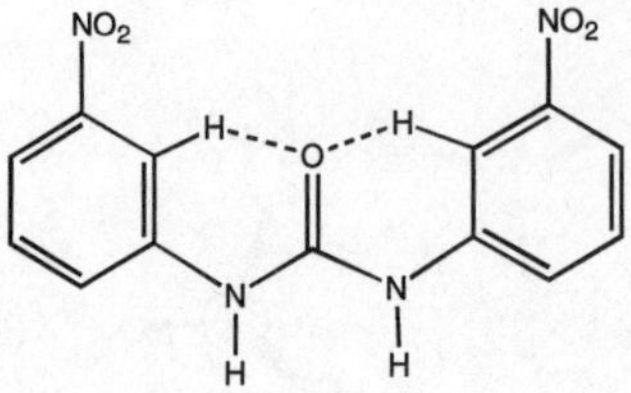

27

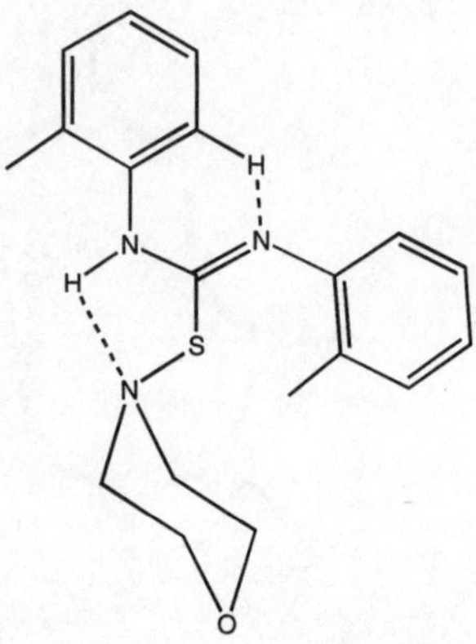

28

(1988) to rationalize the fact that the carbonyl group does not accept an $N-H$ group in a conventional hydrogen bonding situation. This particular compound was described as one of the poorest of hydrogen bond acceptors. In support of their claim, these authors stated that in this group of diphenyl ureas with electron withdrawing substituents, the intramolecular d separations are in the range 2.23–2.29 Å while the two d separations in N,N'-diphenylurea itself are 2.49 and 2.66 Å. While such an intramolecular bond might indeed exist in these compounds, the justification provided for its existence is insufficient because the short d separations could be a consequence of molecular planarity that has arisen for independent reasons, such as maximization of resonance.

A somewhat related case was observed in isothiourea **28** (Sudha *et al.* 1996) and is shown in Fig. 2.36. An unusually short intramolecular $C-H\cdots N$ contact ($d = 2.24$ Å) is accompanied by appreciable widening of the internal angle at the amino N atom. This interaction could be considered to be a hydrogen bond because if it were not, the molecular conformation might have easily given it up to avoid the strain at the N atom. On the other hand, one could argue that the particular conformation is adopted only to maximize resonance between the phenyl and isothiourea groups, and that the $C-H\cdots N$ contact is more a disturbing than a supporting factor. The overall situation is therefore again quite unclear.

The value of the θ angle need not also be very helpful. In part, an intramolecular hydrogen bond is weak if the hydrogen bond angle is constrained to be rather low. Proper angular ranges for intramolecular $C-H\cdots O$ hydrogen bonds have not yet been established and the present situation is confusing. Quite a number of short intramolecular $C-H\cdots O$ contacts that are claimed to be hydrogen bonds in the literature do not pass muster on closer inspection and should be characterized as 'forced contacts'. In general, independent supporting evidence is necessary before an intramolecular $C-H\cdots O$ contact

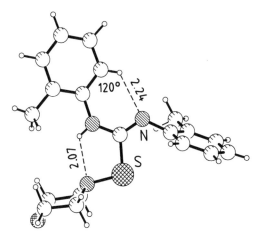

Fig. 2.36. 1,3-Bis(2-methylphenyl)-2-(4-morpholino)-isothiourea, **28**, with two relevant intramolecular interactions, N—H···N and C—H···N (Sudha *et al.* 1996). Geometries are normalized.

29

can be considered to be a hydrogen bond. For a number of examples, such indicators have actually been obtained from other structural parameters. Even for these cases, however, the bonding situation is not as unambiguous as for typical intermolecular C—H···O interactions.

The 11-ethyl and 11-methoxy derivatives of the fused ring aromatic 15, 16-dihydrocyclopenta[*a*]phenanthren-17-one, **29**, present an interesting contrast

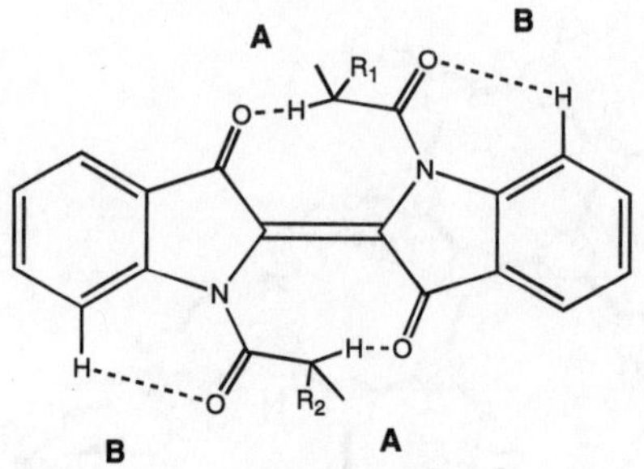

30

(Desiraju *et al.* 1993). In both polymorphs of the former, the anticipated steric hindrance in the bay region of the steroid results in torsional angles between 5° and 20°. Such distortion is absent in the latter which is close to planar because of a possible intramolecular $C-H \cdots O$ bond ($d = 1.98$ Å).

Smith *et al.* (1993) have used an insightful choice of compounds to demonstrate the existence of intramolecular $C-H \cdots O$ hydrogen bonds in N,N'-diacylindigos, **30**. The crystal structures of N,N'-diacetylindigo ($R_1 = R_2 = H$) and its mono ($R_1 = H$, $R_2 = Cl$) and dichloro ($R_1 = R_2 = Cl$) derivatives all exhibit short intramolecular $H \cdots O$ distances. In each molecule, both the N-acyl $C-H$ groups donate to the indigo $C=O$ groups (hydrogen bonds A) while the N-acyl carbonyl groups accept a $C-H$ group each from the indigo aryl groups (hydrogen bonds B) forming a total of four distinct contacts. The A contacts are shorter for the dichloro derivative ($D = 2.98, 2.91$ Å) than they are for the parent compound (3.04, 2.94 Å). This was attributed to the greater acidity of the relevant $C-H$ groups in the dichloro derivative. Chloro-substitution also affects $C=O$ group basicity. So, one of the B contacts is longer in the dichloro derivative (2.90 Å) than in the unsubstituted derivative (2.85 Å). The contacts in the monochloro derivative are therefore of special interest. The more acidic N'-chloroacetyl $C-H$ forms a stronger interaction (2.87 Å) than the N-acetyl $C-H$ (2.93 Å), while the less basic N'-chloroacetyl $C=O$ forms a weaker interaction (2.91 Å) than the N-acetyl $C=O$ (2.85 Å). That the relative ordering of $C \cdots O$ distances may be predicted by simply considering the $C-H$ acidities and O atom basicities shows convincingly that one is dealing with hydrogen bonding interactions.

A number of other studies may be mentioned. The work of Harlow, Sammes and co-workers (1984*a,b*) on intramolecular $C-H \cdots N$ bonds in a series of 2-aminoalkyl-1,3-dithiamine 1,1,3,3-tetraoxides has already been noted in Section 2.1. An intricate system of intramolecular $C-H \cdots N$, $N-H \cdots O$ and $N-H \cdots N$ hydrogen bonds stabilizes the planar conformation of N-*p*-tolyl-5-benzoylamino-2-chloro-4-*p*-tolylamino-6-pyrimidinecarboxamide,

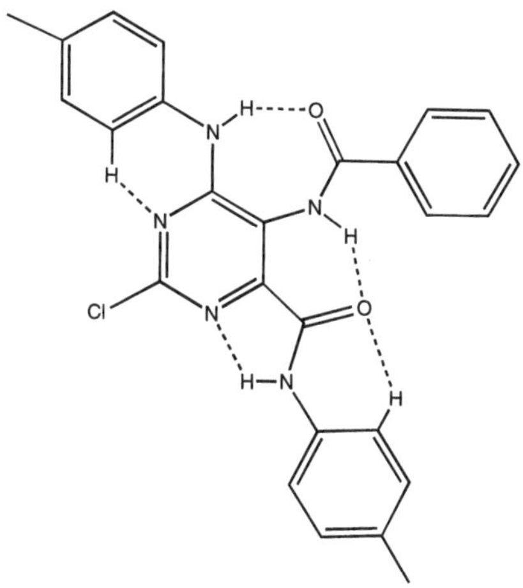

31

31 (Mazurek *et al.* 1995). Wiberg *et al.* (1991) have noted changes in the internal ring conformations of nonanolactone, cyclodecane, cyclodecanone and
nonanolactam that can be rationalized on the basis of an intramolecular
C—H···O hydrogen bond in nonanolactone (d = 2.34 Å; θ = 109.5°). Other
cyclic ketones have been examined by the group of Scheffer (T. J. Lewis *et al.*
1996). These workers (Fu *et al.* 1994) also report a number of very short
intramolecular C—H···O contacts in *N*-(*tert*-butyl)succinimide (four contacts
in the d range 2.17–2.24 Å, θ range 116–124°) that may enhance intramolecular H atom photoabstraction reactions. In an Ru(II) complex of a chiroporphyrin (Mazzanti *et al.* 1996), a methylene group conformation has been
accounted for on the basis of an intramolecular C—H···O interaction (D =
3.18 Å, θ = 142°). It has been noted that the Co(III) complex of bis-*N,N*-
carboxymethyl-L-phenylalanine can coordinate a bidentate amino acid in the
cis-N and *trans-N* configurations (Jitsukawa *et al.* 1994). The former contains
an intramolecular C—H···O bond (D = 2.83 Å) that is confirmed with NMR,
while the latter contains no such interaction. Finally, Taube and co-workers
(Ghosh *et al.* 1995) have rationalized the large distortion from linearity of the
O—V—O angle (173.5°) in V(pyridine)$_4$(O$_3$SCF$_3$)$_2$ based on an intramolecular C—H···O hydrogen bond. In some of these cases, there is enough
flexibility in the molecular framework to make them borderline to quasi-
intermolecular interactions.

Two related theoretical studies should also be mentioned. In a conformational analysis of 2-propoxyethanol, the global energy minimum conformation
was calculated to be as shown in **32** (Gil *et al.* 1995). The hydroxy as well as
the terminal methyl group are oriented towards the central ether O atom and
form O—H···O and C—H···O hydrogen bonds with strongly bent θ angles.

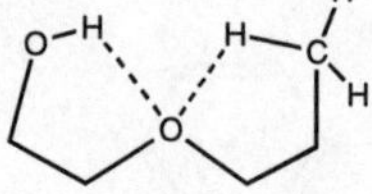

32 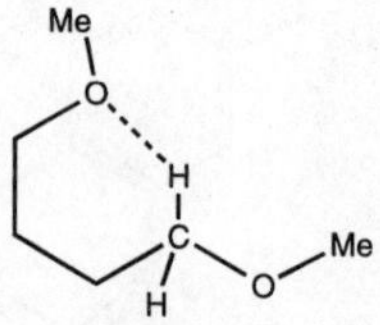   **33**

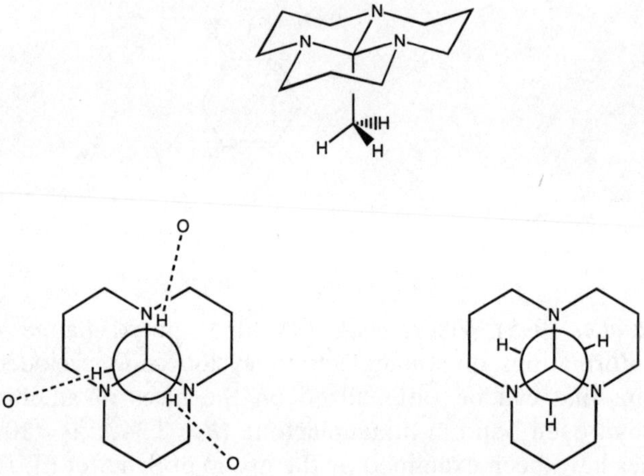

34

The total energy is about 0.37 kcal/mol lower than in the conformer with a fully extended propyl chain, where no $C-H\cdots O$ interaction is formed. In a conformational analysis of dimethoxy ethers, it was calculated that the folded conformation of $Me-O-(CH_2)_4-O-Me$ which contains an intramolecular $C-H\cdots O$ contact, **33**, is more stable by 0.43 kcal/mol than the extended conformation (Law and Sesamuna 1996).

2.2.9.2   Conformational changes

If intermolecular $C-H\cdots O$ hydrogen bonds are comparable in energy to intramolecular conformational barriers, then it should be possible to observe unusual molecular conformations that are mediated by these weak interactions. Although such effects should occur frequently, they have been discussed in depth only for very few examples. A well-cited case described by Seiler and Dunitz (1989) involves the cyclic orthoamide **34**. X-Ray analyses were performed for two crystalline modifications of the compound. In the cubic trihydrate (*Pa*3, *Z* = 8), the molecule has crystallographic threefold rotational

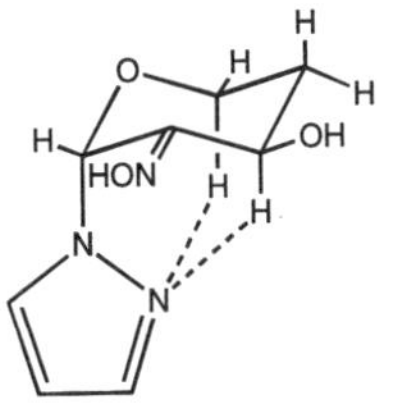

35

symmetry. In the anhydrous monoclinic form ($P2_1/c$, $Z = 8$) the two symmetry independent molecules have different configurations, one all-*trans* (as in the cubic trihydrate) and the other *cis*, *cis*, *trans*. In the cubic trihydrate, each orthoamide molecule is attached to a triad of water molecules by $C-H\cdots O$ bonds ($d = 2.67$ Å, $\theta = 170°$). An unusual feature of this structure is the nearly eclipsed conformation about the central $C-CH$ bond. In the anhydrous form, both types of molecule have the normal staggered orientation of their methyl groups. The authors attributed the reversal of the methyl group orientation (staggered to eclipsed) to $C-H\cdots O$ hydrogen bonding. The minimum energy of the three intramolecular bonds taken together was estimated computationally to be -2.5 kcal/mol (Chao and Chen 1996). A related case of a methyl group that balances on top of a rotational barrier in order to facilitate $C-H\cdots O$ hydrogen bonding has been found in crystalline *N*-methyl-2-pyrrolidone (Müller *et al.* 1996). A very different example of how intermolecular $C-H\cdots O$ bonds can stabilize an otherwise unfavourable conformation has been provided by Leiserowitz (1976) in his exemplary study of fumaric acid. Crystallographic and computational studies by Ciunik (1997) suggest that intramolecular $C-H\cdots N$ hydrogen bonds are important in fixing an equatorial or 'anomeric' conformation of 1-(2-hydroxy-iminopyranosyl)pyrazoles, **35**.

2.2.10 *Influence on crystal packing*

The electrostatic nature of the $C-H\cdots O/N$ hydrogen bond determines its role in influencing crystal packing. Electrostatic interactions between point charges (separated by a distance r) are relatively long range, falling off as $-r^{-1}$. For comparison, the short range van der Waals interactions fall off as $-r^{-6}$. It may be thus inferred that even incipient $C-H\cdots O$ bonds have an orienting effect on crystallizing molecules, and before the effects are felt of the van der Waals interactions that will ultimately determine the close-packing characteristics. For this reason, crystal structures may be viewed as a complex mélange of isotropic and anisotropic interactions. The hydrogen bonds (weak and strong) determine general connectivity patterns of molecules, while the isotropic interactions determine both intramolecular conformations and intermolecular close-packing within the basic scaffolding established by the hydrogen bonds. For this reason it is particularly hard to ascertain which type of interaction is structure-determining. For the typical organic molecule that contains a small number of O and/or N atoms (this is a large sub-group in the CSD), the entire range of interaction hierarchy may be observed. Crystal structures are seen wherein the role of the weak hydrogen bond could be either innocuous, or vary from supportive to intrusive. In the first (very large) category, $C-H\cdots O$ hydrogen bonds are found but these are neither distinctive nor significant. They merely exist within a structure that is almost wholly determined by other interactions. When the $C-H\cdots O$ bonds play a supportive role, their orientational requirements are in consonance with those of the other interactions. A typical example would be 1,4-benzoquinone in which $C-H\cdots O$ bonds and stacking interactions are neatly matched. Many other related examples have been discussed by Berkovitch-Yellin and Leiserowitz (1984). Intrusive $C-H\cdots O$ hydrogen bonds would appear to direct the packing in unexpected ways, in other words the packing simply cannot be understood until and unless these bonds are actively invoked (Desiraju 1991a, 1996a).

A few prototypical studies may be mentioned. The early work of the Israeli school (Bernstein et $al.$ 1974; Leiserowitz 1976) on the packing characteristics of quinones and carboxylic acids has been discussed in Section 2.1, and the importance of the 1,4-benzoquinone structure has been mentioned. The relationship of these structures to that of 2-methoxy-1,4-benzoquinone was discussed by Keegstra et $al.$ (1994) in the context of molecular recognition (see also Section 4.1.2). In a series of aromatic steroids, the cyclopenta[a]phenanthrenes, there are no hydrogen bond donors other than $C-H$ groups (Desiraju et $al.$ 1993). The observed patterns are complex and often contain $C=O$ groups accepting four and even five $C-H\cdots O$ hydrogen bonds. Gavezzotti (1991a) analysed the packing of 590 oxohydrocarbons and reported the frequent occurrence of $C-H\cdots O$ contacts mostly to $C=O$ groups. He concluded that $C-H\cdots O$ hydrogen bonds are secondary interactions with weak directionality.

2.2.10.1 *Supportive and intrusive roles for weak hydrogen bonds*

In structures where the $C-H\cdots O$ bonds play a supportive role, their directional preferences are satisfied within the geometrical constraints of the stronger $O-H\cdots O$ and $N-H\cdots O$ hydrogen bonds, the $\pi\cdots\pi$ stacking, herringbone and donor–acceptor interactions. However, when they become intrusive they are able to significantly disturb the effects of either the close-packing interactions or of the strong hydrogen bonds. The latter case is particularly dramatic because it seems counterintuitive to expect that weak hydrogen bonds can affect the disposition and arrangements of strong hydrogen bonds. Indeed, most of the important studies of the effects of strong hydrogen

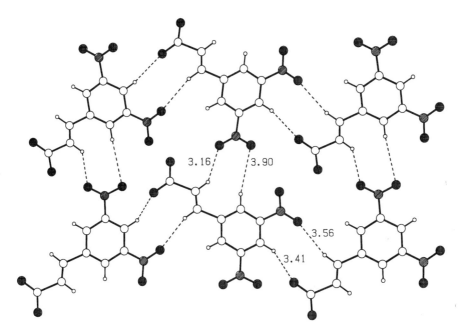

Fig. 2.37. Network of C—H···O hydrogen bonds in 3,5-dinitrocinnamic acid. *D* values are given. Molecules related by O—H···O hydrogen bonding are not shown (adapted from Desiraju and Sharma 1991).

bonding on crystal packing patterns are silent on the question of weak hydrogen bonds in general and intrusive weak hydrogen bonds in particular. In reality, C—H···O hydrogen bonds are able to discriminate between alternative O—H···O or N—H···O networks, which though geometrically reasonable are structurally quite distinct. In such cases, weak hydrogen bonds may be termed steering or 'tugboat' interactions which, though small in energy, are sufficient to select a particular crystallization pathway from among several possibilities. The energy of an individual C—H···O/N hydrogen bond may be small, but if their number crosses a critical threshold, the structure may focus into an unconventional packing.

A good example is provided by 3,5-dinitrocinnamic acid, a molecule that contains many activated C—H groups and acceptor O atoms (Desiraju and Sharma 1991). This acid crystallizes as an O—H···O dimer where the hydrogen bonded molecules are related, not by an inversion centre as is usual, but by a twofold rotation axis (space group *C2/c*). Both inversion and twofold dimers are equienergetic and, in principle, both are reasonable possibilities. However, the crystal structure (Fig. 2.37) shows an extensive network of C—H···O hydrogen bonds. These C—H···O bonds dominate the structure and their combined directional requirements appear to be incompatible with the O—H···O inversion dimer. The formation of all these C—H···O bonds depends in some measure on optimal conformational freedom of the molecule, and the isomeric 2,4-dinitrocinnamic acid with its conformationally locked *ortho* nitro group adopts the normal centrosymmetric $O-H\cdots O$ dimer.

The absence of a critical number of $C-H\cdots O$ hydrogen bonds may also disturb the $O-H\cdots O$ topology. In (4-chlorophenyl)propiolic acid, the complete absence of $C-H\cdots O$ bonds leads to an unexpected catemer motif rather than the dimer, such as is found in (4-chlorophenyl)cinnamic acid (Desiraju *et al.* 1990). The $O-H\cdots O$ catemer is intrinsically more stable than the dimer (Leiserowitz 1976). Therefore, if the dimer is to be formed, it must be stabilized additionally by $C-H\cdots O$ bonds, or in some other manner. Alternatively, if the $C-H\cdots O$ bond-forming ability is reduced, the catemer is favoured. Arguably this happens in the propiolic acid above relative to the corresponding cinnamic acid because of the absence of the critical alkenic $C-H$ groups. These and other studies show that $C-H\cdots O$ bonds need to be considered as important contributors in the formation of hydrogen bond patterns and their presence or absence could well determine the manner of networking stronger $O-H\cdots O$ and $N-H\cdots O$ hydrogen bonds.

Nitro compounds. The nitro group, though not inherently very basic, acts as a $C-H\cdots O$ acceptor in the crystal structures of a large number of unsaturated compounds (Sharma and Desiraju 1994). Conjugation of the nitro group enhances its acceptor ability in a $C-H\cdots O$ bonding situation via cooperativity effects. The formation of the unexpected *homo*dimer in the molecular complex of 4-(*N*,*N*-dimethylamino)benzoic acid and 3,5-dinitrobenzoic acid has been rationalized on the basis of poor $C-H\cdots O$ hydrogen bonding (Sharma *et al.* 1993) and is mentioned also in Section 2.2.3.3. This complex is unique among a group of nine complexes formed by nitrobenzoic and nitrocinnamic acids because the eight others studied form the expected heterodimers. In solution, both homo and heterodimers probably exist but because of difficulties in $C-H\cdots O$ bond formation, stacking interactions become significant. It was shown that these interactions are much better for homodimers than for heterodimers; in the latter case, atom charges within an $O-H\cdots O$ hydrogen bonded dimer get dissipated because of resonance-assistance (Gilli *et al.* 1989). Therefore stacking of aromatic rings can prevail over other types of interactions like $C-H\cdots O$ or even $O-H\cdots O$ and such effects should be included when modelling the structures of biomolecules.

2.2.10.2 *Relative importance of intermolecular interactions*

Considering that crystal packing results as the sum of many different contributions of directional and non-directional intermolecular interactions, it is important that different types of interactions be considered jointly in structure analysis. There is a growing need to simultaneously assess different interactions and to establish their hierarchy. For molecules containing only C, H,

O and/or N atoms, it is erroneous to think that forces like $O-H\cdots O$ and $N-H\cdots O$ can solely control crystal packing. Formulation of rules may achieve a certain superficial simplification but in the long run, one may encounter more exceptions than examples. Against such a backdrop, it is pertinent to consider an interesting method, NIPMAT, for the visualization of organic crystal structures that has been developed in the Cambridge Crystallographic Data Centre. A pictorial matrix is formed using the atoms of a molecular skeleton $(A_1, A_2 \ldots A_m \ldots A_n)$ and the matrix element $A_m A_n$ which is defined by the shortest intermolecular contact $A_m \ldots A_n$ is shown in terms of a grey scale. The shorter the contact, the darker the square which represents that particular contact. Therefore, the plot obtained is a simultaneous visual representation of all the short intermolecular interactions. It is possible that two atoms A_m and A_n have more than one short intermolecular contact, and the inability to show more than one $A_m \ldots A_n$ contact is a limitation of the method.

In the context of $C-H\cdots O$ hydrogen bonding, the NIPMAT plots for 1,4-benzoquinone and fluoranil are shown in Fig. 2.38. In benzoquinone, the $C-H\cdots O$ bonds are revealed as dark regions in the upper and right extremities. However, there is an overall greyness to the plot especially in the carbon region, revealing that stacking interactions are also important. The utility of this sort of analysis is confirmed on inspecting the NIPMAT plot of fluoranil. Here, and in the absence of H atoms, the dominant feature is the dipolar $C\cdots O$ interaction, the geometry of which leads to the skewed arrangement of rings, which is revealed as a pale area in the carbon region (no stacking). A comparison of these NIPMAT plots shows that the packing of benzoquinone and fluoranil are fundamentally different. Consideration of a single NIPMAT plot allows a simple but elegant visual assessment of intermolecular interactions in a crystal that is almost impossible otherwise.

2.2.10.3 Replacement of strong by weak hydrogen bonds

Some cases of structural isomorphism have been reported where an $O-H\cdots O$ or $N-H\cdots O$ bond in one structure is structurally and functionally replaced by a $C-H\cdots O$ bond in the other. Well-known examples are the complexes **36** and **37** of barbital with urea and acetamide. Because of the identity of the contact

36   **37**

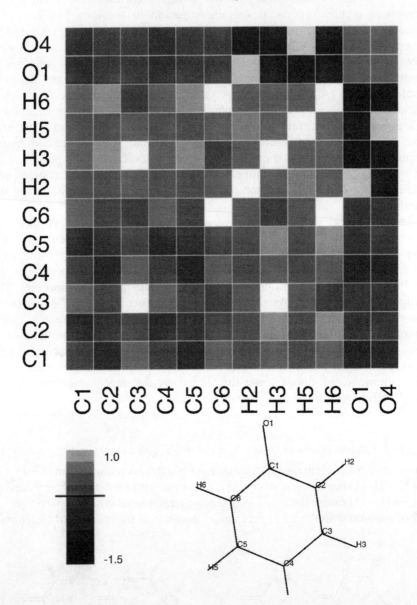

Fig. 2.38. Examples of NIPMAT plots. Grey tones represent intermolecular distances, with the darker shades representing shorter distances. (a) Intermolecular contacts in 1,4-benzoquinone. (b) Intermolecular contacts in fluoranil. Notice the C—H···O contacts in benzoquinone.

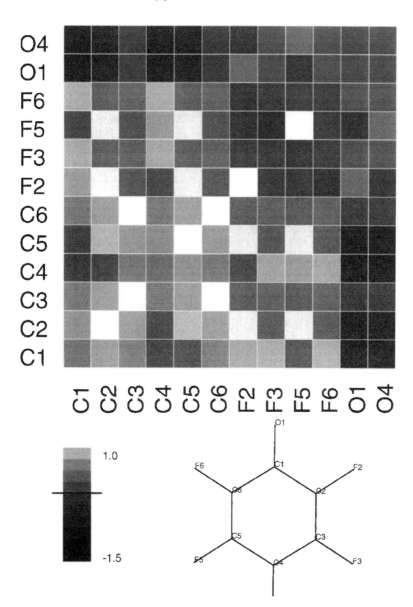

Fig. 2.38. (cont.)

38   **39**

patterns in both structures, Berkovitch-Yellin and Leiserowitz (1984) interpreted the $C-H\cdots O$ geometry here as a hydrogen bond. In the context of
crystal engineering (Chapter 4), the interaction patterns in these two structures
would be said to represent equivalent supramolecular synthons. Similar observations were reported by Jaskólski (1984) who found an $N-H\cdots Cl^-$ bond in
an adenosine derivative replaced by a $C-H\cdots Cl^-$ bond in an isomorphous
structure of a related derivative. Recently, such an observation was made for a
larger molecular assembly with the isomorphous crystal structures of β-
cyclodextrin (β-CD) complexed with diethanolamine and with pentane-1,5-
diol (Steiner *et al.* 1995*b*). This is discussed in greater detail in Section 4.2.

Isostructurality is common in steroids and in a recent interesting case
(Anthony *et al.* 1998*a*), it was found that the crystal structures of 2-oxa-4-
androstene-3,17-dione, **38**, and its 6α-hydroxy analogue, **39**, are very similar.
Both these compounds adopt the same monoclinic space group, $P2_1$, and the
values of the *a*-axis parameters are nearly equal (6.232 and 6.221 Å). This is the
direction of the hydrogen bond interactions and a $C-H\cdots O$ interaction in the
former ($d = 2.67$ Å) is replaced by a $C-O-H\cdots O$ hydrogen bond in the latter
($d = 1.90$ Å) without other major differences (Figs 2.39 and 2.40). The closeness
in the hydrogen bond patterns in these structures is evidenced by the formation
of a binary solid solution of the two components. Binary solid solution formation is the most stringent criterion for isostructurality (Kálmán and Párkányi
1997), and this appears to be the first reported instance of solid solution formation in pairs of compounds related by weak/strong hydrogen bond equivalence.
This example shows again that $O-H\cdots O$ and $C-H\cdots O$ hydrogen bonding
can influence crystal packing in very similar ways and is underpinned by the fact
that the C6 H atoms in Δ^4-steroids are allylic in nature and, as such, activated as
$C-H$ donors. Isostructural replacement of strong by weak hydrogen bonds
occurs frequently upon suitable mutation of amino acid side-chains in proteins,
as is shown for several examples in Section 5.2.

2.2.10.4 *Miscellaneous examples*

As the concept of the weak hydrogen bond has become increasingly accepted
by crystallographers, there has been a steady rise in the number of papers
where this interaction has been explicitly identified and discussed. In many of

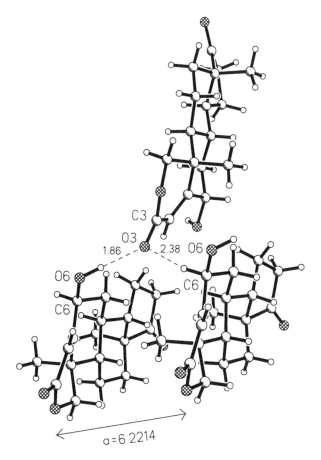

Fig. 2.39. O—H···O and C—H···O hydrogen bonds in steroidal hydroxy lactone **39** involving the C6 and O3 centres. The *a*-axis direction is indicated. O-atoms are shaded (adapted from Anthony *et al.* 1998*a*).

these publications, the role of the C—H···O/N hydrogen bonds ranges from a passive to a supportive one. As might be expected, such discussions have become less detailed in recent times, and one appears to be at the stage now where authors feel confident enough to be able to invoke the presence of a weak hydrogen bond without a specific reference citation to earlier papers and reviews of the subject. This circumstance in itself is telling and indicates that the weak hydrogen bond, and especially the C—H···O/N hydrogen bond, is now established as an important associative interaction in molecular crystals. It is impractical, even impossible to list these papers here and a casual browse through any current issue of say *Acta Crystallographica Section C, Zeitschrift für Krystallographie, Inorganic Chemistry* or even say, *Tetrahedron Letters* will reveal several papers where the presence of weak hydrogen bonds has been

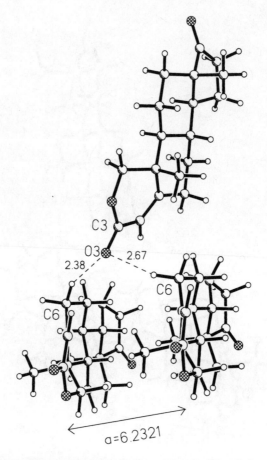

Fig. 2.40. C—H···O hydrogen bonds in lactone **38** involving the C6 methylene group. Notice the great similarity to Fig. 2.39 along the *a*-axis direction. A C—O—H···O hydrogen bond in **39** is replaced by a C—H···O hydrogen bond. Compounds **38** and **39** form a solid solution (adapted from Anthony *et al.* 1998*a*).

noted. Different authors seem to have different criteria for terming a particular C—H···O/N geometry as a hydrogen bond. Regrettably, many still use the Taylor–Kennard distance criterion ($d < 2.4\,\text{Å}$) and even more inexplicably so for intramolecular geometries. Others prefer more relaxed criteria ($d < 2.8$ Å). We would definitely advocate the latter, with hopefully some spectroscopic or auxiliary evidence for the longest of contacts ($d < 3.0\,\text{Å}$). The angular cut-offs most often used are in the 100–120° range, which is reasonable. The following structures are only representative and give the reader a flavour of the wide variety of organic and organometallic structures where the role of weak hydrogen bonds in crystal packing has been mentioned or discussed.

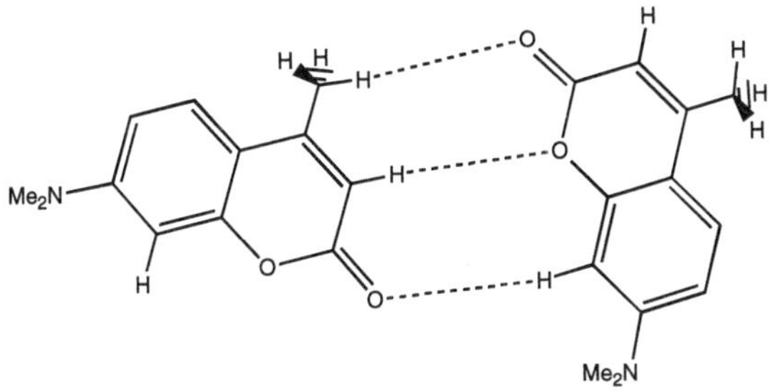

40

In ferulic acid (3-methoxy-4-hydroxycinnamic acid), intermolecular D values of 3.39 and 3.18 Å have been noted (Nethaji *et al.* 1988). A particularly detailed justification for $C-H\cdots O$ ($D = 3.49$ Å) and $C-H\cdots N$ hydrogen bonding is presented for (Z)-4-benzylidene-2-methyl-5(4H)-oxazolone (Souhassou *et al.* 1986). D values in the range 3.30–3.39 Å have been reported for $C-H\cdots O$ bonds to anionic oxygen in crown ether complexes (Reddy and Chacko 1993). Blaschette *et al.*(1993) report a $C-H\cdots O$ bond ($D = 3.22$, $d = 2.48$ Å) in another crown ether derivative. Planar heterocycles are particularly prone to the formation of $C-H\cdots O$ hydrogen bonds as is seen in the crystal structure of a furoxan trimer (Gasco *et al.* 1993) and in an isoxazole carboxamide (Jaulmes *et al.* 1993). Ratajczak-Sitarz and Katrusiak employ the Taylor–Kennard criterion in 3-chloro-5-tosylmethylpyridazine (1994a) and in phenylsulphonylmethyl bromide (1994b). A relatively short contact ($D = 3.21$ Å) is reported in L-mannono-1,4-lactone by Shalaby *et al.* (1994), and an elaborate discussion of $C-H\cdots O$ bonds in an auxin related xylopyranoside has been presented by Tomic *et al.* (1995). The presence of two symmetry independent molecules in 4-methyl-(7-N,N-dimethylamino)coumarin, **40**, may be rationalized by $C-H\cdots O$ hydrogen bonding which is responsible for their mode of association (Yip *et al.* 1995). The absence of disorder in a host–guest complex of cobaltocenium fluoride in a nonasil cavity has been ascribed to a dense system of $C-H\cdots O-Si$ interactions (Behrens *et al.* 1995). It is noteworthy that in this structure, all the $C-H$ groups of the guest species are thus tethered.

Strong and weak interactions ($N-H\cdots O, C-H\cdots O, C-H\cdots N$ and $\pi\cdots\pi$ stacking) have been discussed in a series of phthalimides by Barrett *et al.* (1995). Davidson (1995) has pointed out that in a phosphonium aryloxide, the $C-H\cdots O$ interactions ($D = 3.20, d = 2.27$ Å, $\theta = 166°$ and others) must be genuine because the donor groups are ideally oriented to participate in the observed hydrogen bonding despite being free to rotate. In methyl-3,5-dinitrocinnamate, $C-H\cdots O$ bonded dimers of energy -8.2 kcal/mol have

been proposed (Sharma *et al.* 1995). Toda *et al.* (1995) have suggested that
$C-H \cdots O$ bonds with D in the range 3.18–3.32 Å could be the cause for the
high melting points of the polymorphic forms of 3,4-dihydro-3,4-bis(diphenylmethylene)—*N*-methylsuccinimide and related compounds. Weak hydrogen
bonds have been suggested in 2-amino-5-chlorobenzophenone by Vasco-
Mendez *et al.* (1996) and in some tetrametavanadates (Wéry *et al.* 1996).
Seyeda *et al.* (1996) refer to $C-H \cdots O$ hydrogen bonding in the context of
the geometry of the ozonide anion. $C-H \cdots O$ networks have been described
in detail for ferrocene derivatives by Ferguson *et al.* (1996). $C-H \cdots O$
and $C-H \cdots \pi$ hydrogen bonds have been invoked to explain the efficient
resolving abilities of some 2,3-di-*O*-(arylcarbonyl)tartaric acids (Tomori *et al.*
1996).

2.2.11   *Repulsive and destabilizing $C-H \cdots O$ contacts*

The entire body of spectroscopic and crystallographic work discussed so far
in this chapter conveys a comprehensive and hopefully convincing picture of
the $C-H \cdots O/N$ interaction as a hydrogen bond, though generally a weak
one. The interaction is largely electrostatic in character, it depends on $C-H$
acidity and O/N basicity, it also has polarization character and as such is
enhanced by cooperativity effects. All these properties account, to a greater
or lesser extent, for its important role in determining intramolecular conformation and intermolecular crystal packing. This then raises the question as to
whether *all* $C-H \cdots O/N$ geometries which conform to usually accepted criteria for a weak hydrogen bond are indeed so. In specific terms, is there any
likelihood of contacts with say $3.0 < D < 4.0$ Å, $2.0 < d < 3.0$ Å, $100 < \theta < 180°$
not being stabilizing and of the hydrogen bond type? If so, what are the conditions for the formation of such destabilizing interactions? And might they,
at least to some degree, justify reservations about what has been said in the
previous sections in this chapter? In this section, it will be shown that destabilizing $C-H \cdots O$ interactions actually do occur in crystal structures, but that
on the whole they do not play disquieting roles.

Whilst considering destabilizing $C-H \cdots O$ interactions, it is useful to distinguish two categories: those which have the chemical potential to be bonding
but are rendered destabilizing by adverse or peculiar structural circumstances;
and those which are destabilizing by their very chemical nature. The latter can
never be hydrogen bonds. We begin with the former and far more common
category, and within this category, we shall concentrate on compressed
$C-H \cdots O$ contacts.

Let us first recall the definition of the terms *repulsive* and *destabilizing* given
in Section 1.1.3. By repulsive, we mean contacts associated with a repulsive
force vector, and by destabilizing we mean contacts with a positive interaction
energy. For $X-H \cdots A$ hydrogen bonds in an optimal geometry, the energy is
at the minimum and the force vector is zero. In a crystal, hardly any hydrogen

bond can adopt this ideal geometry and almost all deviate from it because of compromises with other inter- and intra-molecular effects. If the hydrogen bond distance is slightly elongated, this leads to an attractive force, and if it is reduced it leads to a repulsive force. Both try to restore the optimal geometry. Because small deviations from optimal geometry lead only to a small energetic disadvantage, slightly attractive *and also slightly repulsive* force vectors will occur frequently on hydrogen bonds in crystals. These deviations are for the most part unremarkable. Of greater concern are large deviations from optimal geometry. Elongation leads to weakening and ultimately to breaking of the bond, but the interaction energy can never become positive. Extreme compression, on the other hand, can in principle lead to a truly destabilizing situation. There are experimental indications that very repulsive intermolecular $C-H\cdots O$ contacts do in fact occur in crystals, but only rarely. Furthermore, it cannot normally be shown if they are above or below the zero energy level. To obtain further insight, one has to rely on computational and statistical studies.

To ascertain if sizable compression can be expected to occur frequently in intermolecular arrangements, the schematic hydrogen bond potential from Fig. 1.3 is drawn again for three kinds of hydrogen bonds that we have chosen to distinguish (Fig. 2.41). The corresponding distributions of distances d

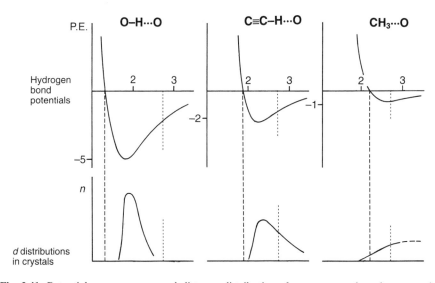

Fig. 2.41. Potential energy curves and distance distributions for strong, weak and very weak hydrogen bonds. The schematic potential energy curves in the upper row are associated with appoximate distributions of distances d shown below them. Because most hydrogen bonds have geometries corresponding to energies within 1 kcal/mol of the minimum, the destabilizing region is accessible to only very weak hydrogen bond types. The $H\cdots O$ van der Waals distance at $2.7\,\text{Å}$ is also indicated, illustrating the adverse consequences of applying the van der Waals cut-off to the weaker hydrogen bonds.

observed in crystals are shown as sketches. For $O/N-H\cdots O$ hydrogen bonds, the energy minimum is about $-5\,kcal/mol$, and at least this energy is needed to compress the interaction into the destabilizing region. Not a single example is known where this happens. The distribution of d in crystals has a width of several tenths of an Å, the maximum is slightly offset towards longer distances. That the whole distribution is *inside* the bonding region has actually been demonstrated for hydrogen bonds between hydroxyl groups in carbohydrates (Kroon *et al.* 1975). For $C-H\cdots O$ hydrogen bonds with acidic donors, the energy minimum is $-2\,kcal/mol$ or deeper (Fig. 2.41 *centre*), and the situation is essentially the same as for $O/N-H\cdots O$ hydrogen bonds. In molecular orbital calculations on the benzoquinone-type dimer linked by a pair of $C-H\cdots O=C$ hydrogen bonds, the energy minimum was calculated to be about $-4\,kcal/mol$ for the pair. In an accompanying database study, all metrics found were well inside the bonding region (Section 2.2.6, Fig. 2.28; van de Bovenkamp *et al.* 1999). No experimental indication has yet been found for a destabilizing $C-H\cdots O$ contact involving one of the stronger $C-H$ donors.

For the weaker $C-H$ donors, the situation is quite different (Fig. 2.41, *right*). The energy minimum is $-1\,kcal/mol$ or shallower, and for non-activated methyl groups no more than $-0.5\,kcal/mol$. Compressing this kind of contact into destabilizing geometries requires little in energetic terms and must be expected to occur frequently. No theoretical or statistical study on this particular matter has been published as yet. However, it is likely that the short-end tail of the distance distribution in crystals, which is notoriously ill-defined and without a clear maximum for weak $C-H$ donors, reaches into the destabilizing region. Though this contention is unproven, it suggests that one examine the shortest contacts of weak donors to O atoms with some caution. These contacts are certainly not 'strong' $C-H\cdots O$ hydrogen bonds; they are in all probability repulsive; they might even be destabilizing.

Some authors have commented on these matters. Let us consider the case of a hexakis-adduct of C_{60} studied with variable temperature X-ray techniques (Seiler *et al.* 1996). The molecule contains $>C(CO_2C_2H_5)_2$ groups and the crystal structure contains seven distinct sets of $C-H\cdots O$ geometries involving these groups. At 270 K, six of these occur in the D and θ ranges 3.26–3.60 Å and 125–155° (neutron corrected). The crystal structure was redetermined at 230 K and again at 180 K, and it was observed that these six $C-H\cdots O$ geometries became shorter with a decrease in temperature. The decrease in D ranged from 0.01–0.09 Å while θ remained practically unchanged. This kind of behaviour is very characteristic of interactions that are both attractive and stabilizing. The seventh interaction was, however, noted to be different. It occurs in the centrosymmetric dimer pattern **41** involving carbethoxy groups. At 270 K, the values of D, d and θ are 3.49 Å, 2.41 Å and 174°. Upon cooling to 230 K, the values become 3.51 Å, 2.41 Å and 177°. Finally at 180 K, they are 3.56 Å, 2.46 Å and 178°. If the H atom positions are uncor-

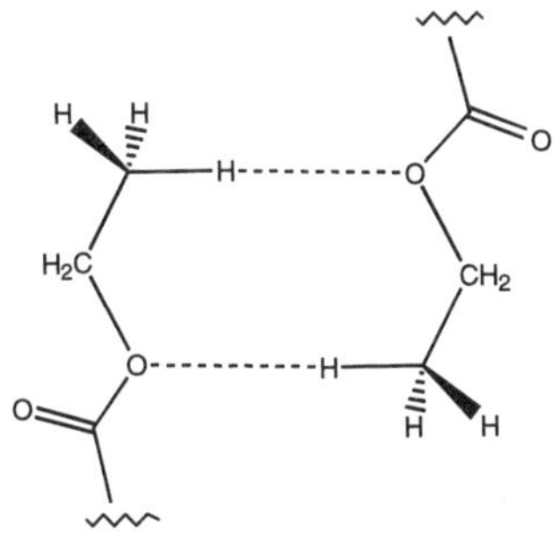

41

rected, however, the d value is seen to rise from 2.40 Å to 2.49 Å to 2.50 Å while the θ value *falls* from 173° to 154° before rising to 163° at the lowest temperature. Though the authors interpreted these observations as being caused by repulsion, we feel that the matter is still ambiguous. The contact is formed by a terminal methyl group of a highly flexible ethoxy group and has plenty of opportunities to avoid this contact. Conformational adjustments, as in the case of 2-nitrobenzaldehyde (see below) could well minimize the repulsion. More significantly, the contact is not so short (2.41 Å) that it could be distinctly repulsive or destabilizing in the first instance. In the absence of other data, it is suggested that the contact is lengthened because some other contact is affected upon cooling.

The study of some pyrimidine nucleosides and their salts by Wiewiórowska *et al.* (1992) may also be noted in this context. These workers have claimed that a short intermolecular contact ($d \cong 2.4$ Å) between the C(6)—H group and the O5′ oxygen is repulsive in nature, on the basis of small hypsochromic shifts (15–25 cm^{-1}) of the C(6)—H stretching vibrations. As an internal check, the short contacts formed by the C(5)—H group appear to be normal C—H···O hydrogen bonds when similarly evaluated. The reader will note, however, that if a d separation of 2.41 Å is not considered to be short enough to constitute a highly repulsive contact for a C(sp^3)—H donor in the Seiler *et al.* case, then it can hardly be considered as being in the repulsive region for a C(sp^2)—H donor. Cases such as 1-methyl-2-methylsulphonyl imidazole (Vampa *et al.* 1995), where there is a short C—H···O geometry (3.27 Å, 2.38 Å, 153°, uncorrected) between the activated methyl C—H group and the sulphonyl O atom, would appear to be good test candidates for further study, while the possibility of a repulsive C—H···O geometry in a contact formed by a *tert*-butyl group has been noted by Meehan *et al.* (1997). Similar behaviour has been noted by Braga *et al.* (1998*a*) for a short and linear C—H···O contact (2.38 Å, 173°) in a dibenzenechromium derivative. Validation of a C—H···O geometry as being repulsive or destabilizing is not easy and requires detailed experimentation. Much effort will be required in the study of this borderline phenomenon.

2.2.11.1   *Forced intramolecular contacts*

For short intramolecular $C-H\cdots O$ interactions, the situation is even more problematic. As outlined in Section 2.2.9, it is difficult to judge reliably whether an intramolecular $C-H\cdots O$ contact stabilizes or destabilizes the molecular conformation. Intramolecular contacts are often associated with steric strain and are then ususually considered as destabilizing 'forced contacts', with the $C-H$ and O moieties only in each other's way. On the other hand, steric strain alone is not a sufficient criterion to disregard what might in reality be a weak hydrogen bond. As shown above, a contact can be repulsive (in terms of forces) and bonding (in terms of energies) at the same time. Only the very strongly repulsive contacts are not bonding. The dilemma is well illustrated by the classical example of 2-nitrobenzaldehyde which has been lucidly discussed by Coppens (1964) and is shown in Fig. 2.42. In this neutron crystal structure, a short intramolecular $C-H\cdots O$ contact is formed between the formyl and nitro groups ($d = 2.38$ Å, $\theta = 94°$). Both groups are rotated out of the aromatic plane, so that the $H\cdots O$ contact is longer than it would be in a planar molecule. For either substituent taken separately, the lowest energy conformation is the one in which it is coplanar with the aromatic group. Therefore, in 2-nitrobenzaldehyde there must be a force that twists the groups out of the plane, and correspondingly a strain energy. This means that the $C-H\cdots O$ contact here is repulsive. However, this does not yet mean that it is destabilizing. Only if the steric strain pushes the contact geometry above the zero-energy contour in the potential energy surface, can one actually speak of a destabilizing contact. Whether this is the case cannot be judged from mere

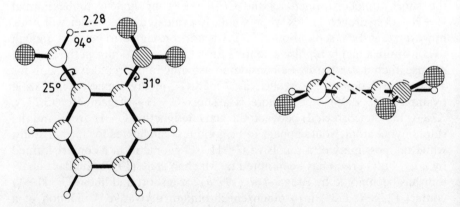

Fig. 2.42. Two views of the repulsive intramolecular $C-H\cdots O$ interaction in the neutron crystal structure of 2-nitrobenzaldehyde (Coppens 1964).

geometry. Simplified calculations are not advisable in a situation where chemical subtleties might be crucial. In summary, a definitive interpretation of the C—H···O contact in 2-nitrobenzaldehyde and related cases is still awaited.

In other cases too, structural evidence is available that shows that an intramolecular contact is repulsive and possibly destabilizing. The axial–equatorial preferences of substituents in cyclohexanes are of classical importance in physical organic chemistry. Generally, axial substituents are disfavoured because of the 1,3-diaxial repulsions. The short C—H···O contacts that occur systematically in axially O-substituted cyclohexanes, **42**, are cases in point. The distances d are around 2.6–2.7 Å, while the angles θ are around 95°, which is not a geometry strongly suggestive of hydrogen bonding. Steric strain is indicated because the C—O bond is slightly twisted away from the cyclohexane molecular axis, and evidence for the repulsive character of this contact comes from the fact that for mono-O-substituted cyclohexanes, the equatorial conformation **43** is statistically preferred over the axial **42** (Steiner and Saenger 1998a). Were the C—H···O interaction to be stabilizing, the situation would have been reversed and a related effect is observed in the family of penams where the axial orientation of the 3α-CO$_2$H group appears to be mediated by an intramolecular C—H···O contact (Nangia and Desiraju 1998c). This is discussed more fully in Section 4.4.

2.2.11.2   *Other borderline phenomena*

We move now to the second category of short C—H···O contacts that are inherently destabilizing for chemical reasons. These are rare occurrences. An example is provided by contacts between C—H groups and O atoms carrying a partial or full positive charge, such as in R$=$O$^+$—H. The primary conditions for hydrogen bonding are obviously not fulfilled here but relatively short contacts of the kind C—H···O$^+$ can occur occasionally with distances d down to about 2.5 Å. Presumably, these short contacts are forced by other intermolecular interactions. A statistical survey shows that these contacts have the opposite directionality behaviour to hydrogen bonds; the shortest contacts are all strongly bent, that is, the H atom is virtually pushed away from O$^+$, and relatively linear C—H···O$^+$ geometries occur only for very long separations d, Fig. 2.43 (Steiner 1999a,b).

Another interesting case is provided by C—H···O interactions involving formyl groups. Unlike most other C—H groups, formyl C—H is slightly

42   **43**

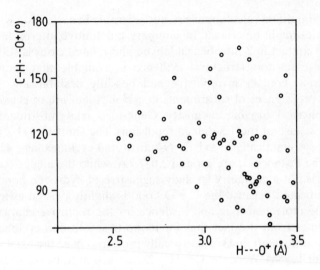

Fig. 2.43. Scatterplot of angles θ against normalized distances d in intermolecular $C-H\cdots O^+$ contacts (acceptors $C=OH^+$ and $S=OH^+$). Although short contacts do occur occasionally, the directionality is reverse to that of hydrogen bonds, in that the shortest contacts are strongly bent (Steiner 1999a,b).

hydridic in character, and the seat of the positive charge may be the C atom rather than the H atom. This leads to a peculiar and clearly anomalous situation in interactions with O atoms; the formal charge distribution is $C^{\delta+}-H^{\delta-}\cdots O^{\delta-}$ (or $C^{\Delta+}-H^{\delta+}\cdots O^{\delta-}$), which is clearly not the usual hydrogen bond situation, and a repulsion between H and O might be anticipated. Nevertheless, if the net charge on the group is positive, $(CH)^{\delta+}$, the total electrostatic interaction between the groups would still be stabilizing. A recent study by Hollingsworth and co-workers suggests that formyl $C-H\cdots O$ contacts are stabilizing interactions which are not of the normal hydrogen bond type; for a series of alkylformamides, structural evidence points at a stabilizing function of such contacts (Chaney *et al.* 1996). On the other hand, IR absorption spectra show a blue shift of the $C-H$ stretching frequency, opposite to the behaviour in hydrogen bonds. Corey *et al.* (1997) have postulated involvement of formyl group $C-H\cdots O$ interactions in determining the stereochemical outcome in some organic transformations (Section 2.2.12). In this light, formyl $C-H$ should be classified as a very special case, but not as being repulsive with respect to the O atom.

At the beginning of this section, we posed the question: are there short $C-H\cdots O$ geometries which do not represent hydrogen bonds? There is an almost trivial situation where the answer is 'yes', but one that is frequently overlooked in the interpretation of crystal structures. A short $C-H\cdots O$ contact can be the second-nearest neighbour part of an even shorter $C-H\cdots H-O$ contact. This is illustrated in **44** to **46** with the example of water

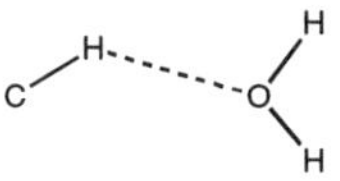

hydrogen bond   **44**

uncertain   **45**

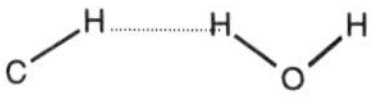

no hydrogen bond   **46**

acceptors, for which this phenomenon is observed in neutron crystal structures (Steiner and Saenger 1993a). Occasionally, H(O) and H(C) atoms approach to within van der Waals separation, and this is associated with a relatively short distance from H(C) to O. Sometimes d is around 2.8 Å or even slightly shorter. Though clearly not hydrogen bonds, such arrangements will slip into hydrogen bond tabulations if only the positions of C, H(C) and O are considered. Intermediate cases such as **45** are really problematic because it is not possible to decide on the hydrogen bond character even upon close inspection, and distances $d(O)$ are observed down to about 2.5 or 2.6 Å. The entire system of atoms should probably be considered as a supermolecule in computational studies, to shed more light on the classification of these geometries.

2.2.11.3   *Most C—H···O contacts are stabilizing*

In the discussion on the roles and the 'truth' of C—H···O hydrogen bonds, repulsive and possibly destabilizing arrangements have often been used as a main argument to play down the importance of the interaction as such, and even to question its existence. In particular, Donohue (1968) used the repulsive intramolecular C—H···O contact in Coppens' neutron crystal structure of 2-nitrobenzaldehyde (Fig. 2.42) in his overall polemics against the C—H···O hydrogen bond. Of course, the argument that if one C—H···O contact is destabilizing, *all* are destabilizing, is wrong. We note that the possible counterargument: if most C—H···O contacts are stabilizing (as is experimentally known), then *all* are stabilizing, is not much better. *When surveying the present experimental and theoretical situation, it seems that the stabilizing nature of intermolecular C—H···O contacts is the rule and that the destabilizing nature is the exception.* The exception occurs more frequently as the hydrogen bond donor under consideration gets weaker. For the strongest C—H donors, no destabilizing contacts to O have been found as yet and the situation is similar to that which is obtained for O—H and N—H donors. Intramolecular C—H···O contacts are, however, very different and forced arrangements that destabilize a molecular conformation certainly occur in many cases. We caution readers from classifying all short intramolecular (C)H···O contacts in a molecule as 'hydrogen bonds' just because they are short.

Some parts of this section touch upon the problem of defining a hydrogen bond itself (see Section 1.2.2). Should the zero energy line of Fig. 2.41 be taken as a delimiter, or in other words does a $C-H···O$ 'hydrogen *bond*' lose this name if compressed sufficiently so as to become destabilizing? One could argue then that it is no longer a 'bond'. Enthalpic considerations are tempting in such a context and linguistic concerns also seem to support such a contention, but this is not such a simple matter. When crossing the zero energy line upwards, there is no fundamental change in the electronic interaction. Again, can one use the same definition of a hydrogen bond for inter- and intra-molecular $C-H···O$ contacts? Is the problem in terming the formyl $C^{\delta+}-H^{\delta-}···O^{\delta-}$ interaction a hydrogen bond a chemical problem, or only a semantic one? All in all, studies of repulsive and especially of destabilizing $C-H···O$ geometries constitute a new frontier in hydrogen bond research and would seem to offer fair pickings to both experimentalists and theoreticians.

2.2.12   Weak hydrogen bonds in liquids and solution

Historically, the study of $C-H···O$ interactions in solution (Section 2.2.1) predates solid state investigations, but these early studies dealt only with interactions donated by activated $C-H$ groups. However, $C-H···O$ interactions formed by less activated $C-H$ groups are weaker than typical solute–solvent interactions and thermal effects in fluids, and because of this, their detection and observation in liquids and solution is difficult. Evidence for their intermediacy in solution processes is therefore obtained only indirectly. Despite these limitations, there has been a growing interest in this subject. In the present section, some recent examples of the study of $C-H···O$ hydrogen bonding in the liquid state and in solution are noted.

In a useful study, Gerothannasis and Vakka (1994) observed that there is a very good linear correlation between the NMR solution chemical shifts $\delta(^{17}O)$ and amide I stretching frequencies $v(CO)$ for amides and peptides in different solvents with varying dielectric constants and solvation abilities. These shifts were found to depend strongly on hydrogen bonding between solute and solvent. These authors found that $CHCl_3$ and CH_2Cl_2 solutions of DMF result in larger or equal shielding (for the carbonyl group O atom) compared to those obtained in CH_3CN, acetone and DMSO (Table 2.11). All these chemical shifts occur in a range (310–330 ppm) that is different from the shifts found in *n*-hexane, CCl_4 and toluene (335–350 ppm). The conclusion is that specific $C-H···O$ hydrogen bonds between solute and solvent play a significant role. More recently, ^{17}O NMR has been used to study possible

Table 2.11 Solvent effects on the ^{17}O chemical shifts of dimethylformamide (Gerothanassis and Vakka 1994)

Solvent	Chemical shift (ppm)	Chemical shift (infinite dilution) (ppm)
n-Hexane	347.2	347.1
CCl_4	337.8	337.8
Toluene	335.3	335.3
CH_2Cl_2	316.5	316.8
$CHCl_3$	310.3	310.5
Acetone	327.4	327.1
CH_3CN	320.2	320.1
DMSO	320.4	—
EtOH	299.3	299.2
MeOH	292.5	292.3
Water	268.7	268.9

*intra*molecular $C(sp^3)-H\cdots O$ interactions in some aliphatic alcohols (Tezuka *et al.* 1997).

Jedlovszky and Turi (1997) have carried out computational studies on weak and strong hydrogen bonds in liquid formic acid. As a result of Monte Carlo simulations, they suggest that $C-H\cdots O$ hydrogen bonds are present in the liquid along with the stronger $O-H\cdots O$ bonds, and that similar to the crystalline phase both types of hydrogen bond play an important role in stabilizing the extended structure. Although the $C-H\cdots O$ bonds are weaker than the $O-H\cdots O$ bonds and though their geometry is somewhat more distorted, especially with regard to deviations from linearity, the basic geometrical preferences are the same for all hydrogen bonds. This is not surprising. We know from Section 2.2.3 that all $C-H\cdots O$ interactions prefer to be linear but that the preference becomes fuzzy with weakening $C-H$ acidity. The liquid structure of formic acid is calculated to be significantly different from the cyclic dimer structure in the gaseous state, and resembles more the solid state structure with an extensive networking of strong and weak hydrogen bonds. The occurrence in the liquid of oligomers constituted with strong hydrogen bonds is predicted, and the oligomers are calculated to be linked with weak hydrogen bonds. In the simulation, only 7 per cent of the liquid structure consists of cyclic dimers. This study is likely to be of importance because it strongly suggests that the common feature in the structures of the two condensed phases of formic acid is the simultaneous presence of $O-H\cdots O$ and $C-H\cdots O$ hydrogen bonds.

The intermediacy of $C-H\cdots O$ hydrogen bonded species during the course of chemical reactions is a largely unexplored subject but one that could well reveal new insights. In the crystal structure of $[Pd(OCH(CF_3)_2)(OPh)(bipy)]$. PhOH, Kapteijn *et al.* (1996) have noted the presence of an intramolecular

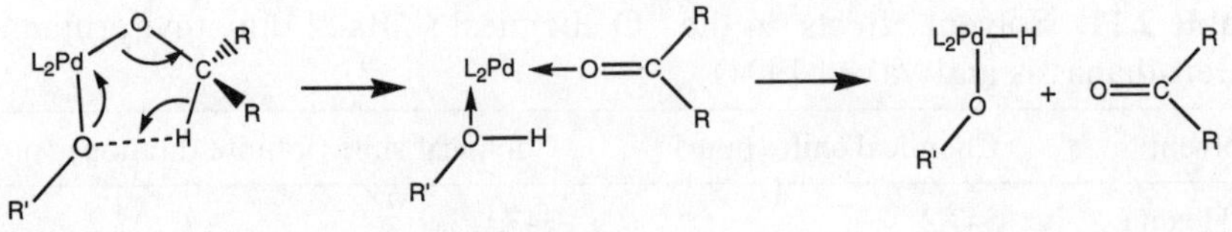

47

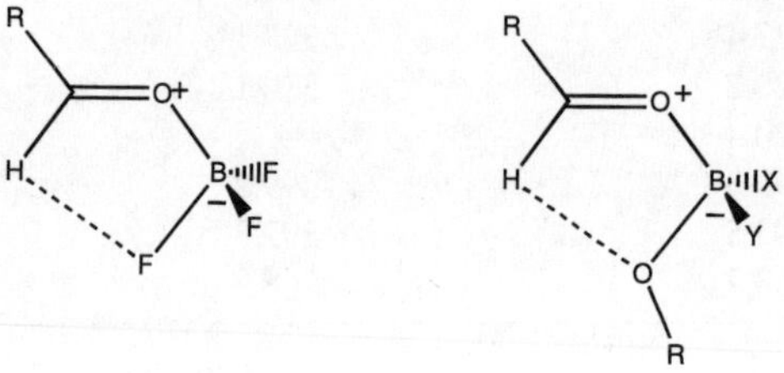

48

C—H···O bond (D = 2.95 Å) between the activated H atom of the OCH(CF$_3$)$_2$ group and the O atom of the phenoxide ligand. It is likely that this intramolecular interaction is able to steer the molecular conformation (see Section 2.2.9). Additionally, this interaction may also be regarded as representing the incipient stage of a base-assisted β-hydrogen elimination involving transfer of the proton from OCH(CF$_3$)$_2$ to the O atom of the phenoxide that would result in a ketone and a phenol coordinated to the Pd-centre as shown in **47**.

Corey *et al.* (1997) have presented a hypothesis for conformational restriction based on formyl C—H···O and C—H···F hydrogen bonds in complexes of formyl compounds with boron Lewis acids. Based on this hypothesis, they account for the observed stereoselectivities in several organic transformations. The X-ray crystal structure of the BF$_3$–dimethylformamide complex was determined and, as in the previously published structures of C$_6$H$_5$CHO·BF$_3$ and H$_2$C=C(CH$_3$)CHO·BF$_3$, was seen to have a conformation wherein a B—F bond is eclipsed with respect to the C=O bond. It has been proposed (Mackey and Goodman 1997) that this conformation is stabilized by an anomeric effect involving the lone pair on the O atom and the σ* orbital of the B—F bond. Corey *et al.* have suggested an alternative explanation, namely that a C—H···F hydrogen bond is favoured (d = 2.35 Å) because the formyl proton (see Section 2.2.11) is greatly acidified by coordination to the Lewis acid BF$_3$, as shown in **48**. In support of their suggestion, they note that in the crystalline complexes of DMF with BCl$_3$, BBr$_3$ and BI$_3$, the coplanar eclipsed

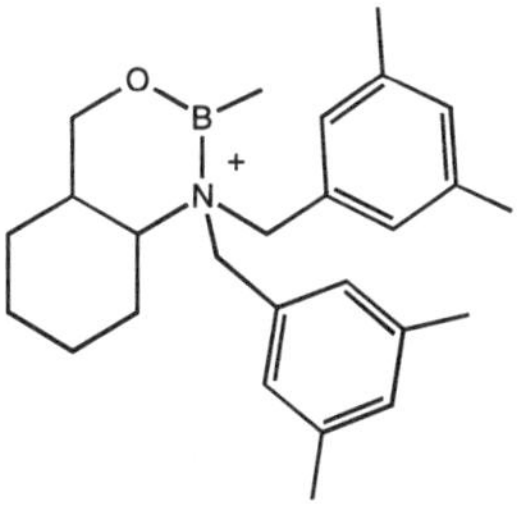

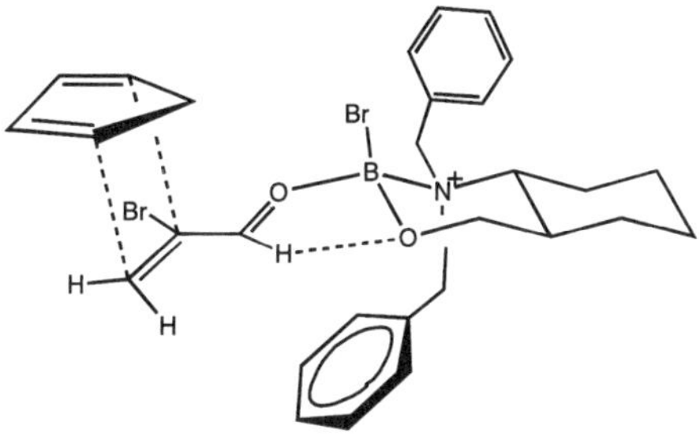

49

conformation is not observed. If the anomeric effect were important, a coplanar conformation would be most likely for $BI_3 \cdot DMF$, and if $C-H \cdots X$ hydrogen bonding were important, it would be the most likely for $BF_3 \cdot DMF$, as is indeed observed. Analogous $C-H \cdots O$ hydrogen bonding to boron species BXY(OR) was also suggested.

The mechanism for stereocontrol in a number of Diels–Alder reactions was explained in terms of these weak hydrogen bonds (Corey and Rohde 1997). For example, the reaction of substituted acroleins with several 1,3-dienes is catalysed by the B-containing chiral Lewis acid. In the transition state, the formyl H atom is placed in close proximity to the O atom substituent on B and held there by a $C-H \cdots O$ hydrogen bond to the equatorial O atom lone pair, as in **49**. This specific arrangement of molecules correctly predicts the absolute configuration of the Diels–Alder product.

The topic of $C-H \cdots O$ hydrogen bonding is closely linked to carbanion chemistry. Based on crystallographic evidence, Reetz *et al.* (1993) have noted that there are very few situations where 'naked' carbanions can exist. Rather, most carbanionic species in solution are really supramolecular aggregates where the negative charges are stabilized (as enolates) with $C-H \cdots O$ hydrogen bonds. A good example is shown by $EtC(CO_2i\text{-}Pr)_2^- \cdot NBu_4^+$. The crystal structure shows several $C-H\cdots O$ hydrogen bonds in the D range 3.26–3.29 Å. These originate from the α-methylene units of the tetrabutylammonium counterions. It is suggested that in solution also the anions and cations interact with one another via weak hydrogen bonds in a highly ordered manner. This self-organization results in a definite supramolecular identity in solution. In another and more recent study, Davidson and Lamb (1997) have characterized the phosphonium amide $(Ph_3PEt)^+(NPh_2)^-$ wherein the association between the cation and anion is through $C-H\cdots N$ hydrogen bonds from both phenyl (3.38 Å, 2.32 Å, 171°) and ethyl (3.54 Å, 2.49 Å, 164°) H atoms of the cation. The N atom is a bifurcated acceptor. The authors have noted that this is the first instance when this anion could be isolated in the solid state and have referred to it as 'ion-separated'.

Other examples of the manifestations of $C-H\cdots O$ hydrogen bonding, especially in the supramolecular and molecular recognition context, may be found in Section 4.4.

2.2.13   Recapitulation

Many conclusions can be drawn from the considerable body of work on the various properties and attributes of weak $C-H\cdots O$ and $C-H\cdots N$ hydrogen bonds. Weaker than conventional hydrogen bonds and less numerous than van der Waals interactions, acceptance of their role and significance has been generally slow in coming. Many authoritative discussions on weak hydrogen bonding, say the review of Allerhand and Schleyer (1963), the Green monograph (1974), the quinone review of Bernstein, Cohen and Leiserowitz (1974), the carboxylic acid paper of Leiserowitz (1976), the Taylor–Kennard paper (1982), and more recently the several papers and reviews written by the present authors (Desiraju 1989a, 1991a, 1992, 1996a; Steiner 1996a, 1997a), provide a nearly complete survey of the subject to date. These weak hydrogen bonds have been invoked to explain many structural, spectroscopic and crystallographic phenomena and they have been handled in computational studies. Yet, it is hard to understand why the solitary (for many years) dissenting note regarding their existence, the book chapter of Donohue (1968), continued to be mentioned, quoted and cited so often and for so many years, even though the author himself recanted in part, years later (1983). Perhaps, the strident and over-forceful tones of his dissent so uncommon to scientific writing, repeated incidentally in a second dissenting note (Cotton et al. 1997), have been remembered only because of the unusual prose style employed. Unlike the 1968 Donohue article, the Cotton et al. paper was swiftly rebutted (Mascal 1998; Steiner and Desiraju 1998) reflecting the much more well-developed state of the subject today. Looking back to the sixties and seventies, however, one senses that the concept of weak hydrogen bonding was so vulnerable to the ravages of a single erroneous paper, because the chemical and biological sciences had just not developed to the point where this concept was really useful or necessary. In this sense, scientists like Sutor, Ramachandran and Sundaralingam were many decades ahead of their times. A secondary factor that prevented a fuller under-standing of weak hydrogen bonds, even after it was recognized that they are important in structural chemistry and biology, was the use of the van der Waals distance cut-off in the enormously influential Taylor–Kennard paper (see Section 2.1).

In summary, however, there is a broad and general consensus that:

1.   An $X-H\cdots A$ interaction may be termed a hydrogen bond if it occurs more frequently than would be expected on the basis of the stoichiometry of the crystals in a statistical analysis.

2.   Length cut-offs should be applied with care, if at all, even as one appreci-ates that d separations less than the van der Waals limit are almost certainly of significance, especially for acidic $C-H$ groups.

3.   Weak hydrogen bonds tend to linearity like strong hydrogen bonds but because they are weak, this directionality is blurred by other crystal packing forces. The stronger the $C-H$ acidity, the clearer the tendency towards lin-earity.

4.   In a weak hydrogen bond $X-H\cdots A$, the $X-H$ stretching frequency experiences a bathochromic shift in the IR spectrum.

5.   Weak hydrogen bonds may be fortified additionally by cooperativity effects.

6.   Weak hydrogen bonds form well-defined and reproducible patterns in crystal structures, and these patterns can be rationalized on the basis of donor and acceptor capabilities of the various functionalities.

Thus $C-H\cdots O$ and $C-H\cdots N$ interactions have emerged as the arche-types of the weak hydrogen bond. Gradually it has also become recognized that other non-conventional donors and acceptors may participate in hydro-gen bond type interactions. These interactions form the subject of the next chapter.

Other weak and non-conventional hydrogen bonds

In the previous chapter we have described the $C-H\cdots O$ hydrogen bond in some detail and have referred to it as the archetypical weak hydrogen bond. While this bond does indeed serve as a point of reference to demarcate the limits between weak hydrogen bonds and their stronger cousins, the scope of this book is wider. In Chapter 1, we have referred to the fact that donors and acceptors in weak and non-conventional hydrogen bonds encompass wide ranges of acidity and basicity and of hardness and softness. In this sense, the $C-H\cdots O$ bond, with its weak acid–strong base and soft acid–hard base combination, exemplifies but only a small region of the overall domain of weak hydrogen bonding. In the present chapter, we wish to extend our discussion to a much larger variety of weak hydrogen bonds.

We shall proceed by systematic variations of the donor and acceptor moieties. We begin with the important aromatic hydrogen bonds that are accepted by the faces of phenyl groups, move through other acceptors of variable efficacies and then through more exotic systems involving metal donors and acceptors before we conclude with $X-H$ groups of reverse polarization, that is with a negative charge on the H atom. It will not really be necessary to justify the hydrogen bond nature *per se* of these various interactions in as great a detail as was done for the $C-H\cdots O$ bond in Chapter 2. However, attempts to verify the hydrogen bond nature of these interactions will be described for some bond types if the situation is significantly different from the $C-H\cdots O$ case. The tenor of this chapter is essentially descriptive and it is hoped that the reader will acquire from it a flavour for the extremely varied nature of weak and non-conventional hydrogen bonds.

3.1 π-Acceptors

There is no good reason to suppose that electron-rich π-systems like aromatic rings and $C\equiv C$ triple bonds would not participate in hydrogen bonding interactions. Indeed they constitute the most common classes of non-conventional acceptors.

3.1.1 *What is a π-acceptor?*

In the hydrogen bonds discussed so far, a donor $X-H$ interacts with a distinct acceptor atom A. In electrostatic terms, this may represent a dipole–dipole or a dipole–charge interaction with the relevant charges being centred more or less around the atomic nucleii. In chemical terms, such a hydrogen bond can be considered as a three-centre four-electron interaction of the kind $X-H\cdots|A$ or $X-H\cdots|A^-$ in which the acceptor atom contributes a sterically accessible electron pair in a filled orbital. However, a donor $X-H$ can also interact with the electrons in a *bonding orbital*, assuming that it is sterically accessible. This then is the situation with the so-called π-acceptors wherein the donor interacts with electrons in a π orbital of a multiple bond. From purely electrostatic considerations, this may be regarded as the inter-action of a donor with the negative charge of the π cloud, and very formalis-tically, as an interaction of a dipole with a quadrupole. The best known π-acceptors are phenyl rings and multiple bonds like $C\equiv C$ and $C=C$. There are also chemical residues which can act as conventional as well as π-acceptors; cyano and pyridyl groups are typical examples.

3.1.1.1 *Historical comment*

A brief comparison of the genesis of thought regarding $X-H\cdots\pi$ and $C-H\cdots O/N$ hydrogen bonds is pertinent. All these hydrogen bonds were dis-covered at almost the same time, that is during the mid-1930s, by physical chemists and published in prominent journals (with priorities possibly: $C-H\cdots N$ by Kumler 1935, $O-H\cdots\pi$(Ph) by Wulf *et al.* 1936, $C-H\cdots O$ by Glasstone 1937). In the following years, pioneering work on $X-H\cdots\pi$ and on $C-H\cdots O/N$ hydrogen bonds was carried out by vibrational spectroscopists, with some support from NMR studies and many of the essential properties of these hydrogen bonds were firmly established by the mid-1960s. However, structural chemists reacted slowly to these developments. A small number of structural papers appeared in the 1950s, but remained almost muffled. In 1962, Sutor published her structural study on $C-H\cdots O$ hydrogen bonds in crystals (Section 2.1), which led to heated discussions, with the critics ini-tially gaining the upper hand. Only after the work of Taylor and Kennard (1982) did structural chemists become increasingly interested in the phenom-enon, subsequently taking the lead from spectroscopists. The development of studies on $X-H\cdots\pi$ hydrogen bonds was far quieter and there was no equiv-alent of the Sutor/Donohue debate. Despite this and also despite the con-vincing spectroscopic data available, crystallographers took little interest in this variety of hydrogen bond as well and accepted them only in the early 1990s, that is around the time when $C-H\cdots O$ hydrogen bonds were also fully accepted.

It is of historical interest to seek the reasons for this very long delay on the part of crystallographers in accepting phenomena for which experimental evi-dence had long since been available from other techniques. Donohue's criti-cism of $C-H\cdots O$ hydrogen bonds was intense, but scientifically not strong enough to delay an entire field for more than a decade. More likely than not, the reasons for the non-acceptance of both $C-H\cdots O/N$ and $X-H\cdots\pi$ interactions as hydrogen bonds by crystallographers lay in the ruling paradigm of the time with regard to van der Waals cut-offs. In this respect the $X-H\cdots\pi$ hydrogen bond was affected in exactly the same way as the $C-H\cdots O$ bond.

3.1.2 Solution and gas phase experiments

Before discussing $X-H\cdots\pi$ hydrogen bonds in the solid state, we present some results of solution and gas phase experiments on simple systems. These results establish the hydrogen bond nature of the interaction and lay a sound basis for crystallographic research. Such independent validation of hydrogen bonding is important because further studies are then possible in more complicated systems that are accessible to crystallographic rather than to spectroscopic methods.

3.1.2.1 Solution infra-red spectroscopy

Intramolecular $O-H\cdots\pi$ bonds.   The first evidence of $O-H\cdots\pi$ hydrogen bonding was provided by the classical work of Wulf, Liddel and Hendricks (1936), who measured the IR absorption spectra of numerous *ortho*-substituted phenols in dilute CCl_4 solutions. When O atoms are present in the *ortho*-position (catechol etc.), intramolecular $O-H\cdots O$ hydrogen bonds are formed leading to strong red shifts of the $O-H$ stretching vibration ν_{OH}. Interestingly, ν_{OH} of *ortho*-phenylphenol is also red shifted, to an extent weaker than for O atom acceptors, but still large enough for a hydrogen bond.

In the following years, intramolecular hydrogen bonds were found in solution for many other π-acceptors, including the pseudo-π-bonded cyclopropane ring. A selection of these hydrogen bonds is shown in Fig. 3.1, with references and ν_{OH} red shifts given in the figure legend. The reader will note that most of the hydrogen bond geometries must suffer severely from steric constraints, but this does not prevent hydrogen bond formation as such. The ν_{OH} red shifts are appreciable, typically in the range 40–90 cm^{-1}, with the exception of the very weak cyclopropane acceptor for which $\Delta\nu_{OH} = -16$ cm^{-1}. With these experiments, the $O-H\cdots\pi$ hydrogen bond phenomenon was established as such at an early stage. It is simple then to conclude that related hydrogen bonds must also occur with donors other than $O-H$, and with π-acceptors other than those shown in Fig. 3.1.

Intermolecular $O-H\cdots\pi$ hydrogen bonds.   One can readily study intermolecular $O-H\cdots\pi$ hydrogen bonds by dissolving small quantities of $R-O-H$ donor molecules in solvents with π-acceptors (benzene, toluene). A more variable method is to dissolve $R-O-H$ donor and π-acceptor molecules in apolar solvents in which they can form hydrogen bonded adducts (say in CCl_4). Many such studies were performed in the 1960s, and the literature was com-

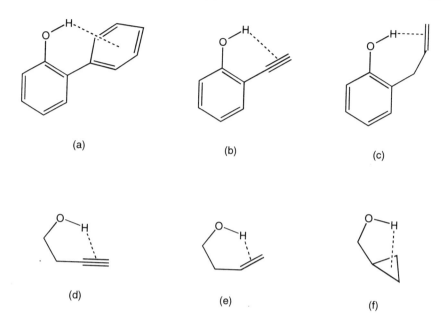

Fig. 3.1. Intramolecular O—H···π hydrogen bonds detected by solution IR spectroscopy. (a) *o*-Phenylphenol, Δv_{OH} = 45 cm^{-1} (Wulf *et al.* 1936). (b) *o*-Ethynylphenol (Prey and Berbalk 1951). (c) *o*-Allylphenol, Δv_{OH} = 63 cm^{-1} (Baker and Shulgin 1958). (d) But-3-yne-1-ol, Δv_{OH} = 42 cm^{-1} (Schleyer *et al.* 1958). (e) But-3-ene-1-ol, Δv_{OH} = 40 cm^{-1} (Schleyer *et al.* 1958). (f) 2-Cyclopropyl-ethan-1-ol, Δv_{OH} = 16 cm^{-1} (Schleyer *et al.* 1958).

prehensively compiled by Joesten and Schaad (1974). As an example, spectroscopic data of phenol–π-base complexes in CCl$_4$ are given in Table 3.1. In general, the bathochromic shifts of v_{OH} are in the range 40–120 cm^{-1}, which overlaps with the range for O—H···O hydrogen bonds (>100 cm^{-1}). It is obvious that O—H···π hydrogen bonds are strongly influenced by substituents on the acceptor. In particular, methyl substitution increases the acceptor strength significantly.

Heterocyclic aromatics. The issue as to whether the π electron system of heteroarenes can act as a hydrogen bond acceptor has been a matter of some controversy. For a convincing study, the reader is directed to the recent work of Chardin *et al.* (1996), who examined hydrogen bonded adducts of O—H donors to a series of pyridine derivatives with increasing steric hindrance at the N atom, Fig. 3.2. Pyridine and 2-*tert*-butylpyridine accept hydrogen bonds associated with strong v_{OH} shifts, suggesting that the basic N atom acts as the acceptor, Table 3.2. If approach to the N atom is blocked by bulky substituents at both sides, a different kind of hydrogen bond is accepted which leads to v_{OH} shifts about an order of magnitude smaller than for pyridine, and very similar to that found for benzene. This can only be explained by hydrogen bonds accepted by the pyridine π bond system.

Table 3.1 Frequency shifts $\Delta\nu_{OH}$ of phenol in $O-H\cdots\pi$ bonded complexes in CCl_4. More complete tabulations can be found in the original publications

π-Base	$-\Delta\nu_{OH}$ (cm^{-1})	Reference
Benzene	49	(a)
Toluene	58	(a)
Mesitylene	78	(a)
Hexamethylbenzene	106	(a)
Naphthalene	48	(a)
Ferrocene	70	(a)
1-Methyl-1,3-butadiene	56	(a)
Cyclopentadiene	70	(a)
1-Octyne	83	(b)
2-Octyne	115	(b)
n-Propylcyclopropane	40	(c)

(a) Yoshida and Osawa 1966. (b) Yoshida et al. 1972. (c) Yoshida et al. 1969. Typical experimental standard uncertainties are ±1–2 cm^{-1}.

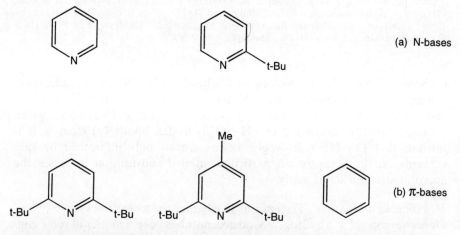

Fig. 3.2. The pyridine bases used by Chardin et al. (1996) to show the π-acceptor potentials of the pyridine skeleton. Spectroscopic data are given in Table 3.2.

Summary. Solution IR spectroscopy shows clearly and undisputably that a large number of π bonded systems can accept intra- and inter-molecular hydrogen bonds. The spectroscopic effects caused by these hydrogen bonds are smaller than for conventional hydrogen bonds, but they are still appreciable. The question as to whether $X-H\cdots\pi$ interactions can be treated as hydrogen bonds may therefore be answered in the affirmative. Because IR

Table 3.2 Frequency shifts Δv_{OH} of 4-fluorophenol when hydrogen bonding with N- and π-bases in CCl_4 (Chardin *et al.* 1996)

Base	Δv_{OH} (cm^{-1})
Pyridine	486
2-*tert*-Butylpyridine	500
2,6-di-*tert*-Butylpyridine	44
2,6-di-*tert*-Butyl-4-methylpyridine	52
Benzene	50

Table 3.3 Bond distances $d(H\cdots\pi)$ in some gas phase complexes with π-base acceptors

Complex	$d(H\cdots\pi)$ (Å)	Reference
Acetylene–HF	2.19	Shea *et al.* 1984
Acetylene–HCl	2.41	Shea *et al.* 1984
Acetylene–HCN	2.60	Aldrich *et al.* 1983
Acetylene–acetylene	2.73	Fraser *et al.* 1988
Propyne–HF	2.18	Shea *et al.* 1984
Ethylene–HF	2.22	Shea *et al.* 1984
Ethylene–HCl	2.44	Shea *et al.* 1984
Ethylene–acetylene	2.78	Fraser *et al.* 1992
Benzene–HF	2.25	Shea *et al.* 1984
Benzene–HCl	2.35	Shea *et al.* 1984

spectroscopy cannot give information on interaction geometries and is difficult in complex systems, a more detailed understanding of $X-H\cdots\pi$ hydrogen bonding can be achieved only with crystallographic techniques.

3.1.2.2 *Gas phase dimers*

A number of $X-H\cdots\pi$ bonded gas phase dimers have been studied by microwave rotational spectroscopy, for donors of various strengths ranging from HF to acetylene. A selection of relevant dimers is shown in Fig. 3.3, and $H\cdots\pi$ bond lengths are given in Table 3.3. For reviews on the method and more complete compilations of results, the reader is referred to the papers by Legon and Millen (1987*a,b*) and Hobza and Zahradnik (1988).

$C\equiv C$, $C=C$ *and cyclopropane.* The gas-phase association of linear molecules $R-X-H$ with acetylene yields planar T-shaped dimers with C_{2v} symmetry. In these dimers, the $X-H$ vector points at the midpoint of the $C\equiv C$ triple bond (Fig. 3.3(a–c)). The H atom vibration amplitude in the

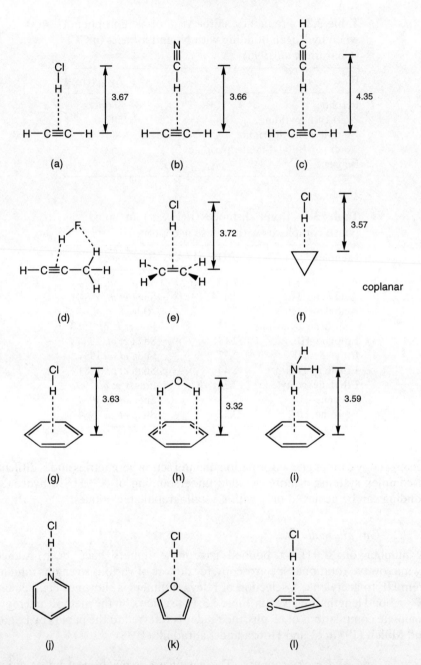

Fig. 3.3. Gas phase dimers studied by rotational spectroscopy: (a) acetylene–HCl (Legon *et al.* 1981); (b) acetylene–HCN (Aldrich *et al.* 1983); (c) acetylene dimer (Fraser *et al.* 1988); (d) propyne–HF (Shea *et al.* 1984); (e) ethylene–HCl (Aldrich *et al.* 1981); (f) cyclopropane–HCl (Legon *et al.* 1982); (g) benzene–HCl (Read *et al.* 1983); (h) benzene–water (S. Suzuki *et al.* 1992); (i) benzene–ammonia (Rodham *et al.* 1993); (j) pyridine–HCl (Cooke *et al.* 1998*a*); (k) furan–HCl (Shea and Kukolich 1983); (l) thiophene–HCl (Cooke *et al.* 1998*b*).

dimer plane is much larger than for the out-of-plane vibrations, showing that the bending potential is broad in the plane but steep perpendicular to it. This is because the H atom does not lose contact with the extended cylindrical π electron cloud when moving in the dimer plane (Legon *et al.* 1981). The acetylene–HCN dimer has the same T-shaped geometry as acetylene–HCl (Aldrich *et al.* 1983), and the acetylene dimer is T-shaped too (Fraser *et al.* 1988). Both are good examples of C—H⋯π hydrogen bonding. The propyne–HF dimer is interesting because the F—H bond points at the middle of the C≡C acceptor, whereas the F atom is displaced towards the methyl group to accept a C—H⋯F hydrogen bond, Fig. 3.3(d) (Shea *et al.* 1984). The ethylene–HCl dimer has internal C_{2v} symmetry, and the Cl—H vector points at the middle of the C=C bond (Aldrich *et al.* 1981), and similar geometries are observed for other donors. Finally, in the cyclopropane–HCl adduct the two molecules are coplanar and the Cl—H vector points at the middle of one of the C—C bonds (Legon *et al.* 1982).

In summary, the complexes of HCl with acetylene, ethylene and cyclopropane can be regarded as being very similar; the X—H donor is placed in the bisecting plane of the C≡C, C=C and C—C bonds, respectively, and is oriented at the bond midpoint.

Benzene. This acceptor has been studied in combination with a variety of donor molecules. In the benzene–HCl dimer, the Cl and H atoms are on time average located on the benzene C_6 axis. However, the H atom performs extensive vibrational motion, indicating a very broad bending potential (Read *et al.* 1983). In a later study on benzene–HBr, Legon and co-workers concluded that the Br—H vector points at the benzene ring itself (not at the midpoint) and undergoes circular motions which allow it to sample the whole π electron density of the ring (Cooke *et al.* 1997).

In the vibrational ground state of the benzene–water dimer, the O atom is located above the centre of the aromatic ring, and both H atoms of the water point towards the π cloud (bidendate geometry). *Ab initio* calculations predict a binding energy of −1.8 kcal/mol (S. Suzuki *et al.* 1992). This geometry is not observed in the crystalline state, where the only examples known have one of the two water H atoms hydrogen bonded to the aromatic group, and the other interacting with a different acceptor. The water molecule in the gas phase dimer undergoes nearly free internal rotation, and the potential well is extremely flat over the benzene surface. The same group of authors also studied the benzene–ammonia dimer (Rodham *et al.* 1993), and found a monodendate geometry with only one H atom forming a hydrogen bond at a time. The ammonia N atom is placed above the ring centre, and the molecule undergoes almost free internal rotation.

Interesting substitution effects were studied for the benzene–HF dimer by Klemperer and co-workers. It was observed that HF binds much stronger to benzene than to the fluorinated analogues $C_6F_3H_3$ and C_6F_6, indicating that the electron-withdrawing F substituents deactivate the π-acceptor potential of benzene (Balocchi *et al.* 1983).

Heterocyclic compounds.   As seen earlier, heterocycles can accept conventional hydrogen bonds at the heteroatoms, and also non-conventional ones at the π-basic centres. In the gas phase complexes of HCl with pyridine and furan, donor and acceptor are coplanar and the HCl molecules donate linear hydrogen bonds to the N and O atom lone-pairs, respectively, Fig. 3.3(j,k) (Cooke *et al.* 1998*a*; Shea and Kukolich 1983). This is as would be expected because N and O atoms are much stronger acceptors than the π electron clouds. The thiophene–HCl dimer has a very different geometry; the molecules are not coplanar but roughly perpendicular, and the hydrogen bond is of the π type. This shows that the S atom of thiophene is a weaker hydrogen bond acceptor than the π electron system. The geometry is such that the Cl atom is placed roughly above the centroid of the thiophene ring, and the Cl—H vector points at the side of the S atom (Cooke *et al.* 1998*b*).

Summary.   Gas phase experiments have significant implications for work done in condensed phases. A number of conclusions may be drawn that are very difficult to derive from crystal structure geometries, given that so many secondary effects prevail in the latter. The most important of these are:

(1)   intermolecular interactions of acids and π-bases are hydrogen bond-like in nature. This is true also for the stronger types of carbon acids like HCN and acetylene;

(2)   for $C\equiv C$, $C=C$ and cyclopropane $C-C$ bonds, the donor points at the bond centre, not at an individual atom;

(3)   the potential energy surfaces of $X-H\cdots\pi$ hydrogen bonds are very flat. This is most pronounced for the benzene acceptor.

3.1.3   *Phenyl groups*

Phenyl groups are by far the most important of π-acceptors. The earliest crystallographic observations of $N-H\cdots Ph$ hydrogen bonds were made by Davies and Staveley (1957) on ammonium tetraphenylborate, while $O-H\cdots Ph$ hydrogen bonding was first reported by McPhail and Sim (1965) for a cyclic peptide. Phenyl groups occur in a vast variety of chemical substances, ranging from organometallic to biological molecules, and this confers a particular importance to $X-H\cdots Ph$ hydrogen bonds. Another term that is commonly used for these interactions is *aromatic hydrogen bonds*.

    In this section, we discuss the general properties of $X-H\cdots Ph$ hydrogen bonds. Let us begin with theoretical calculations which are helpful during the subsequent interpretation of experimental data.

3.1.3.1 *Theoretical calculations*

Geometrical parameters. The literature on X—H···Ph hydrogen bonding is confusing with respect to the use of geometrical parameters. Unlike in X—H···O hydrogen bonds, it is not immediately obvious which geometrical parameters are the most relevant, and different geometrical descriptors have found favour. As the 'hydrogen bond distance', some authors use the distance of H (or X) to the centroid of the phenyl ring, some use the perpendicular distance to the Ph plane, while others prefer to give distances to individual C atoms. The situation is complicated because some X—H···Ph hydrogen bonds point towards the aromatic centroid, some are directed at the midpoint of a particular C—C bond, and some point at an individual C atom. For these situations, differing sets of geometrical parameters could be appropriate. The conventions followed in this book are given in Fig. 3.4. The point of reference is the aromatic centroid M, and distances are given with respect to this point, $d(M) = H···M$ and $D(M) = X···M$. Distances and angles to relevant C atoms are also given if necessary. While referring to the work of other authors who use different descriptors, their system is adopted if appropriate and adequate.

Theoretical background. Numerous computational studies on X—H···Ph hydrogen bonds have been performed at different levels of theory, and only a selection can be given here. Several computations which have been performed in connection with experimental work will be referred to in later sections.

Levitt and Perutz (1988) performed simple potential energy calculations on a model of N—H···Ph interactions, where the donor is represented by a two-atom fragment. Only electrostatic and van der Waals terms were used. In the equilibrium geometry, the N—H donor then points perpendicularly at the centre of the aromatic ring. The energy minimum occurs at $D = 3.40$ Å ($d = 2.40$ Å), and is 3.35 kcal/mol deep, Fig. 3.5. This energy is composed of about 2/3 electrostatic and 1/3 van der Waals contributions. When rotating the donor with respect to the acceptor, the energy falls only slowly, Fig. 3.5(b,c).

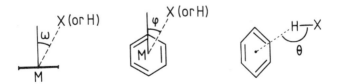

Fig. 3.4. Definition of geometrical X—H···Ph hydrogen bond parameters. ω(H) is the angle between the line H···M and the C_6 axis. φ(H) is the angle between the projection of H···M on the Ph plane and the line C—M, where C is the C atom closest to H. ω(X) and φ(X) are defined analogously. Distances are defined as $d(M) = H···M$, $D(M) = X···M$. Sometimes, it is helpful to give contact distances and angles with respect to individual C atoms.

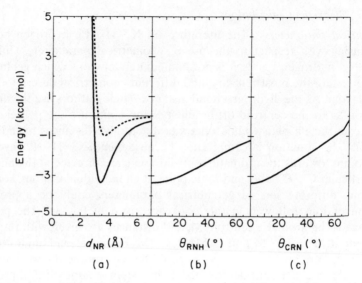

Fig. 3.5. Potential energy calculations of the N—H···benzene hydrogen bond. (a) Variation of the energy with D in perpendicular geometry. The continuous line shows the total energy and the dotted line the van der Waals contribution. (b) Variation as N—H is rotated by changing the R···N—H angle at a fixed N position. (c) Variation as the entire N—H group is rotated while keeping it pointed towards the ring centre R (from Levitt and Perutz 1988).

The authors state that the interaction is so strongly dominated by electrostatic and van der Waals forces that there is no need to invoke lone-pair electrons or delocalized electron clouds. Using a similar structural model but a different set of empirical parameters, Worth and Wade (1995) performed force-field calculations to determine the potential energy surface shown in Fig. 3.6, and found the global minimum for a perpendicular geometry at $d(M)$ = 2.3 Å and an energy of –2.2 kcal/mol. This potential energy surface is shallow over a very large region, and the donor can be moved over the entire face of the acceptor with only small changes in energy, and with considerable variations of the contact angle. The reader will note that that the limits of the map correspond to N—H groups far outside the ring. These studies are discussed again in Section 5.2.3.

Compared to the studies mentioned above, *ab initio* molecular orbital calculations represent a higher level of theory. However, only if the full apparatus of the method is used (high level basis sets, BSSE corrections, electron correlation considered), can the results be regarded as reasonable. We do not attempt to fully survey the earlier literature, and only a few recent studies are mentioned.

A well-studied model system is the benzene–HF dimer. Tang *et al.* (1990) calculated the global energy minimum to be at $d(M)$ = 2.41 Å with the ideal C_6 geometry. At the highest level of theory used, a bond energy of

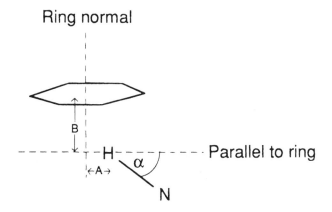

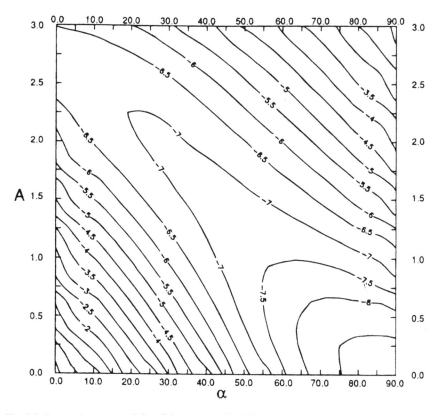

Fig. 3.6. Interaction energy (kJ/mol) between an N—H group and an aromatic ring when varying the coordinates A and α and optimizing B as obtained in force field calculations. The optimal distance B varies only slightly upon variation of A and α, and is around 2.4 to 2.5 Å for all positions above the ring face (from Worth and Wade 1995).

−4.8 kcal/mol was obtained. Brédas and Street (1988) calculated the energies
for different distance optimized geometries, and obtained −5.2 kcal/mol for the
centred geometry, −4.8 kcal/mol if the donor points at the midpoint of a particular $C-C$ bond, and −4.5 kcal/mol if it points at a single C-atom. This means
that moving the donor 1.4 Å from the global minimum costs only 0.7 kcal/mol.
In contrast to these results, Rozas *et al.* (1997*a*) obtained the global minimum
at the geometry where the $F-H$ molecule points at the midpoint of a $C-C$
bond, with $d(\pi) = 2.41$ Å and an interaction energy of about −3.2 kcal/mol.

For the benzene–water dimer, an off-centred minimum energy geometry
was obtained in an *ab initio* study by Rablen *et al.* (1998). These authors report
disquietingly large discrepancies between results obtained at different levels
of theory. In earlier force-field calculations, Jorgensen and Severance (1990)
had found the global minimum at a centred geometry with $D(M) = 3.11$ Å and
$\Delta E = -3.8$ kcal/mol.

Malone *et al.* (1997) have performed *ab initio* calculations on a number of
$X-H\cdots Ph$ dimers with the donors in optimized geometries above the aromatic midpoint and above the edge of the ring. The results in Table 3.4 indicate that the contacts above the ring centre are slightly more favourable, but
that the energy differences between the different positions are very small.
The reader should notice the very large difference in bond energies for
the uncharged donor NH₃ (*ca.* −1.0 kcal/mol) and the ammonium ion (*ca.*
−16 kcal/mol).

Summary.   A theoretical consensus model for the $X-H\cdots Ph$ hydrogen
bond is yet to emerge. In particular, the bond geometry that represents the
global energy minimum is still to be established. Earlier methods favoured the
ideal ring centroid contact, whereas more recent computations prefer contacts
to the centres of individual $C-C$ bonds. There is consensus, however, that the
potential energy surface is very flat and the energies of different contact
geometries are very similar. By implication, the $X-H\cdots Ph$ hydrogen bond
can be easily distorted by other interactions. In this sense, the current uncertainties about the global energy minimum are relatively unimportant in struc-

Table 3.4   Calculated energies of for $X-H\cdots Ph$ bonded dimers (*ab initio*
calculations, 6–31G**, Malone *et al.* 1997)

Donor	Geometry	Binding energy (kcal/mol)
MeOH	Above centroid	−2.81
MeOH	Above C	−2.63
NH₃	Monodentate above centroid	−0.97
NH₃	Bidentate above C	−0.79
NH₄⁺	Monodentate above centroid	−16.3
NH₄⁺	Bidentate above C	−9.0

tural chemistry and biology—what is of importance is that the entire face of the Ph ring can serve as a hydrogen bond acceptor and that typical bond energies with strong donors are in the range −2 to −4 kcal/mol.

3.1.3.2 Crystal structure statistics

The most comprehensive CSD analysis of $X-H\cdots Ph$ hydrogen bonds has been published by Malone *et al.* (1997). A large geometrical variability of $X-H\cdots Ph$ hydrogen bonds was found in organic and organometallic crystal structures. To distinguish different contact geometries, categories as defined in Fig. 3.7 were introduced. The frequencies of interactions in these geometries

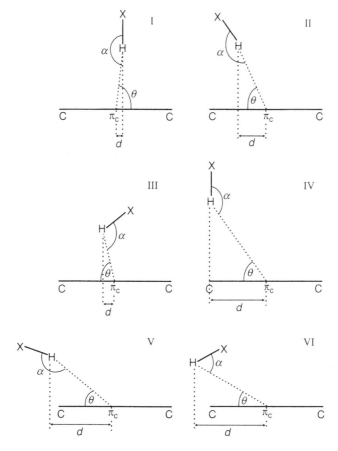

Fig. 3.7. Geometrical categories of $X-H\cdots Ph$ interactions as defined by Malone. Type I: $d(M) \leq 3.05$ Å, $\theta \geq 53°$, $150° \leq \alpha \leq 180°$, $d \leq 0.5$ Å. Type II: $d(M) \leq 3.05$ Å, $\theta \geq 53°$, $150° \leq \alpha \leq 180°$, $0.5 < d \leq 1.4$ Å. Type III: $d(M) \leq 3.05$ Å, $\theta \geq 53°$, $\alpha < 150°$, $d \leq 1.4$ Å. Type IV: $d(M) \leq 3.05$ Å, $40° \leq \theta \leq 60°$, $130° \leq \alpha \leq 150°$, $1.4 \leq d \leq 1.5$ Å. Type V: $d(M) \leq 4.0$ Å, $\theta \leq 90°$, $90° \leq \alpha \leq 180°$, $d > 1.4$ Å. Type VI: $d(M) \leq 4.0$ Å, $\theta \leq 90°$, $\alpha \geq 90°$, $d > 1.4$ Å (Malone *et al.* 1997).

Table 3.5 Frequencies of $X-H \cdots Ph$ interactions in the geometrical categories defined in Fig. 3.7; CSD results of Malone *et al.* (1997)

Donor	Total n	Type I	Type II	Type III	Type IV	Type V	Type VI
O—H	401	5	17	66	4	306	7
N—H	224	18	28	55	4	123	0
N$^+$—H	71	11	2	19	1	39	0
≡CH	10	2	4	4	0	0	0
S—H	3	0	0	1	0	2	0

are given in Table 3.5 for the donor types $O-H, N-H, N^+-H, \equiv C-H$ and $S-H$. Inter- and intra-molecular interactions were not distinguished. Remarkably, the heavily distorted geometries of types III and V occur most frequently. In this context, it must be remembered that:

(1) sterically constrained intramolecular contacts are included in these statistics;

(2) secondary components of bifurcated hydrogen bonds are included; and

(3) the frequencies are not cone-corrected as is usual for conventional hydrogen bonds (see Fig. 2.16).

In fact, the solid angle covered by type I contacts is much smaller than for II, and it is the largest for V. Also type IV contacts correspond to only a small solid angle. Still, these data show that very strongly distorted $X-H \cdots Ph$ bonds occur commonly in crystals.

Geometrical preferences of $X-H \cdots Ph$ hydrogen bonds in crystals are best investigated with water molecules, because these can orient more freely with respect to the acceptor than all other donor types. In a CSD analysis of $O_W-H \cdots Ph$ hydrogen bonds, a distribution of angles $\omega(O_W)$ was found as shown in Fig. 3.8(a). The distribution shows a maximum in the interval 10–20°, but after cone correction (weighting by $1/\sin \omega$, see Section 2.2.3), the maximum occurs in the interval $\omega < 10°$, Fig. 3.8(b). This shows that the preferred location of O_W is actually perpendicular to the ring and over the aromatic centroid. With the limited number of water molecules and the relatively inaccurately refined H atom positions, a related analysis of $\omega(H_W)$ was not feasible.

A CSD analysis of $X-H \cdots Ph$ hydrogen bonds in organometallic crystals ($X = O, N, C$) was performed by Braga *et al.* (1998*b*). The cut-off criterion used, $d(M) < 3.0$ Å, is much more restrictive than in the analysis of Malone *et al.* (1997). Based on this cut-off value, the mean geometries listed in Table 3.6 are obtained, showing the expected sequence of donor strengths $O-H > N-H > C-H$. Scatterplots of angles $\theta(M)$ against $d(M)$ show a preference of $X-H \cdots Ph$ hydrogen bonds for linearity, and that the shorter contacts are on

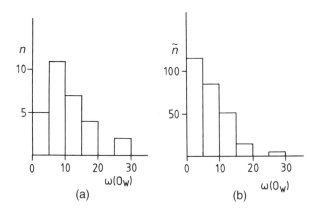

Fig. 3.8. CSD analysis of angles $\omega(O)$ in O—H···Ph hydrogen bonds formed by water molecules. (a) Unweighted distribution. (b) Distribution after $1/\sin\omega$ cone-correction.

Table 3.6 Mean geometries of X—H···Ph hydrogen bonds (X = O, N, C) in organometallic crystal structures (Braga *et al.* 1998*b*)

Donor	n	$d(M)$ (Å)	$D(M)$ (Å)	$\theta(M)$ (°)
O—H	49	2.66(3)	3.41(3)	143(3)
N—H	103	2.71(1)	3.50(1)	145(1)
C—H	24	2.79(2)	3.69(2)	142(2)

Distances d normalized, distance cut-off $d(M) < 3.0$ Å.

average more linear, Fig. 3.9. Furthermore, there is a clear effect of charge assistance, that is hydrogen bonds formed by charged groups are shorter on average than those between neutral molecules, Fig. 3.10. A very detailed statistical survey of N^+—H···Ph—B^- hydrogen bonds in tetraphenylborate salts (Bakshi *et al.* 1994) is discussed in Section 3.1.3.3.

Mean distances of different C—H groups to the faces of phenyl groups have been determined by Nishio and co-workers in a CSD study, Table 3.7. The distances follow the same trend as for conventional O, N and Cl⁻ acceptors, i.e. the donor strengths are ranked as $CHCl_3 > C\equiv C$—H > $CH_2Cl_2 > ClCCH_2$ (Umezawa *et al.* 1998). This means that the influence of C—H donor acidity parallels that for conventional acceptors, and suggests definitely that C—H···O and C—H···Ph interactions are of a related nature.

3.1.3.3 *Intermolecular X—H···Ph hydrogen bonds (X = O, N, Cl)*

In the previous sections, we have surveyed theoretical and statistical studies on X—H···Ph hydrogen bonds. This provides an overall description

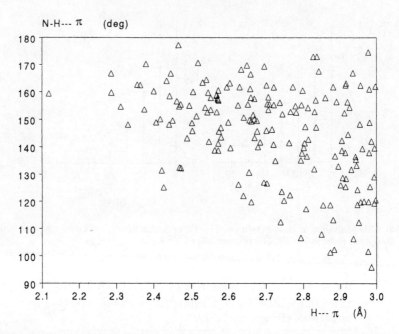

Fig. 3.9. Scatterplot of angles $\theta(M)$ against distances $d(M)$ of $N-H\cdots Ph$ hydrogen bonds in organometallic crystal structures. The behaviour for $O-H\cdots Ph$ and $C-H\cdots Ph$ hydrogen bonds is similar (from Braga *et al.* 1998*b*).

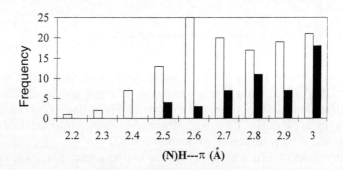

Fig. 3.10. Distribution of distances $d(M)$ of $N-H\cdots Ph$ hydrogen bonds in organometallic crystal structures. *Empty bars*: charged compounds; *solid bars*: neutral compounds (from Braga *et al.* 1998*b*).

of the interaction. We now proceed to an examination of individual crystal structures.

Centred and off-centred $X-H\cdots Ph$ bonds. $X-H\cdots Ph$ hydrogen bonds occur with widely differing geometries in crystal structures. The centred geometry occurs relatively rarely (Table 3.5), but it is occasionally observed in an

Table 3.7 C—H···Ph contact distances for different types of C—H groups. The distance d is here not
defined as H···M, but as the normal distance of the
H atom to the Ph plane (cut-off: 2.9 Å) (Umezawa
et al. 1998)

Donor	n contacts	d (Å)
Cl_3CH	14	2.38 ± 0.16
$C\equiv CH$	26	2.63 ± 0.14
Cl_2CH_2	52	2.65 ± 0.15
$ClCCH_2$	43	2.73 ± 0.14
C_2CH_2	5518	2.74 ± 0.74
CCH_3	6772	2.75 ± 0.10

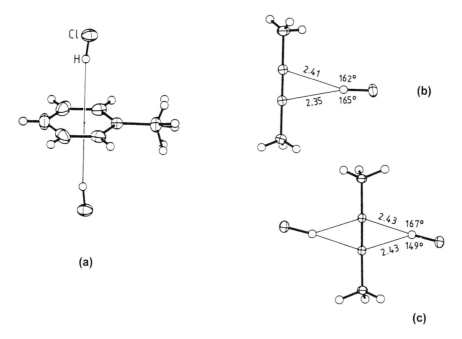

Fig. 3.11. Cl—H···π hydrogen bonded complexes. (a) 1:2 Toluene–HCl (from Deeg and Mootz
1993). (b) 1:1 and (c) 1:2 but-2-yne–HCl (from Mootz and Deeg 1992*a*).

almost ideal form. An impressive example has been found by Deeg and Mootz
(1993) in the crystalline 1:2 complex toluene–HCl, Fig. 3.11(a). The hydrogen
bond distance of $D(M) = 3.52$ Å is somewhat shorter than in the gas phase
dimer benzene–HCl, 3.63 Å (Fig. 3.3). In the same figure, related crystal structures of the 1:1 and 1:2 complexes but-2-yne–HCl are also shown (Mootz and
Deeg 1992*a*).

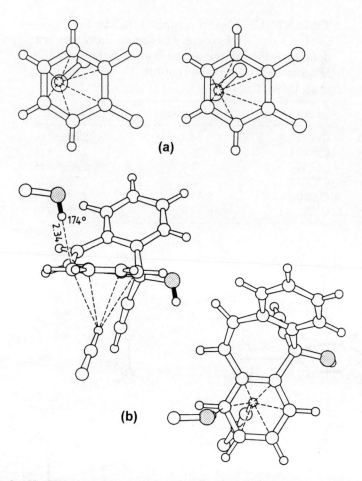

Fig. 3.12. O—H⋯Ph hydrogen bonds with different geometries. (a) Close to centred hydrogen bonds in (S)-2,2,2-trifluoro-1-(9-anthryl)ethanol, **50**, with normalized d(M) = 2.27 and 2.36 Å, and θ(M) = 161 and 170° for the left and right hydrogen bonds, respectively (Rzepa *et al.* 1991). (b) Neutron structure of 5-ethynyl-5H-dibenzo[a,d]cyclohepten-5-ol, **51**, at 20 K, showing a hydrogen bond directed almost linearly at an individual C atom (from Steiner *et al.* 1997c).

X—H⋯Ph hydrogen bonds that deviate from perpendicularity are more frequent. Two relevant examples are shown in Fig. 3.12. In the structure of the trifluoroethanol **50**, pairs of molecules are linked by mutual aromatic hydrogen bonds with a relatively well-centred geometry, and with d(M) = 2.27 and 2.36 Å, and θ(M) = 161° and 170°, respectively (Fig. 3.12(a); Rzepa *et al.* 1991). This can be considered as a good hydrogen bond geometry. A more irritating geometry is found in the low-temperature neutron crystal structure of alkynol **51**, where a hydroxyl group points almost linearly at an individual carbon atom of the acceptor group with d(C) = 2.339(6) Å, and θ(C) = 174.4(5)°. The

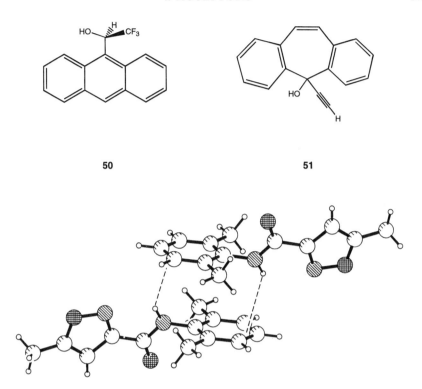

Fig. 3.13. Crystal structure of the anticonvulsant drug 'D2624', **8** (Jaulmes *et al.* 1993), exhibiting intermolecular N—H···Ph hydrogen bonding. Normalized geometry: $d(C)$ range = 2.65–3.86, $d(M)$ = 3.02, $D(M)$ = 3.57 Å, $\theta(M)$ = 116°, $\omega(H)$ = 30° and $\omega(O)$ = 17°. The hydrogen bond is directed roughly at the middle of a C—C bond with $d(M)$ = 2.64 Å. IR spectroscopic measurements of Lutz *et al.* (1996) show a ν_{NH} red shift of 30 cm^{-1} compared to solution in CCl$_4$.

geometry is highly distorted with respect to the aromatic centre with $d(M)$ = 2.98, $\theta(M)$ = 152.6(5)°, $\omega(H)$ = 41.7° and $\theta(O)$ = 35.3°. The hydrogen bond nature of this interaction is confirmed by IR spectroscopy ($\Delta\nu_{OH}$ = −61 cm^{-1}). Interestingly, the other side of the phenyl group accepts a second and well-centred hydrogen bond which is donated by an ethynyl group. In *ab initio* calculations on the O—H···Ph geometries in Fig. 3.12, energies of −2.4 kcal/mol were obtained for those in (a) and −1.3 kcal/mol for the one in (b) indicating that the centred geometry is more favourable (Steiner *et al.* 1996*b*). Despite the energetic disadvantage, however, the neutron structure in Fig. 3.12(b) proves that X—H···Ph hydrogen bonds can be directed at individual C atoms.

In a relatively frequently occurring geometry, X—H···Ph hydrogen bonds are directed approximately at the midpoint of a C—C bond. An example with amide N—H donors is shown in Fig. 3.13 in the crystal structure of the anticonvulsant drug **8**. IR spectroscopic measurements have shown that the N—H···Ph contact is indeed a hydrogen bond (Lutz *et al.* 1996).

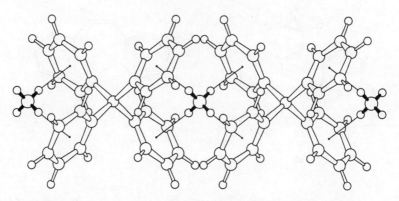

Fig. 3.14. Neutron structure of ammonium tetraphenylborate at 20 K. Notice the symmetrically disposed N—H⋯Ph hydrogen bonds (Steiner and Mason 1999).

X—H⋯Ph hydrogen bonds involving ions. The strength of X—H⋯Ph hydrogen bonds depends strongly on whether charged groups are involved or not (Fig. 3.10, Table 3.4). This is analogous to conventional hydrogen bonding where, say, N^+—H⋯O^- hydrogen bonds are much stronger than the N—H⋯O bonds in uncharged compounds, and can play much more dominant roles in crystal structures. X—H⋯Ph interactions involving ions occur mainly in the forms X^+—H⋯Ph, X—H⋯Ph—Y^-, X^+—H⋯Ph—Y^-, and occasionally also as X^+—H⋯Ph—Y^+.

The quintessential example of an ionic π-acceptor is the tetraphenylborate anion, $[BPh_4]^-$. Bakshi *et al.* (1994) have calculated that in this anion, the negative charge at the central B atom is 0.23 *e* and that each phenyl group therefore carries a delocalized partial charge of 0.19 *e*, making them very good aromatic acceptors. In consequence, tetraphenylborate salts have been extensively used as model systems for studies of hydrogen bonding. Ammonium tetraphenylborate, NH_4^+ $[BPh_4]^-$ is representative (Davies and Staveley 1957; Westerhaus *et al.* 1980; Steiner and Mason 2000) and in its crystal structure, chains are formed in which each ammonium ion donates pairs of N—H⋯Ph hydrogen bonds to two neighbouring anions (Fig. 3.14). The hydrogen bonds are very short and directed almost ideally at the centres of the phenyl rings: $d(M) = 2.154(6)$, $D(M) = 3.085(3)$, $d(C)$ range = 2.47–2.55 Å, $\theta(M) = 162.9(3)°$. This is a classical case of a compound which exhibits *only* non-conventional hydrogen bonds.

Knop, Cameron and co-workers have determined the crystal structures of many ammonium and iminium tetraphenylborate salts, and have surveyed their results together with those of other authors (Bakshi *et al.* 1994; Robertson *et al.* 1998). Extensive N^+—H⋯Ph^- bonding is a common feature of these compounds, often involving bi- and tri-furcated interactions. The hydrogen bond configurations that are observed are listed in Fig. 3.15, showing that the

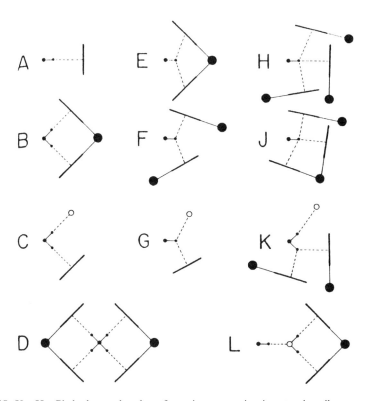

Fig. 3.15. X—H···Ph hydrogen bond configurations occurring in tetraphenylborate salts. The phenyl groups are seen in side-on projection, the B atoms are drawn as large black dots. Small black dots represent N atoms and small circles O atoms (from Bakshi *et al.* 1994).

52

system allows for considerable complexity. An interesting bidendate motif, **52**, involving the guanidinium ion was reported by the same group of authors (Bakshi *et al.* 1994). The donors in Fig. 3.15 are mostly of the N^+—H kind, but some originate from co-crystallized water molecules. The average

Table 3.8 X—H···Ph hydrogen bond geometries in tetraphenylborate salts (Bakshi *et al.* 1994)

	N^+—H, normal	N^+—H, bifurcated	O—H, normal
Mean $d(M)$ (Å)	2.34	2.92	2.58
Mean $D(M)$ (Å)	3.25	3.51	3.34
Mean $\theta(M)$ (°)	156	119	135
$d(M)$ range (Å)	2.11–2.81		2.29–2.80
$D(M)$ range (Å)	3.04–3.76		3.12–3.68

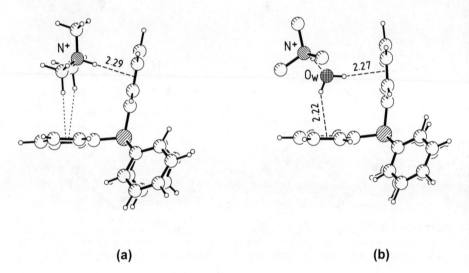

(a)   **(b)**

Fig. 3.16. Two examples of ammonium tetraphenylborates. (a) Trimethylammonium tetraphenylborate (Bakshi *et al.* 1994). Notice that the cation also forms a pair of C—H···Ph interactions. (b) Section of $(n\text{-Bu})_3\text{NH}^+$ $[\text{BPh}_4]^-$ monohydrate (Aubry *et al.* 1977). The alkyl chains of the cation are omitted for clarity. Geometries are normalized.

hydrogen bond distances in these tetraphenylborate salts are given in Table 3.8, and two examples of crystal structures are shown in Fig. 3.16.

X—H···Ph hydrogen bonds occur in many other charged systems. In proteins, for example, hydrogen bonds from charged N—H donors to aromatic side-chains are common (Section 5.2.3.1). An interesting example in a synthetic receptor molecule was found in $\text{Na}_4[\text{calix}[4]\text{arene sulfonate}]\cdot 13.5\ \text{H}_2\text{O}$, Fig. 3.17 (Atwood *et al.* 1991). A water molecule is enclosed in the calixarene cavity and forms two hydrogen bonds with aromatic groups of the host with very short $D(M)$ distances of 3.16 and 3.19 Å, respectively. Because each of the calixarene phenyl rings carries a sulfonate group, the molecule as a whole is a tetraanion with excellent π-acceptor potential.

In an interesting correlation between solid state and solution properties,

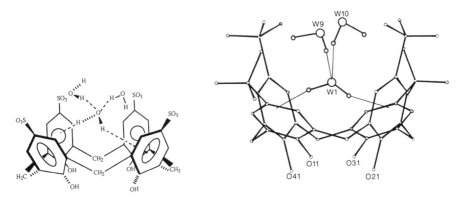

Fig. 3.17. Structure of and hydrogen bonding in Na$_4$[calix[4]arene sulfonate]·13.5 H$_2$O (from Atwood *et al.* 1991).

53   54

Boche and co-workers have identified N—H···Ph hydrogen bonds between amine and carbanion fragments in fluorenyllithium 2-ethylenediamine, **53**, and ammonium 1,2,4-tricyanopentatrienide, **54** (Buchholz *et al.* 1989a,b). In **53**, there are three short H···C distances (2.50, 2.57, 2.59 Å), while in **54** it is 2.58 Å ($\theta = 144°$). The structure of **53** is of relevance because it is a model of a solvent-bridged ion-pair, while the geometry in **54** has been invoked to rationalize observations on racemization of 9-fluorenyl derivatives.

3.1.3.4   *Intramolecular X—H···Ph hydrogen bonds*

Unlike the C—H···O hydrogen bond, the *intra*molecular variation of the X—H···Ph bond has been relatively well investigated. Because the steric constraints on such bonds increase with a reduction in the number of covalent bonds between donor and acceptor, we classify these intramolecular hydrogen bonds according to ring sizes.

Ortho-phenylphenols and related compounds. Five-membered rings.   The first X—H···Ph hydrogen bond reported occurs in *ortho*-phenylphenol and is of the intramolecular kind (Wulf *et al.* 1936). The hydrogen bond nature of this O—H···Ph interaction was confirmed in several seminal studies, in particular and in great detail by Oki and co-workers (summarized by Oki and Iwamura 1967). The crystal structure of the model compound 2,6-diphenylphenol as determined by Ueji and co-workers is shown in two projections in Fig. 3.18 (Nakatsu *et al.* 1978). The hydroxyl groups of neighbouring molecules are far apart from each other so that there can be no O—H···O hydrogen bonds, but the IR stretching frequency of $\nu_{OH} = 3524\,cm^{-1}$ is clearly indicative of hydrogen bond formation. The O—H group is in proximity to two C atoms ($d_{C1} = 2.41, d_{C2} = 2.36$ Å), and the distance to the C—C bond midpoint is even shorter, $d = 2.28$ Å. The contact angles are considerably bent with $\theta_{C1} = 111°$ and $\theta_{C2} = 128°$. The distances to all other C atoms are much longer, >3.3 Å, and the whole range of distances d_C is 2.31–4.21 Å. The geometry with respect to the aromatic midpoint is extremely off-centred with $d(M) = 3.11$, $D(M) = 3.89$ Å, $\theta(M) = 137°$, $\omega(H) = 52°$ and $\omega(O) = 60°$ (for normalized H atom position). This is a geometry for which hydrogen bond character it not easy to judge intuitively, and so IR spectroscopic evidence is clearly necessary.

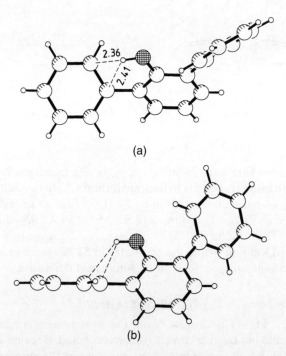

(a)

(b)

Fig. 3.18. 2,6-Diphenylphenol, an early example of an O—H···Ph hydrogen bond (Nakatsu *et al.* 1978). To visualize this hydrogen bond more clearly, the acceptor group is drawn (a) in the plane of paper, and (b) perpendicular to the plane of paper. The geometry is normalized.

Many *ortho*-phenylphenols exhibit intramolecular O—H···Ph hydrogen bonding. If the phenyl acceptor is itself unsubstituted at the 2- and 6-positions, as in **55**, the two phenyl rings are rotated by about 60° with respect to each other. If both phenyl rings carry hydroxyl groups at the 2-positions, as in say **56**, mutual intramolecular O—H···Ph hydrogen bonds are often formed with the phenyl rings being roughly perpendicular.

Because the intramolecular O—H···Ph interactions in *ortho*-phenylphenols are so bent, they leave the head-on direction of the hydroxyl group accessible for intermolecular interactions. In fact, the hydroxyl groups of many *ortho*-phenylphenols form bifurcated hydrogen bonds with one inter- and one intra-molecular component as in **57**. If the intermolecular acceptor A is strong, the intramolecular O—H···Ph interaction is only a secondary effect to a dominant O—H···A hydrogen bond. However, the second acceptor may also be a phenyl group, leading to bifurcated O—H···Ph hydrogen bonds. A good example was found by Ueji *et al.* (1982) in crystalline 4-nitro-2,6-diphenylphenol. The molecules form dimers that are connected by intermolecular O—H···Ph hydrogen bonds formed in addition to the intramolecular O—H···Ph interactions, Fig. 3.19.

An N—H analogue of the intramolecular hydrogen bonds in *ortho*-phenylphenols has been studied by Drew and Willey (1986) in a series of benzophenone hydrazones. In the crystal structure shown in Fig. 3.20, there is an intramolecular N—H···Ph interaction with almost identical geometry

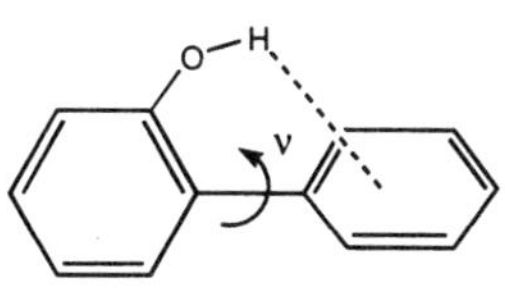

v = 60°

55

v = 90°

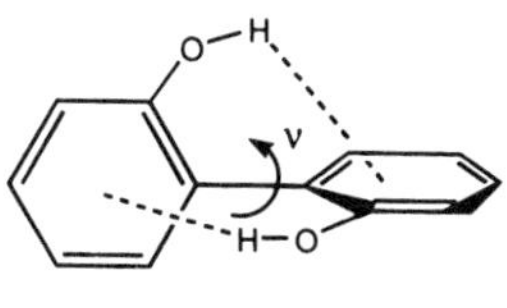

56

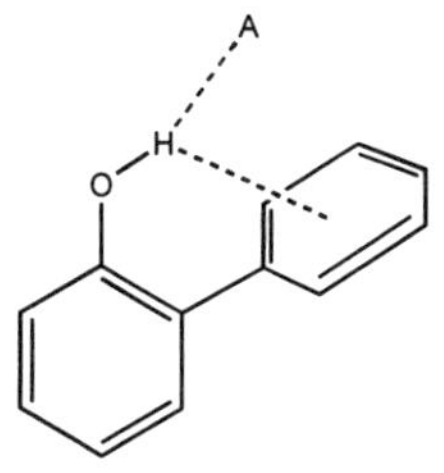

A = O, N, Ph

57

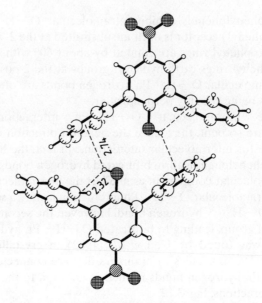

Fig. 3.19. Molecular dimer in the crystal structure of 4-nitro-2,6-diphenylphenol (Ueji *et al.* 1982). The intramolecular hydrogen bond is very similar to that in 2,6-diphenylphenol, the intermolecular hydrogen bond has normalized geometry of $d(M) = 2.74$, $D(M) = 3.27$ Å, $\theta(M) = 114°$, $\omega(H) = 20°$ and $\omega(O) = 4°$.

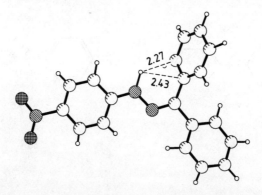

Fig. 3.20. Intramolecular $N—H\cdots Ph$ contact in benzophenone (4-nitrophenyl)hydrazone (Drew and Willey 1986). The normalized geometry is: $d(C1) = 2.43$, $d(C4) = 2.27$, $d(C—C) = 2.25$, $d(M) = 3.09$, $\theta(M) = 126°$, $\omega(H) = 54°$ and $\omega(O) = 61°$.

as the $O—H\cdots Ph$ bonds discussed above. Weak hydrogen bonding character was inferred by comparison of IR spectra of crystals and a CCl_4 solution.

Six-membered rings. The ring size $n = 6$ occurs frequently for intramolecular $X—H\cdots Ph$ hydrogen bonds, and appears to be a favourable one. The

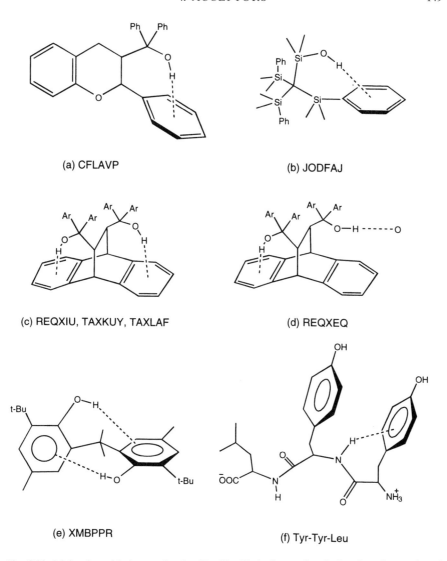

Fig. 3.21. Molecules with intramolecular X—H···Ph hydrogen bonds forming six-membered rings; see Table 3.9. (a) *cis*-Diphenyl(2-phenyl-3-chromanyl)methanol (Baert *et al.* 1983). (b) [Tris(dimethylphenylsilyl)methyl]-methylsilanol (Al-Juaid *et al.* 1991). (c) Roof-shaped hydroxy host compounds (REQXIU: Csöregh *et al.* 1996; TAXKUY and TAXLAF: Weber *et al.* 1996). (d) Roof-shaped hydroxy host compound (Csöregh *et al.* 1996). (e) 2,2-Bis-(2-hydroxy-5-methyl-3-*t*-butylphenyl)propane (Hardy and MacNicol 1976). (f) L-Tyrosyl-L-tyrosyl-L-leucine monohydrate (Steiner 1998*d*).

first example in crystals was reported by Hardy and MacNicol (1976), and many examples were published later by others (Fig. 3.21). All these have similar overall geometries, which are obviously governed by constraints within the six-membered ring (Table 3.9). As a representative example, the

Table 3.9 Geometries of intramolecular X—H···Ph hydrogen bonds which are part of six-membered rings. The molecular structures and literature references are given in Fig. 3.21

CSD code	Donor	d_C-range (Å)	d(M) (Å)	D(M) (Å)	θ(M) (°)	ω(H) (°)	ω(O) (°)
CFLAVP	O—H	2.32–3.42	2.57	3.49	156	26	33
JODFAJ	O—H	2.45–3.39	2.62	3.49	148	23	29
REQXEQ	O—H	2.21–3.14	2.34	3.31	168	25	27
REQXIU (a)	O—H	2.26–3.45	2.56	3.52	165	29	33
REQXIU (b)	O—H	2.26–3.52	2.61	3.53	155	31	34
TAXKUY(a)	O—H	2.12–3.30	2.41	3.35	159	29	29
TAXKUY(b)	O—H	2.26–3.39	2.56	3.40	147	28	28
TAXLAF (a)	O—H	2.21–3.23	2.40	3.35	163	25	27
TAXLAF (b)	O—H	2.13–3.20	2.35	3.31	167	27	27
XMBPPR	O—H	2.03–3.65	2.61	3.54	158	41	39
Tyr–Tyr–Leu	N—H	2.41–3.73	2.82	3.78	156	32	38

(a) and (b) refer to symmetry-independent molecules.

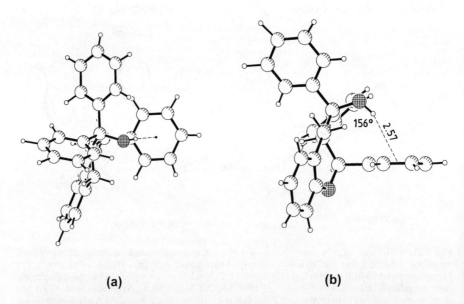

(a)   **(b)**

Fig. 3.22. Intramolecular six-membered ring containing an O—H···Ph hydrogen bond in *cis*-diphenyl(2-phenyl-3-chromanyl)methanol, shown in projections parallel and perpendicular to the acceptor plane (Baert *et al.* 1983).

O—H···Ph hydrogen bond in *cis*-diphenyl(2-phenyl-3-chromanyl)methanol, published by Baert *et al.* (1983), is shown in Fig. 3.22. The donor group is located roughly above the edge of the Ph acceptor, and the O—H vector points roughly at the aromatic centroid. The distance of the H atom to the

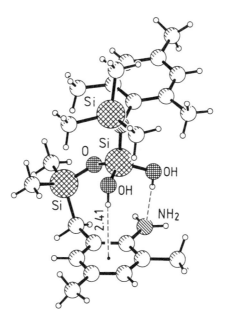

Fig. 3.23. Intramolecular seven-membered ring containing an O—H⋯Ph hydrogen bond. The normalized geometry is $d(M) = 2.41$, $D(M) = 3.35\,\text{Å}$, $\theta(M) = 161°$, $\omega(H) = 2°$, $\omega(O) = 7°$ (Murugavel *et al.* 1996).

pivot atom of the Ph group is short (2.32 Å), and the distance to the aromatic centre M is also short (2.57 Å). The displacement of the hydroxyl group from the acceptor C_6 axis is moderate with $\omega(H) = 26°$ and $\omega(O) = 33°$ (compared to 52° and 60°, respectively, in 2,6-diphenylphenol).

Table 3.9 contains only one N—H⋯Ph hydrogen bond. It is unlikely that this interaction is less favourable than for O—H donors. More likely than not, it is frequently overlooked. The single example shown occurs in a peptide and is discussed in greater detail in Section 5.2.3.2.

Ring sizes n ≥ 7. The larger ring sizes are observed only relatively rarely. However, the first O—H⋯Ph hydrogen bond ever reported in a crystal structure happens to be an intramolecular one with the ring size $n = 7$ (McPhail and Sim 1965). A different example with $n = 7$ occurs in the silanediol molecule shown in Fig. 3.23; this O—H⋯Ph interaction is almost ideally centred with $d(M) = 2.41$ Å (Murugavel *et al.* 1996). A very aesthetic example with $n = 8$ occurs in hexahelicene **58**, Fig. 3.24(a) (van Meerssche *et al.* 1984), and another is shown in Fig. 3.24(b) (Ishi-i *et al.* 1996). With a $d(M)$ of 2.15 Å, the latter is one of the shortest X—H⋯Ph hydrogen bonds ever reported.

With further increase in ring size, intramolecular hydrogen bonds approach the quasi-intermolecular situation where donor and acceptor can orient almost freely with respect to each other. Such large ring sizes do not occur in small organic molecules, but are not unusual in biological structures and in synthetic

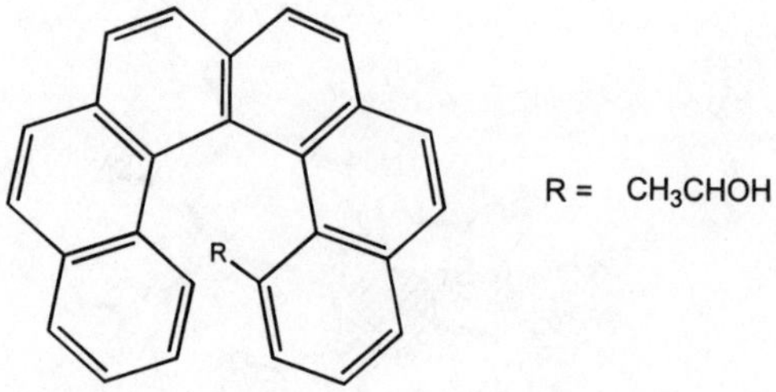

R =   CH₃CHOH

58

macrocyclic and oligomeric compounds. Toniolo and co-workers have reported such hydrogen bonds in a whole series of $C_{\alpha,\alpha}$-disubstituted glycines; an example is shown in Fig. 3.25(a) (Valle *et al.* 1988; Crisma *et al.* 1997). A different example occurs in the neutron structure of coenzyme vitamin B_{12}, Fig. 3.25(b). For this hydrogen bond, an energy of –4.0 kcal/mol has been calculated (Starikov and Steiner 1998).

3.1.3.5   *C—H···Ph hydrogen bonds*

In the previous section, Ph acceptors have been considered only along with strong hydrogen bond donors. Based on the principles elaborated in Chapters 1 and 2, one must expect that related hydrogen bond effects can occur with weaker donors too, with the only difference lying in their magnitude. The donor strengths of C—H groups have a very wide range that spans the scale from zero to the strengths of weaker types of O—H groups (Section 2.2). In consequence, the more acidic C—H types can form clean hydrogen bonds to Ph acceptors. With falling C—H acidity, the hydrogen bond nature of these C—H···Ph interactions falls, and blurs into the van der Waals region. In a CSD study on intermolecular C—H···Ph interactions, Umezawa *et al.* (1998) observed a pronounced dependence of the mean distances $d(M)$ on the C—H donor strength (Table 3.7). The mean distance of the H atom to the Ph plane is 2.38 Å for chloroform donors, but 2.75 Å for methyl groups (cut-off used: $d < 2.9$ Å).

Strongly acidic C—H groups.   For the most acidic C—H groups, C—H···Ph interactions have a distinct hydrogen bond nature. An example has already been shown in Fig. 3.12(b) where the terminal alkyne residue of **51** acts as the donor. The contact geometry is close to centred at M [$\omega(H) = 6.8°$], the distance $d(M) = 2.59$ Å is similar to O/N—H···Ph hydrogen bonds (compare Tables 3.8 and 3.9), and the red shift of the acetylenic C—H stretching frequency compared to CCl₄ solution is 41 cm⁻¹. In *ab initio* calculations, a binding energy of –1.3 kcal/mol was obtained (Steiner *et al.* 1996b). Similar interactions occur in related alkynes, and in a CSD search, six further examples were

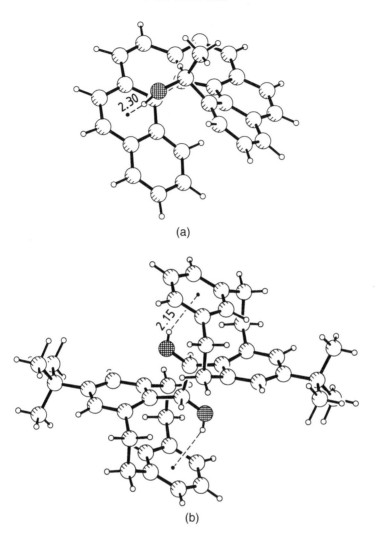

Fig. 3.24. Intramolecular eight-membered rings containing an O—H···Ph hydrogen bond. (a) (Hydroxy-1-ethyl)-1-hexahelicene, **58**, with $d(M) = 2.30$, $D(M) = 3.21$ Å, $\theta(M) = 153°$ (van Meerssche *et al.* 1984). (b) A metacyclophane with a pair of symmetry-equivalent hydrogen bonds, $d(M) = 2.15$, $D(M) = 3.03$ Å, $\theta(M) = 148°$ (Ishi-i *et al.* 1996). Geometries are normalized.

found with $d(M)$ distances ranging from 2.51 to 2.84 Å. The angles $\theta(M)$ range from 147 to 171°, indicating a preference for linearity. In *ab initio* computations on different C≡C—H···Ph configurations, bond energies in the range −1.0 to −2.0 kcal/mol were calculated (Steiner *et al.* 1995a).

Hydrogen bonds from haloforms to phenyl rings have also been reported. Atwood *et al.* (1992) have described a dichloromethane molecule hydrogen bonded to two aromatic rings of a calixarene (Fig. 3.26). The stabilization of

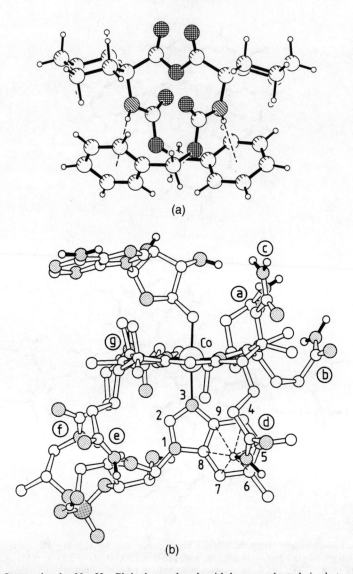

(a)

(b)

Fig. 3.25. Intramolecular N—H···Ph hydrogen bonds with long covalent chains between donor and acceptor. (a) In a synthetic peptide, $d(M) = 2.70$, $D(M) = 3.71$ Å, $\theta(M) = 179°$ (Valle *et al.* 1988; Crisma *et al.* 1997). (b) In the 15 K neutron crystal structure of coenzyme vitamin B_{12}, donated from an amide to a benzimidazole group, $d(M) = 2.58$, $D(M) = 3.42$ Å, $\theta(M) = 143°$ (adapted from Starikov and Steiner 1998, using the structure of Bouquiere *et al.* 1993).

the guest molecule is very similar to that of the water molecule in a related host shown in Fig. 3.17. In these two inclusion complexes, binding of the guest is facilitated by the hydrogen bond motif Ph···H—X—H···Ph, where X=O in Fig. 3.17 and X=C in Fig. 3.26.

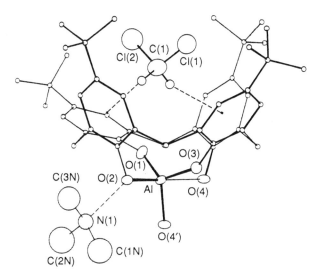

Fig. 3.26. Inclusion of dichloromethane in aluminium-fused bis-p-*tert*-butylcalix[4]arene (from Atwood *et al.* 1992).

59

A good example for intramolecular C—H⋯Ph hydrogen bonding by a strongly activated donor has been found by Cavaglioni and Cini (1997) in the organorhodium compound **59**. Here, a purine C(8)—H donor forms a short and close to centred interaction with a Ph ligand.

Weakly acidic C—H groups. With falling C—H acidity, the hydrogen bond nature of C—H⋯Ph interactions becomes increasingly questionable. Although the interactions are still directional, that is they do not represent van der Waals interactions, one could argue that a degree of long-range nature is not reached that would allow their classification as hydrogen bonds (Section 1.2.2). Furthermore, there is still no consensus among theoreticians about the relative contributions of electrostatic, dispersion, charge transfer and repulsion phenomena to the various kinds of hydrogen bond interaction. The well-known edge-to-face Ph···Ph interaction, which clearly has a directional character, is often called a 'herringbone interaction' (Desiraju and Gavezzotti 1989) as distinct from a hydrogen bond. Ph···Ph herringbone interactions can be a major driving force in the organization of crystal structures of benzenoid and phenyl-substituted compounds (Section 4.3), and the multiple Ph···Ph interactions between XPh$_3$ groups ('sixfold phenyl embrace', Dance and Scudder 1995, 1996, 1998) are well-known. Nishio calls all these 'aromatic CH/π interactions' and avoids a distinction with hydrogen bonds proper (Nishio and Hirota 1989, Nishio et al. 1995, Nishio et al. 1998). In any event, there is no well-defined border between the herringbone interaction and the hydrogen bond and in some cases, it is difficult to categorize a given contact properly. This would seem to be particularly challenging for C—H···Ph contacts with a structure-stabilizing function that goes far beyond what is usual for herringbone interactions.

Structure determining roles of C—H···Ph interactions from C(sp^2)—H donors have been documented in several cases. Desiraju, Nangia and coworkers have observed variations in C—H···Ph interactions formed in the nearly isostructural solvates of 2,3,7,8-tetraphenyl-1,9,10-anthyridine, **60**, with toluene and chlorobenzene (Madhavi et al. 1997). These variations adduce support for the hydrogen bond character of these very weak interactions. The contact of a heterocyclic C—H group of the anthyridine to the π cloud of a neighbouring toluene molecule is 2.54(2) Å but this is lengthened to 2.61(2) Å in the corresponding chlorobenzene solvate where the π-basicity is less (Chapter 6). A nice example of a cooperative ring of six C—H···Ph hydrogen bonds is provided by the activated γ-picoline in the molecular complex it forms with phloroglucinol (Biradha and Zaworotko 1998).

C—H···Ph interactions from methyl groups are even weaker, but may still be directional. An example of specific methyl–Ph interactions are antiparallel stacked arrangements of tolyl moieties in which the methyl groups are placed above the aromatic ring centres (Dastidar and Goldberg 1996). Another recent and related example has been reported by Hancock and Steed (1998) for a cyclotriveratrylene inclusion compound.

General motifs of this kind have been analysed by Ciunik and Jarosz (1998), who find two kinds of systematic 'hybrid interactions' (stacking plus

60 R = CH$_3$, Cl

C—H···Ph) of benzyl groups. In the first kind, **61**, parallel benzyl groups form columns with interplanar distances in the range 3.4–3.7 Å (mean 3.57 Å) and a lateral offset between neighbouring groups of 2.9–3.8 Å (mean 3.3 Å). The C—H···Ph interactions have geometries in the range $d(M) = 2.5–3.0$ Å and $\theta(M) = 120–160°$. In the second type of interaction, **62**, antiparallel benzyl groups form diads with interplanar distances, lateral offsets and C—H···Ph geometries very similar to those seen in **61**.

These arrangements certainly do have a stabilizing and possibly structure-determining function, so that they can be considered as being a part of the grey area between hydrogen bonds and the van der Waals interaction. The occurrence of related interactions in biological structures was reported by Umezawa and Nishio (1998) and by Ciunik *et al.* (1998). These weakest kinds of C—H···π interactions are of importance in structural chemistry and in structural biology and the field has been reviewed by Nishio *et al.* (1998).

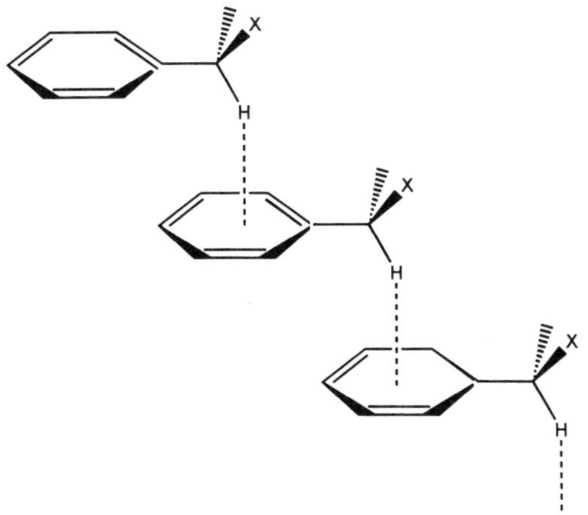

61

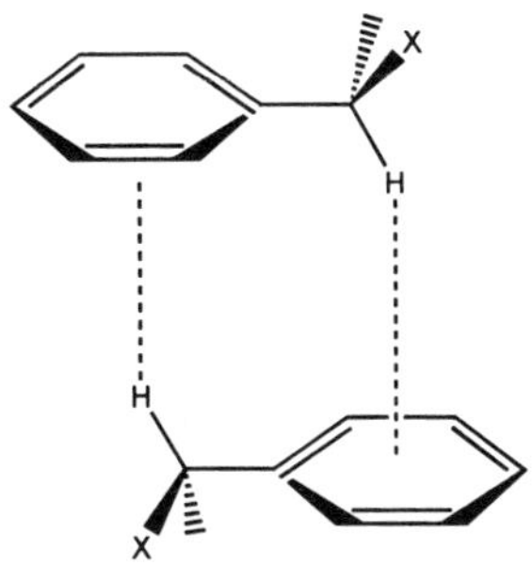

62

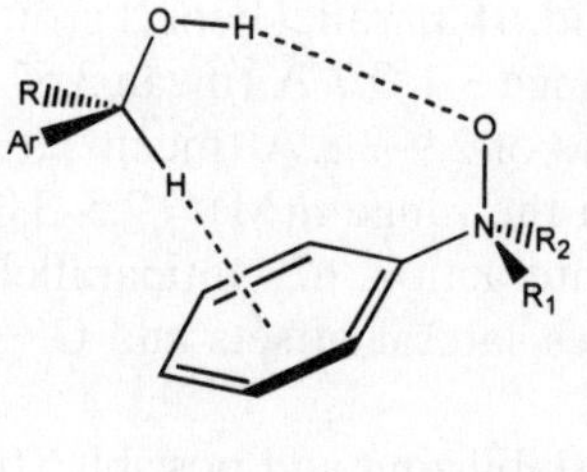

63 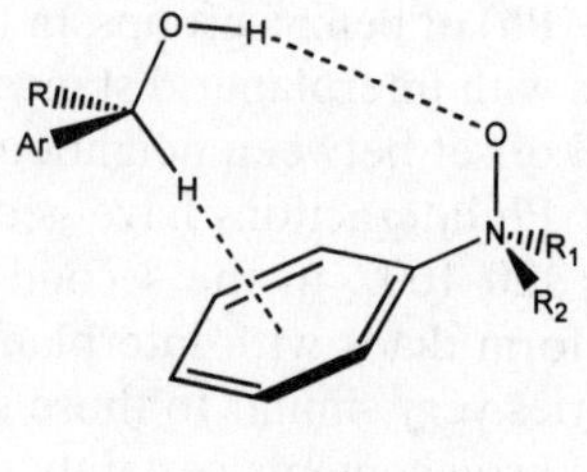   64

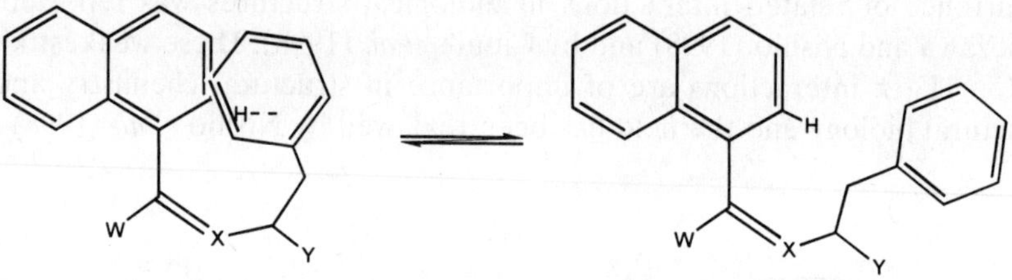

65

Examples from solution chemistry. Two interesting examples from solution chemistry are pertinent here. As far back as 1976, Pirkle and Hauske observed the non-equivalence of NMR chemical shifts in enantiomeric tertiary amine oxides **63** and **64** in the presence of a chiral benzyl alcohol. They rationalized this difference to the anchoring effect of $C-H\cdots Ph$ interactions that cause the groups R_1 and R_2 to be shielded differently in the two diastereomeric conformations. The authors have clearly noted that the carbinyl H atom in the alcohol is similar to the methine H atom in chloroform in that both have electron-withdrawing groups in the α-position.

In a more recent study, Boyd *et al.* (1996) have shown that in an entire family of alkenes, nitrones and imines, **65**, a congested conformation with an attractive intramolecular $C-H\cdots Ph$ interaction is preferred to a more open conformation without this interaction. On the basis of NMR data, this interaction has been assigned an energy in the range -0.3 to -0.6 kcal/mol.

3.1.3.6   The cation$\cdots$Ph interaction

For weakly acidic $C-H$ donors of uncharged molecules, contacts with Ph acceptors fall into the region intermediate between hydrogen bonds and van der Waals interactions. $X-H\cdots Ph$ hydrogen bonds have also a transition region to a different kind of interaction, one which is much stronger – the

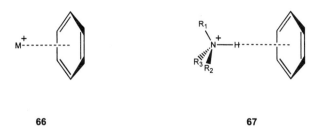

66   **67**

Table 3.10 Mean $M^+ \cdots Ph$(midpoint) and $M^+ \cdots O_W$ distances for interactions of alkali ions with phenyl rings and water molecules. CSD data compiled for this book

	Mean $M^+ \cdots Ph$ [n] (Å)	Mean $M^+ \cdots O_W$ [n] (Å)
Li^+	2.07(9) [5]	1.963(7) [81]
Na^+	2.84(6) [19]	2.423(3) [1143]
K^+	3.16(4) [24]	2.865(7) [433]
Rb^+		3.13(7) [11]
Cs^+	3.51(5) [17]	3.40(5) [33]

so-called cation–π interaction. This interaction plays an important role in supramolecular chemistry (MacGillivray and Atwood 1996) and in biological structures (Sussman *et al.* 1991), and comprehensive reviews are available (Dougherty 1996; Ma and Dougherty 1997). The classical cation–π interaction is that of alkali ions with benzene. In these complexes, the cation is positioned on the C_6 axis of benzene and interacts with the negative charge at the π-face of the molecule, **66**. Related complexes are formed also with other cations, in particular with cations that carry X—H groups such as NH_4^+, Me_4N^+, with distances and energies as given in Tables 3.10 and 3.11. Notably, the energy of the complexes $NH_4^+ \cdots Ph$ and $K^+ \cdots Ph$ are very similar, around -19 kcal/mol.

In crystals, interactions of ammonium groups with phenyl rings are clearly directional, **67**, and possess the IR spectroscopic features of hydrogen bonds (Westerhaus *et al.* 1980). Examples have already been shown in Figs 3.14 and 3.16(a). On the other hand, the binding energy is strongly dominated by the purely electrostatic interaction of the positive charge centre and the π electron cloud, largely exceeding energies of X—H $\cdots$ Ph hydrogen bonds in neutral species (Table 3.4). This suggests that the N^+—H $\cdots$ Ph interaction lies in between a pure cation–π interaction (as in $K^+ \cdots Ph$) and a pure aromatic hydrogen bond (as in Cl—H $\cdots$ Ph).

If quarternary ammonium ions such as R—N^+Me_3 interact with phenyl rings, they form short contacts of the methyl C—H groups with the ring edge,

Table 3.11 Experimental gas-phase cation-base binding energies (Meot-Ner (Mautner) and Deakyne 1985; Ma and Dougherty 1997)

Complex	Energy (kcal/mol)
Cation⋯Ph	
$Na^+⋯Bz$	−28.0
$K^+⋯Bz$	−19.2
$Bz⋯K^+⋯Bz$	−14.5
$NH_4^+⋯Bz$	−19.3
$NH_4^+⋯(1,3,5,-Me_3)Bz$	−21.8
$Me_3NH^+⋯Bz$	−15.9
$Me_4N^+⋯Bz$	−9.4
Cation⋯O/N	
$K^+⋯H_2O$	−17.9
$Me_4N^+⋯H_2O$	−9.0
$Me_4N^+⋯MeNH_2$	−8.7

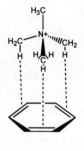

68

68 (Lee *et al.* 1995; see also Fig. 3.16(a)). In crystals, this is associated with typical $N^+⋯M$ separations of around 4.5 Å (Verdonk *et al.* 1993). Are these $C–H⋯Ph$ contacts hydrogen bonds? According to Nishio *et al.* (1998), they would be considered as a particular kind of stabilizing $C–H⋯\pi$ interaction, whereas others would ascribe their occurrence merely to a minimization of the $N^+⋯M$ separation.

A survey of such phenomena indicates that the hydrogen bond is an interaction with very open borders – on the weak side, there is a gradual transition into the van der Waals continuum and on the strong side there are similarly gradual transitions to the covalent bond, to the purely ionic interaction and also to the less well-studied but strong cation–π interaction.

3.1.3.7   *Role in crystal packing*

When considering the role of X—H···Ph hydrogen bonds in determining crystal packing, it is important to recall that:

1. The interaction energies are smaller than in conventional hydrogen bonds, but are still of the same order of magnitude. This allows limited competition with X—H···O/N hydrogen bonds.

2. The entire face of the aromatic ring can serve as hydrogen bond acceptor, including the individual C atoms. This makes the aromatic acceptor a 'target that is very easy to hit' (S. Suzuki *et al.* 1992). Consequently, X—H···Ph hydrogen bonds can be formed even in sterically very adverse situations. The donor can be moved laterally from the optimal position by as much as 3.0 Å, and the interaction still may remain a hydrogen bond (Fig. 3.6).

A simple function of phenyl acceptors in crystal lattices can be to compensate for a local or global lack of stronger acceptors (Hanton *et al.* 1992). If there are too many donors, they will tend to satisfy their hydrogen bond potentials with weaker acceptors, if these are available. A good example is aromatic molecules carrying amino groups. In larger systems, there may be a local lack of strong acceptors even if the total numbers of donors and acceptors are roughly balanced. In this case too, X—H···Ph hydrogen bonds may be formed as a last resort, and the large target size of the phenyl group makes it particularly useful for this purpose. A very typical example is amine–aromatic interactions in proteins (Section 5.2.3), and a representative example in a small molecule crystal structure is shown in Fig. 3.27.

Nevertheless, one should not be tempted to take the last-resort function as a predictive tool; there are many crystal structures where X—H···Ph hydrogen bonds are formed despite the presence of many conventional acceptors. In the structures shown in Fig. 3.12, strong donors and acceptors are balanced (Rzepa *et al.* 1991; Steiner *et al.* 1996*b*). In the calixarene complex in Fig. 3.17 (Atwood *et al.* 1991), there are strong SO_3^- acceptors immediately neighbouring the weak Ph acceptors, and in the structure shown in Fig. 3.13 (Lutz *et al.* 1996), there is even a clear surplus of strong acceptors. In the latter structure, the formation of the cooperative pattern O=C—N—H···O =C—N—H···is in principle possible, but the strong amide N—H donor forms a hydrogen bond with the phenyl group instead. This structure is an example of the complete failure of structure predictions based on hierarchies of donor and acceptor strengths (Etter 1990). Crystal packing is not determined by a small number of strong interactions, but by the cooperation and competition of a large number of strong, weak and very weak interactions. The strongest possible interaction does not in any event prevail.

Additionally, other general properties of X—H···Ph hydrogen bonds lead

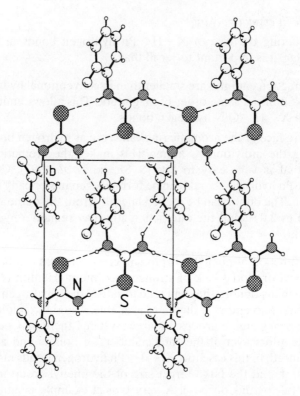

Fig. 3.27. N—H⋯Ph hydrogen bonds are formed in 2-fluorophenylthiourea to compensate for an imbalance of conventional donors and acceptors (from Steiner 1998*e*).

to unpredictable behaviour. The large size of the acceptor and its very flexible directionality hardly allow rational use for crystal engineering purposes. The X—H⋯Ph hydrogen bond is not weak, but it is too flexible to allow predictive control. Whilst comparing different crystal structures containing aromatic hydrogen bonds, one gets the initial impression that each is a particular case with little structural relation to the others. Unlike with conventional or with C—H⋯O hydrogen bonds, arrangements that deserve to be called *supramolecular synthons* do not show up immediately (a requirement for a supramolecular synthon is robustness and repetitivity in crystal structures, see Section 4.3). A notable exception are the tetraphenylborate salts, for which Knop, Cameron and co-workers could specify a number of N—H⋯Ph synthons (Fig. 3.15). The chelation of a bidendate donor between two phenyl rings of the anion, as is seen with ammonium ions and water molecules, is a repetitive motif (Figs 3.14 and 3.16). A different X—H⋯Ph synthon which occurs repeatedly in crystals is the formation of mutual hydrogen bonds between benzylic Ph—C—X—H units, **69** (Subramanian *et al.* 1999). Here, X—H can be N—H as well as O—H (Fig. 3.28).

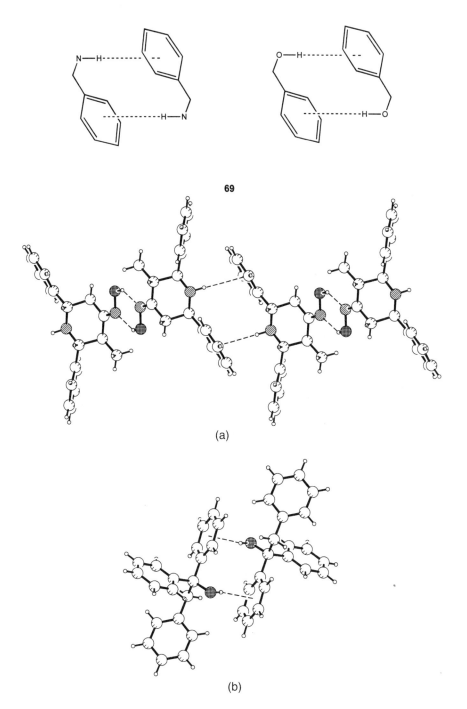

Fig. 3.28. Mutual X—H⋯Ph hydrogen bonds formed by Ph—C—X—H units. (a) Molecular chain in which molecules are linked by oxime dimers and by N—H⋯Ph dimers (Subramanian *et al.* 1999). (b) Dimer linked by mutual O—H⋯Ph hydrogen bonds in 1,1,2-triphenylethanol (Ferguson *et al.* 1994).

Limited progress in rationalizing the roles of $X-H\cdots Ph$ bonds in crystal packing has also been made in some other cases. In a careful analysis of lattice inclusion compounds with the host shown in Fig. 3.21(c) (with Ar = Ph), Csöregh *et al.* (1996) have noted the existence of an intramolecular $O-H\cdots Ph$ hydrogen bond when a 'non-hydrogen bonding' guest like nitroethane and benzene is present. Upon inclusion of a hydrogen bonding guest like dioxane, one of the intramolecular bonds is opened so that an intermolecular $O-H\cdots O$ hydrogen bond can be formed (Fig. 3.21(d)). The authors have stated correctly that the orientation of the $O-H$ group is a 'soft' parameter.

To conclude, $X-H\cdots Ph$ hydrogen bonds are extremely flexible. Therefore, they are difficult to utilize in crystal engineering. Typically, there are no clear trends for the involvement of $X-H\cdots Ph$ hydrogen bonds in specific and extended supramolecular arrays. However, it is observed that $X-H\cdots Ph$ bonds can form repetitive patterns if covalent geometries and $X-H\cdots Ph$ geometries match favourably. The phenomenon is not well understood and clearly more work is required.

3.1.4 Alkynes

Alkynes are much simpler π hydrogen bond acceptors than aromatic groups. The acceptor is composed of only two atoms, the distance between these atoms is short (around 1.18 Å), and ideally the system exhibits cylindrical symmetry. Gas phase experiments (Fig. 3.3) and theoretical studies on $X-H\cdots C\equiv C$ hydrogen bonds find that the optimal geometry is T-shaped with the donor pointing linearly at the centre of the triple bond, and not at one of the individual alkynic C atoms. This T-shaped geometry is clearly illustrated in the 1:1 crystalline complex but-2-yne–HCl, Fig. 3.11(b) (Mootz and Deeg 1992a). In the corresponding 1:2 complex, the geometry is somewhat distorted due to crystal packing effects. Gas phase bond energies for acetylene acceptors were calculated as −4.5 and −3.0 kcal/mol for the donor HF, depending on the computational method used (Tang *et al.* 1990; Rozas *et al.* 1997a), and −1.9 kcal/mol for HCN (Fan *et al.* 1996). All this hints at a generally good hydrogen bond acceptor potential of the $C\equiv C$ group.

3.1.4.1 Intermolecular $O-H\cdots C\equiv C$ hydrogen bonds

The geometry of hydrogen bonds to $C\equiv C$ groups can be described much more unambiguously than for aromatic acceptors. As there are only two atoms in the group, it is straightforward to specify the geometries with respect to these two atoms [i.e. $d(C)$, $D(C)$ and $\theta(C)$], and this is indeed done in many publications. Alternatively, one can take the centre M of the triple bond and the bisecting plane as references, and give the values of $d(M)$, $D(M)$, $\theta(M)$

70

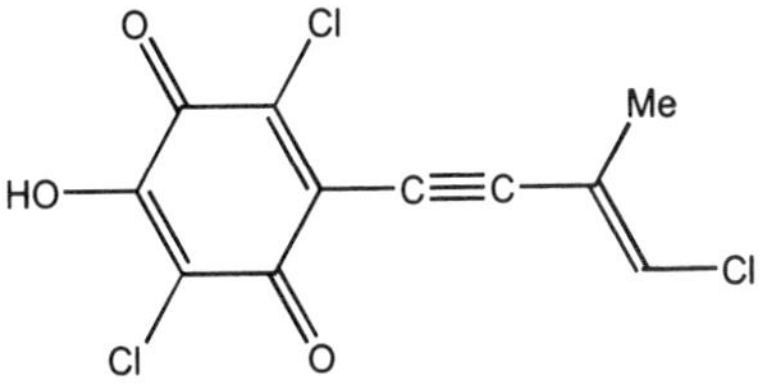

71

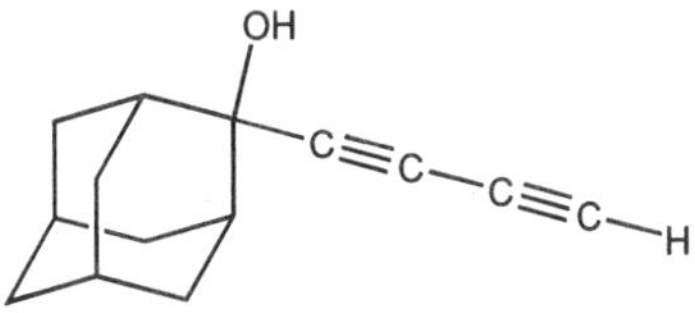

72

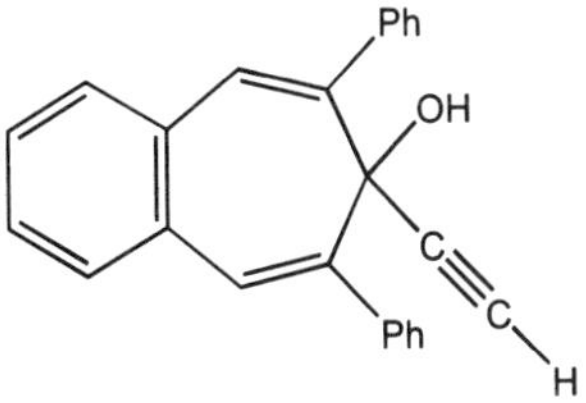

73

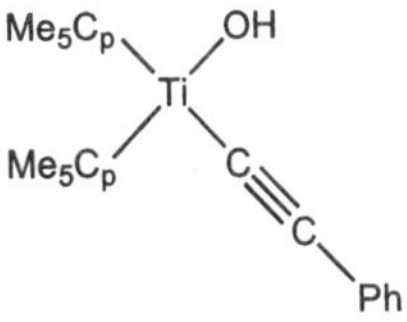

74

and ω(X/H), **70**. The angle ω is the elevation angle of the H/X$\cdots$M line from the bisecting plane of C≡C (easily measured as 90° minus the angle C−M$\cdots$X/H). For terminal alkynes, one can introduce a defined orientation of ω, such as $\omega > 0$ for displacement towards the terminal C atom, and $\omega < 0$ for displacement away from it.

O/N−H$\cdots$C≡C hydrogen bond geometries in crystal structures are given in Table 3.12 for intermolecular bonds without bifurcation. The number of hydrogen bonds is far too small to perform detailed statistical analysis, but some structural characteristics can still be seen. The hydrogen bond distances cover a wide range, d(M) = 2.26–2.80, mean value = 2.51 Å, and D(M) = 3.22 to 3.64, mean value = 3.42 Å. The shortest distance d to an individual C atom is 2.27 Å. Some of the hydrogen bonds are directed almost exactly at M, such as in 2-ethynyladamantan-2-ol, **21**, with θ(M) = 179° (Fig. 2.32). Others point more towards one of the C atoms, such as in danazole, **25** (Fig. 2.33). The

Table 3.12 Intermolecular O—H···C≡C hydrogen bonds without bifurcation. Terminal alkynes are labelled C2≡C1—H

Compound	Ref.	Acceptor	d(C1) (Å)	d(C2) (Å)	d(M) (Å)	D(M) (Å)	θ(C1) (°)	θ(C2) (°)	θ(M) (°)	ω(H) (°)	ω(O) (°)
21	(b)	C≡C—H	2.29	2.38	2.26	3.22	164	166	179	4	4.7
71	(c)	C—C≡C—C	2.78	2.76	2.70	3.62	159	147	156	1	4.5
72	(d)	C≡C—H	2.27	2.55	2.34	3.31	167	165	178	14	13.9
26	(e)	C≡C—H	2.59	2.86	2.66	3.64	158	175	170	13	15.7
73	(f)	C≡C—H	2.51	2.49	2.43	3.37	158	153	159	−1	−2.1
73	(f)	C≡C—H	2.42	2.47	2.38	3.25	153	138	147	4	−1.7
25	(g)	C≡C—H	2.40	2.71	2.49	3.34	144	140	143	16	13.6
74	(h)	Ti—C≡C—H	2.77	2.97	2.80	3.59	131	140	137	9[i]	12.9[i]
Mean values[θ]			2.50	2.65	2.51	3.42	154	153	159	8(3)[k]	7(3)[k]

(a) Distances d normalized. (b) Allen et al. 1996b. (c) Hautzel et al. 1990. (d) Keller et al. 1995. (e) Steiner et al. 1997a. (f) Steiner et al. 1996c. (g) Viswamitra et al. 1993. (h) Polse et al. 1995. (i) Displaced away from the Ti atom. (j) Standard uncertainties of mean values are about 0.07 Å for d and 5° for θ. (k) Only for terminal alkynes.

variation of the angle ω is small compared to the variation seen in hydrogen bonds to aromatic acceptors, where values of $\omega > 30°$ are not uncommon. This means that the T-shaped geometry is relatively well maintained.

For the six hydrogen bonds to terminal alkynes in Table 3.12, the $\omega(O)$ values range from -2.1 to $15.7°$ with a mean value of $7(3)°$. The mean angle $\omega(H)$ is $8(3)°$. This means that on average, the donor group is slightly displaced from the bisecting plane of $C\equiv C$ in the direction of the terminal C atom. On average, $d(C1)$ is $0.16\,\text{Å}$ shorter than $d(C2)$. This slight displacement could simply be due to steric effects because the terminal atom is more easily accessed in intermolecular contacts. There are possibly also electronic reasons caused by the chemical inequivalence of the two C atoms.

$O-H\cdots C\equiv C$ hydrogen bonds can be part of bifurcated arrangements, with an O atom normally acting as the major acceptor. Then, hydrogen bond geometry can deviate much more from the ideal values than in non-bifurcated bonds, and this must be taken into account in structure interpretation. Occasionally, however, bifurcated bonds are observed where the alkyne group acts as more than a secondary player. Bifurcated bonds donated to a hydroxyl group and a geminal ethynyl fragment, **75**, occur repeatedly. In hydrogen bonds of this kind, the hydroxyl donor points roughly midway between the two acceptors, but the donor O atom is still positioned approximately on the bisecting plane of the $C\equiv C$ bond [i.e. $\omega(O) \approx 0°$]. Rzepa et al. (1994) have drawn attention to a water molecule donating a pair of such bifurcated bonds in the organoplatinum compound **76**. These types of hydrogen bonds are also formed in the ethynyl steroid **77**, where they are interconnected to an infinite chain (Fig. 3.29). Note that the bifurcated bond in **77** is almost symmetrical in the sense that the two $d(A)$ distances are very similar, as are the two $\theta(A)$ angles.

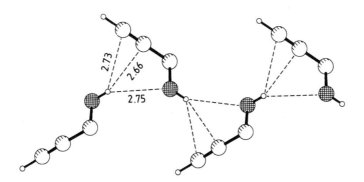

Fig. 3.29. Chain of interconnected bifurcated hydrogen bonds in steroid **77** (Rohrer et al. 1976a). For clarity, only the relevant fragment $H-O(17)-C(17)-C(20)\equiv C(21)-H$ is drawn. Normalized geometries: $d(O) = 2.75\,\text{Å}$, $\theta(O) = 135°$, $d(M) = 2.63\,\text{Å}$, $\theta(M) = 128°$, $\omega(O) = 1°$.

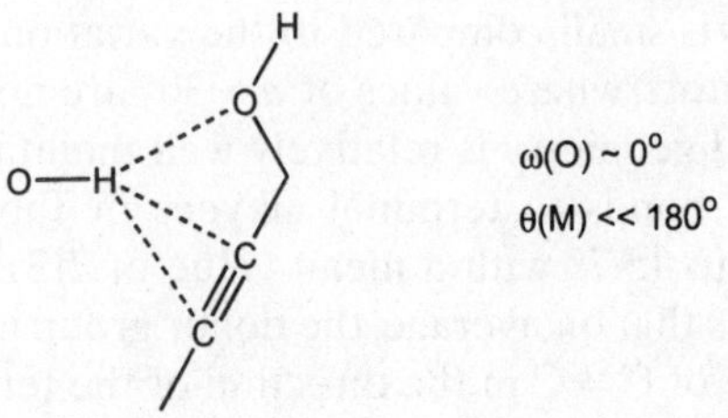

$\omega(O) \sim 0°$

$\theta(M) \ll 180°$

75

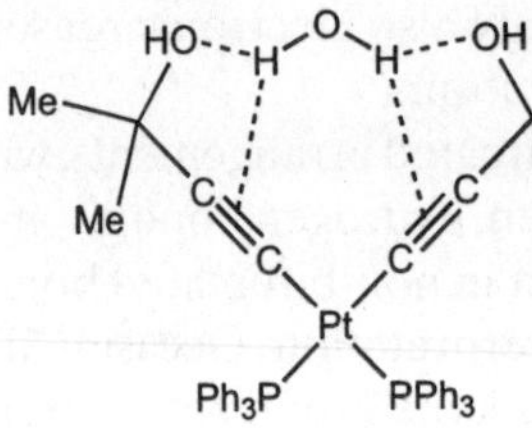

76

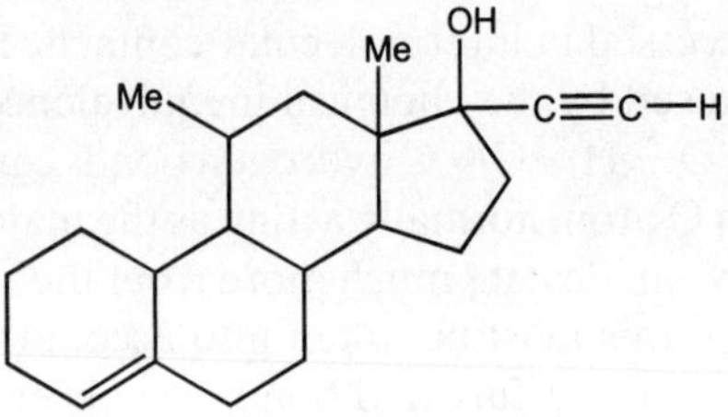

77

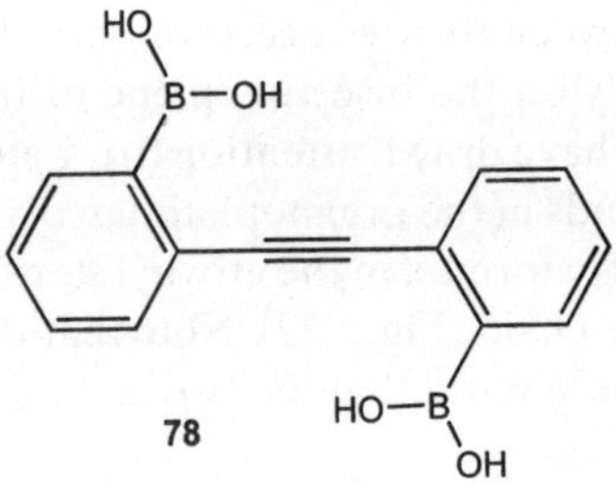

78

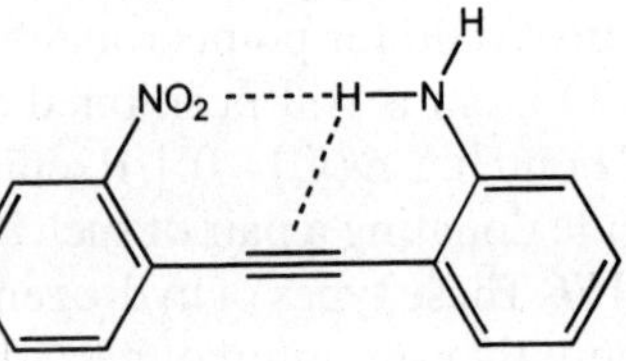

79

3.1.4.2 Intramolecular O/N−H···C≡C hydrogen bonds

Intramolecular hydrogen bonds to alkynes are only rarely observed. An interesting example is the diboronic acid **78** shown in Fig. 3.30(a). Pilkington *et al.* (1995) reported that the central C≡C bond accepts a pair of relatively short intramolecular hydrogen bonds as part of a bifurcated arrangement. The other component is intermolecular with an O atom acceptor. The intramolecular hydrogen bonds form six-membered rings. An intramolecular N−H···C≡C hydrogen bond with the ring size $n = 5$ has been found in the nitro amine **79**, Fig. 3.30(b) (Pilkington *et al.* 1996). This interaction is part of a bifurcated arrangement with the major component directed at a nitro O atom, and it represents the N−H analogue of the intramolecular hydrogen bond in *ortho*-ethynylphenol, Fig. 3.1(b). For the latter, hydrogen bonding character had been clearly shown by solution IR spectroscopy.

Intramolecular O−H···C≡C bonding in but-3-yne-1-ol was established early on by solution IR spectroscopy (Fig. 3.1(d)). This hydrogen bond is necessarily distorted from optimal geometry. The relevant fragment is a

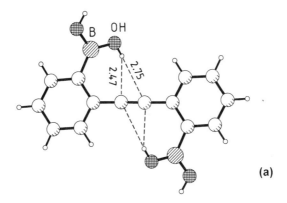

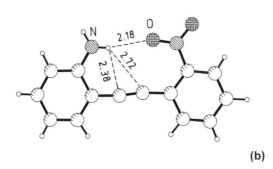

Fig. 3.30. X—H···C≡C hydrogen bonds (a) in compound **78**, $d(C1) = 2.47$, $d(C2) = 2.75$, $d(M)$ = 2.55, $D(M) = 3.14$ Å, $\theta(M) = 118°$, $\omega(H) = 13°$, $\omega(O) = 18.6°$ (Pilkington *et al.* 1995); (b) in compound **79**, $d(C1) = 2.38$, $d(M) = 2.49$ Å, $\theta(M) = 117°$, $\omega(H) = 17°$ $\omega(O) = 33.9°$ (Pilkington *et al.* 1996). Geometries are normalized.

common one in organic molecules, and in crystals, several of them adopt the conformation required to obtain an intramolecular O—H···C≡C inter-action. However, these are always minor components of bifurcated arrange-ments such as **80**, and the geometries are very distorted indeed. A CSD search revealed four hits with the following mean geometry: $d(C1) = 2.74$, $d(M) = 2.96$, $D(M) = 3.32$ Å, $\theta(M) = 103°$, $\omega(H) = 27°$, $\omega(O) = 44°$.

3.1.4.3   *C—H···C≡C hydrogen bonds*

Short C—H···C≡C contacts are observed in many crystal structures but it is questionable if they can be routinely classified as hydrogen bonds. The matter is unambiguous only for acidic C—H donors. With falling C—H acidity, the interaction enters the grey area between hydrogen bonds and van der Waals interactions. The strength of C—H···C≡C interactions depends

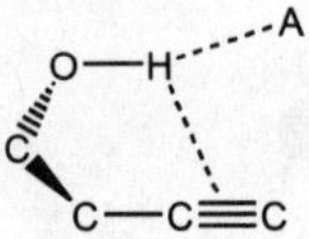

80

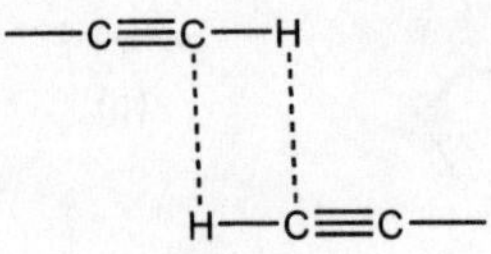

d (C) < d (H) d (C) > d (H)

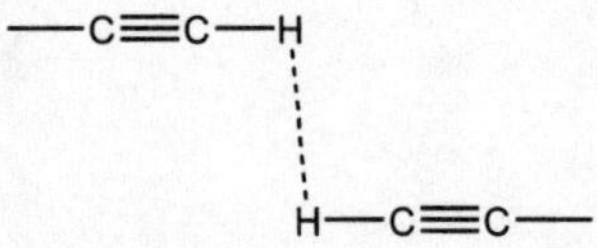

81   **82**

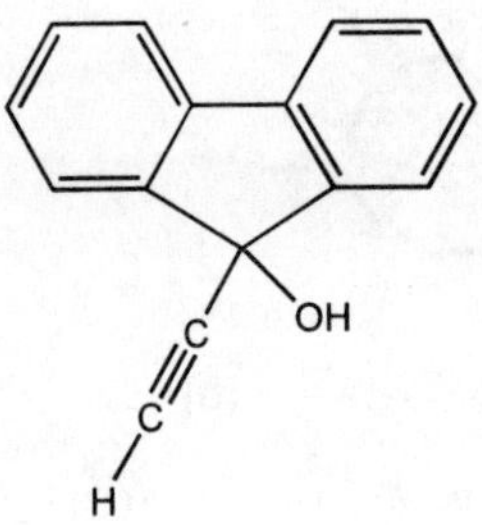

83 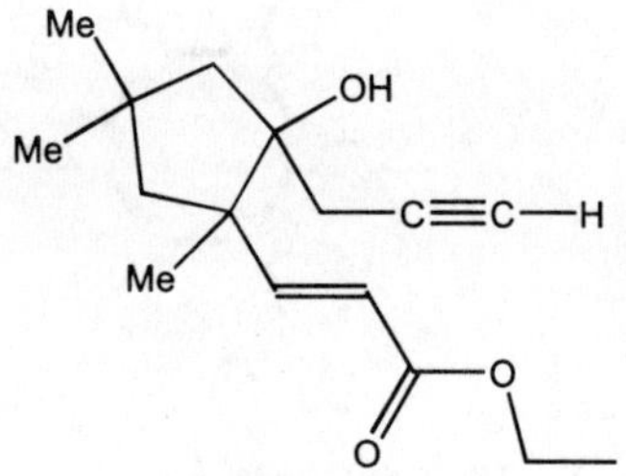   **84**

not only on the donor, but also on the acceptor. In *ab initio* calculations on a series of $C-H\cdots C\equiv C$ complexes, Fan *et al.* (1996) obtained the energy and distance values given in Table 3.13. The contact geometry is T-shaped in all cases. When varying the donor in the sequence CH_4, CH_3Cl, CH_2Cl_2, $CHCl_3$ and with the acceptor kept constant (acetylene), the bond energy increases from −0.1 to −1.8 kcal/mol. This is what might have been expected from donor acidity considerations. A much larger variation is obtained if the acceptor is changed from acetylene to metallated species; because the metal atoms donate electrons to the $C\equiv C$ moiety, the acceptor strength is dramatically enhanced. Metallation of the π-acceptor can turn a weak hydrogen bond into a strong one. In purely organic compounds, $C-H\cdots C\equiv C$ hydrogen bonds will always be weak to very weak.

By far the most abundant $C-H\cdots C\equiv C$ hydrogen bonds are those formed

Table 3.13 Optimized bond energies and distances $d(M)$ for a series of C−H···C≡C complexes. *Ab initio* MP2 computations by Fan *et al.* (1996)

Donor	Acceptor	$d(M)$ (Å)	Energy (kcal/mol)
CH_4	H−C≡C−H	3.08	−0.1
CH_3Cl	H−C≡C−H	2.85	−0.6
CH_2Cl_2	H−C≡C−H	2.62	−1.3
$CHCl_3$	H−C≡C−H	2.49	−1.8
$CHCl_3$	H_3P−Au−C≡C−Au−PH_3	2.17	−6.0
CH_4	Na−C≡C−Na	2.82	−0.5
CH_3Cl	Na−C≡C−Na	2.50	−3.5
CH_2Cl_2	Na−C≡C−Na	2.10	−7.6
$CHCl_3$	Na−C≡C−Na	2.02	−10.1

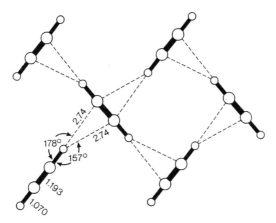

Fig. 3.31. Neutron structure of the orthorhombic low-temperature phase of acetylene (McMullan *et al.* 1992; from Steiner 1998*b*).

between terminal alkynes. It is known that in the gas phase, acetylene molecules form T-shaped dimers, Fig. 3.3. Bond energies are calculated in the range −1.0 to −2.0 kcal/mol, depending on the computational method (Scheiner 1997). In the orthorhombic (*Acam*) low-temperature phase of acetylene, an arrangement is formed which can be understood as a two-dimensional tiling of these T-shaped dimers, Fig. 3.31 (McMullan *et al.* 1992). Every acetylene molecule donates and accepts two C−H···C≡C interactions of close to optimal geometry. Notably, in the cubic (*Pa3*) high temperature phase, no such C−H···C≡C contacts are formed. Simple calculations on the orthorhombic structure, point at a weak hydrogen bond nature of the C≡C−H···C≡C interactions (Grabowski 1995). Spectroscopic evidence for hydrogen bond nature was obtained for a chain C≡C−H···C≡C−H···C≡C−H occurring in pentynoic acid with $d(M) = 2.67$ Å. The red shift of v_{CH} compared

Table 3.14 Geometry of $C\equiv C-H\cdots C\equiv C-H$ hydrogen bonds in crystals; atom labelling is $C2\equiv C1-H$

	Range	Mean
$d(C1)$ Å	2.54–3.03	2.76(2)
$d(C2)$ Å	2.52–3.20	2.82(3)
$\theta(C1)$ (°)	>117	153(3)
$\theta(C2)$ (°)	>125	154(2)
$d(M)$ (Å)	2.51–3.00	2.72(2)
$D(M)$ (Å)	3.55–3.95	3.72(2)
$\theta(M)$ (°)	>131	155(2)
$\omega(H)$ (°)	−19 to +29	+3(1)
$\omega(O)$ (°)	−15.5 to +37.8	+4(1)

CSD data retrieved for this book, organic crystal structures, distances d normalized, $d(M) < 3.0$ Å, 32 contacts in 25 crystal structures.

to CCl_4 solution is $33\,cm^{-1}$, which is indicative of weak but noticeable hydrogen bonding (Steiner *et al.* 1995a). Further spectroscopic measurements on $C\equiv C-H\cdots C\equiv C-H$ interactions showed that the distance range is very long, extending far beyond the van der Waals separation. Even for alkyne–alkyne contacts with $d(C)$ distances ≥ 2.9 Å, effects on the vibrational spectrum could be observed (Steiner *et al.* 1996d). This is as it must be for a hydrogen bond.

$C\equiv C-H\cdots C\equiv C-H$ hydrogen bonds are so abundant in crystal structures of terminal alkynes that their geometries can be statistically analysed with good significance. The distribution of distances $d(M)$ shows a clear maximum in the interval 2.6–2.7 Å, with a mean value of 2.72(2) Å, and falls steeply for longer distances (Fig. 3.32 and Table 3.14). This shape of distance distribution is typical for bona fide hydrogen bonds. This distribution is a good example to show the catastrophic consequences of using the van der Waals cut-off criterion as an indicator of hydrogen bonding. A cut-off at 2.4 Å would neglect *all* these hydrogen bonds and a cut-off at 2.7 Å would cut through the maximum of the distribution. Obviously, this histogram calls for a long $d(M)$ cut-off of at least 3.0 Å, if one is to be used at all. Also the hydrogen bond angle $\theta(M)$ has a distribution that is typical of hydrogen bonding, with the maximum for linear angles (after cone-correction), and a pronounced fall-off for bent angles. The distribution of the angle $\omega(H)$ is slightly skewed with the average value of 3(1)°. The donor can approach the acceptor from different directions, but on average it is slightly displaced in the direction of C1. These statistical data show that the geometrical behaviour of the $C\equiv C-H\cdots C\equiv C-H$ hydrogen bond is relatively simple, and is quite unlike that seen for hydrogen bonds to aromatic acceptors.

There is another mode for contacts between terminal alkynes in crystals. The

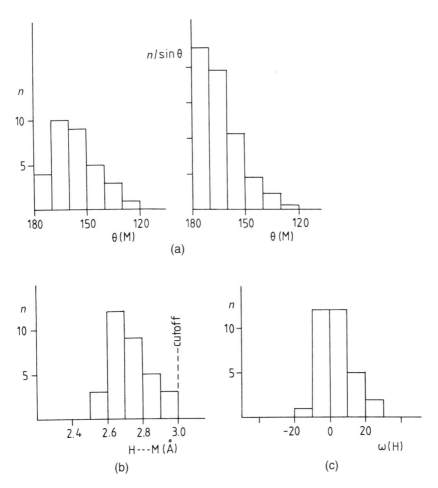

Fig. 3.32. Normalized geometry of C≡C—H···C≡C hydrogen bonds for which numerical data are given in Table 3.14. (a) Distribution of the angle $\theta(M)$: *left* original data, *right* data after cone correction. (b) Distribution of $d(M)$. (c) Distribution of the angle $\omega(H)$ defined in **70**.

groups can be arranged not only perpendicularly, but also antiparallel to each other. Depending on the angular diplacement, the shortest contacts can be between the H and C1 atoms, **81**, or between the terminal H atoms, **82**. The former case represents antiparallel dipoles placed side by side, so that it might be weakly bonding. For an example found in alkynol **83** with d(C1) = 2.97 Å, however, an interaction energy of only −0.1 kcal/mol has been calculated (Steiner *et al.* 1995a). The geometries **82** with d(H) < d(C1) represent dipoles arranged head-to-head, and may be destabilizing. An extreme example with d(H) = 2.26 Å was found in **84** (Lakshmi *et al.* 1995b). The intermediate cases with d(H) ≈ d(C1) could have close to zero interaction energies. Notably, attractive dispersion forces can balance small electrostatic repulsions, so that interpretation of contact geometries **81** and **82** is not straightforward.

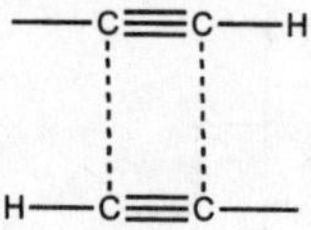

85

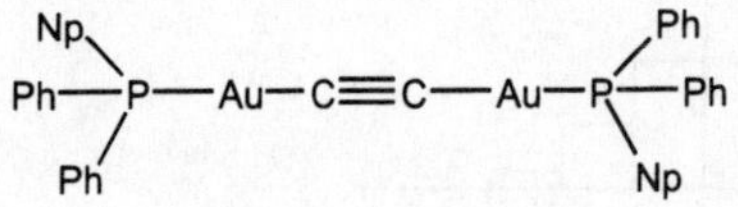

86

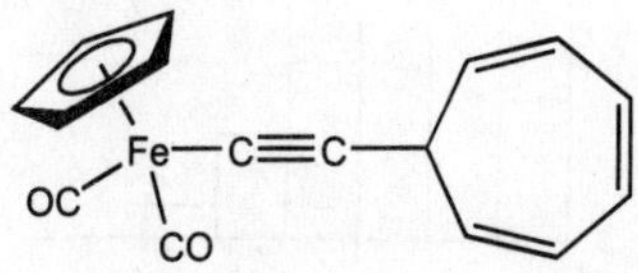

87

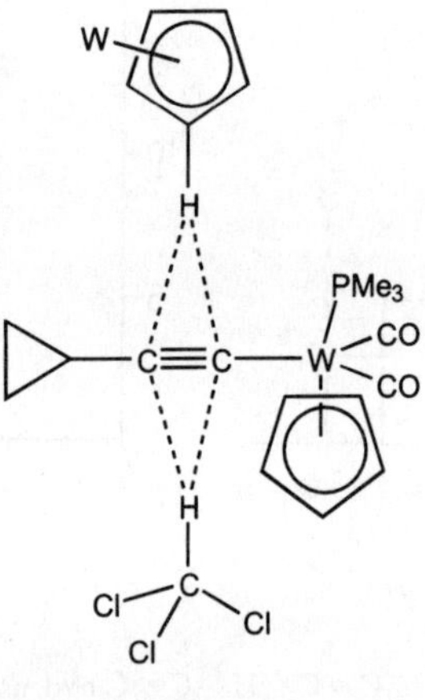

88

In this context, one can draw a connection to the field of solid-state re-activity. When shifting the terminal alkyne groups of geometry **81** further with respect to each other, one arrives at geometry **85** where the two C≡C moieties form lateral contacts with C···C separations of around 3.5 Å. This is a topochemically reactive motif that Foxman and co-workers have exploited effectively for solid-state reactions triggered by ^{60}Co γ-rays (Booth *et al.* 1987).

There are C—H···C≡C hydrogen bonds which are stronger than those between terminal alkyne groups, and others which are weaker. Examples of stronger bonds are seen in the 1:2 and 1:4 chloroform adducts of **86** (Fig. 3.33; Müller *et al.* 1994). The Au atoms donate electrons into the triple bond,

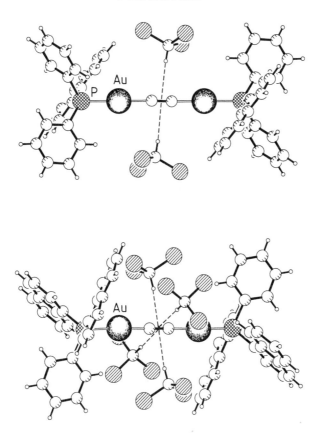

Fig. 3.33. Crystal structures of the 1:2 and 1:4 complexes of **86** with CHCl₃. The normalized distances $d(M)$ are 2.30 Å for the 1:2 complex, and 2.46 and 2.48 Å for the 1:4 complex (Müller *et al.* 1994).

enhancing its acceptor potential. In the 1:2 complex, $d(M)$ is 2.30 Å; this would be considered as very short even for an O—H···C≡C hydrogen bond (Table 3.12). Donors weaker than chloroform also form relatively short hydrogen bonds with metal-bonded C≡C groups. An example with the Cp donor in compound **87** is given in Fig. 3.34 and similar interactions occur in **88** (Steiner and Tamm 1998). The nature of C—H···C≡C interactions with weakly acidic C—H and purely organic C≡C groups is of general interest, but has not yet been investigated in detail.

3.1.4.4 *Hydrogen bonded networks*

Most π-acceptor moieties have either no hydrogen bond donor potential (C—C≡C—C, etc.) or they can act only as very weak donors (Ph, aryl). Therefore, their roles in hydrogen bond patterns are usually very simple. They

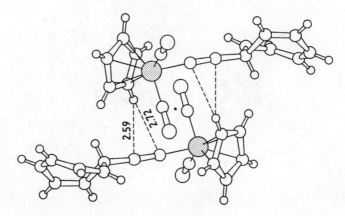

Fig. 3.34. Short (Cp)C—H···C≡C—Fe interaction in compound **87** with the normalized geometry $d(M) = 2.58$, $D(M) = 3.60$ Å, $\theta(M) = 154°$ (from Steiner and Tamm 1998).

accept either single or pairs of isolated hydrogen bonds, **89**, or they act as chain stoppers, **90**. Notably, the double-acceptor function has a negative cooperativity (or anti-cooperativity), that is the bond energy is smaller than the sum of two isolated bonds (Bone *et al.* 1989), so that it is disfavoured compared to single acceptors.

The terminal alkyne group is quite different, because it is at the same time one of the strongest C—H donors and one of the best π-acceptors. This duality of relatively strong donor and acceptor potential immediately reminds one of the hydrogen bond capabilities of the hydroxyl group. Terminal alkynes allow for the formation of hydrogen bond chains which are analogous to chains formed by hydroxyl groups. The simplest possible chains consist of only two hydrogen bonds as in **91**. Here, X—H can be any donor and A can be any chain stopper. Two such examples have been discussed in the literature. The first is a chain C≡C—H···C≡C—H···Ph in the benzyl alcohol **92** (Steiner *et al.* 1995a), and the second is a pair of chains O—H···C≡C—H···Ph in **73**. The latter is shown in Fig. 3.35.

The most frequently occurring hydrogen bond chains of terminal alkynes are not finite, but infinite of the kind···C≡C—H···C≡C—H···C≡C—H. Because the angle between a donated and accepted hydrogen bond is preferably rectangular, these chains typically form planar or close to planar zigzag lines, **93**. This motif is not only favourable geometrically, but also nicely allows for cooperativity effects. Chains of this kind need not adopt only planar arrangements, but can also be twisted into helices or closed into cycles. In 1,3,5-triethynylbenzene (Weiss *et al.* 1997a), there is a very complex system of C≡C—H···C≡C hydrogen bonds, which also contains a helix with roughly fourfold screw symmetry, shown schematically in **94**.

Four-membered cycles are found in the orthorhombic phase of acetylene

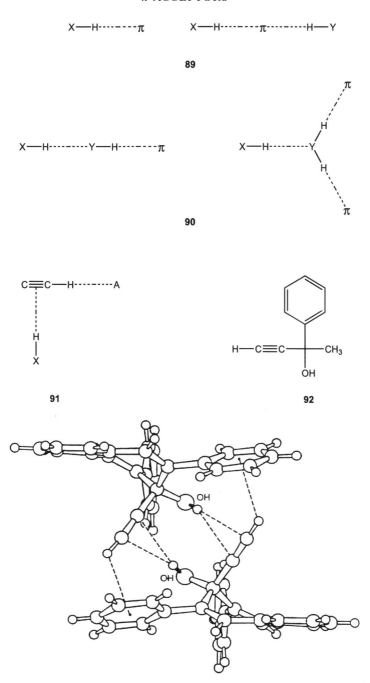

Fig. 3.35. Hydrogen bonded dimer of compound **73**. The geometry of the O—H···C≡C bonds is given in Table 3.12. The C≡C—H···Ph bonds have normalized distances $d(M)$ = 2.58 and 2.59 Å, respectively (from Steiner *et al.* 1996c).

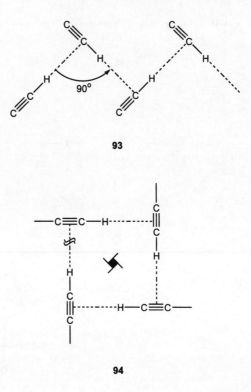

93

94

(Fig. 3.31), and in more distorted geometry in tetrakis-(4-ethynylphenyl)-methane (Galoppini and Gilardi 1999). This structure is the ethynyl equivalent of the corresponding tetrakis-(4-bromophenyl)-methane (Reddy *et al.* 1996) and demonstrates the topological and chemical relationship between ethynyl···ethynyl and Br···Br interactions. A six-membered puckered ring is found in ethynylferrocene, **95**, Fig. 3.36. All these chains, helices and rings are topologically equivalent to the chains, helices and rings that are frequently formed by $O-H\cdots O$ hydrogen bonding hydroxyl groups, and one may conclude that the synthons $O-H\cdots O-H$, $C\equiv C-H\cdots C\equiv C-H$ and $Br\cdots Br$ are supramolecular equivalents.

In **96**, the ethynyl groups form antiparallel zigzag chains of hydrogen bonds, Fig. 3.37(a) (Ahmed *et al.* 1972, repeated by Weiss *et al.* 1997*a*) and a very similar interaction pattern occurs in 1,4-diethynylcubane, **97** (Fig. 3.37(b)). The structure of pentynoic acid is revealing (Steiner *et al.* 1995*a*). The carboxylic acid groups form the expected dimers **98**, leading to a supramolecular species which carries two ethynyl groups that point in opposite directions, analogous to **96** and **97**. The crystal structure is likewise similar (Fig. 3.37(c)) and its relationship to the lower homologues butynoic acid and propynoic acid may be noted (Benghiat and Leiserowitz 1972; Leiserowitz 1976). These structures constitute good examples as to how different molecules can have very similar

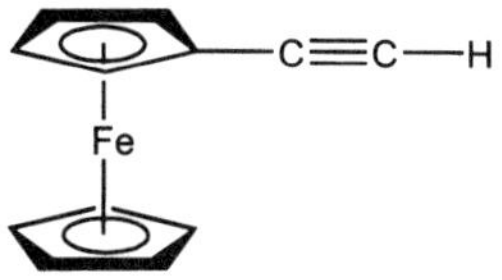

95

96

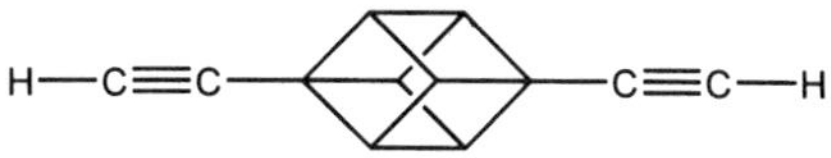

97

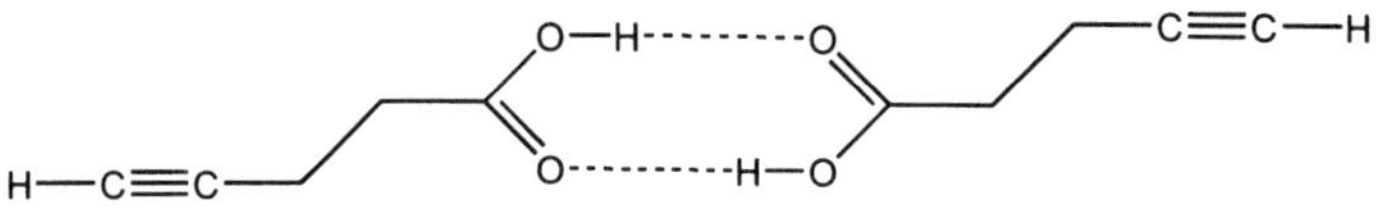

98

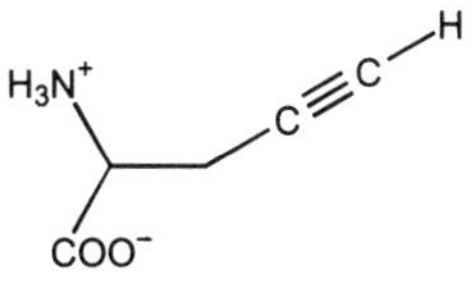

99

crystal structures provided there is an equivalence of at least some molecular and supramolecular synthons. The matter is taken up in greater detail in Chapter 4. In keeping with such trends, the reader may note that the same pattern topology is formed by $O-H\cdots O-H\cdots O-H$ hydrogen bonds in γ-hydroquinone and by $Cl\cdots Cl$ interactions in monoclinic 1,4-dichlorobenzene.

   Alkyne chains constitute an exceptionally robust hydrogen bond motif. They are formed not only with molecules that carry no other hydrogen bonding

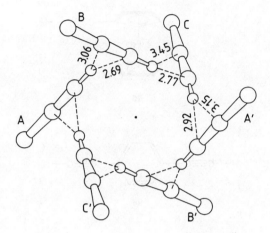

Fig. 3.36. Puckered six-membered ring of C≡C—H···C≡C hydrogen bonds in ethynylferrocene, **95**. For clarity, only the ethynyl groups and the carrier atom of the Cp ring are shown
(Steiner *et al.* 1996*d*).

groups, but also with those that carry strong donors and acceptors. In prop-2-
ynylglycine **99**, for example, the zigzag chain is nicely formed within a framework of much stronger ionic hydrogen bonds (Fig. 3.38).

Zigzag chains of alkynes can also be part of bifurcated hydrogen bonds.
Then, however, they are necessarily distorted from ideal geometry. An impressive example with O atoms as the second acceptors has been found by
Moloney and Foxman (1995) in the coordination polymer dimethyl(propynoato)thallium, Fig. 3.39. Bifurcated bonds with both acceptors being alkyne
groups, **100**, occur in the crystal structure of diyne **101** (Steiner *et al.* 1996*d*).

Terminal alkynes are sufficiently similar in size to hydroxyl groups to allow
the formation of mixed networks. Infinite chains with alternating hydroxyl and
ethynyl groups have been shown already in Section 2.2.7 in the context of
hydrogen bond cooperativity (Figs 2.32 to 2.34). Even more complex patterns
involving ethynyl groups, hydroxyl groups and water molecules have been
found in the hydrated dialkyne **102** (Subramanian *et al.* 1996), Fig. 3.40. In
these mixed networks, the different sizes and directionality characteristics of
the hydroxyl and the ethynyl groups seem to be levelled out by the general
flexibility of the hydrogen bonds. This is not trivial if one considers that for
hydroxyl groups, the preferred angle between donated and accepted hydrogen bonds is tetrahedral, whereas for the ethynyl group, it is rectangular.

Molecular fragments containing an ethynyl and a hydroxyl group in geminal
position often form dimers such as seen in Figs 2.32, 3.35 and 3.40(b). Such
dimers are found in two different geometries; in one, the fragments are related
by a centre of symmetry, and in the other by a twofold axis, Fig. 3.41. Such a
dimer also occurs also in **74**, where it is formed by a fragment that is chemically different from those in Fig. 3.41, **103**.

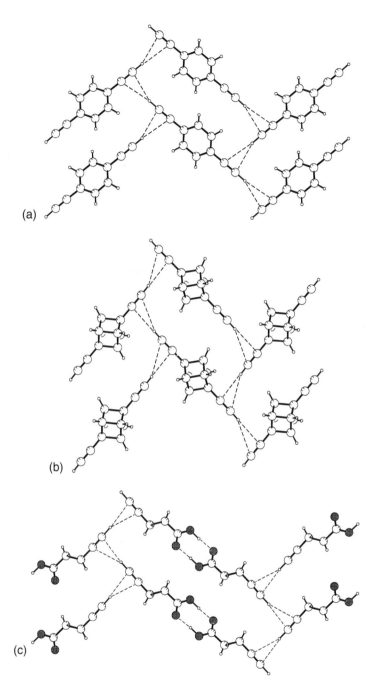

Fig. 3.37. Three structures with antiparallel chains of C≡C−H···C≡C hydrogen bonds. (a) Diethynylbenzene, **96**, $d(M) = 2.60$ Å (Ahmed *et al.* 1972). (b) 1,4-Diethynylcubane, **97**, $d(M) = 2.60$ Å (Eaton *et al.* 1994). (c) Pentynoic acid, **98**, $d(M) = 2.67$ Å (Steiner *et al.* 1995a). Distances are normalized. Notice the topological resemblance between these structures.

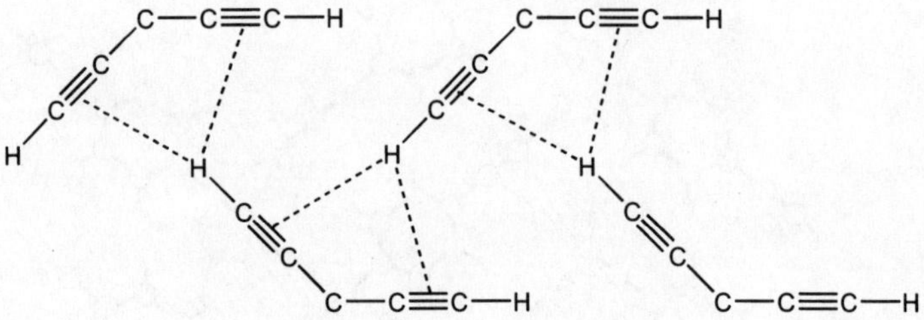

100

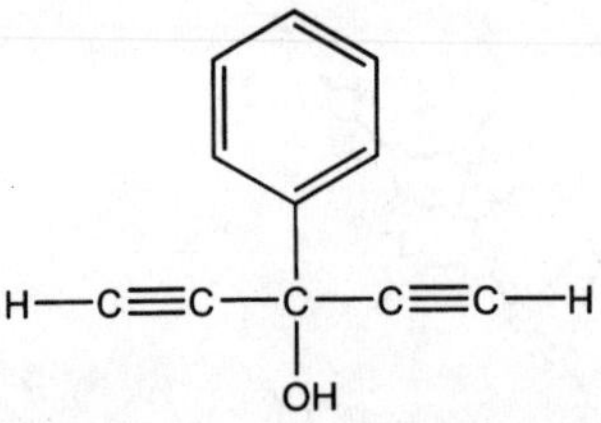

101

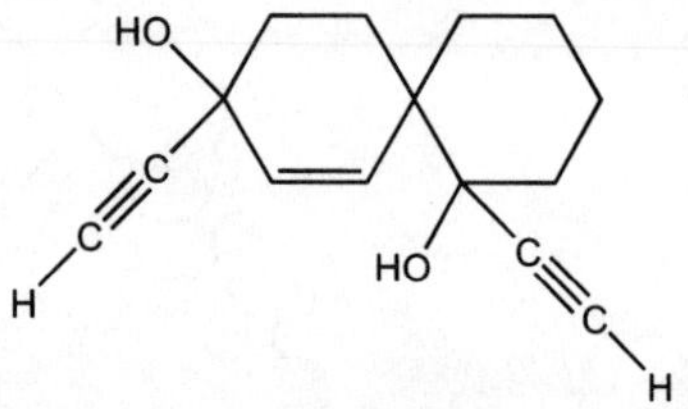

102

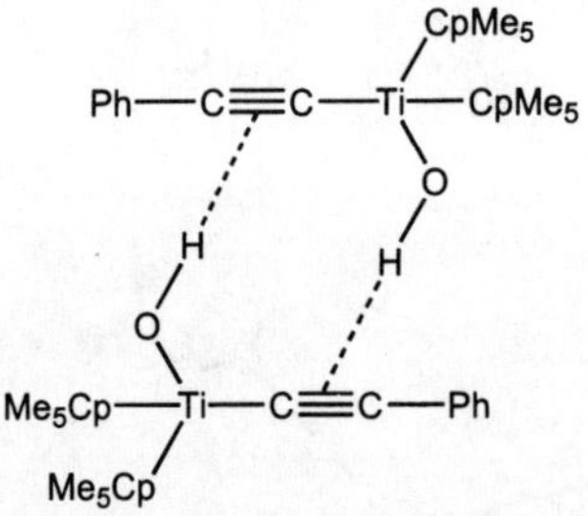

103

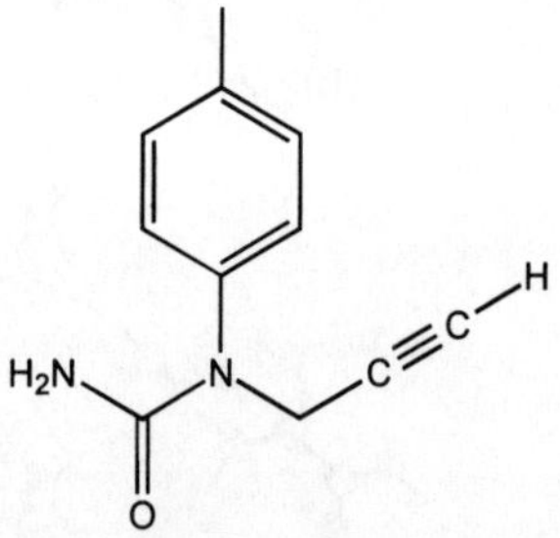

104

The ethynyl group can form mixed hydrogen bond networks with $N-H$ groups also. An example has been reported for the urea derivative **104**, Fig. 3.42 (Kumar *et al.* 1998*a*). In this crystal structure, a hydrogen bond pattern is formed which consists of $N-H \cdots O = C$, $C \equiv C-H \cdots O = C$ and $N-H \cdots C \equiv C-H$ interactions.

In mixed hydrogen bond networks involving ethynyl groups, the individual

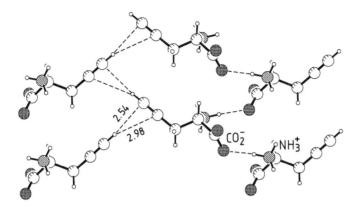

Fig. 3.38. Prop-2-ynylglycine, **99**. A structure where chains of $C\equiv C-H\cdots C\equiv C$ hydrogen bonds are formed within a framework of much stronger ionic hydrogen bonds (Steiner 1995c). Compare this with Fig. 3.37.

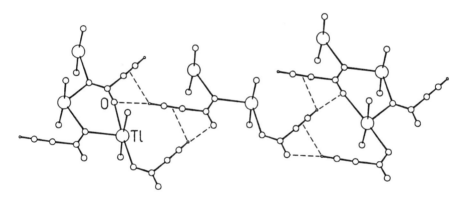

Fig. 3.39. Infinite chains of bifurcated $C\equiv C-H\cdots A$ hydrogen bonds with the acceptors $A = O$ and $C\equiv C$ in the coordintion polymer dimethyl(propynoato)thallium (Moloney and Foxman 1995).

constituents typically have similar or identical functions. It is then not possible to distinguish between primary and secondary interactions, or between the strong bond framework and the supporting or bystander weak bonds. In the ring structures shown in Fig. 3.40, none of the constituents can be removed without leading to collapse of the entire network. The classification into strong and weak is incomplete or even misleading in these cases. It may be noted, however, that these observations should not be extrapolated to $C-H\cdots A$ and $X-H\cdots\pi$ hydrogen bonds in general. They only show what weaker types of hydrogen bonds *can* do, not that such effects will be seen always. Most ethynyl groups in crystals do not act as hydrogen bond acceptors but rather form

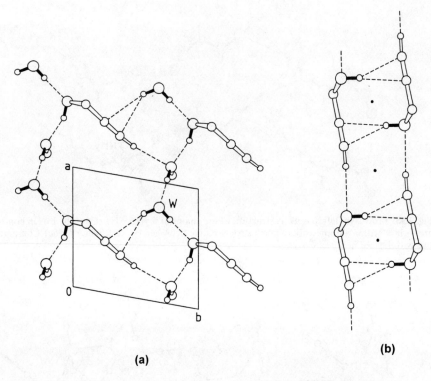

Fig. 3.40. Mixed hydrogen bond patterns in the hydrated dialkyne **102**. (a) Section showing a co-crystallized water molecule. (b) Section showing fused circular arrangements. (Subramanian *et al.* 1996).

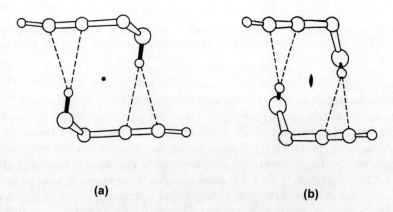

Fig. 3.41. The two kinds of O—H···C≡C hydrogen bonded dimers formed by the geminal hydroxyl-ethynyl fragment. (a) Symmetry relation by an inversion centre in **21**. (b) Relation by a twofold or pseudo-twofold axis as in **73**.

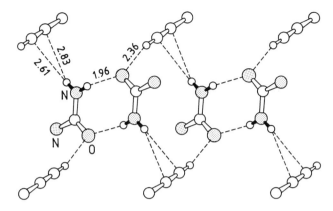

Fig. 3.42. Mixed hydrogen bond network in the urea derivative **104**. Geometries are normalized (from Kumar *et al.* 1998*a*).

105   **106**

relatively simple C≡C—H···O/N hydrogen bonds, several examples of which have been described in Chapter 2.

3.1.5 *Alkenes*

The hydrogen bond acceptor properties of the C=C group have been investigated far less intensively than those of the Ph and C≡C groups. In gas phase experiments (Fig. 3.3) and theoretical calculations (Tang *et al.* 1990; Rozas *et al.* 1997*a*), it was found that in the optimal geometry, the donor is placed on the axis normal to the >C=C< moiety, pointing at the midpoint of the C=C bond. Because the acceptor has two π faces, it can accept two hydrogen bonds, X—H···C=C···H—Y. From CSD analyses, Viswamitra *et al.* (1993) and Rzepa *et al.* (1994) found a number of relevant crystal structures. For intermolecular hydrogen bonds, the crystal structures of compounds **105** and **106** were alluded to. Further examples **108–110** related to the 17-ethynyl steroid

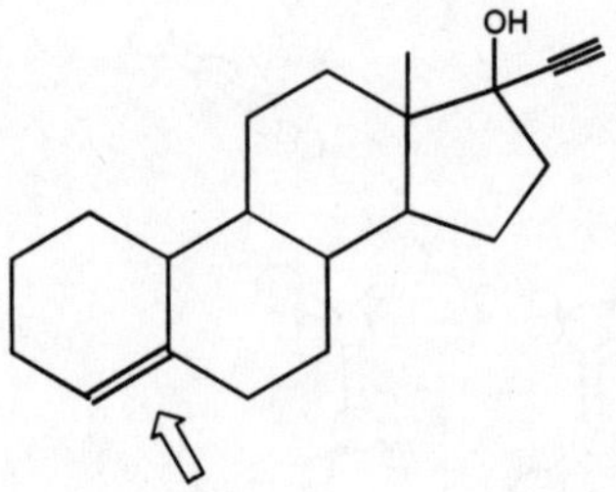

107

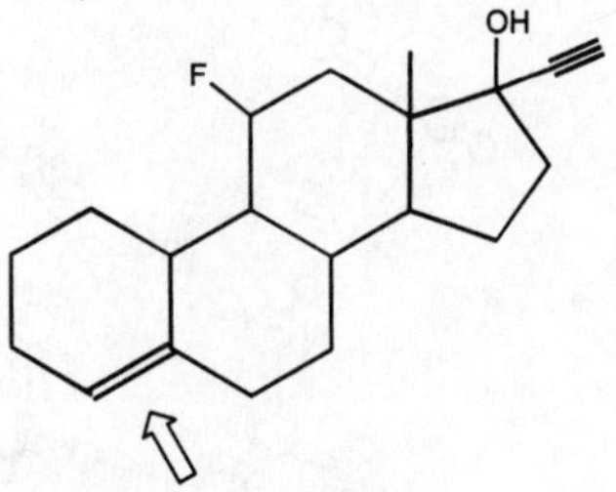

108

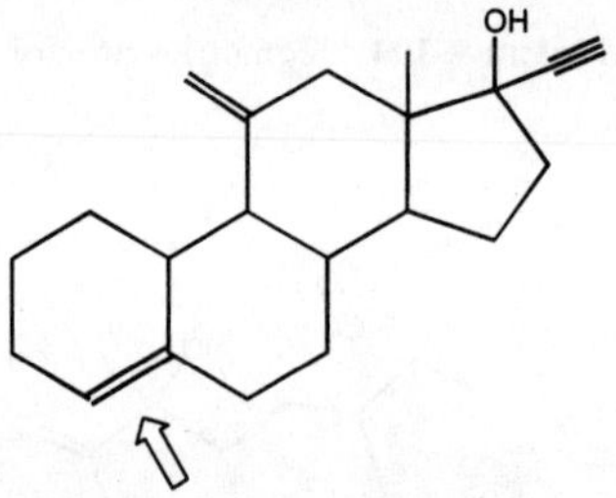

109

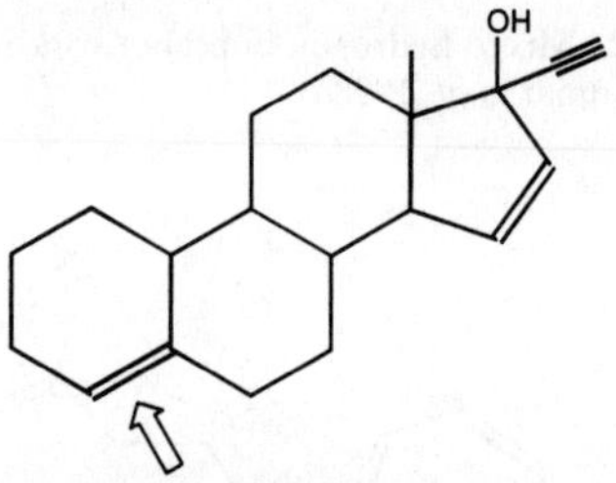

110

Table 3.15 Geometry of intermolecular $O-H \cdots C \equiv C$ hydrogen bonds

Compound	Ref.	d(C1) (Å)	d(C2) (Å)	d(M) (Å)	D(M) (Å)	θ(M) (°)
105	(*a*)	2.34	2.65	2.41	3.30	150
106	(*b*)	2.82	2.40	2.51	3.47	164
107	(*c*)	2.67	2.78	2.64	3.52	149
108	(*d*)	2.65	2.70	2.59	3.48	152
109	(*e*)	2.53	2.59	2.47	3.40	157
110	(*f*)	2.75	2.74	2.66	3.63	167

(*a*) Bowden *et al.* 1982. (*b*) Coleman *et al.* 1980. (*c*) Lynestrenol; Rohrer *et al.* 1976*b*. (*d*) Rohrer *et al.* 1978. (*e*) Van Geerestein 1987. (*f*) Van Geerestein and Leeflang 1988. Distances *d* normalized.

lynestrenol, **107**, were later compiled by Lutz *et al.* (1998). Geometrical hydrogen bond parameters for these examples are listed in Table 3.15, and two of the structures are shown in Figs 3.43 and 3.44.

The data in Table 3.15 do not provide a complete survey of $O-H \cdots C = C$ hydrogen bonds in crystals, but some characteristics of these interactions are revealed. The distances d(M) are in the range of about 2.4 to 2.7 Å, with the shortest being 2.34 Å. Four of the six examples have a symmetrical geometry

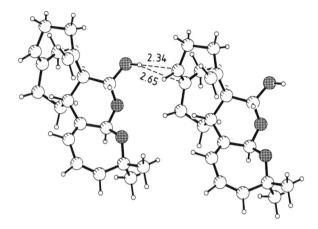

Fig. 3.43. The intermolecular O—H···C=C hydrogen bond in **105** with normalized geometry (Bowden *et al.* 1982; Rzepa *et al.* 1994).

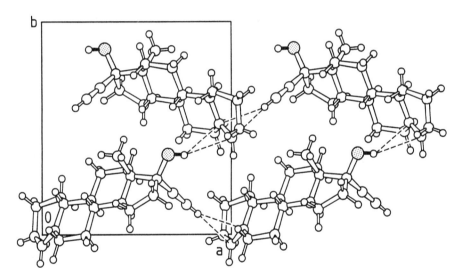

Fig. 3.44. Intermolecular O—H···C=C and C≡C—H···C=C hydrogen bonds in lynestrenol, **107**. No conventional hydrogen bonds are formed (from Lutz *et al.* 1998).

with $d(M)$ being shorter than the $d(C)$ values. O—H···C=C hydrogen bonds can be secondary components of bifurcated arrangements, and are then typically longer and more distorted than the examples given in Table 3.15.

C≡C—H···C=C hydrogen bonding has been described for lynestrenol, Fig. 3.44 (Lutz *et al.* 1998). Note that the C=C bond in the A-ring accepts two hydrogen bonds, one from an ethynyl and one from a hydroxyl group. Related

Table 3.16 Geometry of $C\equiv C-H\cdots C=C$ hydrogen bonds in ethynyl-steroids. The acceptor is in all cases a $C4=C5$ bond of the steroid A-ring (Lutz *et al.* 1998)

	Range	Mean
$d(C4)$ (Å)	2.62–3.18	2.88
$d(C5)$ (Å)	2.57–2.95	2.75
$d(M)$ (Å)	2.52–2.98	2.74
$D(M)$ (Å)	3.55–4.01	3.75
$\theta(M)$ (°)	>142	156

Distances d normalized.

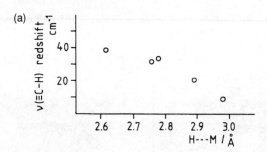

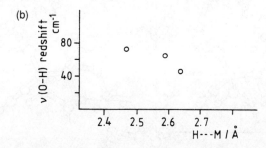

Fig. 3.45. (a) Correlation of the acetylenic ν_{CH} red shift with $d(M)$ in $C\equiv C-H\cdots C=C$ hydrogen bonds. (b) Correlation of the ν_{OH} red shift with $d(M)$ in $O-H\cdots C=C$ hydrogen bonds. Data for ethynyl steroids from Lutz *et al.* (1998).

interactions could be identified in nine other steroid structures with the geometries listed in Table 3.16. IR absorption spectra were measured for several of these compounds, and showed appreciable ν_{CH} red shifts up to $39\,\mathrm{cm^{-1}}$. Even a correlation with the distance $d(M)$ could be established, Fig. 3.45. This smooth correlation shows the long range nature of the $C\equiv C-H\cdots C=C$ hydrogen bond, and once more illustrates that the use of rigid distance cut-off criteria is faulty. A cut-off at $d \leq 2.6\,\mathrm{Å}$ would miss the

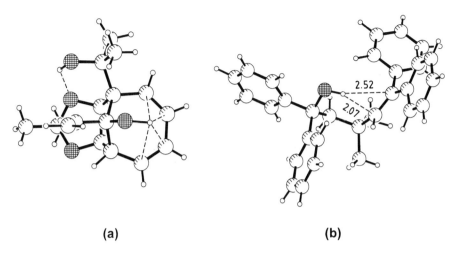

(a)   (b)

Fig. 3.46. Intramolecular O—H···C=C hydrogen bonds discussed by Rzepa *et al.* (1994). (a) Compound **111** (Parvez *et al.* 1987) two hydrogen bonds with intramolecular ring sizes $n = 5$ and normalized geometries of $d(C) = 2.46$ and 2.34, $d(M) = 2.31$, $D(M) = 2.86$ Å, $\theta(M) = 115°$, and of $d(C) = 2.36$ and 2.41, $d(M) = 2.31$, $D(M) = 2.86$ Å, $\theta(M) = 117°$. (b) Compound **112** (Zimmermann and Zuraw 1989) intramolecular ring size $n = 6$, $d(C) = 2.52$ and 2.07, $d(M) = 2.21$, $D(M) = 3.12$ Å, $\theta(M) = 154°$.

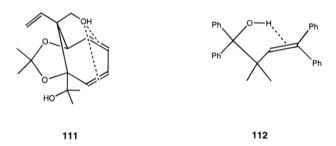

111   **112**

entire correlation shown and the experimental IR frequency shifts would remain completely mysterious. A cut-off at $d = 2.8$ Å is less catastrophic because it does at least recognize the phenomenon of C≡C—H···C=C hydrogen bonding. Still, it cuts right through the middle of the distribution. The related correlation for O—H···C=C hydrogen bonds in the same compounds contains only three data points, and is also shown in Fig. 3.45. Wave number shifts up to $72 \, \text{cm}^{-1}$ are found, indicating respectable hydrogen bonding.

In their survey, Rzepa *et al.* (1994) identified a number of intramolecular O—H···C=C hydrogen bonds. Two examples are shown in Fig. 3.46 with intramolecular ring sizes of $n = 5$ and 6 for **111** and **112**, respectively.

In principle, C=C groups should be able to participate in cooperative hydrogen bond arrays, if at least one of the substituents is a hydrogen atom: X—H···C=C—H···A. However, C=C—H is a relatively weak C—H donor so that the cooperativity effect is presumably only small. No experimental evidence for this effect has been provided.

Of all the crystal structures of molecules that contain both C=C bonds and O—H groups, the number that has O—H···C=C hydrogen bonds is very small. This means that their intrinsic chance of formation is low. From spectroscopic experiments, on the other hand, it is established that the C=C group is a good hydrogen bond acceptor. The reason for this discrepancy is presumably that C=C lacks certain properties of the more successful Ph and C≡C acceptors. One feature that makes Ph an important acceptor is its large size. Ph can accept hydrogen bonds from donors in a very large volume of space and, furthermore, the directionality requirements are very flexible. Therefore, it can easily and conveniently act as a second-choice acceptor if there are no other acceptors present. In crystal structures, C=C also acts as a second-choice acceptor, but it is a much smaller target and its π faces are accessible from only a restricted volume of space. Furthermore, it is often sterically shielded by substituents. The C≡C group is about the same size as C=C but because of its cylindrical symmetry, it can be approached from a much larger volume of space than the latter. Furthermore, and quite unlike the C=C group, it can also be involved in cooperative hydrogen bond patterns that make it quite attractive as a hydrogen bond partner. The C=C group, with its small size and lacking the capability to participate in strong cooperative arrays, is in effect the least important of the classical π-acceptors in crystal structures. Despite its limited importance though, it would be misleading to disregard the C=C acceptor completely. When formed, an X—H···C=C bond can play an important role in structure stabilization.

3.1.6 *Heterocycles*

As in carbocyclic aromatics, the π faces of heterocyclic compounds can accept hydrogen bonds. Solution IR spectroscopy reveals that hydrogen bonds directed to the π systems of pyridine and benzene are similar (Section 3.1.2.1). In fully corrected *ab initio* calculations on the dimer pyrrole–HF, Jiang and Tsai (1997) calculated a bond energy of –3.1 kcal/mol for the F—H···π bonded dimer, and a very flat potential energy surface of the interaction was obtained. This allows hydrogen bonding to the entire face of the ring, including the side of the N atom. According to these calculations, the global energy minimum is over the C3—C4 bond but the position over the N atom still has 65 per cent of the optimal bond energy.

In the crystal structure of pyrrole, Goddard *et al.* (1997) have actually observed an N—H···π hydrogen bond directed at the midpoint of C3—C4 with $D(M) = 3.30$ Å, Fig. 3.47(a). The N—H bond forms an angle of 70° with

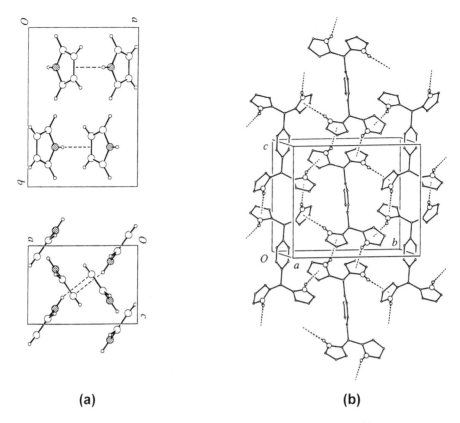

(a)   **(b)**

Fig. 3.47. Interconnected N—H···π hydrogen bonds between pyrrole rings. (a) Pyrrole at 103 K (Goddard *et al.* 1997). (b) Compound **113** (Bennis and Gallagher 1998).

the pyrrole ring. Because pyrrole can simultaneously act as a donor and an acceptor, the observed arrangement is suggestive of cooperativity. In 1,4-bis(di-2-pyrrolyl-methyl)benzene **113**, Bennis and Gallagher (1998) found chains of N—H···π hydrogen bonds between the pyrrole rings, Fig. 3.47(b). For the two symmetry-independent hydrogen bonds, the geometries are $D(M) = 3.22$ and $3.32\,\text{Å}$, and $\theta(M) = 154$ and $156°$, respectively.

Pyridine is an important heterocyclic π-acceptor, but no structural studies with a focus on X—H···π(pyridine) hydrogen bonds have been published. A search through the CSD, though, reveals two very interesting model structures, 4-aminopyridine and 2-amino-5-methylpyridine (Fig. 3.48). In the former, N—H···N hydrogen bonded chains are formed, while the second amino H atom is involved in N—H···π bonds which link the chains. In the latter, molecules form N—H···N dimers and the second amino H atom is again is involved in an N—H···π hydrogen bond. In both cases, the presence of N—H···π bonding can be rationalized by a mismatch of conventional donor

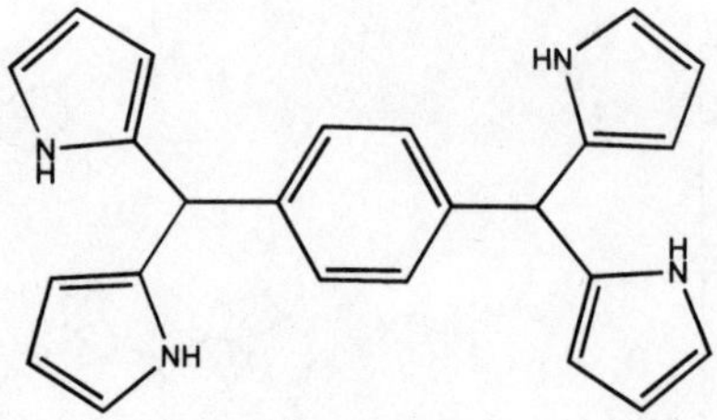

113

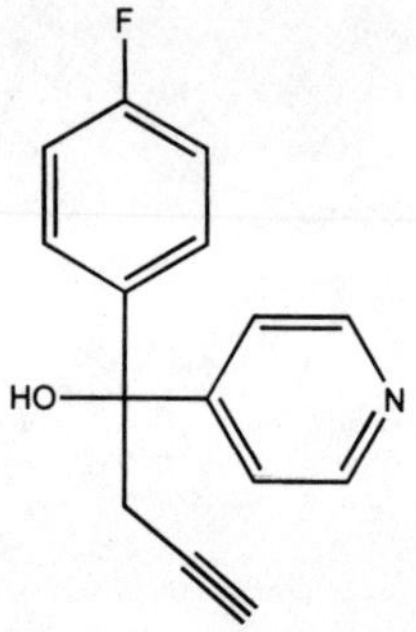

114

and acceptor sites (see also Section 4.3.1 for the related crystal structures of the three isomeric aminophenols).

For the imidazole acceptor, a possible N—H···π hydrogen bond with a histidine side-chain as the acceptor has been identified by Gilliland and co-workers in the crystal structure of human carboxyhemoglobin (Vásquez *et al.* 1998). C—H···π hydrogen bonding with heterocyclic acceptors has also been reported. In a crystal structure containing co-crystallized furan molecules, Hunter *et al.* (1994) observed purportedly structure-stabilizing C—H···π(furan)···H—C interactions. Because the C—H donors are from a methyl and a phenyl group, this interaction is presumably in the borderline region of the hydrogen bond phenomenon.

In some instances, hydrogen bonds to heterocycles are directed neither at the heteroatom lone pair, nor at the π face, but rather in between the two. An example is found in the terminal alkyne **114**, where a pyridyl residue accepts an O—H···N hydrogen bond directed at the N atom lone-pair, and an additional C≡C—H···N interaction which is almost perpendicular to the ring plane (Fig. 3.49). It is expected that the latter interaction contains a significant contribution to the pyridyl π electron system. Numerous other heterocycles

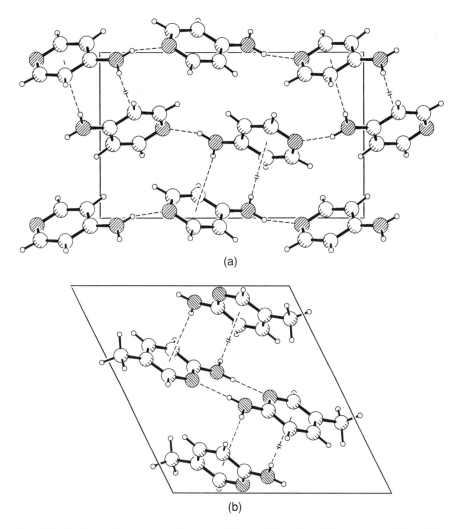

(a)

(b)

Fig. 3.48. N−H···π hydrogen bonds accepted by pyridine rings. (a) 4-Aminopyridine (Chao and Schempp 1977) $d(M) = 2.58$ $D(M) = 3.46$ Å, $\theta(M) = 145°$. (b) 2-Amino-5-methylpyridine (Nahringbauer and Kvick 1977) $d(M) = 2.55$, $D(M) = 3.41$ Å, $\theta(M) = 144°$. Geometries are normalized.

should be capable of accepting π hydrogen bonds, but have not yet been discussed in this context. As the general awareness of these interaction increases, one can surely expect to see numerous findings.

3.1.7 Other π-acceptors

In the previous sections, we have shown that a large number of π-bonded moieties can act as hydrogen bond acceptors. The main requirement seems to be

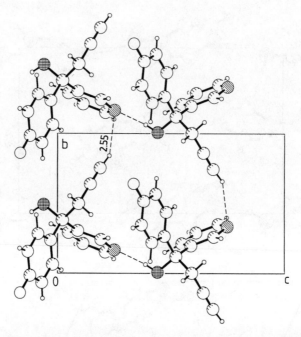

Fig. 3.49. The C≡C—H···N hydrogen bond in **114** approaches the pyridyl group almost per-pendicular to the ring plane, and would appear to have π hydrogen bond character (from Kumar *et al.* 1998*b*, based on Albinati *et al.* 1985).

that filled π bond orbitals are sterically available for the setting-up of inter-molecular interactions. This requirement is fulfilled by groups other than those discussed above. In this section, we present some π hydrogen bond acceptors which have been mentioned only fleetingly in the literature, but where at least some experimental data is available. Presumably it is only a question of time before the list of known π hydrogen bond types grows further.

3.1.7.1 *Cyclopropane and related compounds*

Cyclopropane rings have long since been recognized by IR spectroscopy as interesting hydrogen bond acceptors (Section 3.1.2). In this moiety, the formal carbon–carbon single bonds have an appreciable π bond character, and behave in many chemical aspects like C=C. The electron density in the bonds is not linear but banana-shaped and this shape can even be seen in experimental charge density studies in crystals (Nijveldt and Vos 1988*a,b*). Because of these singular properties, the cyclopropane ring is sometimes called 'pseudo-π-bonded'.

Two principal geometries can be expected for hydrogen bonds to cyclo-propane, edge-on (the donor is roughly in the ring plane and points at a C—C bond), and face-on. In the gas phase dimer cyclopropane–HCl, the edge-on

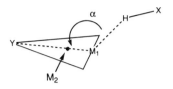

115

Table 3.17 O/N—H···π hydrogen bonds with cyclopropane, aziridine and oxirane acceptors; given are the number of examples found in the CSD, and the mean geometries (Allen *et al.* 1996*a*)

	Edge-on	Intermediate	Face-on
Donor type			
O—H	18	3	10
N—H	5	2	3
Acceptor type			
Cyclopropane	16	3	11
Aziridine	2	1	1
Oxirane	5	1	1
Inter/intramolecular			
Intermolecular	14	2	4
Intramolecular	9	3	9
Geometry			
d(M) (Å)	2.81(4) to M_1	2.82(5) to M_1	2.68(4) to M_2
H···M_1—Y (°)	158(3)	114(3)	111(3)

Distances d normalized, cut-off d(C) < 3.0 Å.

hydrogen bond is formed (Fig. 3.3). Also IR spectroscopic experiments (Joris *et al.* 1968) and *ab initio* calculation (Rozas *et al.* 1997*a*) favour this geometry. In a database survey on intermolecular interactions of the three-membered rings cyclopropane, aziridine and oxirane, Allen *et al.* (1996*a*) analysed the C—H···X donor properties, and also the hydrogen bond acceptor characteristics of these compounds. To distinguish edge-on and face-on geometries, they introduced the angle α as defined in **115**. For edge-on contacts, α is close to 180° and if the donor H atom is placed above the face centre M_2, α is around 80°. In the analysis of O/N—H···α contacts, a bimodal distribution is found for α, with the maxima corresponding to edge-on and face-on geometries. Intermediate cases also occur. The frequencies and mean distances of these hydrogen bonds are given in Table 3.17. Most of the O/N—H···π(cyclopropane) hydrogen bonds are secondary components of bifurcated arrangements and presumably play only a minor role in structure stabilization. Some

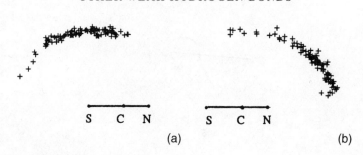

Fig. 3.50. H atom approach to thiocyanate anions (O—H and N—H donors), (a) around the S-acceptor, (b) around the N-acceptor. The interface between these regions represents the π-acceptor region (from Tchertanov and Pascard 1996).

of the intramolecular interactions represent ring sizes $n = 4$ and are very distorted from ideal geometries.

3.1.7.2   Thiocyanate anions

The thiocyanate anion is a strong hydrogen bond acceptor. The chemical constitution is often given as $^{-}S—C\equiv N$, but the negative charge can be imagined as being localized at either of the terminal atoms, or delocalized over the whole ion, $(SCN)^{-}$. The angular preferences of the hydrogen bonding to this ion have been determined by Tchertanov and Pascard (1996) in a CSD analysis as shown in Fig. 3.50. The preferred approach to the S atom is roughly perpendicular to the S—C bond with an average X···S—C angle of 99°, and the preferred approach to the N atom is in the direction of the ion axis with a mean X···N—C angle of 145°. A donor can approach the ion also in the direction of the π electrons, and simultaneous short contacts to S and C, or to N and C, are not uncommon. The distance to C may be even shorter than to S or N. There is a gradual transition from hydrogen bonding with the N atom lone pair to interaction with the π electrons, and many intermediate cases are observed. A number of crystal structures are known where the ion accepts several hydrogen bonds to S, N, and also to π, such as is shown in Fig. 3.51. There are also examples of C—H···π hydrogen bonding to thiocyanate ions.

Lommerse and Cole (1998) performed quantum-chemical calculations on the hydrogen bonding to thiocyanate ions. For the model donor MeOH, they obtain the total energy contour plot shown in Fig. 3.52. There are energy minima of −8.9 and −12.8 kcal/mol for approach to S and N, respectively, at optimal geometries of $d(S) = 2.5$ Å, H···S—C = 105°, and $d(N) = 1.9$ Å, H···N—C = 180°, respectively. In the direction of the π electron system, there is no minimum but a saddle point with an energy of −6.6 kcal/mol, 2.5 Å away from the C atom. The thiocyanate ion can accept hydrogen bonds in effect from *all* directions. The theoretical partitioning of the bond energies is given

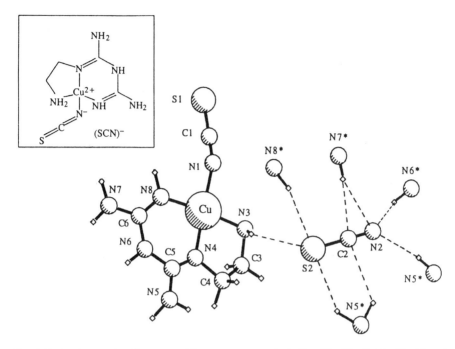

Fig. 3.51. Hydrogen bonding to a thiocyanate anion, including N—H···S, N—H···N and N—H···π hydrogen bonds (from Tchertanov and Pascard 1996, based on Andreetti *et al.* 1971).

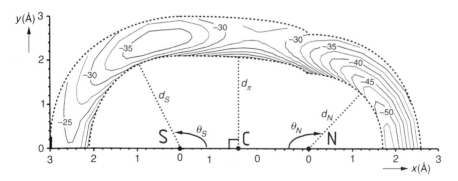

Fig. 3.52. Potential energy contour plot for hydrogen bonds of MeOH to (SCN)⁻ as obtained in *ab initio* calculations. The contour levels are given in kJ/mol (Lommerse and Cole 1998).

in Table 3.18 for optimal geometries to the three acceptor types (S, N and π). The thiocyanate ion is a very strong π-acceptor because of the direct charge support; the O—H···π bond energy is much larger than for uncharged π-acceptors, and is about half as large as that for strong conventional O—H···N hydrogen bonds directed to the N atom of the same ion.

Table 3.18 Energy contributions to the total energy in the thiocyanate–methanol system at optimal geometries. IMPD calculations by Lommerse and Cole (1998)

	E_{es} (kJ/mol)	E_{er} (kJ/mol)	E_{pol} (kJ/mol)	E_{ct} (kJ/mol)	E_{disp} (kJ/mol)	E_{total} (kJ/mol)
N-acceptor	−64.0	+40.9	−11.4	−8.9	−10.1	−53.5
S-acceptor	−43.5	+22.4	−6.0	−5.6	−4.7	−37.3
π-acceptor	−24.0	+7.4	−4.3	−2.5	−4.3	−27.7

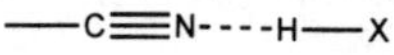

116 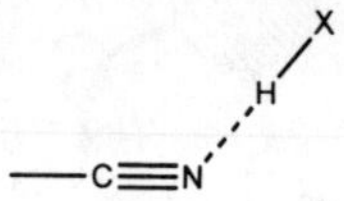   117 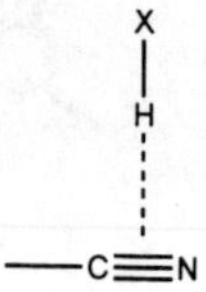   118

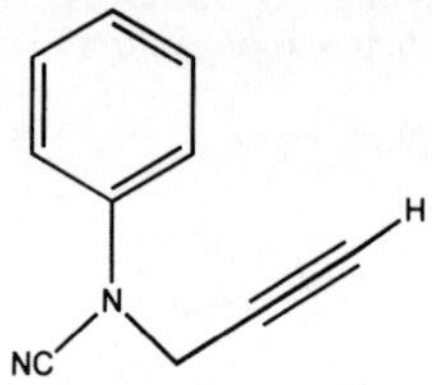

119 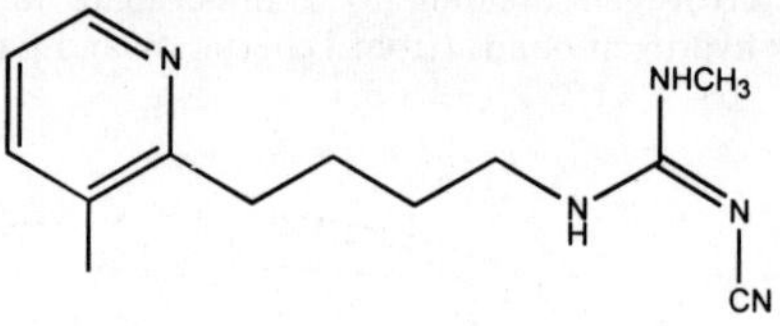   120

3.1.7.3 *Cyano groups*

For cyano groups, one should expect that hydrogen bonding to the π electron system would be possible. One should also expect that there is a gradual transition between the X—H···N and the X—H···π bonding modes, **116** and **118**. However, there is no charge support for the X—H···π bonding. Therefore the relative energy difference between the X—H···N and X—H···π modes should be much larger than in thiocyanate ions, favouring the conventional bonding to the nitrogen electron lone pair.

Despite the abundance of cyano groups in organic crystal structures, no reliable example of an O/N—H···π(cyano) hydrogen bond has been reported. However, interaction geometries of the intermediate case **117** are observed. An example with an alkynyl donor has been found by Kumar *et al.* (1998*b*) in

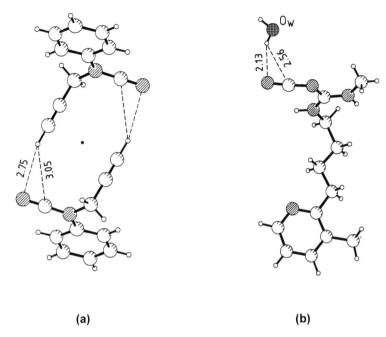

(a)   **(b)**

Fig. 3.53. (a) C≡C—H···N≡C hydrogen bonds in **119**. The geometry is intermediate between 'normal' C—H···N and C—H···π(N≡C) hydrogen bonding (from Kumar *et al.* 1998*b*). (b) O—H···N≡C hydrogen bond in **120** (Prout *et al.* 1993) with a similar geometry as in (a); $d(N)$ = 2.13, $d(C)$ = 2.56 Å, $\theta(C)$ = 154°, angle C—N···H = 97°, angle C—N···O = 89°. Geometries are normalized.

the crystal structure of alkyne **119**, Fig. 3.53(a). With an H···N≡C angle of 93°, the hydrogen bond geometry is closer to **118** than to **116**. In the IR absorption spectrum, the alkynyl ν_{CH} band is centred at 3280 cm^{-1}, indicating a relatively weak but noticeable hydrogen bond (compare Fig. 2.8 in Section 2.2.1). A related hydrogen bond of a water donor may be identified in the histamine antagonist **120**, Fig. 3.53(b).

If cyano groups are coordinated to metal atoms, M—N≡C—R, the nitrogen electron lone pair is no longer available for hydrogen bonding but the π electron system is still accessible for intermolecular interactions. Because the metal atom donates electrons into the cyano group, better acceptor potentials than for organic cyano groups may be expected.

3.1.7.4 *Cyclopentadienyl ligands*

Infra-red spectroscopy has shown that Cp rings can accept hydrogen bonds (Yoshida and Osawa 1966) and some structural data has been tabulated by Braga *et al.* (1998*b*). Glidewell *et al.* (1996*a*) discussed intramolecular O—H···π(Cp) hydrogen bonding in a series of α-ferrocenyl alcohols of the

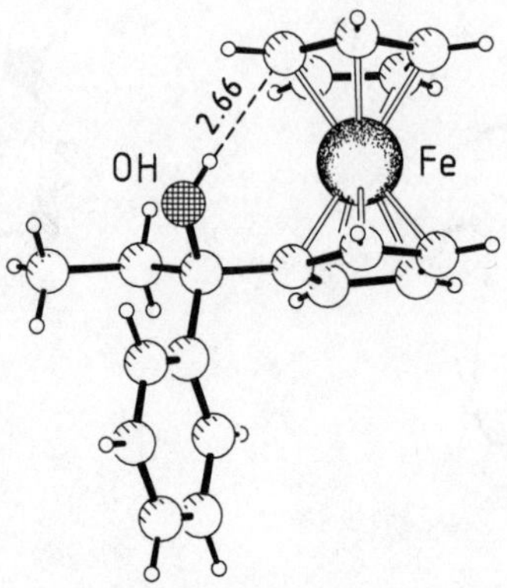

Fig. 3.54. Intramolecular $O-H \cdots \pi(Cp)$ hydrogen bond in compound **121** with $R_1 = Et$, $R_2 = Ph$. Normalized geometry: $d(C) = 2.67$, $D(C) = 3.55$ Å, $\theta(C) = 149°$ (Glidewell *et al.* 1996*a*).

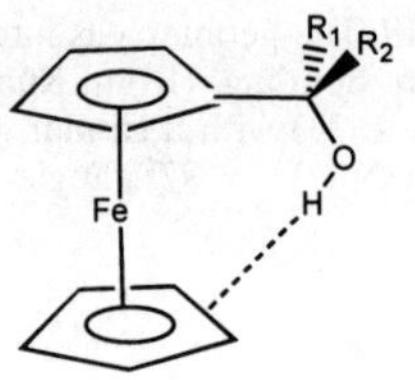

121 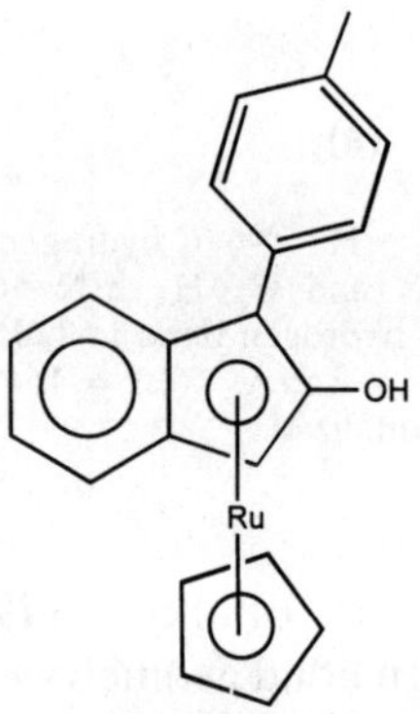   **122**

type **121**. In these compounds, a systematic intramolecular hydrogen bond is donated by the hydroxyl group to a carbon atom of a Cp ring, as shown for a typical example in Fig. 3.54. Intermolecular hydrogen bonding to Cp ligands has not been discussed explicitly, but the CSD contains an interesting example for **122** which is shown in Fig. 3.55.

Glidewell *et al.* (1996*b*) have invoked intramolecular $C-H \cdots \pi$ interactions between Cp ligands, i.e. $C(Cp)-H \cdots \pi(Cp)$. This interaction might be placed at the lower end of hydrogen bonding. It might also represent the $Cp \cdots Cp$ analogue of the $Ph \cdots Ph$ herringbone interaction.

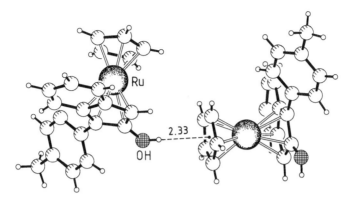

Fig. 3.55. Intermolecular O—H···π(Cp) hydrogen bond in compound **122** with normalized geometry $d(M) = 2.33$, $D(M) = 3.26$ Å, $\theta(M) = 159°$ (Yang *et al.* 1994).

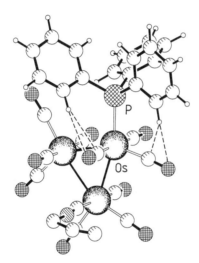

Fig. 3.56. Organometallic compound exhibiting intramolecular C—H···π(OC—M) hydrogen bonding with normalized geometry $d(\pi) = 2.27, 2.42$ Å, $\theta(\pi) = 165, 156°$ (Lu *et al.* 1992; Brammer 1999).

3.1.7.5 *Carbonyl ligands*

Based on structure geometries retrieved from the CSD, Brammer (1999) suggested that the $\pi(C\equiv O)$ bonds of terminal metal carbonyls can serve as hydrogen bond acceptors. X—H···π(OC—M) hydrogen bonds are found to be uncommon for X = N and O, but relatively abundant for X = C. There are even examples where C—H···π(OC—M) hydrogen bonds appear to help in defining the conformation of organometallic complexes. An example is shown in Fig. 3.56, where the donors are the *ortho*-C—H groups of a triphenylphosphine ligand.

3.2 Weak atomic acceptors

In this section, we discuss hydrogen bonds to atomic acceptors other than O
and N. We begin with covalently bonded halogen atoms, proceed to group VI
elements other than O, then to group V elements other than N, and finally to
group IV elements. Some of these acceptors are not weak, but certainly non-
conventional in the sense defined in Chapter 1. The conventional S-acceptor
is a borderline case between strong and weak, and is included for reasons of
completeness. Hydrogen bonds to transition metal atoms are regarded as a
separate case and are discussed in Section 3.5.

3.2.1 *Group VII elements – covalent halogen*

This section is concerned with covalently bonded halogen atoms as hydrogen
bond acceptors. We consider the cases $Hal-X$ where $X = C$, H or a transition
metal atom. The bihalide ions $[Hal-H-Hal]^-$ and inorganic anions like
BF_4^-, PF_6^- and $ZnCl_4^{2-}$ are conventional and strong acceptors, and unless taken
in combination with $C-H$ donors, are not within the scope of this book. We
shall consider halides in Section 3.3 and there too only in combination with
the $C-H$ donor.

In view of the high electronegativity of the halogens, in particular that of F,
one might expect that covalent $C-Hal$ bonds should be strongly polar and
that organic halogen atoms would be excellent hydrogen bond acceptors. In
reality, however, the hydrogen bonding propensities of organic halogen is
poor, and the the reasons for this are far from satisfactorily explored or
explained. $O-H \cdots Hal-C$ hydrogen bonding was discovered in 1936 by Wulf
et al. in spectroscopic studies on *ortho*-halogenophenols but the IR frequency
shifts are small (Baker 1958; Baker and Shulgin 1958). For intermolecular
$O-H \cdots Hal-C$ hydrogen bonds too, infra-red spectroscopy detects only
small shifts of ν_{OH}, about a factor of five smaller than for oxygen bases (West
et al. 1962). Because of this weakness, hydrogen bonds to covalently bonded
halogen atoms have attracted little attention. Only in recent years have
more strenuous efforts been made to investigate these interactions in greater
detail.

3.2.1.1 *Fluorine*

$O-H \cdots F-C$ and $N-H \cdots F-C$ hydrogen bonds.   Hydrogen bonds from
the strong donors $O-H$ and $N-H$ to organic F occur in crystal structures
only rarely. Nevertheless, this phenomenon is of general interest and has been
the subject of several database analyses (Murray-Rust *et al.* 1983; Shimoni and
Glusker 1994; Howard *et al.* 1996; Dunitz and Taylor 1997). In their CSD analysis, Dunitz and Taylor (1997) retrieved 5947 $F-C$ bonds occurring in crystal
structures with at least one potential $O-H$ or $N-H$ donor. Of these $F-C$

groups, only 37 are involved in possible $O/N-H\cdots F-C$ hydrogen bonds with distances $d < 2.3$ Å, i.e. 0.6 per cent of the potential $F-C$ acceptors. Even of the 37 possible hydrogen bonds, only a fraction can be regarded as unambiguous. The very small percentage of hydrogen bonding organic F atoms leaves no doubt that $F-C$ is a weak hydrogen bond acceptor indeed. The authors describe a number of published crystal structures with $O/N-H\cdots F-C$ hydrogen bonds, a selection of which is given in Fig. 3.57. In 2-fluoro-1,1,2-triphenylethanol, molecular dimers are formed which are linked across an inversion centre by a pair of $O-H\cdots F$ hydrogen bonds with $d = 2.02$ Å, Fig. 3.57(a). It is of interest to note that in the related molecule 1,1,2-triphenylethanol, which differs only in that the F atom is replaced by H, a pair of $O-H\cdots Ph$ hydrogen bonds is formed (Section 3.1.3, Fig. 3.28(b)). In calcium 2-fluorobenzoate, Fig. 3.57(b), the Ca-coordinated water molecule

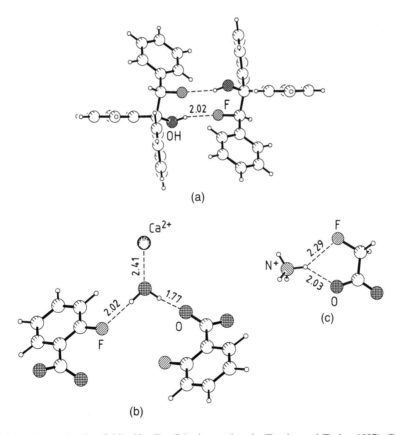

Fig. 3.57. Intermolecular $O/N-H\cdots F-C$ hydrogen bonds (Dunitz and Taylor 1997). Geometries are normalized. (a) 2-Fluoro-1,1,2-triphenylethanol, $d = 2.02$ Å, $\theta = 152°$ (DesMarteau *et al.* 1992). (b) Calcium 2-fluorobenzoate dihydrate, $d = 2.02$, $D = 2.99$ Å, $\theta = 170°$ (Karipides and Miller 1984). (c) Ammonium monofluoroacetate, $d = 2.29$ Å, $\theta = 141°$ (Wei and Ward 1976).

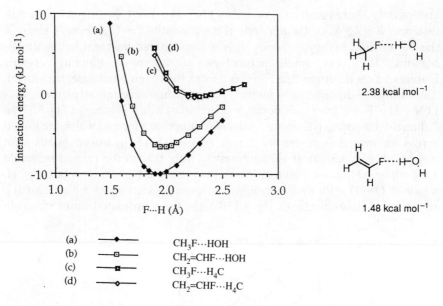

Fig. 3.58. *Ab initio* calculations on C—H···F hydrogen bonds. Potential energy curves are calculated for the donors H_2O and CH_4 (from Howard *et al.* 1996).

donates one hydrogen bond to a carboxylate O atom, and one to an F atom. Because the F atom is part of an anion, it is presumably electron-rich and the conditions for hydrogen bonding are favourable (Karipides and Miller 1984). And finally, a bifurcated hydrogen bond is formed in ammonium monofluoroacetate, with one component donated to a carboxylate O atom, and one component to the F atom, Fig. 3.57(c); bifurcated hydrogen bonds of this kind occur frequently with fluorocarboxylate anions (Murray-Rust *et al.* 1983).

In a different database analysis, Howard *et al.* (1996) considered only monofluorinated functional groups, i.e. they disregarded CF_2 and CF_3 systems which are expected to introduce statistical bias. Using the distance cut-off $d <$ 2.35 Å, they found that 40 out of 1163 organic F atoms possibly accept a hydrogen bond from O—H or N—H (=3.4 per cent). This is a larger fraction than that reported by Dunitz and Taylor, but it includes 25 intramolecular interactions many of which have very unfavourable geometries. Poor directionality characteristics are found for O/N—H···F hydrogen bonds in general. In the statistical data, hydrogen bonds to $C(sp^3)$—F are somewhat more frequent than to $C(sp^2)$—F, and in *ab initio* calculations, lower energies are found for fluoromethane than for fluoroethene acceptors, Fig. 3.58. It appears from these observations that hydrogen bonds to $C(sp^3)$—F are stronger than those to $C(sp^2)$—F. The shortest H···F contact in the data set of Howard *et al.*, however, is intramolecular and directed to a $C(sp^2)$—F group in molecule **123**, Fig. 3.59(a). These authors have also identified a short intramolecular

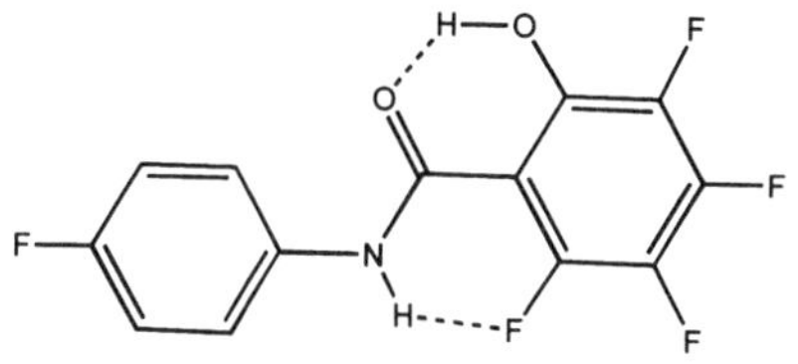

123

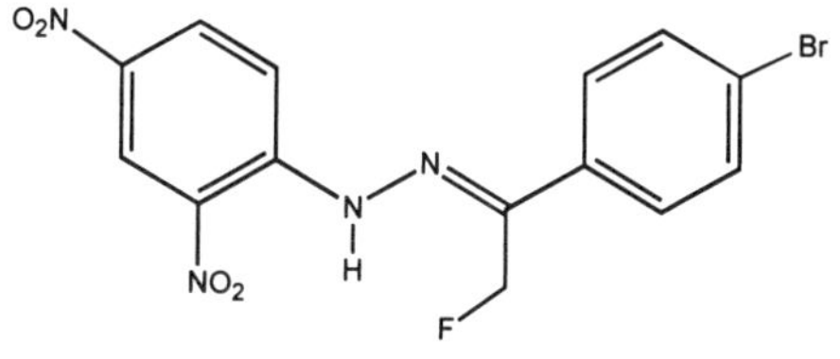

124

N—H···F hydrogen bond in their study of hydrazone **124** (Fig. 3.59(b)). Here, the acceptor is of the type C(sp^3)—F (Borwick *et al.* 1997).

The database analyses mentioned above can neither provide tools that allow one to predict under which circumstances O/N—H···F—C hydrogen bonds are preferably formed, nor do they help to control these interactions in practical structural chemistry. They only permit one to conclude that these interactions are very rare. The amount of available structural data on O/N—H···F hydrogen bonds is so small that statistical analysis of geometries can hardly be performed. For the question as to why F is such a poor hydrogen bond acceptor despite its high electronegativity, Dunitz and Taylor (1997) mention a few reasons. In particular, they attribute the poor acceptor potential to the combination of two factors: one is the low proton affinity and hardness of F (low basicity, tightness of the electron shell), and the inability to modify this by intramolecular electron delocalization or intermolecular cooperative effects.

C—H···F—C hydrogen bonds. According to the CSD analysis of Howard *et al.* (1996), short C—H···F—C contacts occur somewhat more frequently than O/N—H···F—C hydrogen bonds, but are still relatively rare. For the 1163 F—C groups analysed, only 57 accept intermolecular C—H···F—C contacts with $d < 2.35$ Å. In their *ab initio* calculations, the authors obtain binding energies that are so low for the donor CH_4 (−0.2 kcal/mol, Fig. 3.58) that they do not consider this kind of contact as being a hydrogen bond at all. With more activated C—H donors, however, binding energies are better (see below).

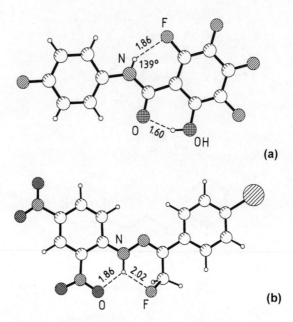

Fig. 3.59. Intramolecular $N-H\cdots F-C$ hydrogen bonds. (a) Compound **123**, $d = 1.86$, $D = 2.70$ Å, $\theta = 139°$ (discussed by Howard *et al.* 1996 in the structure of Banks *et al.* 1995). Note also the resonance-assisted $O-H\cdots O$ hydrogen bond. (b) Compound **124**, $d = 2.02$, $D = 2.73$ Å, $\theta = 125°$ (Borwick *et al.* 1997). Geometries are normalized.

Shimoni and Glusker (1994) have stated that the poor competition of the $C-F$ group with O and N atom acceptors extends to $C-H$ donors. They have suggested accordingly that in the evaluation of the acceptor capabilities of the $C-F$ group, $C-H\cdots F-C$ geometries in compounds containing *only* C, H and F atoms are better candidates. This indeed was the approach used by one of us in a study of a number of simple substituted fluorobenzenes, **125** (Thalladi *et al.* 1998*a*). All these compounds are liquid at room temperature and single crystals for X-ray diffraction were grown *in situ*. The analysis of the $C-H\cdots F$ interactions that are found in the crystal structures takes the form of comparisons with related $C-H\cdots O/N$ situations. Fluorobenzene, for instance, bears a close relationship to the pyridine–HF adduct (see also below), pyridine-1-oxide and benzonitrile at the level of individual interactions, showing that the characters of the structure-determining intermolecular interactions in these four structures are the same (Fig. 3.60). Similarly, 1,4-difluorobenzene and 1,4-benzoquinone are related, the $C-H\cdots F$ interactions in the former playing the same structural role as the $C-H\cdots O$ interactions in the latter (Fig. 3.61). With an increase of the F content of the molecules, the $C-H$ acidity increases and the $C-H\cdots F$ interactions in 1,3,5-trifluorobenzene and 1,2,4,5-tetrafluorobenzene become shorter and more important. These compounds are isostructural with 1,3,5-triazine and 1,2,4,5-

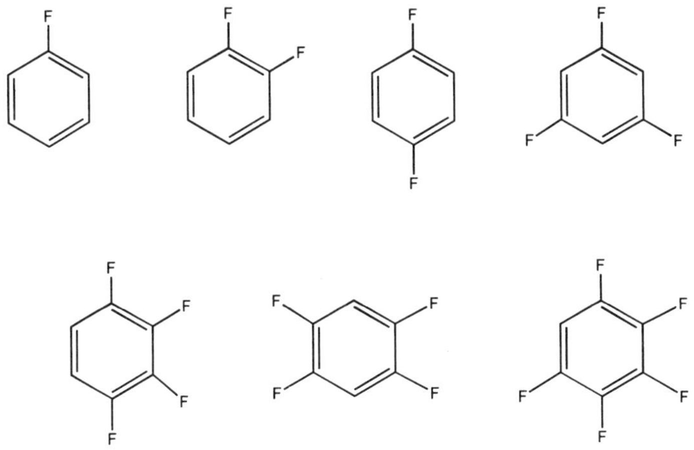

125

tetrazine, structures that are heavily dominated by C—H···N hydrogen bonds. This isostructurality only strengthens the argument that C—H···F interactions resemble C—H···N interactions, and provides more evidence for their description as weak hydrogen bonds.

It is pertinent to compare the C—H···F interactions in these crystal structures with all the others in the CSD that contain only C, H and F atoms. Figure 3.62 shows scatterplots of H···F distances d against the C—H···F angles θ (H atom positions normalized). Interactions between all types of C—H and C—F groups are shown in Fig. 3.62(a), while Fig. 3.62(b) includes only interactions between $C(sp^2)$—H and $C(sp^2)$—F groups. There is no real difference between these scatterplots. Generally, some kind of inverse correlation between d and θ may be seen, but there are many points in the top right hand corner of the plots that simply add to the crystallographic noise. Some of these correspond to bifurcated interactions but no specific conclusion may be drawn on this or on any other basis. Figure 3.63 in contrast, which is the corresponding d–θ scatterplot for the simple substituted fluorobenzenes, shows a strong negative correlation that is very characteristic of hydrogen bonding. The top right hand corner is now completely empty, suggesting that when a C—H···F contact is present in these compounds, it is there for chemical reasons. These scatterplots show that only when the carbon acidity is enhanced and only in the absence of competing acceptors, is the hydrogen bond nature of the C—H···F interaction even revealed. Once these conditions are met, however, C—H···F contacts seem to display all the characteristics of weak hydrogen bonds.

A relatively large number of individual structure publications reporting C—H···F—C hydrogen bonds is also available. A convincing example with a strong C—H donor occurs in 4-fluoroethynylbenzene, Fig. 3.64 (Weiss *et al.*

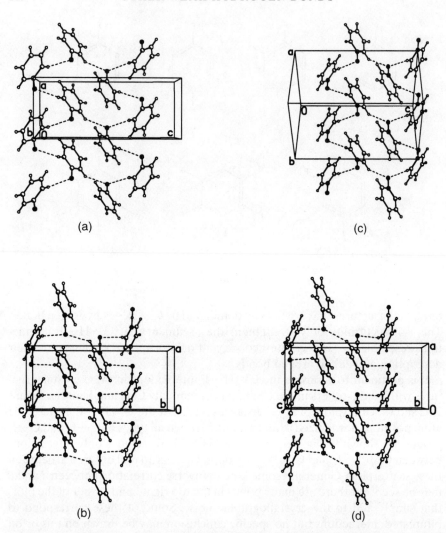

Fig. 3.60. Structural similarity between fluorobenzene, the pyridine–HF adduct, pyridine-1-oxide and benzonitrile. (a) C—H···F mediated helices in crystalline fluorobenzene. (b), (c) and (d): C—H···F⁻, C—H···O and C—H···N mediated helices in pyridine–HF, pyridine-1-oxide and benzonitrile, respectively (from Thalladi *et al.* 1998a).

1997*b*). The molecules are aligned head-to-tail in chains with $d = 2.26$ Å and $\theta = 140°$, which is a good indication of C≡C—H···F—C hydrogen bonding. In addition, there are long C—H···C≡C interactions with different molecules (not shown in the figure). An aesthetically pleasing example is reported for a benzene solvent molecule which co-crystallized with a heterobimetallic fluoroalkoxide, Fig. 3.65 (Teff *et al.* 1997). The benzene molecule is embedded in the crystal lattice forming a very large number of C—H···F—C interactions with CF₃ groups which are, despite relatively long distances d

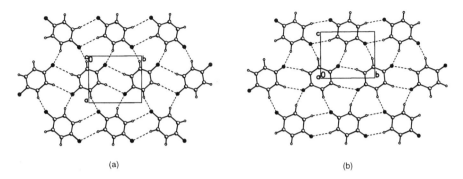

(a)   (b)

Fig. 3.61. Structural similarity between 1,4-difluorobenzene and 1,4-benzoquinone. (a) Corrugated layer structure of 1,4-difluorobenzene, stabilized by $C-H\cdots F$ hydrogen bonds. (b) Flat layer structure of 1,4-benzoquinone, stabilized by $C-H\cdots O$ hydrogen bonds (from Thalladi *et al.* 1998*a*). Notice the similarity between these two structures.

between 2.59 and 3.05 Å, reasonably interpreted as hydrogen bonds. $C-H\cdots F-C$ interactions of weak $C-H$ donors have also been discussed in some detail for a partially fluorinated cyclopenta[*a*]phenanthren-17-one (Shimoni *et al.* 1994).

Methyl fluoride, CH_3F is a simple and interesting molecule in which the $C-H$ donor stength is substantially enhanced compared to CH_4. In the crystal structure, numerous $C-H\cdots F-C$ interactions are formed with distances d down to 2.48 Å, Fig. 3.66 (Ibberson and Prager 1996). In earlier *ab initio* calculations on CH_3F clusters, interaction energies around -0.8 kcal/mol were obtained for optimal geometries with $d = 2.1$ Å (Oi *et al.* 1983). These calculations obtain larger hydrogen bond energies in clusters than in dimers, suggesting some kind of cooperative effect.

HF as an acceptor.   Hydrogen fluoride is a classical hydrogen bond acceptor (Kumler 1935) and is expected to be much stronger than $F-C$. Only a few crystal structures that contain this molecule have been published, mainly by Mootz and co-workers, but some are of great beauty and deserve closer discussion. In crystalline pyridine–HF, referred to above, molecular adducts are formed by very strong $F-H\cdots N$ hydrogen bonds ($D = 2.47$ Å). These adducts are aligned in chains linked by short $C-H\cdots F$ hydrogen bonds, leading to the hydrogen bond motif $C-H\cdots F-H\cdots N$, Fig. 3.67 (Boenigk and Mootz 1988). A more intricate example is provided by the complex of formic acid with HF, Fig. 3.68 (Wiechert *et al.* 1997). In this compound, strong $O-H\cdots O$ and $F-H\cdots O$ hydrogen bonds are formed, but also much weaker $C-H\cdots O$ and $C-H\cdots F$ interactions, showing structure stabilization by a complicated interplay of strong and weak hydrogen bonding.

Metal-bonded F atoms.   The hydrogen bond acceptor potentials of metal-bonded F have not been investigated systematically, but there are several indi-

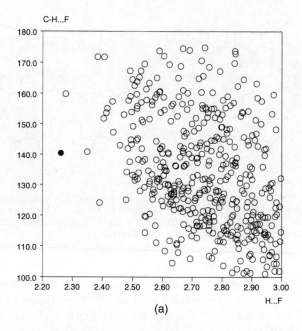

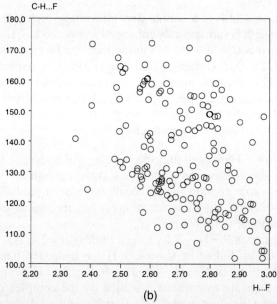

Fig. 3.62. Scatterplots of normalized distances d against angles θ in $C-H\cdots F-C$ interactions. CSD results: (a) interactions between all types of $C-H$ and $F-C$ groups. The $C\equiv C-H\cdots F-C$ interaction in 4-fluoroethynylbenzene (Weiss *et al.* 1997*b*) is marked as •; (b) interactions between $C(sp^2)-H$ and $F-C(sp^2)$ groups (from Thalladi *et al.* 1998*a*).

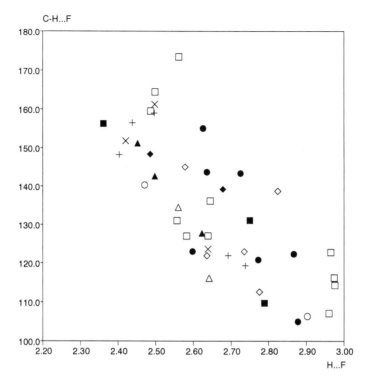

Fig. 3.63. Scatterplots of distances d against angles θ for $C-H\cdots F-C$ interactions in fluoroben-zenes, **125**. Different symbols mark the different crystal structures in the scheme. Ordinary and bifurcated bonds are included. Notice the contrast between the wide scatter of data points in Fig. 3.62 and the narrow band here. The absence of points in the high d-high θ region may also be noted (from Thalladi *et al.* 1998*a*).

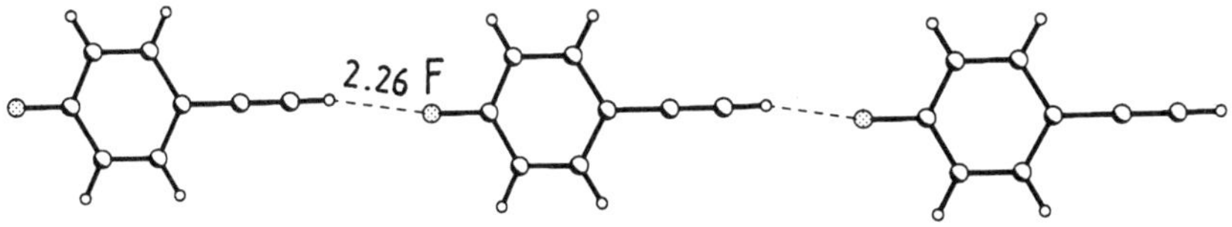

Fig. 3.64. 4-Fluoroethynylbenzene, exhibiting short $C\equiv C-H\cdots F-C$ hydrogen bonds (Weiss *et al.* 1997*b*).

vidual reports on relatively strong hydrogen bonds. It can be assumed that analogous to Cl (Aullón *et al.* 1998, see below), metal-bonded F is a much better acceptor than F—C. A good example is provided by the crystal struc-ture of **126**, where a tungsten-coordinated cyanoacetylene molecule forms a short $C\equiv C-H\cdots F-W$ hydrogen bond, Fig. 3.69 ($d = 2.16$, $D = 3.14\,\text{Å}$, $\theta = 149°$). The acetylenic stretching frequency of $3174\,\text{cm}^{-1}$ indicates a hydro-gen bond of considerable strength (Kiplinger *et al.* 1995). An example with a

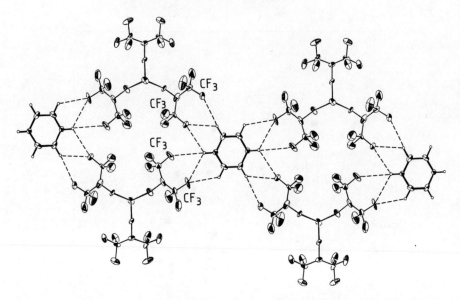

Fig. 3.65. Section of the heterobimetallic compound $[Pb_2Li_2(HFIP)_6\ (C_6H_6)]$ with HFIP = $OCH(CF_3)_2$. The benzene $C-H$ groups form numerous directional interactions with the peripheral trifluoromethyl F atoms of the organometallic compound (adapted from Teff *et al.* 1997).

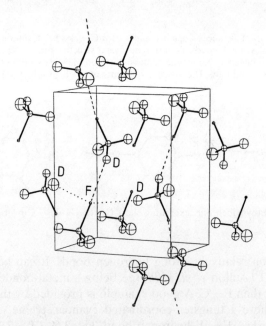

Fig. 3.66. Neutron powder diffraction structure of CD_3F. The F atoms are drawn as small dots. Only some of the many short $C-D\cdots F$ contacts are indicated (adapted from Ibberson and Prager 1996).

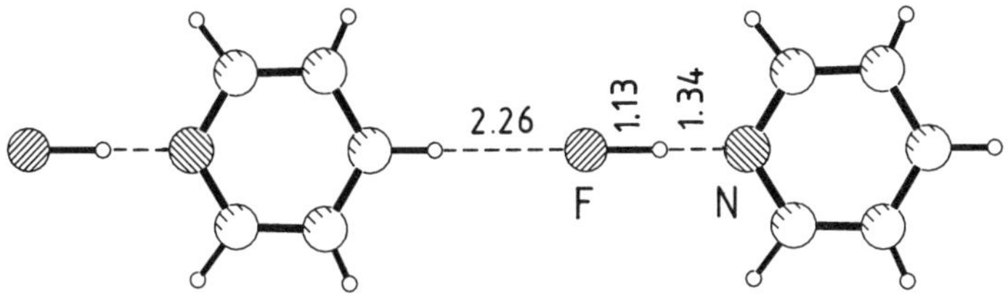

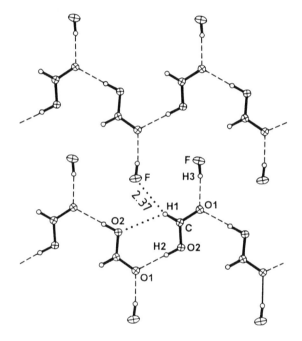

Fig. 3.67. The pyridine–HF adduct at –50°C; $D = 3.34$ Å (Boenigk and Mootz 1988). See also Fig. 3.60.

Fig. 3.68. The formic acid–HF adduct at 123 K. Notice the O—H⋯O, F—H⋯O, C—H⋯O and C—H⋯F hydrogen bonds (adapted from Wiechert *et al.* 1997).

less acidic C(sp^2)—H donor and a W—F acceptor has been reported by Osterberg *et al.* 1990 ($d = 2.03$, $D = 3.10$ Å, $\theta = 170°$).

F in inorganic anions. There are many inorganic anions with peripheral F atoms, such as BF_4^-, PF_6^-, AsF_6^-, SiF_6^{2-}, AlF_6^{3-}, UF_6^-, and so on. These represent conventional anionic hydrogen bond acceptors. It is trivial to note that for all kinds of X—H donors, hydrogen bonds to these anions are much stronger than hydrogen bonds to uncharged organic F, and have much shorter hydrogen bond distances.

Of the anions listed above, BF_4^- and PF_6^- are particularly popular in organometallic chemistry and occur regularly as counterions to organometal-

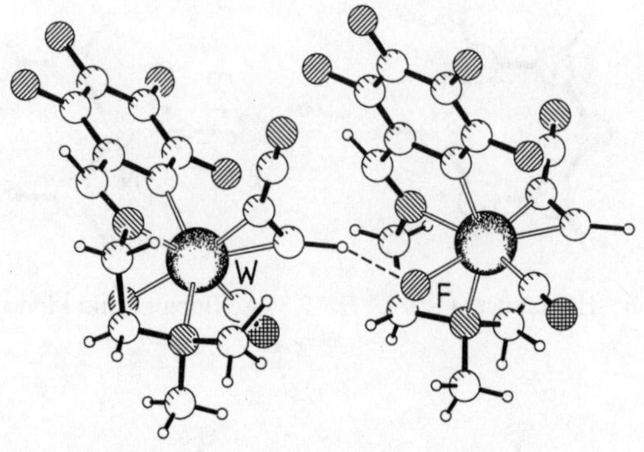

Fig. 3.69. Short C—H···F hydrogen bond in compound **126** ($D = 3.14$ Å) formed by a cyano-acetylene ligand and a tungsten-bonded F atom (Kiplinger *et al.* 1995).

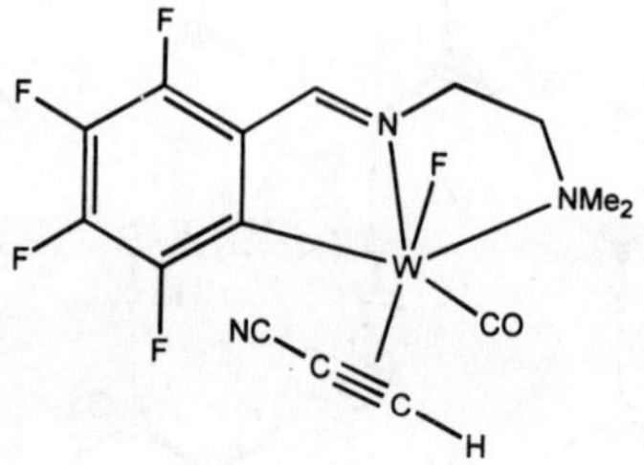

126

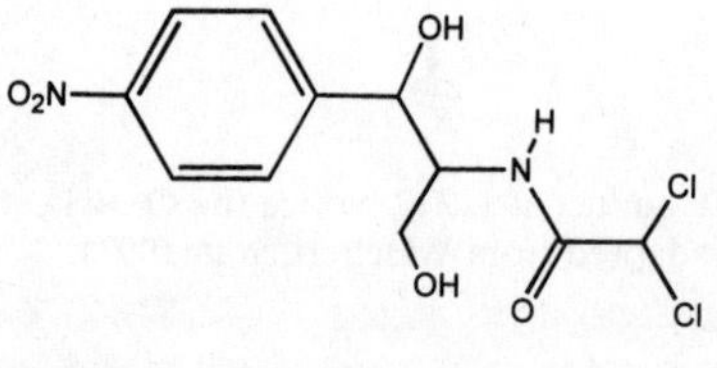

127

lic cations. It is not surprising that numerous short C—H···F contacts occur in compounds of this kind. An example where a chloroform molecule forms a hydrogen bond to a PF_6^- anion is shown in Fig. 3.70(a) (Xu *et al.* 1996). An example for extensive C—H···F hydrogen bonding around a BF_4^- anion has been discussed by Brammer *et al.* (1996), Fig. 3.70(b). In this neutron-determined crystal structure, the shortest distance d is 2.08 Å. A CSD study on these matters has been provided by Grepioni *et al.* (1998).

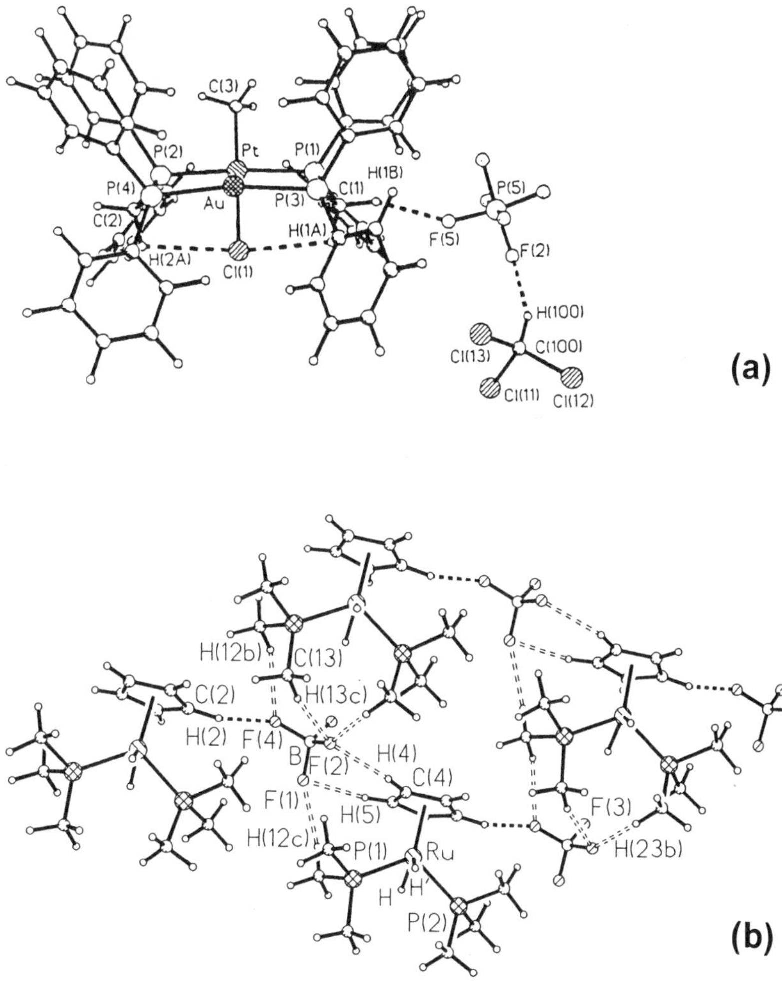

Fig. 3.70. C—H···F hydrogen bonds donated to inorganic anions. (a) With a $CHCl_3$ donor and PF_6^- anion acceptor, $d = 2.33$ Å (from Xu *et al.* 1996). (b) C—H···F hydrogen bonds around a BF_4^- anion in the neutron diffraction study of $[Cp(PMe_3)_2RuH_2]BF_4$ at 20 K. The shortest distance d is 2.08 Å (Brammer *et al.* 1996).

3.2.1.2 *Chlorine*

The hydrogen bond acceptor potentials of different kinds of Cl have been investigated by Aullón *et al.* (1998) in a CSD study. The authors considered hydrogen bonds from O—H and N—H donors to the acceptors Cl⁻, M—Cl (M = transition metal) and C—Cl. The histograms of distances d for the resulting six donor/acceptor combinations are shown in Fig. 3.71. The histograms for chloride ions are narrow and show distinct peaks at intervals of $d = 2.15$–2.20 Å for O—H as well as for N—H donors. The histograms for

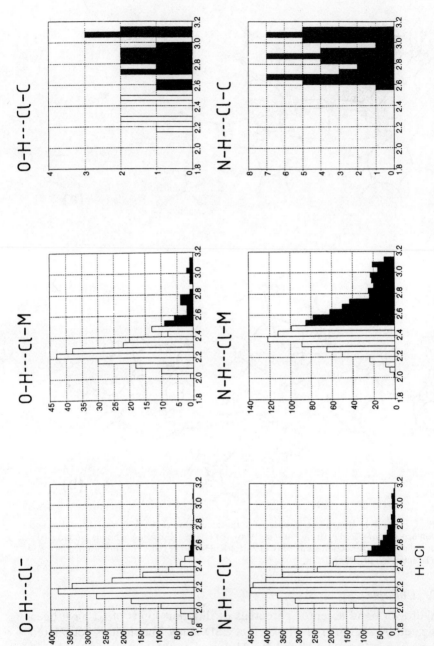

Fig. 3.71. Distances d of hydrogen bonds between the donors O—H and N—H and the acceptors Cl⁻, M—Cl (M = transition metal) and C—Cl, showing that the sequence of acceptor strengths is Cl⁻ > M—Cl ≫ C—Cl (from Aullón *et al.* 1998). The shaded parts of the histograms show distances $d > 2.5$ Å.

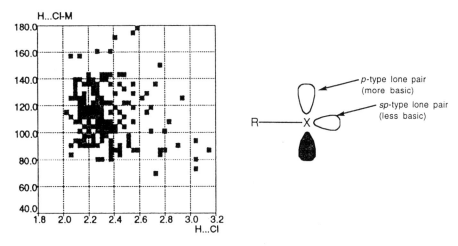

Fig. 3.72. Scatterplot of H···Cl—M angles against the distance d in O—H···Cl—M hydrogen bonds. The data show preference for angles in the range 90 to 120° (from Aullón *et al.* 1998). *Right:* Proposed basicity characteristics of the Cl lone pairs explaining the observed acceptor directionality (from Yap *et al.* 1995).

M—Cl acceptors are slightly broader, but also show clear peaks which are at intervals of $d = 2.20$–2.25 Å for O—H and of $d = 2.35$–2.40 Å for N—H donors. For C—Cl acceptors, only very few short H···Cl contacts are observed and the histograms show no recognizable maxima. The shapes and locations of the histograms clearly suggest that O/N—H···Cl—M interactions are good hydrogen bonds, whereas interactions with organic Cl are much weaker. The hydrogen bond acceptor strength of Cl follows the sequence Cl⁻ > Cl—M ≫ Cl—C.

Aullón *et al.* (1998) also studied the acceptor directionality at M—Cl and C—Cl groups. A typical scatterplot of H···Cl—M angles is shown in Fig. 3.72 (for O—H···Cl—M hydrogen bonds). The angles are clustered around values of 100 to 120°, whereas close to linear H···Cl—M geometries hardly ever occur. This is statistical support to earlier views (Yap *et al.* 1995) that in hydrogen bonds to M—Cl acceptors, there should be a pronounced angular preference for close to perpendicular approach because halogen *p*-type lone pairs are more basic than the *sp*-type lone pair (Fig. 3.72, *right*). The latter statement is based on observed coordination geometries of M—Hal—C complexes which tend to be close to rectangular at Hal.

Organic chlorine. O/N—H···Cl—C hydrogen bonds occur very rarely in crystal structures, but examples are available which show the typical situations in which these interactions are formed. Relatively common are *intra*molecular hydrogen bonds which are typically formed because the donor cannot find a stronger acceptor. The classical examples are the *ortho*-halogenated

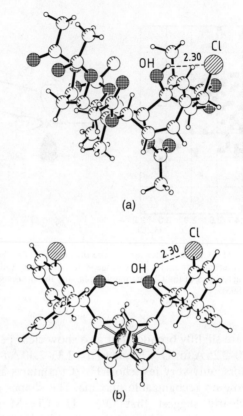

(a)

(b)

Fig. 3.73. Intramolecular O—H···Cl—C hydrogen bonds. (a) The natural diterpenoid briarein A (Burks *et al.* 1977). (b) A ferrocene derivative (Toda *et al.* 1997). Distances *d* are normalized.

phenols in dilute apolar solution (Wulf *et al.* 1936). In a crystal lattice, this situation can occur in molecules with hydroxyl groups that are not easily accessible for intermolecular contacts, or in molecules generally lacking strong hydrogen bond acceptors. An impressive example occurs in the diterpenoid briarein A, Fig. 3.73(a) (Burks *et al.* 1977). Here, a hydroxyl group is located roughly at the bottom of the bowl-shaped molecule, and is more or less inaccessible for the formation of intermolecular hydrogen bonds. Although the outer surface of the molecule carries a number of strong carbonyl acceptors, the hydroxyl group can only form an intramolecular hydrogen bond with a Cl—C group ($d = 2.30$, $D = 3.11$ Å, $\theta = 139°$). In Fig. 3.73(b), a ferrocene derivative is shown; this is a potential host compound carrying two tertiary hydroxyl groups. Because of the awkward molecular shape, the molecules cannot arrange themselves in a fashion that would allow favourable intermolecular hydrogen bonding. Therefore, a remarkable intramolecular hydrogen bond chain O—H···O—H···Cl—C is formed ($d_{Cl} = 2.30$, $D_{Cl} = 3.09$ Å, $\theta = 139°$; Toda *et al.* 1997).

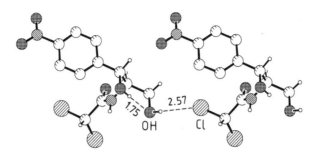

Fig. 3.74. Intermolecular O—H···Cl—C hydrogen bonds in chloramphenicol, **127** (Acharya *et al.* 1979). Distances *d* are normalized.

Although most examples with O—H···Cl—C hydrogen bonds involve sterically hindered donors, this is not true in all cases. As an example, the crystal structure of the antibiotic chloramphenicol **127** is shown in Fig. 3.74 (Acharya *et al.* 1979). In this compound, the relevant donor is a primary hydroxyl group which in principle has a number of acceptors stronger than Cl—C available (hydroxyl, carbonyl and nitro groups). Nevertheless, a very unexpected hydrogen bond scheme is formed, that is a chain O—H···O—H···Cl—C involving an intra- and an intermolecular hydrogen bond ($d_{Cl} = 2.57$, $D_{Cl} = 3.26$ Å, $\theta = 127°$). This example serves as another warning to simplistic views – prediction of hydrogen bond schemes from the nature of the functional groups involved can be dramatically misleading.

Metal-bound Cl atoms. The statistical analysis of Aullón *et al.* (1998) follows a number of structure publications pertaining to clear-cut X—H···Cl—M hydrogen bonds (e.g. Brammer *et al.* 1987; Yap *et al.* 1995). In general, hydrogen bonds of strong donors to Cl—M acceptors are too close to conventional to come within the scope of this work. A closer look, however, should be taken at weak donors, in particular C—H. Good examples have been published with the most acidic of C—H donors, such as chloroform, Fig. 3.75 (Xu *et al.* 1996). Very interesting examples are provided by two closely related organoplatinum compounds **128** and **129** from the laboratory of van Koten. In keeping with the analogy between hydroxyl and ethynyl groups, the two compounds form similar supramolecular arrays, molecular chains linked by X—H···Cl—Pt hydrogen bonds (Fig. 3.76). For **128**, $d = 2.19$ and $D = 3.13$ Å, and the red shift of ν_{OH} compared to CCl$_4$ solution is 314 cm^{-1} (Davies *et al.* 1996). In **129**, $d = 2.48$ and $D = 3.50$ Å, and the red shift of ν_{CH} is 94 cm^{-1} (James *et al.* 1996). Both frequency shifts are indicative of appreciable hydrogen bonding.

Examples of intramolecular C—H···Cl—M hydrogen bonds have been provided by Xu *et al.* (1996) for the compounds shown in Figs 3.70(a) and 3.75, and by Spaniel *et al.* (1998) for compound **130**, Fig. 3.77 ($d = 2.57$ Å, $\theta = 137°$). Needless to say, the phenomenon of C—H···Cl—M hydrogen bonding should also occur with less polar C—H groups. However, no examples are available.

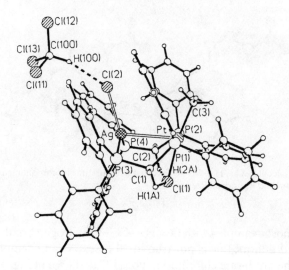

Fig. 3.75. C—H···Cl—Ag hydrogen bond donated by a chloroform molecule with the normalized geometry $d = 2.47$, $D = 3.52\,\text{Å}$, $\theta = 163°$ (Xu *et al.* 1996).

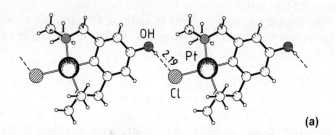

(a)

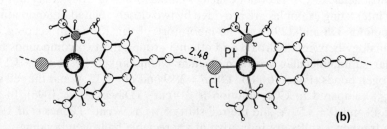

(b)

Fig. 3.76. Molecular chains connected by X—H···Cl—Pt hydrogen bonds in the two related organoplatinum compounds **128** and **129**: (a) with an O—H donor (Davies *et al.* 1996); (b) with a C≡C—H donor (James *et al.* 1996). Distances d are normalized.

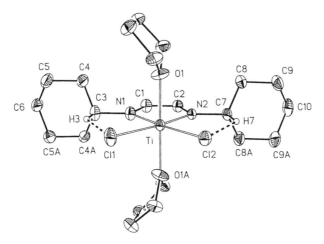

Fig. 3.77. Intramolecular C—H···Cl—Ti hydrogen bonds in compound **130**, d = 2.59 Å (from Spaniel *et al.* 1998).

128 R = OH

129 R = ethynyl

130 R = thf

3.2.1.3 *Bromine and iodine*

No statistical data on Br-acceptors have been published, so we carried out a CSD analysis during the progress of this work. The mean distances of O—H···Br hydrogen bonds are given in Table 3.19 for the acceptors Br⁻, M—Br and C—Br, respectively, and the corresponding histograms are shown in Fig. 3.78. As might have been expected, the sequence of acceptor strengths

Table 3.19 Mean geometries of O—H···Br hydrogen bonds with the acceptors Br⁻, Br—M (M = transition metal) and Br—C

Acceptor	n	H···Br (Å)	O···Br (Å)	O—H···Br (°)	H···Br—X (°)
Br⁻	257	2.372(8)	3.294(6)	156.6(8)	
Br—M	45	2.54(3)	3.40(2)	150(3)	108(3)
Br—C	25	2.75(3)	3.47(4)	134(4)	117(5)

CSD data retrieved for this text ($R < 0.08$, $d < 3.0$ Å), distances d normalized.

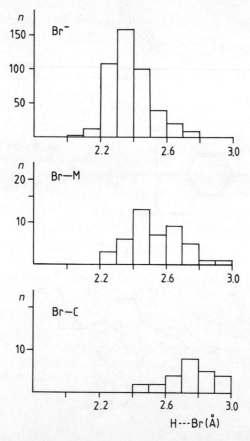

Fig. 3.78. Histograms of normalized distances d in O—H···Br hydrogen bonds with the acceptors Br⁻, Br—M (M = transition metal) and Br—C. CSD data retrieved for this book, numerical data in Table 3.19.

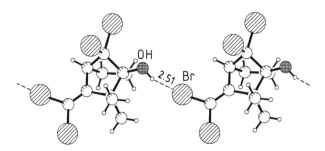

Fig. 3.79. Intermolecular O—H···Br—C hydrogen bonding in a brominated norbornane derivative (Kubicki *et al.* 1995). The distance *d* is normalized.

is Br^- > M—Br > C—Br. For Br^- and M—Br acceptors, the distance histograms have clear peaks, indicating relatively strong hydrogen bonds. The histogram for C—Br acceptors also appears to have a distinct maximum (in the interval 2.7 to 2.8 Å), and resembles the classical picture of hydrogen bonding more closely than the corresponding histograms for C—Cl (Fig. 3.71). This could be an effect caused by higher polarizability of C—Br. The hydrogen bond directionality clearly decreases within the sequence Br^-, M—Br, C—Br, as is shown by the mean angle θ falling from 156.6(8) for Br^- to 134(4)° for C—Br. The mean H···Br—X angles of 108(3) and 117(5)° for Br—M and Br—C, respectively, indicate preference for side-on approach of hydrogen bond donors, as reported similarly for Cl—X (Fig. 3.72).

Conventional hydrogen bonds bonds of O—H and N—H donors to strong acceptors like M—Br and $CuBr_4^{2-}$ are not discussed here. Hydrogen bonds to Br^- ions are discussed in Section 3.3. (numerical data in Table 3.25). Hydrogen bonds to organic bromine, C—Br, even those formed by strong donors, deserve a closer look. As for organic chlorine acceptors, a number of relevant crystal structures with sterically hindered strong donors are available, in particular tertiary hydroxyl groups. A typical example is shown in Fig. 3.79. Here, heavily brominated norbornane molecules form chains linked by short O—H···Br—C hydrogen bonds with $d = 2.51$, $D = 3.17$ Å, $\theta = 124°$ (Kubicki *et al.* 1995). It is noted as a curiosity that O—H···Br—C hydrogen bonds appear to occur in quite a number of marine natural compounds carrying Br atoms; the typical hydrogen bond motif is the isolated O—H···Br—C interaction discussed above.

Available structural data on covalently bonded I-acceptors hardly allows for any kind of statistical analysis. No behaviour fundamentally different from that of Br is expected. As a single literature example of N—H···I—C hydrogen bonding, the crystal structure of 3-iodo-L-tyrosine is shown in Fig. 3.80. Molecules related by a 2_1 axis are linked by a pair of O—H···O and an N$^+$—H···I—C hydrogen bonds ($d_I = 2.85$, $D_I = 3.78$ Å, $\theta = 153°$; Okabe and Suga 1995).

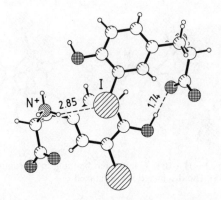

Fig. 3.80. Intermolecular $N^+ - H \cdots I - C$ hydrogen bonding in 3-iodo-L-tyrosine (Okabe and Suga 1995). Distances d are normalized.

3.2.2   Group VI elements – S, Se and Te

The atoms of group VI, that is O, S, Se and Te, can act as hydrogen bond acceptors in a falling sequence of strengths (the Pauling electronegativities are respectively 3.5, 2.5, 2.4 and 2.1). Whereas O is a very strong acceptor, S and Se are normally weak. Te has a similar or even lower electronegativity than H, and therefore cannot accept hydrogen bonds except in some cases that are discussed below. O is the best studied of all hydrogen bond acceptors. Much less is known about S and the information on Se and Te as hydrogen bond acceptors is scanty.

Despite the differences in electronegativity, chemical and structural similarities within the group are obvious. This leads to the occurrence of hydrogen bond patterns that are common to the group O, S, Se. For instance Fig. 3.81 shows isostructural molecular ribbons formed by molecules of urea, thiourea and selenourea. Ribbons of this kind are formed in many complexes of urea (Li and Mak 1997a, 1998; Thaimattam *et al.* 1998), thiourea (Li and Mak 1997b) and selenourea, and also in the pure compounds. Urea itself (Sklar *et al.* 1961) has an exceptional crystal structure and is not discussed here.

For a closer scrutiny of the hydrogen bond geometries for the group from O to Se, the geometries of the cyclic amide dimer motif and its thioamide and selenoamide analogues, **131**, are listed in Table 3.20. Hydrogen bond distances increase by about 0.5 Å from O to S acceptors, but only by a further 0.2 Å from S to Se. This is a structural consequence of the fact that the electronegativity difference between O and S is much larger than that between S and Se. It is also of interest that the acceptor directionality changes within the series; $N - H \cdots O = C$ hydrogen bonds have an average angle of 123.4° with the $O = C$ bond, whereas for $S = C$ and $Se = C$ acceptors a more perpendicular approach is preferred (107.6 and 105°, respectively).

The sequence of acceptor strengths O ≫ S > Se is also verified from

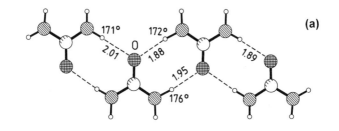

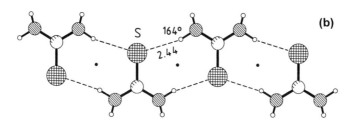

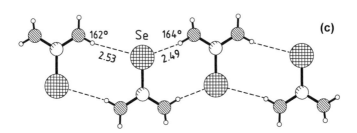

Fig. 3.81. Molecular ribbons formed by (a) urea, (b) thiourea and (c) selenourea. The geometries are normalized. Examples (a) and (c) are from lattice inclusion compounds (Li *et al.* 1995, Wright and Meyers 1980, respectively), and (b) shows the pure compound (Truter 1967). See also, Thaimattam *et al.* (1998) for related examples.

Table 3.20 Mean geometry of the amide dimer and its S and Se analogues

	C=O	C=S	C=Se
n	1255	309	7
C=X (Å)	1.234(1)	1.689(1)	1.85(1)
H⋯X (Å)	1.921(3)	2.457(7)	2.64(8)
N⋯X (Å)	2.901(3)	3.415(7)	3.58(6)
N—H⋯X (°)	166.9(3)	161.7(7)	157(3)
C—N⋯X (°)	114.2(1)	124.1(3)	119(2)
H⋯X=C (°)	123.4(2)	107.6(4)	105(3)
N⋯X=C (°)	121.9(2)	110.0(4)	108(3)

CSD, $R < 0.08$, distances d normalized.

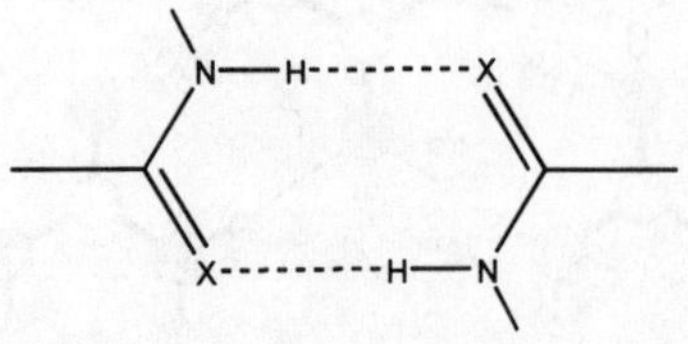

131 X = O, S, Se

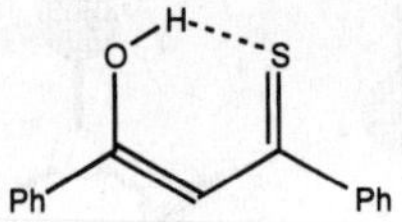

132 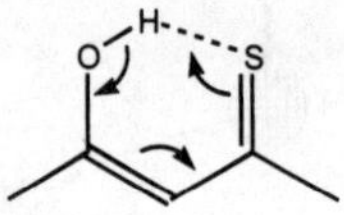   **133**

Table 3.21 Calculated hydrogen bond energies and distances of gas-phase dimers with F—H donors and hydrides of group VI and V elements as acceptors (DZP basis set)

Acceptor	ΔE (kcal/mol)	$d(F \cdots X)$ (Å)	Reference
H_2O	−9.0	2.72	Hinchliffe 1984
H_2S	−3.5	3.67	Hinchliffe 1984
H_2Se	−3.4	3.89	Hinchliffe 1984
H_3N	−11.8	2.76	Hinchliffe 1985
H_3P	−3.7	3.53	Hinchliffe 1985
H_3As	−3.5	3.60	Hinchliffe 1985

theoretical calculations. As an example, the calculated energies and distances for hydrogen bonds between the strong donor HF and the acceptors H_2O, H_2S and H_2Se are provided in Table 3.21 (Hinchliffe 1984).

3.2.2.1 *Sulfur*

Sulfur can be regarded as a conventional hydrogen bond acceptor, and is therefore not taken up here for detailed discussion. Nevertheless, this acceptor has been neglected in classical overviews of hydrogen bonding and so we have provided some structural information below. The role of S as an acceptor in homonuclear S—H···S hydrogen bonds will be discussed in Section 3.4.1.3.

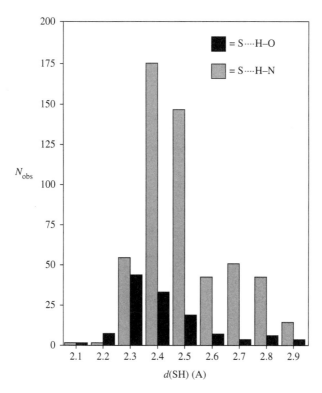

Fig. 3.82. Distributions of normalized distances d in intermolecular $N-H\cdots S=C<$ (grey bars) and $O-H\cdots S=C<$ (black bars) hydrogen bonds (from Allen *et al.* 1997a).

O/N−H⋯S hydrogen bonds. Database studies on the S atom acceptor have been published by Allen *et al.* for the $C=S$ and $-CS_2^-$ systems (1997a), and for divalent sulfur, $Y-S-Z$ and $R-S-H$ (1997b). These studies are based on the van der Waals cut-off definition $d < 2.9$ Å. For $O-H$ and $N-H$ donors in combination with $>C=S$ acceptors, histograms of distances d are obtained as shown in Fig. 3.82. These histograms have the shape that is typical for conventional hydrogen bonds of moderate strengths. The hydrogen bond angles are also normal for moderate hydrogen bonds, with the most probable values between 160 and 170°. Mean values averaged over all types of $O-H$ and $N-H$ donors are given in Table 3.22 in comparison with data for $>C=O$ acceptors. For the $>C=S$ acceptors, mean hydrogen bond distances are more than 0.5 Å longer. This is larger than the difference of the S and O van der Waals radii (1.80 and 1.52 Å, respectively, Table 1.8), reflecting that S is not only larger, but also a much weaker acceptor than O.

Lone-pair directionality at $>C=S$ acceptors is very pronounced, with in-plane contacts preferred, Fig. 3.83. The preferred contact angle with respect to the $C=S$ bond is somewhat below 110°, as compared to the $>C=O$ acceptor

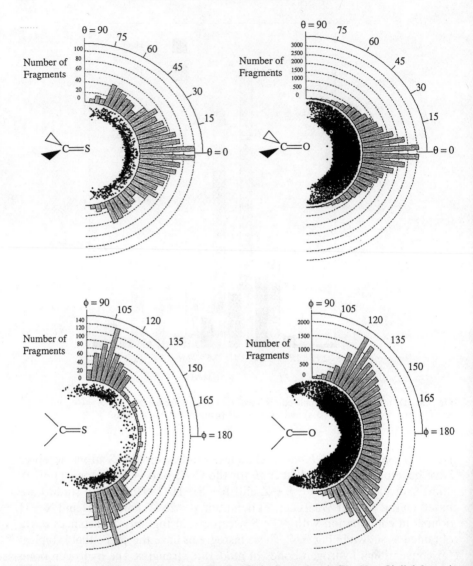

Fig. 3.83. Lone-pair directionality of $X-H \cdots S=C<$ hydrogen bonds ($X = N$ or O) (*left* figures) as compared with the $O=C<$ analogues (*right* figures). Symmetrized polar scattergrams in views perpendicular to and onto the $S/O=C<$ planes (from Allen *et al.* 1997*a*).

where the most frequent angle is greater than 120°. The sharpening of acceptor directionality and the preference for more perpendicular approach with respect to the $S=C$ bond have been observed also by Platts *et al.* (1996*a*) in an independent CSD analysis, and were corroborated with *ab initio* calculations on complexes of HF with O^- and S^- bases.

The polarity of the $C=S$ bond is dramatically influenced by the residues to

Table 3.22 Mean geometries of intermolecular $O-H\cdots S$ and $N-H\cdots S$ hydrogen bonds as compared with $>C=O$ acceptors (Allen *et al.* 1997*a,b*)

Type	d (Å)	D (Å)	θ (°)	n data
S-acceptors				
$O-H\cdots S=C<$	2.41(1)	3.32(1)	157(1)	119
$N-H\cdots S=C<$	2.51(1)	3.43(1)	157(1)	530
$O-H\cdots S<$	2.67(5)	3.39(1)	134(1)	39
$N-H\cdots S<$	2.75(2)	3.58(2)	142(2)	47
O=C-acceptors				
$O-H\cdots O=C<$	1.884(2)	2.790(2)	158.2(2)	5861
$N-H\cdots O=C<$	2.011(2)	2.931(2)	156.2(2)	7014

Distances d normalized, distance cut-off $d < 2.9$ Å.

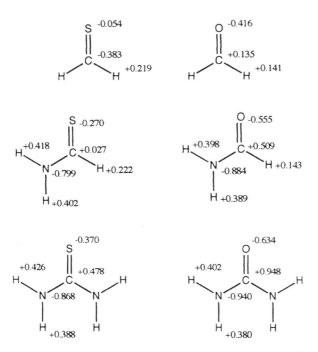

Fig. 3.84. Calculated Mulliken partial atomic charges for a series of model molecules containing the $>C=S$ fragment and their oxygen analogues. *Ab initio* MO calculations using the 6-31G* basis set (Allen *et al.* 1997*a*).

which it is bonded. This effect is much more pronounced when compared to $>C=O$ which has a strong inherent dipole moment due to the large electronegativity difference of C and O. This circumstance is illustrated in Fig. 3.84 where calculated atomic partial charges are shown for some molecules

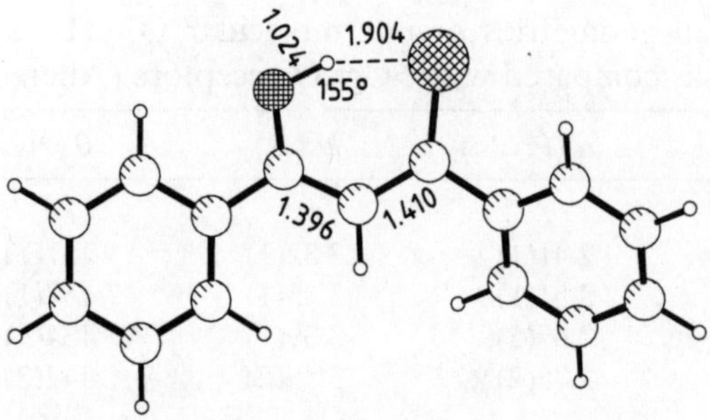

Fig. 3.85. Strong resonance assisted $O-H\cdots S$ hydrogen bond in the neutron structure of compound **132** (Power *et al.* 1976; Steiner 1998*a,f*).

containing the $>C=S$ fragment and their $>C=O$ analogues. Whereas the partial charge on S increases from $0.05e$ in thioformaldehyde to $0.37e$ in thiourea, it increases only from $0.42e$ in formaldehyde to $0.63e$ in urea. This means that in general, substituent effects concern $X-H\cdots S$ hydrogen bonds to a stronger degree than the O atom analogues.

Although S is a weak acceptor in general, strong $O-H\cdots S$ hydrogen bonding is possible in principle. By far the shortest $O-H\cdots S$ hydrogen bonds known are intramolecular and occur in very special fragments such as monothiodiketone enols, and related arrangements. In Fig. 3.85, an example is shown for compound **132** (Power *et al.* 1976). These hydrogen bonds have been identified as *resonance assisted*, that is strong hydrogen bonding is facilitated by charge flow through the system of conjugated double bonds, **133** (Steiner 1998*f*). Note that in **132**, the formal $C=C$ and $C-C$ bonds have almost equal lengths, i.e. there is almost complete π-delocalization at $C=C-C$. For related compounds, IR spectroscopy has shown ν_{OH} red shifts of around $450\,\mathrm{cm^{-1}}$, also an indication of strong hydrogen bonding (Steinwender and Mikenda 1990).

Divalent sulfur $X-S-Y$ is a much weaker acceptor than $S=C<$, and is only rarely involved in hydrogen bonding. If hydrogen bonds are formed, distances with $O-H$ and $N-H$ donors are on average longer by *ca.* $0.2\,\text{Å}$ when compared to $S=C<$ acceptors, Table 3.22 (Allen *et al.* 1997*b*). In thioethers, S may even carry a small positive partial charge, and is then unsuitable as a hydrogen bond acceptor. In this situation, S may engage in a completely different kind of electrostatic intermolecular interaction, that is $>S^{\delta+}\cdots X^{\delta-}$ (Allen *et al.* 1997*b*). Exceptions are provided by the thiol group, $C-S-H$, and by H_2S, which can operate as donor and acceptor simultaneously (see Section 3.4.1). Intramolecular $N-H\cdots S$ hydrogen bonds involving divalent S have been invoked as being of importance in determining conformational preferences in the penam class of antibiotics (Nangia and Desiraju 1999).

$C-H\cdots S$ hydrogen bonds. The above section has considered hydrogen bonds from the strong donors $O-H$ and $N-H$ to the weak acceptor S.

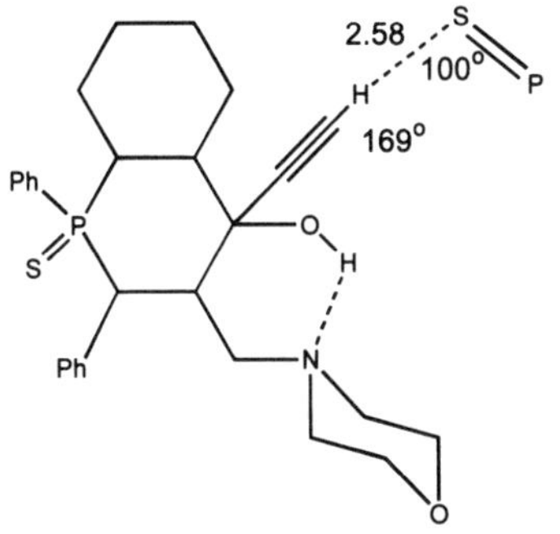

2.58
S
100°
H
P
169°

Ph
P
S

O
H

Ph

N

O

134

H
O

Se

t-Bu

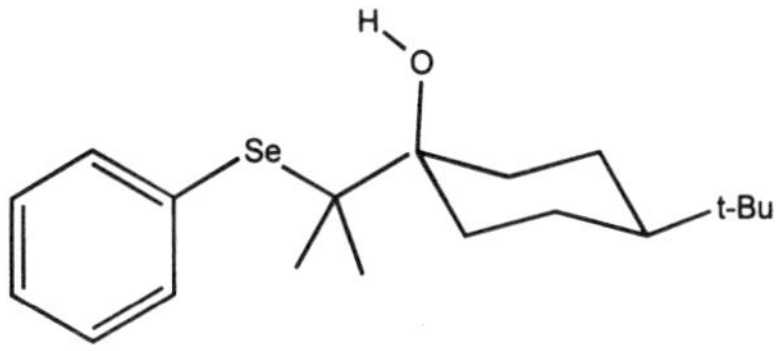

135

Hydrogen bonds with the weak C—H donors have not yet been the subject of systematic investigations. Although the point was addressed by Taylor and Kennard (1982), little convincing literature has appeared since. Because C—H···S interactions must be considerably weaker than C—H···O hydrogen bonds, only minor roles in structure stabilization can normally be expected. In exceptional cases, however, structure determining roles also occur.

A clearcut example of a C—H···S hydrogen bond with a strong alkynic donor and an S=P acceptor occurs in **134** (Shalamov *et al.* 1990). In the close to linear interaction, the distance d is only 2.58 Å, not much longer than the mean value for N—H···S=C< hydrogen bonds (Table 3.22). C—H···S interactions with weakly acidic C—H donors and the strong acceptors S=P have been described in a series of organophosphorus compounds and the hydrogen bond nature of these interactions was confirmed by ^{31}P NMR spectroscopy (Potrzebowski *et al.* 1998). An unusual example of a hydrogen bond from a chloroform molecule to an S=Ge acceptor is discussed in Section 3.2.2.2. The possibility of intramolecular C—H···S hydrogen bonding, despite a strongly bent angle ($\theta = 105°$), has been discussed for a push–pull butadiene (Surange *et al.* 1997).

3.2.2.2 Selenium

O—H···Se hydrogen bonds. Only two unambiguous examples of O—H···Se hydrogen bonds are found in the CSD. In the non-ionic **135**, a tertiary hydroxyl group forms an intermolecular O—H···Se hydrogen bond with

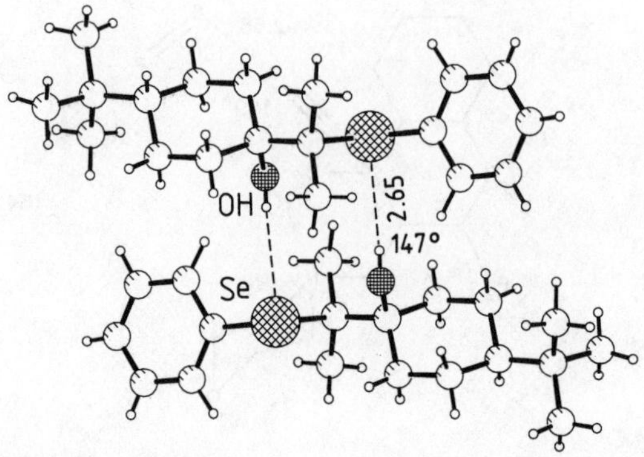

Fig. 3.86. Intermolecular O—H···Se hydrogen bonds in compound **135** (Labar *et al.* 1985). The geometry is normalized.

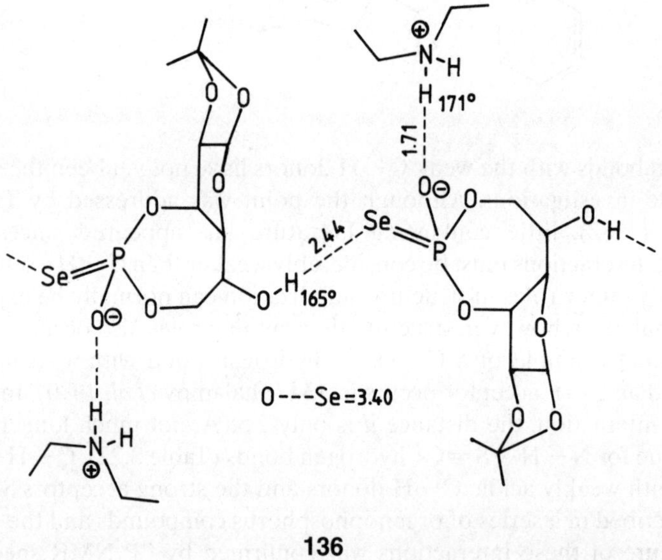

136

$d = 2.65$ Å, Fig. 3.86 (Labar *et al.* 1985). Although the C—Se—C group must be a weaker acceptor than C=Se, the hydrogen bond distance is comparable with those in the cooperative N—H···Se=C interactions listed in Table 3.20. It is not unusual that in tertiary alcohols, the hydroxyl group cannot engage in O—H···O hydrogen bonding (Brock and Duncan 1994). This is due to steric hindrance, and the O—H group resorts to weaker acceptor types, in this case, Se.

In the ionic compound **136**, the situation is very different. Here, Se is bonded to a deprotonated (i.e. negatively charged) phosphinic acid group. The Se atom

Table 3.23 Mean geometries of intermolecular $N-H\cdots Se$ hydrogen bonds

Acceptor	$n_{data}/n_{struct.}$	d (Å)	D (Å)	θ (°)
Se^{2-}	9/2	2.38(3)	3.30(2)	153(4)
$P=Se$	4/4	2.63(7)	3.58(6)	158(4)
$C=Se$	17/8	2.64(4)	3.55(3)	153(3)

CSD, $R < 0.08$, distances d normalized, distance cut-off $d < 3.0$ Å.

accepts a short $O-H\cdots Se$ hydrogen bond from a hydroxyl donor with $d =$ 2.44 Å. The relevance of the $O-H\cdots Se$ interaction is strongly supported by the fact that the S analogue is isostructural with a clear $O-H\cdots S=P$ hydrogen bond, $d = 2.37$ Å (Magomedova *et al.* 1991).

Hydrogen bonds to Se atoms at the periphery of inorganic ions like $GeSe_4^{4-}$ are not within the scope of this book, because they represent conventional (albeit highly exotic) ionic hydrogen bonds. An example of this kind of $O-H\cdots Se$ interaction has been found in $Na_4GeSe_4\cdot14\ H_2O$ with distances D between 3.28 and 3.63 Å (Krebs and Jacobsen 1976).

$N-H\cdots Se$ hydrogen bonds. For hydrogen bonds between $N-H$ donors and Se acceptors, a relatively large quantity of structural data is available, mostly from selenoamides, $Se=C-N-H$, and the more activated $Se=P-N-H$. The pronounced acceptor strength of Se in these configurations can be explained by contribution of the polar resonance form **137**, which provides surplus negative charge at Se (Peng *et al.* 1994). For $Se=C$ and $Se=P$ acceptors, mean hydrogen bond geometries are given in Table 3.23 in comparison with data for the selenide ion, Se^{2-}. The hydrogen bond distances for $Se=C$ and $Se=P$ are very similar, and are about 0.25 Å longer than for the doubly charged selenide anions.

By far the most frequent motif of $N-H\cdots Se$ hydrogen bonds is the cooperative cyclic selenoamide dimer, **131**. Analogous dimers are formed by the $Se=P-N-H$ group. Examples of both kind of dimers occur in **138** and **139** and are shown in Fig. 3.87. Like amides, selenoamides and $Se=P-N-H$ groups form not only cyclic patterns, but also cooperative hydrogen bonded chains $Se=C-N-H\cdots Se=C-N-H\cdots Se=C-N-H$. Examples of cycles and chains of this kind have been reported for a series of closely related selone molecules by Peng *et al.* (1994).

Particularly rich systems of interconnected $Se=C-N-H\cdots Se=C-N-H$ hydrogen bonds are formed in crystalline selenourea and in binary crystals containing this molecule. In tris(acetylacetonato) Co(III)–selenourea, the selenourea molecules form molecular ribbons as already shown in Fig. 3.81(c) (Wright and Meyers 1980). The crystal structure of selenourea itself is very peculiar with nine molecules per asymmetric crystal unit. The molecules form ribbons related to those shown in Fig. 3.81, though not in a planar

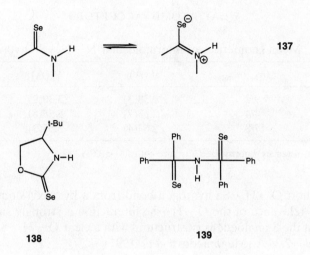

137

138

139

140 E = S, Se, Te

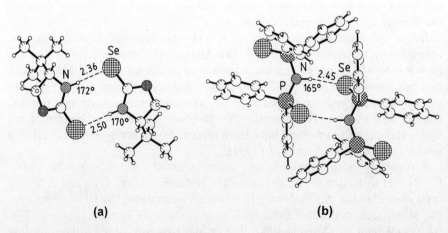

(a)   (b)

Fig. 3.87. Cyclic N—H···Se hydrogen bond dimers: (a) with the Se=C—N—H fragment of the selone **138** (Peng *et al.* 1994); (b) with the Se=P—N—H fragment of compound **139** (Bhattacharyya *et al.* 1995). Geometries are normalized.

arrangement. Rather, consecutive molecules are twisted by 120° along the ribbon axes. In the nine independent ribbons, the hydrogen bonds have very similar hydrogen bond distances D around 3.51 Å (Rutherford and Calvo 1969).

For reasons of completeness, conventional $N—H \cdots Se$ hydrogen bonds with Se^{2-} anions may also be mentioned. Hydrogen bond distances are short (d around 2.34 Å, Table 3.23) and are similar to those found with chloride acceptors (Section 3.3, Table 3.25).

C—H⋯Se hydrogen bonds. A good example of $C—H \cdots Se$ hydrogen bonding is found in a series of crystal structures of the terminal chalcogenido complexes $(\eta^4\text{-Me}_8\text{taa})GeE$ with E = S, Se and Te, **140** (Kuchta and Parkin 1994; Steiner 1998g). In these crystal structures, chloroform molecules are arranged around the terminal GeE moieties, forming hydrogen bonds with E, Fig. 3.88 and Table 3.24. The Ge ≈ E bond lengths suggest that the bonding between Ge and E is best described as an intermediate between the $Ge^+—E^-$ and the Ge=E resonance forms. This means that E carries a partial negative charge, $E^{\delta-}$, and is therefore able to accept hydrogen bonds. The sulfido and selenido compounds are isostructural and contain a 4:1 ratio of chloroform. The tellurido complex is differently shaped and contains only a 2:1 ratio of chloroform. It is of interest to compare the hydrogen bond distances with the average value for $Cl_3C—H \cdots O$ interactions (Steiner 1994a): mean separations d are 2.22 Å for O-acceptors, 2.45 Å for the sulfido, 2.55 Å for the selenido and 2.80 Å for the tellurido compound. This lengthening of distances when going down group VI is very similar to that observed for hydrogen bonds to halide ions of the corresponding rows (Cl^- to I^-), see Table 3.25 below.

Intramolecular $C—H \cdots Se$ hydrogen bonding in **141** has been investigated in great detail by crystallographic, IR and NMR spectroscopic methods. Intramolecular $H \cdots Se$ contacts with $d = 2.92$ Å are observed in the crystal

Table 3.24 $C—H \cdots E$ hydrogen bonds of the chloroform donors in the chalcogenido complexes $(\eta^4\text{-Me}_8\text{taa})GeE$ with E = S, Se and Te, **152** (Kuchta and Parkin 1994; Steiner 1998g); see Fig. 3.88

Acceptor	Donor	$H \cdots E$ (Å)	$C \cdots E$ (Å)	$C—H \cdots E$ (°)
Ge=S	CHCl$_3$, Mol. 1	2.38	3.44	167
	CHCl$_3$, Mol. 2	2.51	3.53	156
Ge=Se	CHCl$_3$, Mol. 1	2.47	3.53	168
	CHCl$_3$, Mol. 2	2.62	3.60	151
Ge=Te	CHCl$_3$, Mol. 1	2.90	3.88	151
	CHCl$_3$, Mol. 2a	2.67	3.74	169
	CHCl$_3$, Mol. 2b	2.84	3.87	158

Distances d normalized.

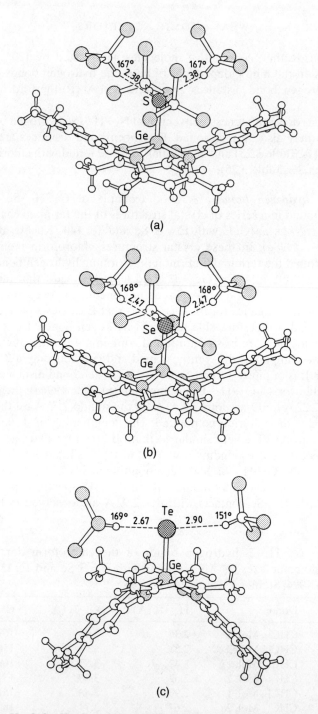

Fig. 3.88. C—H···S, C—H···Se and N—H···Te hydrogen bonds donated by chloroform molecules. Compounds **140** with (a) E = S, (b) E = Se, and (c) E = Te. Geometries are given in Table 3.24 (Kuchta and Parkin 1994; Steiner 1998g).

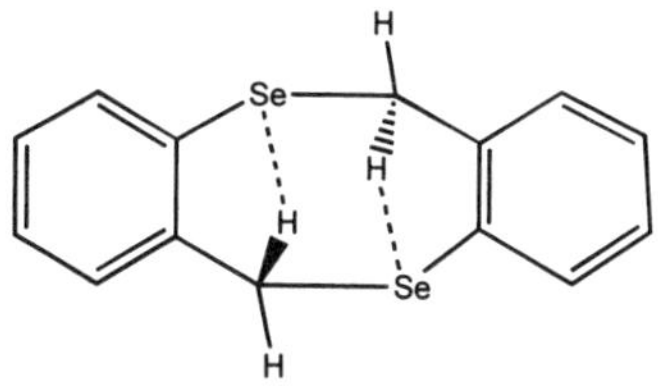

141

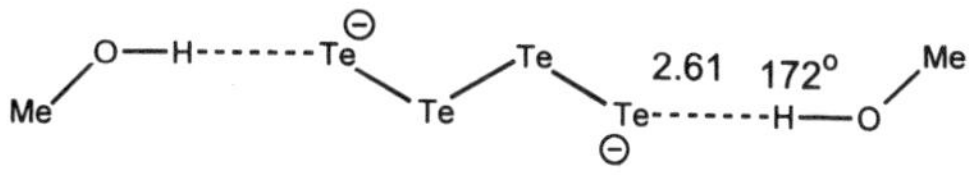

142

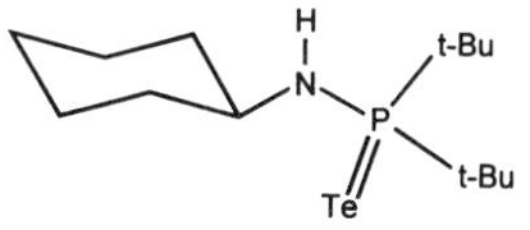

143

structure (Iwaoka and Tomoda 1994). Despite the strongly bent angle $\theta = 101.7°$, the weakly hydrogen bonding nature of this contact was inferred from the IR wavenumber shift of the relevant $C-H$ oscillator ($\Delta\nu = 52\,cm^{-1}$), and a deuterium-induced ^{77}Se NMR upfield isotope shift of $\Delta\Delta\delta = 0.25\,ppm$ (Iwaoka et al. 1996).

3.2.2.3 Tellurium

Because of its low electronegativity, Te does not normally accept hydrogen bonds. In some chemical situations, however, surplus negative charge can be provided to Te atoms, making them suitable as hydrogen bond acceptors. Only few examples are known from crystal structures, and these are discussed below.

An $O-H\cdots Te$ hydrogen bond has been found in bis(tetraphenylphosphonium) tetratelluride methanol solvate, **142** (Huffman and Haushalter 1984). The tetratelluride dianion accepts two $O-H\cdots Te^-$ hydrogen bonds from methanol molecules. In this case, Te is made suitable for hydrogen bonding by the double negative charge on the tetratelluride chain. Notably, the hydrogen bonds are directed at the terminal Te atoms, which carry the formal charges.

An example of $N-H\cdots Te$ hydrogen bonding occurs in the tellurophosphinic amide **143**, which appears to have a charge distribution suitable for hydrogen bond formation, $Te=P-N-H\cdots Te=P-N-H$, Fig. 3.89 ($d = 2.96$, $D = 3.87\,Å$, $\theta = 150°$; Bochmann et al. 1995). A more conventional

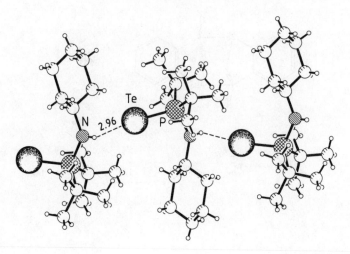

Fig. 3.89. Intermolecular $N-H\cdots Te$ hydrogen bonding in compound **143** (Bochmann *et al.* 1995). The distance d is normalized.

situation occurs in a structure reported by Warren *et al.* (1994), where diaminoethane molecules, $H_2N-CH_2-CH_2-NH_2$, donate hydrogen bonds to Te atoms of the complex tetraanion $[Sb_6Te_9]^{4-}$ ($d = 2.94$, $D = 3.83\text{Å}$, $\theta = 146.3°$).

Hydrogen bonds of chloroform molecules to a polar $Ge=Te$ group have already been discussed above (Fig. 3.88; Kuchta and Parkin 1994; Steiner 1998g). This is the only documented case of a $C-H\cdots Te$ hydrogen bond.

3.2.3 *Group V elements – P, As and Sb*

The group V elements can act as hydrogen bond acceptors if their bonding situation provides an electron lone pair, Fig. 3.90. For N, there are several configurations with different N hybridization states. The chemical variability of N leads to a very broad spectrum of acceptor properties of this atom, but this is not the subject of the present work. The other group V elements also have a very variable chemistry, but the number of configurations which are capable of accepting hydrogen bonds is reduced, and even these are uncommon. Fig. 3.90 shows configurations for which there is experimental evidence for hydrogen bond capability. The inorganic chemist can easily suggest more group V atom configurations which only await experimental realization as hydrogen bond acceptors.

Much of our knowledge on the acceptor potentials of P and As is based on studies on the hydrides PH_3 (phosphine) and AsH_3 (arsine). The subject has been reviewed by Sennikov (1994). In the gas phase, PH_3 and AsH_3 form complexes of the kind $Hal-H\cdots XH_3$ with hydrogen halides. $C-H\cdots P$ hydrogen

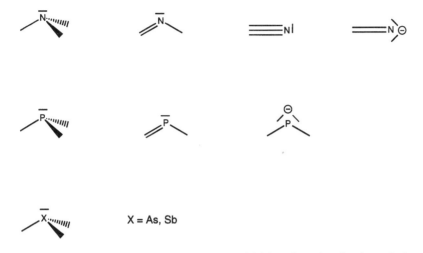

X = As, Sb

Fig. 3.90. Covalent configurations of Group V atoms which have been found acting as hydrogen bond acceptors.

bonding with $D = 3.91\,\text{Å}$ was observed in the gas phase dimer $N\equiv C-H\cdots PH_3$ (Legon and Willoughby 1984). Hydrogen bonded complexes of HF with PH_3, AsH_3 and SbH_3, and also with trimethylphosphine, $P(CH_3)_3$, have been identified spectroscopically in solid argon matrices (Arlinghaus and Andrews 1984). $C-H\cdots XH_3$ hydrogen bonds formed by terminal alkynes have also been found with the matrix isolation technique (Jeng and Ault 1990). In dilute solutions of the strong donors H_2O and EtOH in liquid PH_3 and AsH_3, the spectral effects of weak $O-H\cdots XH_3$ hydrogen bonding are observed (Sennikov 1994). Theoretical calculations on $Hal-H\cdots XH_3$ dimers indicate a sequence of acceptor strengths $NH_3 \gg PH_3 > AsH_3$, which is as would be expected, Table 3.21 (Hinchliffe 1985). As for the group VI elements, the acceptor strengths drop dramatically from the first to the second row, and only slightly from the second to the third row.

There are exceedingly few crystal structure publications where $O/N-H\cdots P$ hydrogen bonding is discussed. Fortunately, however, there is at least one literature example for each of the P atom acceptor types in Fig. 3.90 so that the field as such is not on shaky ground. The shortest hydrogen bond reported is formed by the phenolic $O-H$ group of **144** to a P atom bonded to three C atoms, Fig. 3.91(a) ($d = 2.30$, $D = 3.19\,\text{Å}$, $\theta = 151°$; Heinicke *et al.* 1996). The closely related **145** is more difficult to pack because of additional *tert*-butanol residues, and a short intramolecular $O-H\cdots P$ hydrogen bond is formed, Fig. 3.91(b) ($d = 2.41\,\text{Å}$; Heinicke *et al.* 1996).

Hydrogen bonds from $N-H$ donors to P atoms with three covalent bonds were also reported. Brauer *et al.* (1986) found an example in the diazadiphospholane **146**, Fig. 3.92(a) The figure does not show that the other $N-H$ donor

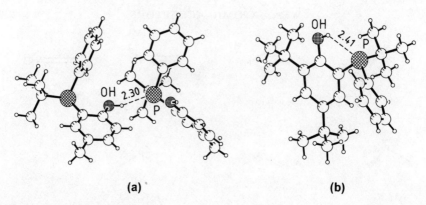

Fig. 3.91. Intra- and intermolecular O—H···P hydrogen bonds in (a) **144** and (b) **145** (Heinicke *et al.* 1996). Distances *d* are normalized.

144

145

146

147

148

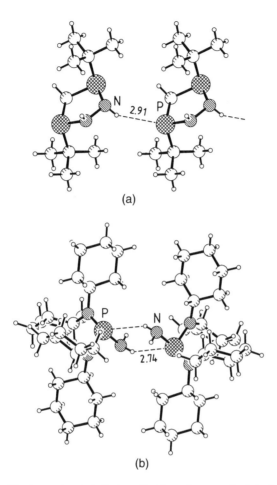

Fig. 3.92. N—H···P hydrogen bonds to P atoms that form three covalent single bonds. (a) In **146**, $d = 2.91$, $D = 3.83$ Å, $\theta = 152°$ (Brauer *et al.* 1986). (b) Aminophosphane dimer in **147**, $d = 2.74$, $D = 3.64$ Å, $\theta = 149°$ (Schick *et al.* 1996).

forms an N—H···N hydrogen bond, resuting in a cooperative chain N—H···N—H···P. A remarkable dimer is formed by aminophosphane **147**, where molecules are joined by mutual and presumably cooperative P—N—H···P hydrogen bonds, Fig. 3.92(b) (Schick *et al.* 1996).

A very interesting hydrogen bond system has been found in the phosphaalkene **148** containing a C—P=C acceptor, Fig. 3.93 (Paasch *et al.* 1995). The amino group donates an intermolecular N—H···Ph hydrogen bond and an N—H···P hydrogen bond to the same molecule, thus forming a chelated arrangement.

Finally, P atoms with a formal negative charge can act as acceptors of hydrogen bonds with partly ionic character. Examples can be found with

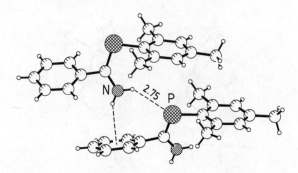

Fig. 3.93. N—H···P hydrogen bond involving a C—P=C acceptor in compound **148**. Normalized geometry $d = 2.75$, $D = 3.56$ Å, $\theta = 138°$. The NH$_2$ donor also forms an N—H···Ph hydrogen bond (Paasch *et al.* 1995).

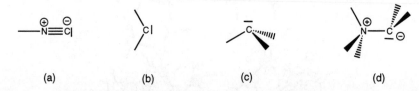

(a)	(b)	(c)	(d)

Fig. 3.94. Carbon atoms with free electron pairs: (a) isonitriles (isocyanides); (b) carbenes; (c) carbanions; (d) nitrogen ylides. All these are strong non-conventional hydrogen bond acceptors (see Table 1.7).

complex ions like P$_{11}^{3-}$, which contains bi- and tri-valent P atoms. Examples of N—H···P$^-$ hydrogen bonds are archived in the CSD but, unfortunately, none have been discussed by the original authors.

It is known from studies on AsH$_3$ and SbH$_3$ that As and Sb can in principle accept hydrogen bonds (Sennikov 1994). In a CSD search, however, no examples for hydrogen bonds in crystals of O—H and N—H donors to these atoms could be clearly identified.

3.2.4 *Group IV elements – isonitriles, carbanions, carbenes and silylenes*

Carbon may act as a hydrogen bond acceptor in two different situations. In the π-acceptors Ph, C≡C, C=C etc., the hydrogen bonds are directed at electron clouds engaged in covalent bonding (Section 3.1). The other kind of C atom acceptors contain free electron pairs. These include isonitriles, carbenes and carbanions (Fig. 3.94). Hydrogen bonds to these acceptors are not weak, but are discussed here because they are clearly non-conventional. Si cannot form π bonds, so Si-based π-acceptors do not exist. However, Si analogues of carbenes, the silylenes, can in principle act as hydrogen bond acceptors.

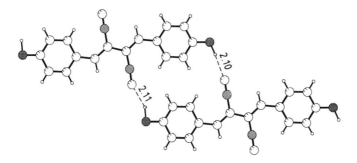

Fig. 3.95. Xanthocillin, **149**, with pairs of O—H···C hydrogen bonds involving isonitrile acceptors (Britton *et al.* 1981). Distances *d* are normalized.

3.2.4.1   *Isonitriles*

Isonitriles (isocyanides) are best represented by the structure R—N≡C| in which formal charges can be assigned as R—N$^+$≡C$^-$. Hydrogen bonds to isonitriles in solution have been discovered with IR spectroscopy by Ferstanding (1962) and Schleyer and Allerhand (1962). Because the N atom is electron-deficient, the hydrogen bonds can only be directed at the C lone pair which is in an *sp*-orbital. For phenol and methanol donors, IR ν_{OH} frequency shifts exceed those with cyano acceptors, showing that the hydrogen bond is strong (phenol–isocyanides: 190 to 250 cm^{-1}, phenol–cyanides: 144 to 174 cm^{-1}; Schleyer and Allerhand 1962). Using the strong C≡C—H donor of phenylacetylene and the acceptor benzyl isocyanide, Ferstanding (1962) could characterize the first C—H···|C hydrogen bond (in solution, ν_{CH} shift = 26 cm^{-1}). In theoretical calculations, the N—H···C hydrogen bond in hydrogen isocyanide, C≡N—H···C≡N—H, was calculated to have an energy of −5.4 kcal/mol (Alkorta *et al.* 1998).

In the crystalline state, hydrogen bonds to isonitriles occur only rarely. In a CSD search on O—H···|C≡N hydrogen bonds, only very few hits were found (*n* = 4, *d* range 2.10 to 2.19 Å, mean *d* = 2.12(2), mean *D* = 3.02(2) Å). A good example occurs in crystalline xanthocillin, **149**, Fig. 3.95 (Britton *et al.* 1981).

3.2.4.2   *Carbenes*

Carbenes, >C|, are very reactive species which are only rarely stable in the solid state. However, a very interesting and unique case of stable C—H···|C< hydrogen bonding has been reported by Arduengo *et al.* (1995). In adduct **150**, an imidazolium C2—H donor forms a short hydrogen bond to a carbene acceptor, Fig. 3.96 (*d* = 2.10, *D* = 3.18 Å, θ = 173°). Note that the donor and acceptor molecules are conjugate acid and base, which is a situation favouring strong hydrogen bonding. Nevertheless, the N—C—N angles are very different, 107.6° at the donor and 102.8° at the acceptor, showing that the imi-

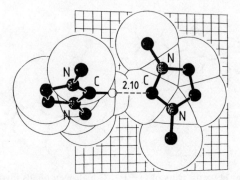

Fig. 3.96. C—H⋯C hydrogen bond with a carbene acceptor in compound **150**. The bulky ligands attached to the N atoms are omitted for clarity (Arduengo *et al.* 1995).

dazolium and carbene units are clearly distinct and that the proton is not equally shared between the two units.

In theoretical calculations on hydrogen bonds of several donors to the simple carbenes H_2C and F_2C, Alkorta and Elguero (1996) have estimated bond energies to be around −5 kcal/mol for the donor H_2O, and −2.5 kcal/mol for the weaker donor NH_3. For a strongly simplified approximation to the

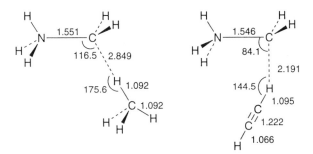

Fig. 3.97. Theoretically proposed C—H···C hydrogen bonding with nitrogen ylides. MP2/6-311++G(d,p) optimized geometries of H_3N—CH_2 in complexes with methane and acetylene. The bent geometry of the acetylene complex indicates a secondary interaction of the C≡C moiety with the ammonium group (from Platts *et al.* 1996b).

experimental situation in Fig. 3.96, that is $H_2C\cdots H_3C^+$, a very strong hydrogen bond with $d = 1.99$ Å and an interaction energy of *ca.* -20 cal/mol is obtained.

3.2.4.3 *Carbanions, ylides*

Hydrogen bonds to carbanions were postulated early on by Ferstanding (1962). With the aim of calculating possibly strong C—H···C hydrogen bonds, Platts *et al.* (1996b) and Platts and Howard (1997) selected very basic C-acceptors of N- and P-ylides in combination with strong acetylenic donors. Particularly in N-ylides, the CH_2 group can formally be regarded as a carbanion bonded to an ammonium counterion, Fig. 3.94(d). For the complexes of the simplest N-ylide, NH_3^+—CH_2^-, with acetylene and methane, C—H···C hydrogen bonds were calculated with interaction energies of -35 kcal/mol and -5 kcal/mol, respectively, and the geometries shown in Fig. 3.97. For the less basic P-ylides (which have a large contribution of the non-ionic resonance form $H_3P{=}CH_2$), lower hydrogen bond energies were calculated.

Because of the high basicity of ylides (and carbanions), the formation of hydrogen bonds with strong donors can only be expected as intermediates leading to proton transfer reactions. Only with donors of low acidity, can stable X—H···$|C^-$ hydrogen bonding occur. Examples for P-ylides were reported by Davidson and co-workers (Batsanov *et al.* 1996). In particular, these authors comment about compound **151**, where phenyl C—H groups donate hydrogen bonds to the ylidic C atom with $d = 2.58$ and 2.81 Å, respectively, Fig. 3.98.

3.2.4.4 *Silylenes*

The Si analogue of carbenes, the silylenes $|Si<$, have been proposed as hydrogen bond acceptors in a theoretical work of Alkorta and Elguero (1996).

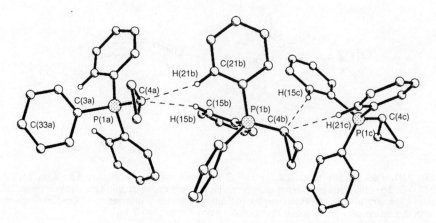

Fig. 3.98. C—H···C hydrogen bonds in the phosphorus ylide **151** with $d = 2.58$ and 2.81 Å, respectively. The structure is unusual because of a very strong pyramidalization of the C atom, rendering it particularly basic (from Batsanov *et al.* 1996, based on Schmidbaur *et al.* 1982).

Interaction energies are predicted to be lower by roughly a factor of three compared to carbenes. Experimental structural data are unavailable.

3.3 Halide anions

Though halide ions are among the strongest hydrogen bond acceptors, they rarely attract specific attention. If incorporated in crystal structures, they play a major role in crystal stabilization, but this circumstance is normally regarded as so trivial that no further discussion is undertaken (and is indeed unnecessary). Hydrogen bonds of O—H and N—H donors to halide ions are conventional and are not discussed here. The hydrogen bonds formed by halide ions with weaker donors are, however, within the scope of the present text and will now be described.

In Table 3.25, average hydrogen bond distances to halide ions are compiled for conventional and some weak hydrogen bond donors (Steiner 1998*h*). For O—H and N—H donors, different acidities clearly reflect in different mean distances d, and the same is true for C—H donors. Note that the average distance d for chloroform (2.39 Å) is not much longer than that for the weakest of the N—H donors, $-N(sp^2)H_2$ (2.35 Å). For weaker C—H donors than $(NC)C(sp^2)-H$, no related data is available as yet. In Fig. 3.99, histograms are shown of H···Cl$^-$ distances d with the donors CHCl$_3$, C≡C—H and $(CN)C(sp^2)-H$. The pronounced maxima of these distributions point at a clear hydrogen bonding character.

Within the group of halides, mean hydrogen bond distances increase by

Table 3.25(a) Mean distances $d(\text{H}\cdots\text{A}^-)$ in $\text{X}-\text{H}\cdots\text{A}^-$ hydrogen bonds to halide ions with $\theta > 140°$ (Steiner 1998*h*)

Acceptor	F⁻	Cl⁻	Br⁻	I⁻
O —H donors				
H—O—H	1.71(2) [14]	2.237(3) [799]	2.400(8) [148]	2.66(1) [47]
C(sp^3)—O—H	1.58 [1]	2.150(5) [299]	2.310(9) [90]	2.55(2) [27]
O=C—O—H	1.50(1) [5]	2.044(8) [65]	2.20(2) [16]	2.42 [1]
P—O—H	—	1.97(2) [13]	—	—
C=O⁺—H	—	1.91(2) [4]	—	—
N —H donors				
—N(sp^2)H$_2$	1.74 [2]	2.350(7) [314]	2.52(2) [77]	2.79(2) [30]
—N⁺H$_3$	1.67 [3]	2.247(5) [467]	2.49(2) [88]	2.72(2) [29]
(CCC)N⁺—H	—	2.079(4) [232]	2.29(1) [90]	2.54(4) [5]
C —H donors				
Cl$_3$C—H	—	2.39(3) [14]	2.62 [1]	2.84(2) [4]
Cl$_2$CH$_2$	—	2.53(3) [17]	2.73 [2]	2.85(3) [12]
C≡C—H	—	2.49(6) [8]	2.70(6) [5]	—
(NN)C(sp^2)—H	—	2.54(2) [48]	2.73(4) [7]	2.90 [3]
(NC)C(sp^2)—H	2.18 [1]	2.64(1) [110]	2.74(2) [56]	2.99(2) [44]
Other donors				
S—H	—	2.23(7) [4]	2.77 [1]	—
P⁺—H	—	2.52 [3]	2.40 [2]	—

Distances *d* normalized, number of contacts given in brackets.

more than 0.5 Å from F⁻ to Cl⁻, by about 0.15 Å from Cl⁻ to Br⁻, and another 0.25 Å from Br⁻ to I⁻. This behaviour is invariant for all donor types, ranging from the strongest, like carboxylic acids, to the weakest in Table 3.25, (NC)C(sp^2)—H. This lengthening of distances corresponds exactly to the increase of the halide ionic radii (1.33, 1.81, 1.96 and 2.20 Å for F⁻ to I⁻, respectively; Shannon 1976).

The strong halide acceptors are often well coordinated with O—H and N—H donors, and are then sterically unavailable for weaker donor types. However, a single Cl⁻ ion, which has approximately the size of a water molecule, can bind about four strong donors (Jeffrey and Saenger 1991), and this is often more than are available in a given structure. Then, the weaker donor types can come into play. There are halide crystal structures which contain no strong donors at all, so that hydrogen bonds can be accepted *only* from C—H donors. An example is crystalline **152**, which possesses a very short C≡C—H⋯Br⁻ hydrogen bond associated with an unusually low acetylenic C—H stretching frequency of 3150 cm⁻¹, Fig. 3.100(a) (Steiner 1996*b*). The distance $d = 2.58$ Å would be more typical for N—H⋯Br⁻ hydrogen bonds with weak N—H donor types (Table 3.25). Different examples are provided by

Table 3.25(b) Mean distances $D(X \cdots A^-)$ in $X-H \cdots A^-$ hydrogen bonds to halide ions with $\theta > 140°$ (Steiner 1998h)

Acceptor	F⁻	Cl⁻	Br⁻	I⁻
O−H donors				
H−O−H	2.68(2) [14]	3.190(3) [799]	3.339(7) [148]	3.60(1) [47]
C(sp^3)−O−H	2.57 [1]	3.100(4) [299]	3.254(8) [90]	3.48(1) [27]
O=C−O−H	2.47(1) [5]	2.997(6) [65]	3.14(1) [16]	3.38 [1]
P−O−H	—	2.94(2) [13]	—	—
C=O⁺−H	—	2.89(1) [4]	—	—
N−H donors				
−N(sp^2)H₂	2.73 [2]	3.299(6) [314]	3.46(1) [77]	3.66(1) [30]
−N⁺H₃	2.65 [3]	3.207(4) [467]	3.44(1) [88]	3.68(2) [29]
(CCC)N⁺−H	—	3.059(3) [232]	3.247(7) [90]	3.50(3) [5]
C−H donors				
Cl₃C−H	—	3.42(2) [14]	3.56 [1]	3.86(4) [4]
Cl₂CH₂	—	3.57(3) [17]	3.74 [2]	3.88(2) [12]
C≡C−H	—	3.51(5) [8]	3.72(5) [5]	—
(NN)C(sp^2)−H	—	3.57(2) [48]	3.72(4) [7]	3.85 [3]
(NC)C(sp^2)−H	3.21 [1]	3.66(1) [110]	3.75(2) [56]	4.00(2) [44]
Other donors				
S−H	—	3.54(8) [4]	4.03 [1]	—
P⁺−H	—	3.93 [3]	3.75 [2]	—

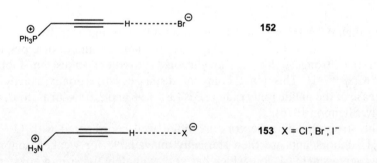

152

153 X = Cl⁻, Br⁻, I⁻

many halide salts of 1,3-dialkylated imidazolium ions. Since both N atoms carry alkyl groups, conventional hydrogen bonding is not possible. An example where discrete dimers are constituted by $C-H \cdots Cl^-$ hydrogen bonds has been discussed by Seddon and co-workers and is shown in Fig. 3.100(b) (Abdul-Sada *et al.* 1990). Another very impressive example is shown for a ternary compound in Fig. 3.100(c), where a Cl⁻ ion satisfies its acceptor potential with three chloroform molecules and the imidazolium C(2)−H donor (Boese *et al.* 1994). It has been shown that imidazolium–halide contacts

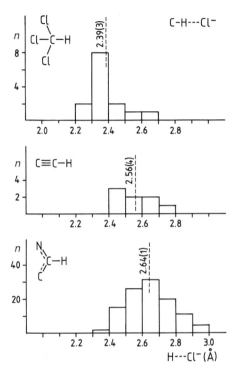

Fig. 3.99. Distances d in $C-H\cdots Cl^-$ hydrogen bonds ($d < 3.0\,\text{Å}$, $\theta > 140°$) for three different kinds of $C-H$ donors (from Steiner 1998h).

of this kind possess the IR spectroscopic characteristics of hydrogen bonds, and therefore cannot be regarded as a simple ion-pair interaction (Elaiwi *et al.* 1995).

In other halide salts, there are one or two strong donors per anion. Then, conventional $O/N-H\cdots Hal^-$ hydrogen bonds satisfy part of the acceptor potential, but a part remains available for weak donors. Examples have been discussed by Aakeröy and Seddon (1993b) for the polymorphs of pyridinium chloride. In one crystal form, the chloride ion is threefold coordinated by one N^+-H and two $(N)C-H$ donors. In the other crystal form, there are two symmetry-independent ions, one of which accepts hydrogen bonds from one N^+-H and four $C-H$ donors, and the other one from one N^+-H and three $C-H$ donors. The latter is shown in Fig. 3.101. $C-H\cdots Cl^-$ hydrogen bonds in an organometallic compound which is deficient in conventional donors have been described by Davidson *et al.* (1995).

In halide salts which have many conventional donors, hydrogen bonds with $C-H$ groups can still be formed. Examples are the isostructural propargyl ammonium halides **153**, in which the halide ions accept hydrogen bonds from three ammonium groups and the propargyl residue, Fig. 3.102 (Steiner 1998i).

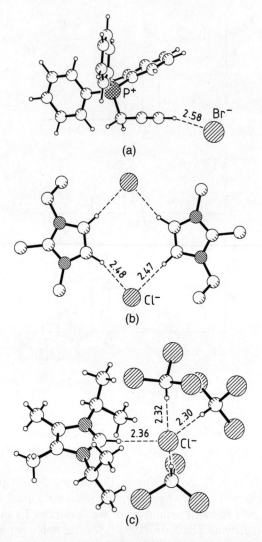

Fig. 3.100. Halide crystal structures devoid of conventional donors: (a) compound **152** (Steiner 1996*b*); (b) 1-ethyl-2,3-dimethylimidazolium chloride (Abdul-Sada *et al.* 1990); (c) 1,3-diisopropyl-4,5-dimethylimidazolium chloride tris (chloroform) with four short C—H···Cl⁻ hydrogen bonds (based on Boese *et al.* 1994). Geometries are normalized.

The C≡C—H···Hal⁻ bonds are longer than average (2.62, 2.75 and 2.96 Å for Cl⁻, Br⁻ and I⁻, respectively), whereas the N⁺—H···Hal⁻ bonds are slightly shorter than average. This shows that in a competitive situation, the ammonium–halide interactions are optimized at the expense of relatively unfavourable C—H···Hal⁻ geometries.

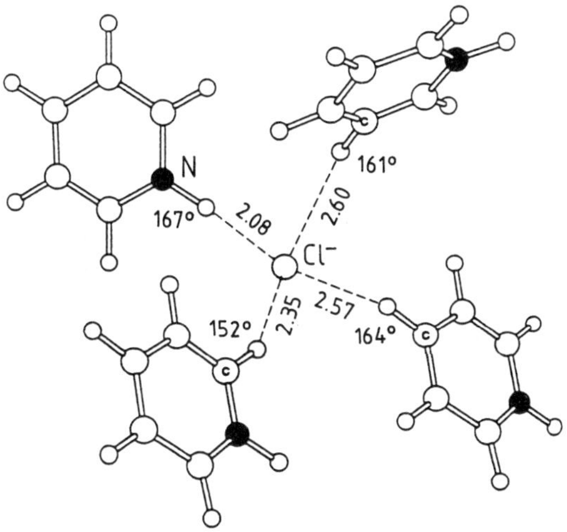

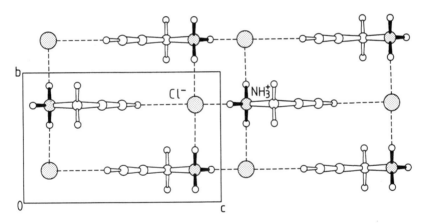

Fig. 3.101. Monoclinic pyridinium chloride, a halide crystal structure with only one conventional donor (adapted from Aakeröy and Seddon 1993b, structure determined by Hensen et al. 1988).

Fig. 3.102. Propargylammonium chloride, **153**. The Cl⁻ ion accepts three hydrogen bonds from ammonium groups and one from the terminal alkyne residue (from Steiner 1998i). The Br⁻ and I⁻ salts are isostructural.

The examples discussed above involve only acidic $C-H$ donors. For these, the hydrogen bond nature of $C-H \cdots Hal^-$ contacts is well established on geometrical and spectroscopic grounds. For less acidic $C-H$ donors, open questions remain. For methyl groups in particular, there are configurations where the hydrogen bond nature of short $N^+-C-H \cdots Hal^-$ contacts may even be questioned. Contacts from halides to $R-NMe_3^+$ groups are typical of the kind

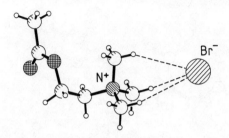

Fig. 3.103. Acetylcholine bromide (Svinning and Soerum 1975). A typical example of a halide ion contact to a quarternary ammonium group. See Table 3.26.

Table 3.26 Mean geometries of $-N^+Me_3\cdots Hal^-$ interactions of the type shown in Fig. 3.103, that is with the anion forming three short contacts to $C-H$ groups

	F^-	Cl^-	Br^-	I^-
n	4	66	41	68
$N^+\cdots Hal^-$ (Å)	3.41(6)	3.96(1)	4.19(1)	4.41(1)
$H\cdots Hal^-$ (Å)	2.26(6)	2.78(1)	2.96(1)	3.16(1)
$C\cdots Hal^-$ (Å)	3.20(6)	3.86(1)	3.95(1)	4.15(1)
$C-H\cdots Hal^-$ (°)	146(2)	149.8(8)	151.8(8)	152.8(6)

CSD, $R < 0.08$, distances d normalized, distance cut-offs $d < d_{max}$, where $d_{max} = 2.5, 3.0, 3.2$ and 3.6 Å for F^-, Cl^-, Br^- and I^-, respectively.

seen in acetylcholine bromide, Fig. 3.103 (Svinning and Soerum 1975). As the halide anion approaches the central charge, it forms short contacts with three $C-H$ groups. In this arrangement, the attractive force is certainly the Coulombic $N^+\cdots Hal^-$ interaction, whereas the $H\cdots Hal^-$ contacts block the anion from further approach to N^+. Although the $C(sp^3)-H\cdots Hal^-$ contacts are short and relatively linear, their geometry is determined by optimization of the ion-pair interaction. This configuration is abundant in crystals and so, mean geometries are given for all halides in Table 3.26.

Halide ions can act as acceptors of bifurcated hydrogen bonds. In these, either both components are directed at halides, or one of the two is directed at a different kind of acceptor. Secondary components of bifurcated bonds typically have strongly bent angles and longer distances than the average values given in Table 3.25. As an example, the hydrogen bond arrangement around the imidazolium group of a histidine residue in a small peptide is shown in Fig. 3.104.

$S-H$ and P^+-H donors can form relatively short hydrogen bonds with halide ions (Table 3.25), but this interaction occurs very rarely. Examples will be shown in the corresponding sections for these donors (3.4.1 for $S-H$ and 3.4.2 for P^+-H).

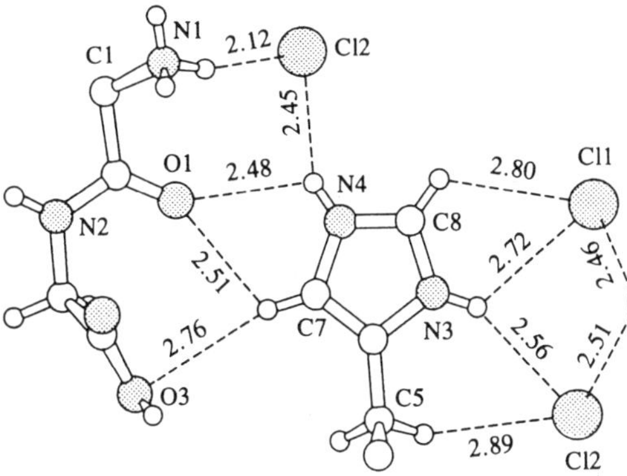

Fig. 3.104. Bifurcated hydrogen bonding involving C—H donors and Cl⁻ acceptors. The hydrogen bond pattern around the histidine side-chain of the dipeptide (L-His-Gly)²⁺·2Cl⁻ (from Steiner 1997b). Distances d are normalized.

3.4 Weak donors

Of all the elements, only N, O, F, Cl and Br are significantly more electronegative than H (Table 1.1). Only for these are X—H bonds distinctively polar as $X^{\delta-}$—$H^{\delta+}$, making them strong hydrogen bond donors. C, S and I are slightly more electronegative than H, so that they are usually weak hydrogen bond donors. By appropriate substitution, they may become moderately strong donors as has been discussed extensively for C—H in Chapter 2. Some other elements have electronegativities similar or slightly smaller than H, that is P, Si, Se, As. For these elements, X—H groups normally cannot donate hydrogen bonds, but in particular cases, may be rendered suitably polar so as to act as hydrogen bond donors. This section is devoted to these donor groups S—H, P—H, Si—H and Se—H. The role of some transition metal hydrides as hydrogen bond donors will be discussed in Section 3.5.2.

3.4.1 S—H

3.4.1.1 General

Though nominally a typical non-metallic element and placed just one position below O in the Periodic Table, S has a much smaller electronegativity (2.5 compared to 3.5 for O). In consequence, the S—H group is a much weaker hydrogen bond donor than O—H. Nevertheless, S—H is one of the classical hydrogen bonding functional groups, and its donor and acceptor potentials have been studied in early times by vibrational spectroscopists (Pimentel and McClellan 1960). For an example, the S—H stretching frequencies ν_{S-H} of thiophenol in different solvents are listed in Table 3.27. The reduction of ν_{S-H} in solvents with suitable acceptors is small, but still indicative of hydrogen

Table 3.27 S—H stretching frequency of dilute thiophenol in various solvents (David and Hallam 1964)

Solvent	Type of hydrogen bond	ν_{S-H} (cm^{-1})	Red shift^a (cm^{-1})
CCl$_4$	None	2590	—
Dioxane	S—H$\cdots$O	2539	51
Acetonitrile	S—H$\cdots$N	2565	25
Benzene	S—H$\cdots\pi$	2574	16

^{a} Compared to dilute CCl$_4$ solution.

bonding. Even S—H$\cdots\pi$ hydrogen bonding with benzene as the acceptor had in this way become apparent by the mid-1960s (David and Hallam 1964).

S—H$\cdots$X hydrogen bonding in crystals, however, is not well documented. In particular for hydrogen bonds donated by 'normal' thiols, relatively few examples have been published. To some degree, this is caused by disorder problems. Since with few exceptions S—H$\cdots$X interactions are weak and the thiol group can rotate, the hydrogen bond can be easily broken; as a consequence, the H atom position often cannot be located. Even in ordered S—H$\cdots$X hydrogen bonds, the H atom position is less well defined than in O/N—H$\cdots$X bonds. In practical terms, an alarmingly large fraction of thiol groups with unrealistic geometries may be found in the CSD. The exact covalent geometry of X—S—H groups in the solid state has been determined by neutron diffraction for only two examples: C(sp^3)—S—H in N-acetyl-L-cysteine has an S—H bond length of 1.338(2) Å and a bond angle of 96.9(2)° (single crystal neutron diffraction at 16 K; Takusagawa *et al.* 1981). In orthorhombic D$_2$S, the mean S—D bond length is 1.33(1) Å and the mean bond angle is 92.9(6)° (neutron powder diffraction at 1.5 K; Cockroft and Fitch 1990).

Still, it is of importance that the S—H group may simultaneously act as hydrogen bond donor and acceptor, analogous to O—H. This means that S—H may participate in complex hydrogen bond arrays, thereby experiencing stabilizing cooperativity effects in a manner very similar to O—H groups.

3.4.1.2 S—H$\cdots$O and S—H$\cdots$N hydrogen bonds

Histograms of distances d and D in S—H$\cdots$O hydrogen bonds are shown in Fig. 3.105 averaged over all types of O atom acceptors. The distributions are broad, but peaks can be identified for d in the interval 2.1–2.3 Å, and D in the interval 3.3–3.6 Å. S—H$\cdots$O hydrogen bonds are mainly formed with carbonyl and carboxylate acceptors, and it is only for these that more detailed statistical analyses are possible. In a CSD study of 15 S—H$\cdots$O=C hydrogen bonds, a mean distance d of 2.34(4) Å was found (for normalized H atom positions; Allen *et al.* 1997a,b). This is much longer than the 1.884(3) Å in O—H$\cdots$O=C and the 2.011(2) Å in N—H$\cdots$O=C hydrogen bonds.

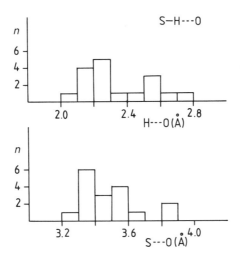

Fig. 3.105. Histograms of normalized distances d and D of intermolecular $S-H\cdots O$ hydrogen bonds in crystals, (CSD, $R < 0.08$, $d < 2.8$ Å, all O atom acceptors).

The most important and best investigated of the $S-H\cdots O{=}C$ hydrogen bonds are those formed by the amino acid cysteine. The donor is of the poorly activated type $C(sp^3)-S-H$, so that these hydrogen bonds clearly fall into the category 'weak'. Three examples are shown as schemes in Fig. 3.106. In the monoclinic form of L-cysteine, there are two molecules per asymmetric unit, and their thiol groups form a short cooperative chain $S-H\cdots S-H\cdots O_2C^-$, Fig. 3.106(a) (Görbitz and Dalhus 1996). In N-acetyl-L-cysteine, the $S-H$ group is involved in a cooperative chain $N-H\cdots S-H\cdots O{=}C$ with a very long $N-H\cdots S$ hydrogen bond, Fig. 3.106(b) (Takusagawa *et al.* 1981), and finally, in glutathione, the $S-H$ group acts only as a donor and forms a hydrogen bond with a carboxylate group, Fig. 3.106(c) (Görbitz 1987). All these all can be considered as very typical $S-H\cdots O$ hydrogen bonds.

Much less data is available on hydrogen bonds of $S-H$ with O atom acceptors other than $O{=}C$. For phenolic acceptors, a remarkable cooperative ring involving two $S-H$ and two $O-H$ groups occurs in a calixarene, Fig. 3.107 (Delaigue *et al.* 1995). This pattern is equivalent to the many hydrogen bond rings reported with four $O-H$ groups. Within the ring, the $S-H\cdots O$ and $O-H\cdots S$ hydrogen bonds are constrained by the calixarene stereochemistry to have almost identical distances D.

A much stronger kind of $S-H$ donor occurs in the monothiocarboxylic acids. As an example, the structure of 2-hydroxythiobenzoic acid is shown in Fig. 3.108 together with the oxygen analog (Mikenda *et al.* 1995). Although there is the similar formation of centrosymmetric cooperative dimers, the difference in $C-X-H$ angles leads to clearly different dimer geometries ($C-S-H = 93°$, $d = 2.05$, $D = 3.39$ Å for normalized H atom position). The short distance d actually indicates relatively strong hydrogen bonding.

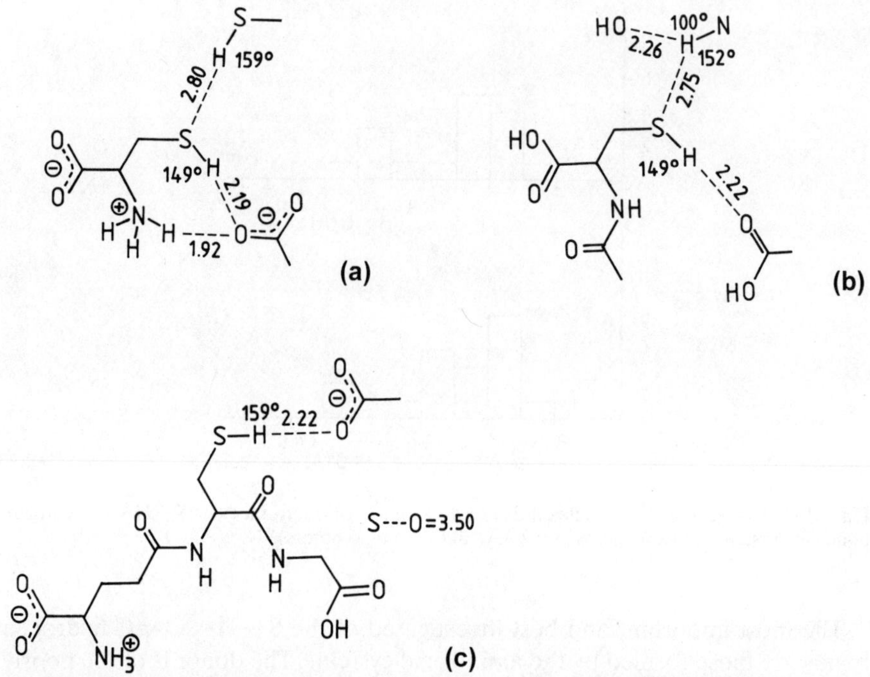

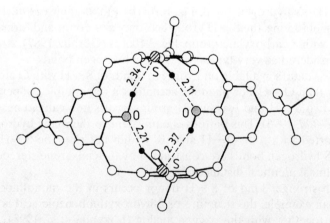

Fig. 3.106. S—H···O hydrogen bonds with carbonyl and carboxylate acceptors. (a) Monoclinic form of L-cysteine with two molecules per asymmetric unit; note the cooperative chain S—H···S—H···O (Görbitz and Dalhus 1996). (b) N-Acetyl-L-cysteine, neutron diffraction study, D = 3.43 Å (Takusagawa *et al.* 1981). (c) γ-L-Glutamyl-L-cysteinylglycine, glutathione (Görbitz 1987). Geometries are normalized.

Fig. 3.107. Cooperative hydrogen bond ring involving two S—H···O and two O—H···S bonds in an uncomplexed calixarene (from Delaigue *et al.* 1995). Geometries are normalized. The S—H···O bonds have distances of D = 3.26 and 3.29 Å, and angles of 144° and 132°, respectively.

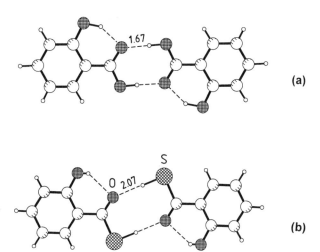

Fig. 3.108. Molecular dimers in (a) 2-hydroxybenzoic acid (Sundaralingam and Jensen 1965) and (b) its O=C−S−H analogue 2-hydroxythiobenzoic acid (Mikenda *et al.* 1995). Distances *d* are normalized.

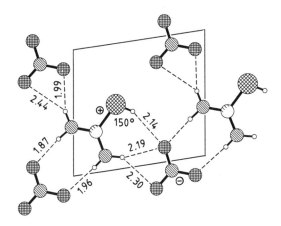

Fig. 3.109. Hydrogen bond network in a layer of isothiouronium nitrate, containing a short S⁺−H···O hydrogen bond (Feil and Loong, 1968). Geometries are normalized.

Similar to N−H and O−H groups, the acidity and therefore the donor strength of S−H is dramatically enhanced by a positive charge, S⁺−H. This can be clearly seen in thiouronium salts, which contain very short S⁺−H···A hydrogen bonds. The example of thiouronium nitrate, which forms a layer-type crystal structure, is shown in Fig. 3.109 ($d = 2.14$ Å; Feil and Loong 1968). In the related salt thiouronium perchlorate, an S⁺−H···O hydrogen bond with $d = 2.15$ Å is formed (Eskudero *et al.* 1992).

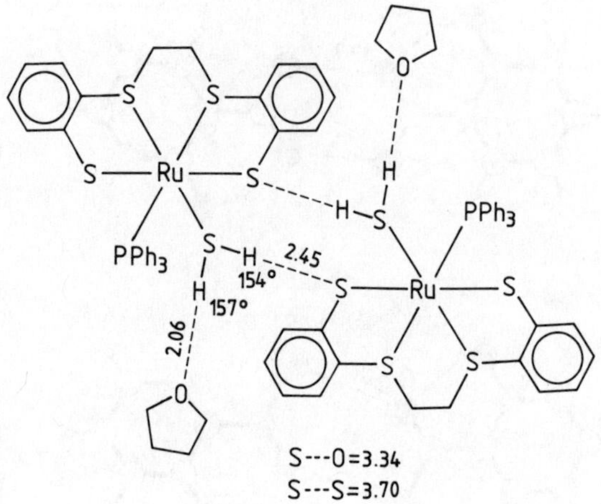

Fig. 3.110. S—H···X hydrogen bonds formed by a ruthenium-coordinated H_2S molecule (Sellmann *et al.* 1992). Geometries are normalized.

A singular hydrogen bond arrangement involving an H_2S molecule has been found in an organometallic crystal structure, Fig. 3.110 (Sellmann *et al.* 1992). The metal-coordinated H_2S molecule donates exceptionally short S—H···O and S—H···S hydrogen bonds, one with a co-crystallized tetrahydrofuran molecule, and one with a symmetry-related organometallic complex.

S—H···N hydrogen bonds occur in crystals very rarely. For an example, we refer to the structure of thiadiazole-2-thiol-5-thione reported by Bats (1976). In this structure fairly linear S—H···N hydrogen bonds are formed with distances $d = 2.29$ and $D = 3.44$ Å.

3.4.1.3. S—H···S hydrogen bonds

The homonuclear S—H···S hydrogen bonds have attracted greater interest than S—H···O and S—H···N interactions. IR spectroscopic data show that S—H···S hydrogen bonds can have very unequal strengths in different types of compounds, Table 3.28. Whereas in the 'normal' thiols, sulfanes and thiophenols, the red shifts due to S—H···S hydrogen bonding amount to only about $20 \, cm^{-1}$, this value can become over $200 \, cm^{-1}$ in thiocarbonic acids and thiophosphinic acids. This means that S—H···S hydrogen bonds are typically very weak, but may become moderately strong in particular compounds.

The smallest molecule in which the S—H group acts as a hydrogen bond donor is the S analogue of water, H_2S, and the hydrogen bonding pattern is also in many ways the most intriguing. Despite the higher molecular weight, hydrogen bonding in H_2S is much weaker than in H_2O as is shown by the lower melting point (MP = 187.6 K). The binding enthalpy of the gas phase H_2S dimer is only $-1.7(3)$ kcal/mol (experimental value of Lowder *et al.* 1994). There are

Table 3.28 Typical IR spectroscopic data on $S-H \cdots S$ hydrogen bonds

Type		ν_{S-H} red shift (cm^{-1})	Reference
Thiols	$C(sp^3)-S-H$	20	(a)
Thiophenols	$Ph-S-H$	20	(b)
Sulfanes	$S-S-H$	20	(c)
Dithiotropolone	$S=C-C-S-H$, cyclic resonant	120	(d)
Trithiocarbonic acid	H_2CS_3	170	(e)
Dithiophosphine acids	$P(=S)-S-H$	170–245	(f)

(a) Spurr and Byers 1958; de Alencastro and Sandorfy 1972. (b) David and Hallam 1965; de Alencastro and Sandorfy 1972. (c) Muller and Hyne 1968. (d) Krebs *et al.* 1984. (e) Tyce and Powell 1965. (f) Allen and Colclough 1957.

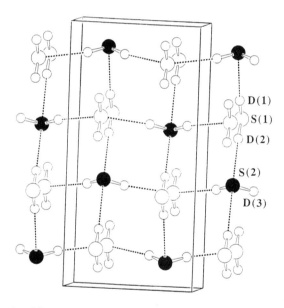

Fig. 3.111. The ordered low-temperature phase of D$_2$S as determined by neutron powder diffraction at 1.5 K (from Cockroft and Fitch 1990).

at least three solid phases of H$_2$S, two cubic phases showing orientational disorder, and an orthorhombic low-temperature phase which is ordered (below 103.5 K, space group *Pbcm*). The crystal structure of the ordered phase is shown in Fig. 3.111, as determined by neutron powder diffraction on D$_2$S at 1.5 K (Cockroft and Fitch 1990). Because the $D-S-D$ angle of 92.9° is close to rectangular, it is not surprising that the molecular arrangement is completely different from that found in any of the water ices. There are two symmetry-independent D$_2$S molecules, each of which donates and accepts two

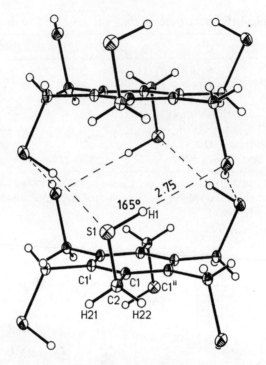

Fig. 3.112. Cooperative cycle of six S—H···S hydrogen bonds in hexakis(mercapto-methyl)benzene (from Mallinson *et al.* 1997*b*). The geometry is normalized.

S—D···S hydrogen bonds. In these hydrogen bonds, the distances D are between 3.985 and 4.027 Å, the distances d between 2.68 and 2.74 Å, and the angles θ between 163.0 and 166.3°. The hydrogen bond situation in the disordered high-temperature phases could not be deduced from the powder diffraction data.

Hydrogen bonds between thiol groups bonded to $C(sp^3)$ are also very long. An example has already been shown in Fig. 3.106(a), for the S—H···S—H···O chain in monoclinic L-cysteine (Görbitz and Dalhus 1996). A crystal structure of exceptional beauty has been found by Mallinson *et al.* (1997*b*) for hexakis(mercaptomethyl)benzene. The six thiol groups in the molecule form cooperative cycles of six symmetry-related S—H···S hydrogen bonds, Fig. 3.112. In both these examples, the hydrogen bond distances are long with distances d and D of around 2.8 and 4.0 Å, respectively.

Much stronger S—H···S hydrogen bonds are formed by activated S—H groups, such as in the thiocarbonic and thiophosphinic acids. Here, the IR bathochromic shift of v_{S-H} exceeds that in thiols by about a factor of ten (Table 3.28), and the hydrogen bond distances are significantly shorter. A series of crystal structures has been published for dithiophosphinic acids

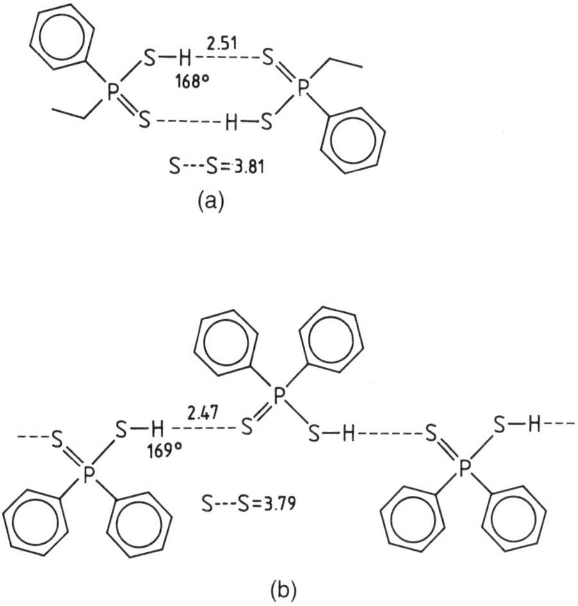

(a)

(b)

Fig. 3.113. Strong S—H···S hydrogen bonding by dithiophosphinic acids. (a) Dimer motif in ethyl-phenyl-dithiophosphinic acid (Krebs 1983). (b) Catemer motif in diphenyl-dithiophosphinic acid (Krebs and Henkel 1981). Geometries are normalized.

(Krebs 1983). The molecules associate either as dimers or in a catemer motif (Fig. 3.113), and the analogy with the hydrogen bond motifs formed by carboxylic acids is obvious. The distances d and D are around 2.50 and 3.80 Å, respectively, about 0.3 Å shorter than for hydrogen bonds between thiol groups.

Trithiocarbonic acid, H_2CS_3, has two crystalline polymorphs. Both have very complicated crystal structures with three and four symmetry-independent molecules (for the β- and α-forms, respectively). In both polymorphs, three-dimensional arrays of S—H···S hydrogen bonds are formed with distances D in the range 3.80–3.97 Å (Krebs *et al.* 1980; Krebs and Henkel 1987). IR spectral shifts of ν_{S-H} approach those in the dithiophosphinic acids (Table 3.28).

Dithiotropolone, **154**, has a very interesting crystal structure (Krebs *et al.* 1984). The S—H···S=C hydrogen bond is extremely short, $D = 3.12$ Å, but quite non-linear because of stereochemical constraints. The IR red shift of ν_{S-H} is about six times that in hydrogen bonds between thiol groups (Table 3.28). The unusually strong S—H···S hydrogen bond can be explained by resonance assistance analogous to related O—H···O systems, that is, as a cooperative effect. As a comparison, the corresponding intramolecular O—H···O=C distance D in tropolone, **155**, is 2.55 Å.

Whereas the strong hydrogen bonds in **154** and **155** may be interpreted as

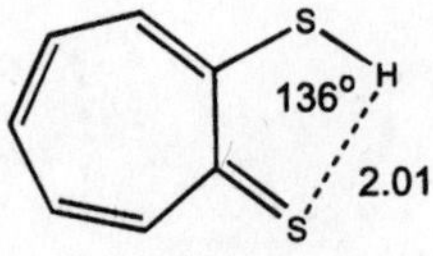

154

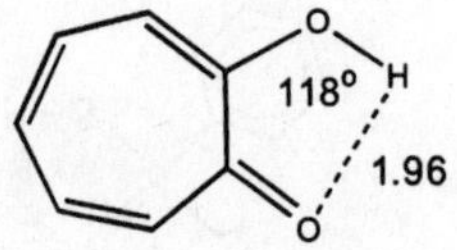

155

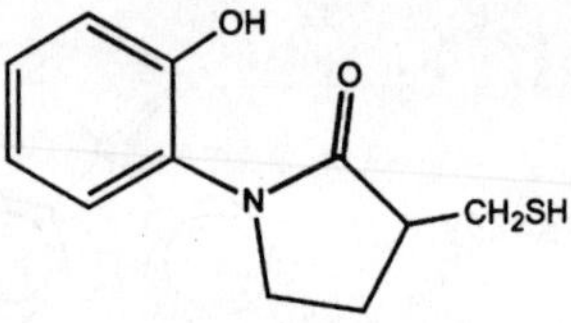

156

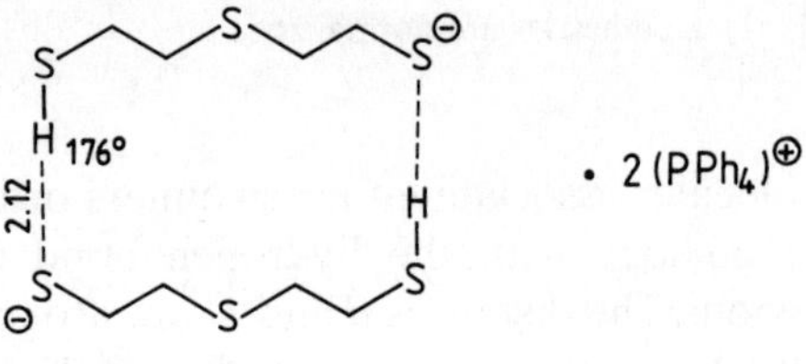

S---S = 3.45

Fig. 3.114. Strong S⁻—H···S hydrogen bonds in a tetraphenylphosphonium thiolate salt (Boorman *et al.* 1992). The geometry is normalized.

resonance assisted, strong S—H···S hydrogen bonding is also possible if mediated by *charge assistance*, that is, if one of the two participants carries a full charge. Such behaviour is well-documented for O—H···O hydrogen bonds (Bertolasi *et al.* 1996). A relevant charge-assisted system of the type S—H···S⁻ has been found by Boorman *et al.* (1992) in the tetraphenylphosphonium thiolate salt shown in Fig. 3.114. With a distance d of only 2.12 Å, this represents the shortest intermolecular S—H···S hydrogen bond reported.

Finally, the anion (SH)⁻ may be mentioned as an extremely weak hydrogen bond donor. An example of S⁻—H···S⁻ interactions formed by this anion is found in the ordered low-temperature phase of CsHS, Fig. 3.115 (neutron diffraction structure of Jacobs and Kirchgässner 1989). The anions form square

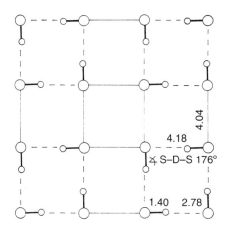

Fig. 3.115. S⁻—H···S⁻ interactions in a layer of the low-temperature phase of CsDS. The Cs ions are coordinated by eight DS⁻ ions (not shown). Neutron powder diffraction at 9 K (Jacobs and Kirchgässner 1989).

arrangements with hydrogen bonds characterized by $d = 2.78$ and $D = 4.18\,\text{Å}$. It is of interest that there are non-bonded S···S contacts between the ions which are even shorter than those involving the hydrogen bonds, 4.04 Å. This crystal structure is possibly highly influenced by the polarizability (softness) of the S atoms.

3.4.1.4 *S—H···Hal⁻ hydrogen bonds*

If the cationic species of a halide salt contains an S—H group, S—H··· Hal⁻ hydrogen bonds can occur. Two different configurations are possible: the positive charge may be placed on S—H itself, i.e. the hydrogen bond is of the type S⁺—H···Hal⁻, or it may be placed elsewhere, (R⁺)—S—H. Clearly, S⁺—H is the more acidic of these two variants, and as such represents the stronger hydrogen bond donor. One example for each of these cases is given in Fig. 3.116. In the chloride salt of cysteine ethyl ester, the S—H group donates a hydrogen bond to the anion, Fig. 3.116(a) (Görbitz 1989). In this case, the positive charge of the cation is localized on the ammonium group, separated from S—H by three σ bonds so that the thiol group can be regarded effectively as uncharged. An example for an S⁺—H···Cl⁻ hydrogen bond is provided by the crystal structure of Me₂S⁺—H·Cl⁻·3 HCl, Fig. 3.116(b) (Mootz and Deeg 1992*b*). Mean hydrogen bond distances to halides averaged over all kinds of S—H donors are given in Table 3.25.

3.4.1.5 *S—H···F—X⁻ hydrogen bonds*

Hydrogen bonds of S—H to the anions BF_4^- and PF_6^- are not at all weak and are mentioned here only for reasons of complete coverage. In a CSD search,

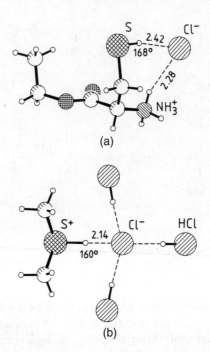

Fig. 3.116. S—H···Cl⁻ hydrogen bonds. (a) L-Cysteine ethyl ester hydrochloride, $D = 3.74\,\text{Å}$ (Görbitz 1989). (b) Me$_2$S$^+$—H·Cl⁻·3 HCl, $D = 3.42\,\text{Å}$ (Mootz and Deeg 1992b). Geometries are normalized.

only four examples were found for S—H···F$_4$B⁻ hydrogen bonds, with mean values of $d = 2.31(6)$ and $D = 3.55(7)\,\text{Å}$ (for normalized H atom position). Only one example is found for an S—H···F$_6$P⁻ hydrogen bond, with $d = 2.35$ and $D = 3.58\,\text{Å}$ (see Section 3.2).

3.4.1.6 S—H···π hydrogen bonds

Spectroscopic evidence for S—H···π hydrogen bonding is long known (Table 3.27), but only three crystal structures are known with unambiguous S—H···Ph hydrogen bonds. In crystalline triphenylmethanethiol (Bernardinelli *et al.* 1991), one of two independent molecules forms a S—H···π bonded dimer across a centre of symmetry, Fig. 3.117(a). The geometry of this interaction is very similar to that of typical O/N—H···Ph hydrogen bonds, and is therefore taken as an indication of hydrogen bonding [d(C) = 2.76–3.09, d(M) = 2.60, D(M) = 3.73 Å, for normalized H atom position]. The symmetry-independent molecule does not form this kind of interaction and appears to have a 'free' thiol group. In **156** (Nishio *et al.* 1996), centrosymmetric dimers are formed too, linked by a pair of S—H···Ph interactions, Fig. 3.117(b) [d(C) = 2.81–3.01, d(M) = 2.60, D(M) = 3.73 Å]. Particularly interesting are the

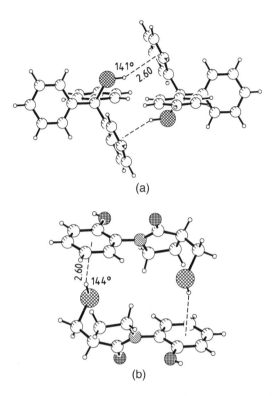

(a)

(b)

Fig. 3.117. Possible intermolecular S—H···Ph hydrogen bonds. (a) Triphenylmethanethiol (Bernardinelli *et al.* 1991). (b) Compound **156** (Nishio *et al.* 1996). Geometries are normalized.

intramolecular hydrogen bonds in compound **157** (Hiller and Rundel 1993). This molecule is closely related to the one shown in Fig. 3.21(e), with the O—H groups substituted by S—H. In both molecules, analogous intramolecular X—H···Ph hydrogen bonds are formed, Fig. 3.118.

3.4.1.7 *Summary*

This section has shown that the hydrogen bond range of the S—H donor is much broader than is commonly assumed. The S—H group is chemically variable and depending on the degree of activation its donor ability can change from very weak to quite strong. Because of its relationship to the O—H group, the S—H group allows for most of the mechanisms that make the former such a variable hydrogen bond functionality. S—H···X hydrogen bonds can experience the same kind of cooperative effects as O—H···X, including resonance and charge assistance. S—H groups in uncharged thiols and thiophenols are the most familiar to organic chemists and may be classified as rather weak donors.

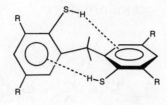

157 R = t-Bu

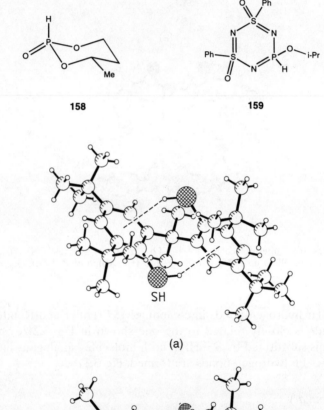

158

159

SH

(a)

OH

(b)

Fig. 3.118. Two related structures with intramolecular X—H···Ph hydrogen bonds. (a) S—H···Ph bonds in compound **157** (Hiller and Rundel 1993). (b) O—H···Ph hydrogen bonds in the compound shown in Fig. 3.21(e) (Hardy and MacNicol 1976).

3.4.2   *P—H and P⁺—H*

The electronegativities of P and H are virtually identical and so P—H bonds
are normally non-polar. Accordingly, it may be questioned as to whether P—H
can act as a hydrogen bond donor at all. However, unambiguous P—H···O
hydrogen bonds are observed if P is bonded to electron-withdrawing groups
or if it carries a positive charge, in other words if the P—H bond is activated.
The amount of structural data available on these systems is very small, so that
only a few examples can be shown below. Statistical information cannot be
given in a satisfactory manner.

3.4.2.1   *P—H···O hydrogen bonds*

Three different cases can be distinguished: P—H···O interactions in
uncharged organic compounds, in charged compounds with P⁺—H donors,
and possibly in organometallic compounds with the weak carbonyl acceptors.
The last case is not discussed here further because there is no original publication that deals with the matter.

In uncharged molecules, P—H is observed as hydrogen bond donor if it is
bonded to strongly electron-withdrawing groups. Two examples discussed in
the literature are shown in Fig. 3.119. In **158**, P—H is bonded to three oxygen
atoms, and forms a hydrogen bond to an O=P acceptor, Fig. 3.119(a) ($d = 2.23$
Å; Saenger and Mikolajczyk 1973). In **159**, P—H is bonded to one oxygen and
two nitrogen atoms, and donates a hydrogen bond to an O=S acceptor with
$d = 2.29$ Å. Pairs of these interactions join adjacent molecules to form
centrosymmetric dimers, Fig. 3.119(b) (Winter *et al.* 1987).

The P⁺—H group in phosphonium ions can easily donate hydrogen bonds
to anions with O atoms. A fine example with a bifurcated P⁺—H···O
hydrogen bond has been discussed by Jones, Blaschette and co-workers for
the triphenylphosphonium salt shown schematically in Fig. 3.120 (Hiemisch *et
al.* 1996). The two hydrogen bond components are donated to different anions,
and the shorter of the two distances D is only 3.30 Å.

3.4.2.2   *P⁺—H···Hal⁻ hydrogen bonds*

Several crystal structures of phosphonium ions with halide counterions have
been published, in which P⁺—H···Hal⁻ hydrogen bonds are formed, Fig. 3.121.
In triphenylphosphonium bromide and in a monohydroxy derivative,
P⁺—H···Br⁻ hydrogen bonds are observed with distances d around 2.4 Å, Fig.
3.121(a,b) (Bricklebank *et al.* 1993; Schmutzler *et al.* 1984, respectively). These
distances are similar to those in water–bromide hydrogen bonds,
O_W—H···Br⁻, Table 3.25 (the distances D are longer than the O_W···Br⁻ distances because of the longer P—H covalent bond, Table 1.3). In methylphosphonium chloride, three P⁺—H···Cl⁻ hydrogen bonds are formed with
distances d around 2.50 Å, Fig. 3.121(c) (Fluck *et al.* 1986).

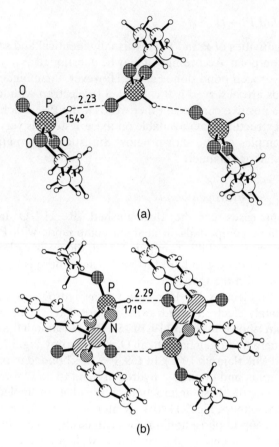

(a)

(b)

Fig. 3.119. P—H···O hydrogen bonds in neutral molecules: (a) compound **158** (Saenger and Mikolajczyk 1973); (b) compound **159** (Winter *et al.* 1987). Geometries are normalized.

3.4.2.3   *P—H···P interactions*

Whereas some unambiguous examples for P—H···O and P⁺—H···Hal⁻ hydrogen bonds are known, significant homonuclear hydrogen bonds of the type P—H···P still await realization in experiment. The unactivated system PH₃ is hardly suitable for this purpose. In *ab initio* calculations on the gas phase dimer, only a very weakly bonded adduct was obtained that is not linked by an H atom placed between the P atoms (Frisch *et al.* 1984). In tetraphosphanylsilane, Si(PH₂)₄, too the P—H groups are more or less unactivated. Despite the large molecular weight, the melting point is low at –25°C, indicating the absence of strong intermolecular interactions. In the interesting crystal structure determined by Driess *et al.* (1998), one of the phosphanyl groups forms relatively linear P—H···P contacts with long distances of $d = 2.97$ and $D = 4.3$ Å, Fig. 3.122. However, the authors note that *ab initio*

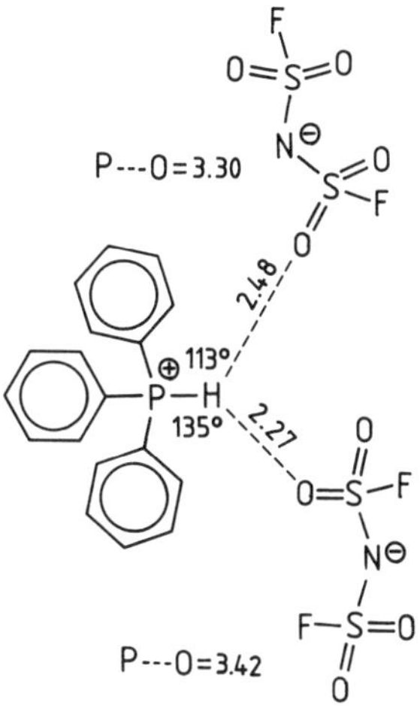

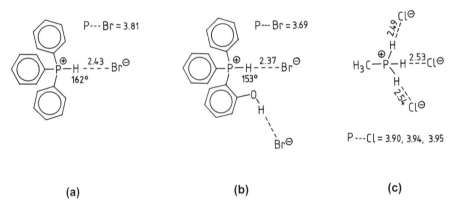

Fig. 3.120. Bifurcated P^+—H$\cdots$O hydrogen bond in a triphenylphosphonium salt (Hiemisch *et al.* 1996). Geometries are normalized.

Fig. 3.121. P^+—H$\cdots$Hal$^-$ hydrogen bonds in (a) triphenylphosphonium bromide (Bricklebank *et al.* 1993), (b) (2-hydroxyphenyl)-diphenylphosphonium bromide (Schmutzler *et al.* 1984), and (c) methylphosphonium chloride (Fluck *et al.* 1986). Geometries are normalized.

calculations on such contacts gave destabilizing energies, and vibrational spectroscopic studies in an argon matrix gave no indication of dimer formation.

3.4.3 *Se*—*H, As*—*H and Si*—*H*

Se has an electronegativity of 2.4 (Table 1.1). This is about the same as S (2.5) and should render the Se—H group a weak hydrogen bond donor. However,

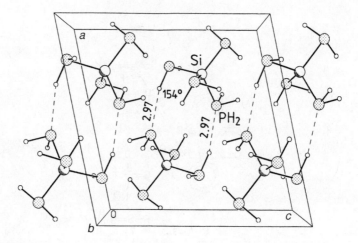

Fig. 3.122. Linear P—H···P contacts in tetraphosphanylsilane. Most P—H groups do not form such interactions (from Driess *et al.* 1998).

almost no structural information is available on this system. The current update of the CSD contains very few Se—H groups, and none of these forms an unambiguous hydrogen bond. For liquid H_2Se, IR spectroscopic data provide some evidence for very weak Se—H···Se interactions, but the estimated binding energy of −0.2 kcal/mol renders classification as a hydrogen bond questionable (Sennikov 1994).

As has an electronegativity of 2.0. This should be sufficient to make As—H a hydrogen bond donor under special circumstances. However, no relevant crystal structures have been reported as yet. The electronegativity of Si is lower than that of H, 1.8 compared to 2.1, so that Si—H groups do not normally donate hydrogen bonds. Exceptions might be possible, if Si carries sufficiently electron-withdrawing substituents such as in trichlorosilane, $SiHCl_3$ (silicochloroform). No crystal structure containing $SiHCl_3$ or a related molecule has as yet been reported and there is no clear example for Si—H···X hydrogen bonding in the solid state.

3.5 Organometallics

We turn now to a discussion of transition metal atoms as hydrogen bond donors and acceptors. The involvement of metal atoms in hydrogen bond arrangements is a fairly recent development (Brammer *et al.* 1995; Braga *et al.* 1998*c*) because the emphasis in hydrogen bond research has remained mostly within the domains of organic and biological compounds, where donors and acceptors are constituted with atoms such as O, N, halogen and S, and in the non-conventional or weak case, with C. In inorganic systems too, attention

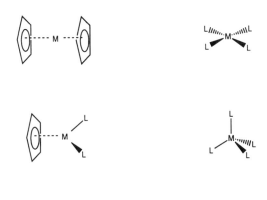

160

has largely been directed towards strong bonds formed by the above main group elements, most commonly as ions (Hamilton and Ibers 1968; Jeffrey 1997). In contrast, hydrogen bonds that involve transition metal atoms are usually weak. They are all certainly non-conventional, at the present time anyway, and in these respects they fall within the scope of the present work.

The peripheries of most organometallic compounds are constituted with organic moieties and the metal atoms are buried deep within the molecule. They are not accessible for the formation of intermolecular interactions and therefore, weak hydrogen bonding in organometallic compounds is mostly a matter of $C-H\cdots O\equiv C$ and $X-H\cdots \pi$ types (Braga *et al.* 1995, 1998*b*). As a result, hydrogen bonds that actually involve the metal centres have generally remained unnoticed. Yet, it has been found in recent times that metal centres do participate both directly and indirectly in hydrogen bonds, $X-H\cdots A$. By direct participation we mean that the metal atom acts directly as a hydrogen bond acceptor or donor, that is $A = M$ or $X = M$. Indirectly, the metal exerts an electronic or steric influence upon the actual hydrogen bond donor or acceptor groups. The latter situation has been detailed elsewhere in the book (Chapters 2 and 4). It is with the former situation that we shall be concerned here. The agostic bond belongs to neither of these categories, but it is an organometallic phenomenon that is clearly related to hydrogen bonding and so it merits discussion.

3.5.1 *Metal atoms as acceptors – $X-H\cdots M$ hydrogen bonds*

Metal centres that can act as hydrogen bond acceptors are typically late transition metals in low oxidation states, containing a sterically accessible filled metal-bound orbital. Following Brammer (1999), pertinent classes of compounds are given in Scheme **160**, with the metal atoms typically being Fe, Co, Ir, Pd, Pt. Database studies by Braga, Desiraju and co-workers (Braga *et al.*

1998c) show that the formation of intermolecular hydrogen bonds to such electron rich metals is a common phenomenon, observed for all traditional and non-traditional hydrogen bonding donor groups. Hydrogen bonding to metal centres in neutral complexes $(X-H\cdots M)$ has been compared with charge-assisted hydrogen bonds $(X^+-H\cdots M^-)$ involving electron rich anionic complexes. For $X = C$, scattergrams of $d(H\cdots M)$ versus $D(C\cdots M)$ and $\theta(C-H\cdots M)$ are given in Fig. 3.123. There are a few examples where even $d < 2.5$ Å. For organic $C-H\cdots X$ contacts, such distances would be taken as indicative of significant hydrogen bonding interaction. The length–angle distributions for $O-H\cdots M$ and $N-H\cdots M$ interactions follow the same trend as the $C-H\cdots M$ interactions. Here too, there are some cases for which $d < 2.8$ Å, $D < 3.5$ Å, and $\theta > 100°$. In $N-H\cdots M$ bonds, the shortest contacts correspond to inter-ionic interactions between organometallic anions and amino cations.

Inter-ionic hydrogen bonds $N^+-H\cdots M^-$ were originally noticed by Calderazzo et al. (1981) with the anion $[Co(CO)_4]^-$ and counterions of the $[NR_3H]^+$ type. As an example, the structure of the $[NMe_3H]^+$ salt is shown in Fig. 3.124(a). In the crystal, there are two symmetry-independent hydrogen bonds with distances d of 2.37 and 2.39 Å, and D of 3.38 and 3.40 Å, and with θ being 179.5° and 179.7°, respectively. A series of related crystal structures have been examined by Brammer and co-workers from the structure correlation viewpoint. In particular, it was found that the angles between the equatorial CO ligands open up by several degrees as the hydrogen bond is formed; a pathway was proposed for the protonation of $Co(CO)_3L^-$ (Brammer 1999). In the N-methylpiperazinium salt of $[Co(CO)_4]^-$ (Fig. 3.124(b)), the average angle between the equatorial CO ligands is 112.1(9)°, whereas the average angle between the axial and equatorial ligands is 106.7(7)° (Brammer and Zhao 1994). The crystal structure of the $[NEt_3H]^+$ salt was determined by neutron diffraction at 15 K, and distances d and D were found to be 2.61 and 3.67 Å, respectively (Brammer et al. 1992).

$N-H\cdots M$ hydrogen bonds have also been observed with the square planar d^8-Pt and d^8-Pd centres. A neutron-determined example is the diplatinium salt $[NPr_4^n]_2[PtCl]_4\cdot cis\text{-}[PtCl_2(NH_2Me)_2]$ reported by Brammer et al. (1991). The dianion of this compound consists of two square-planar d^8-Pt centres held together by intermolecular $N-H\cdots Pt$ and $N-H\cdots Cl$ interactions, Fig. 3.125. The latter interaction is readily identified as an $N-H\cdots Cl-Pt$ hydrogen bond with $d = 2.32$ Å. The very short $N-H\cdots Pt$ contact, $d = 2.26$ Å, is also interpreted as a hydrogen bond because the presence of the filled Pt d_{z^2} orbital oriented towards the $N-H$ group suggests a three-centre four-electron interaction. Related examples occur also with $C-H$ donors, and an example has been explicitly discussed by Alyea et al. (1998) for dimers of $cis\text{-}[PdCl_2(TPA)_2]$, where TPA stands for 1,3,5-triaza-7-phosphaadamantane, Fig. 3.126. The d separation of 2.86 Å is some 0.6 Å shorter than the van der Waals sum. A different example with a chloroform molecule donating a hydrogen bond to a

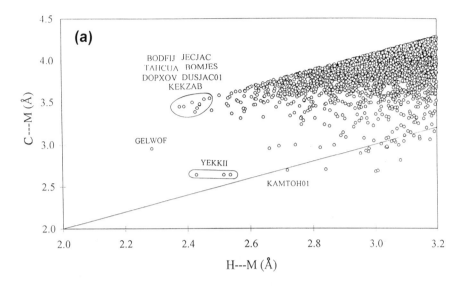

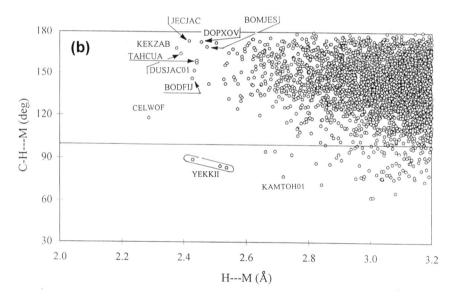

Fig. 3.123. Distances and directionality of C—H···M interactions with M = any transition metal. Scatterplots of (a) D against d, and of (b) θ against d. The diagonal line in (a) indicates $D = d$. Notice that there are a few interactions where $d > D$. The horizontal line in (b) represents $\theta = 100°$ (from Braga *et al.* 1997a).

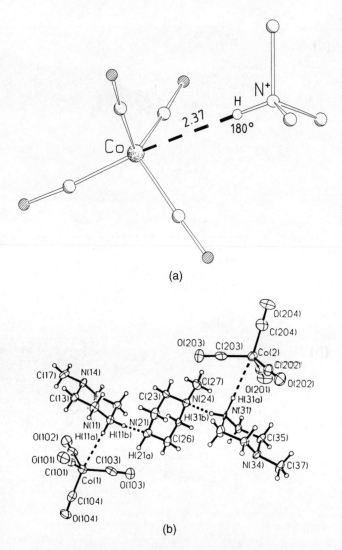

(a)

(b)

Fig. 3.124. $N^+ - H \cdots Co$ hydrogen bonds with Co in $[Co(CO)_4]$ anions: (a) in the $[NMe_3H]$ salt (Calderazzo *et al.* 1981; figure from Braga *et al.* 1998c); (b) in the *N*-methylpiperazinium salt (from Brammer and Zhao 1994).

square-planar Pt centre, $d = 2.38$ and $D = 3.45$ Å, has been mentioned by Braga *et al.* (1997a).

An example of an intermolecular $O - H \cdots M$ hydrogen bond is provided by the square-planar copper complex shown in Fig. 3.127, namely $[CuL_2L_2']\cdot2[H_2O]$ (L = *N*-acetyl-α-alaninato, L' = *N*-methylimidazole) (Battaglia *et al.* 1982). Two water molecules are hydrogen bonded to each complex, completing the coordina-

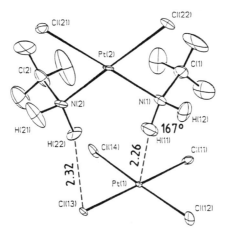

Fig. 3.125. The dianion in the neutron structure of $[NPr_4^n][PtCl_4] \cdot cis\text{-}[PtCl_2(NH_2Me)_2]$, showing short $N-H \cdots Pt$ and $N-H \cdots Cl$ hydrogen bonds (from Brammer *et al.* 1995).

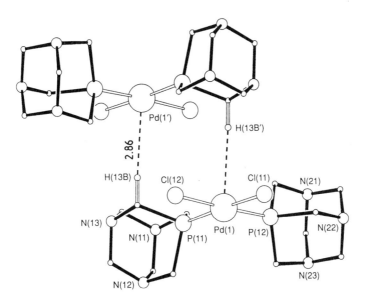

Fig. 3.126. $C-H \cdots Pd$ hydrogen bonds in $cis\text{-}[PdCl_2(TPA)_2]$, where TPA = 1,3,5-triaza-7-phosphoadamantane (from Alyea *et al.* 1998).

tion around the metal. The relevant water H atom forms a bifurcated link with the Cu atom and an O atom of the ligand, constituting a five-membered metallacycle. The *d* separations are 2.57 and 2.03 Å for the $O-H \cdots Cu$ and $O-H \cdots O$ bonds, respectively.

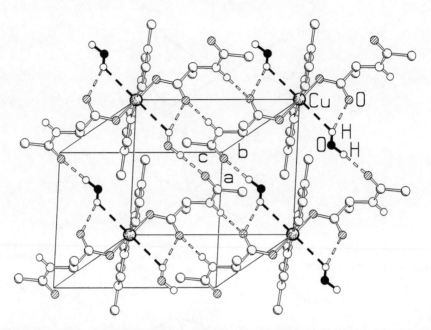

Fig. 3.127. O—H···M hydrogen bonds in crystalline [CuL$_2$L'$_2$]·[H$_2$O] (L = *N*-methylimidazole). Notice the bifurcated interactions (adapted from Braga *et al.* 1997*a*).

X—H···M three-centre four-electron interactions are also observed in their intramolecular variation. A good example is the N—H···Pt bond in the zwitterionic complex [PtCl(1-C$_{10}$H$_6$NMe$_2$-8-C,N)(1-C$_{10}$H$_6$NHMe$_2$-8-C,H)] and its bromo analogue **161** studied by Wehman-Ooyevaar *et al.* (1992*a,b*). Weakening of the N—H bond could be quantified by NMR spectroscopy, indicating the hydrogen bond nature of the interaction.

We should also mention that O—H···M hydrogen bonding involving uncharged donors and acceptors has been detected by IR solution spectroscopy for an extensive series of compounds (η^5-C$_5$R$_5$)ML$_2$, where R = H, Me; M = Co, Rh, Ir; L = CO, C$_2$H$_4$, N$_2$, PMe$_3$ (Kazarian *et al.* 1993). Using perfluoroalcohols as the donors, red shifts of the OH stretching vibrations in the range 260–510 cm^{-1} were measured, corresponding to bond energies roughly in the range −4.9 to −6.9 kcal/mol, depending on the basicity of the metal centre.

To summarize then, X—H···M hydrogen bonding interactions are common when donors such as N—H, O—H and C—H are able to approach a nucleophilic metal centre. Enhanced basicity is usually associated with late transition metals in low oxidation states that can supply a sterically accessible filled *d* orbital. When the metal atom carries a ligand with electronegative atoms, multifurcated hydrogen bonds can be formed involving the metal atom and the ligands as acceptors.

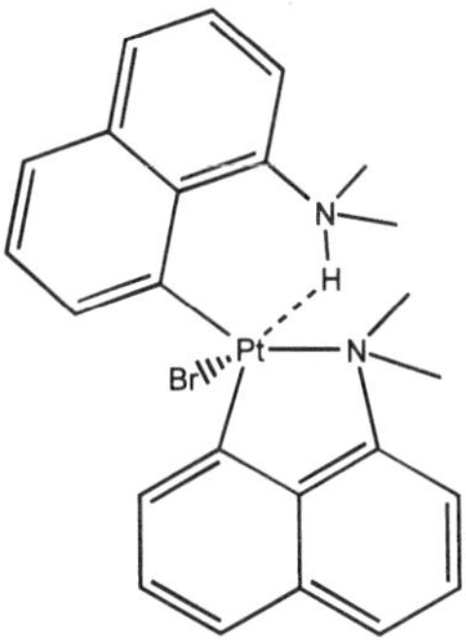

161

3.5.2 *Metal atom groups as donors – M—H···A hydrogen bonds*

Metal atoms in coordination complexes and clusters are amphoteric. It was Pearson (1985) who pointed out that the bond between a transition metal and a hydrogen atom is a highly polarizable one, and that hydrogen can undergo reactions as H^+, $H^•$ and H^-, depending on the relative stability of the resulting species. This variable behaviour depends on a number of factors, such as electronic configuration, oxidation state and electronegativity of the metal, as well as the type and distribution of ligands. This chameleon-like character is reflected in the capacity of an H atom bound to a metal to donate a hydrogen bond despite its nominal classification as a 'hydridic' species. Thus the M—H group can sometimes act as a non-conventional hydrogen bond donor.

It has been demonstrated that metal-bound H atoms in metal clusters, the so-called metal hydrides, can form hydrogen bonds with suitable bases, usually though not necessarily CO, provided that the hydrogen ligand is not sterically hindered (Teller and Bau 1981; Braga *et al.* 1996*a*). When approach is not forbidden by the encapsulation of the H ligand within the ligand shell, the H atoms can form intermolecular bonds comparable in length with those of the C—H···O≡C type discussed in Sections 2.2.2 and 2.2.3. Needless to say, the electronic nature of the metal plays a fundamental role in tuning the polarity of the M—H system. In general, an accumulation of positive charge on the metal-bound H atom is observed in neutral polynuclear cluster complexes where the H atom is often present in a μ_2 or μ_3 bonding fashion. Actually, polynuclear hydrides are most often obtained by protonation of carbonylate anions of group VIII transition metals with acids (Bau *et al.* 1979). A good example with μ_2-H atoms is afforded by $[(\mu_2\text{-H})(\mu_2\text{-NCHCF}_3)\text{Os}_3 (\text{CO})_{10}]$, the structure of which has been determined by neutron diffraction (Fig. 3.128; Dawoodi and Martin 1981). Each cluster molecule participates in two types of hydrogen bonds. The first is of the Os—H···O type ($d = 2.59\,\text{Å}$) while the other is a more traditional C—H···O bond ($d = 2.57\,\text{Å}$).

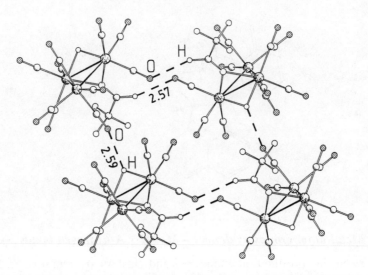

Fig. 3.128. $Os-H\cdots O\equiv C$ hydrogen bonds in crystalline $[(\mu_2\text{-}H)(\mu_2\text{-}NCHCF_3)Os_3(CO)_{10}]$. There are also $C-H\cdots O\equiv C$ hydrogen bonds in the stucture (adapted from Braga *et al.* 1996a).

Intramolecular   Intermolecular

162

Crystalline $[(\mu_2\text{-}H)_3Os_3Ni(CO)_9(C_5H_5)_2]$ also shows a nice pattern of $Os-H\cdots O$ and $C-H\cdots O$ hydrogen bonds, the former again involving μ_2-bonded H atoms. The $Os-H\cdots O$ bonds link molecules in a chain-like fashion, and the $C-H\cdots O$ bonds link centrosymmetrically related molecules into dimers via interactions between $(Cp)C-H$ groups and terminal CO ligands, Fig. 3.129 (Churchill and Bueno 1983). An example of an intramolecular $M-H\cdots A$ bond with $d = 2.33\,\text{Å}$ was reported for M = W and A = O by Fairhurst *et al.* (1995) based on X-ray and NMR data.

An interesting variant of $M-H\cdots A$ hydrogen bonds involves molecular hydrogen, that is H_2 molecules, bonded to metal atoms. Such hydrogen bonds can be formed intra- and inter-molecularly, **162**, and can be readily understood because H_2 functions as σ-donor to the metal and is therefore formally electropositive. A good example was found by Albinati *et al.* (1993) in the neutron crystal structure of $Ir(H_2)(H)(Cl)_2\{P(i\text{-}Pr)_3\}_2$, Fig. 3.130. The Ir-coordinated H_2

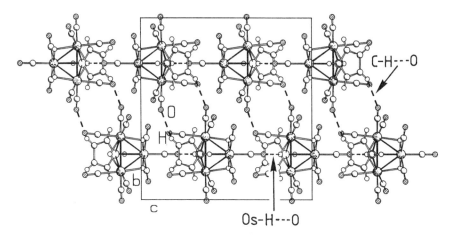

Fig. 3.129. Os—H···O≡C hydrogen bonds in crystalline [(μ₂-H)₃Os₃Ni(CO)₉(C₅H₅)₂]. There are also C—H···O≡C hydrogen bonds in the stucture (adapted from Braga *et al.* 1996*a*).

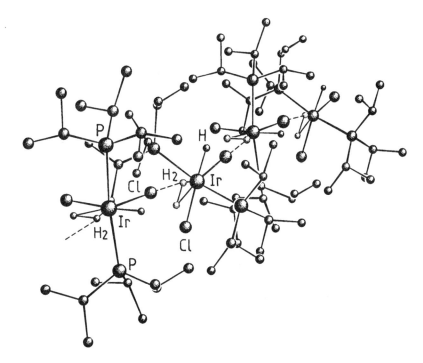

Fig. 3.130. Chain of Cl—Ir—H···Cl—Ir—H···Cl—Ir—H hydrogen bonds in the neutron structure of the trihydride Ir(H₂)(H)(Cl)₂{P(*i*-Pr)₃}₂ (Albinati *et al.* 1993; figure adapted from Koetzle 1996). Here, *d* = 2.64 Å.

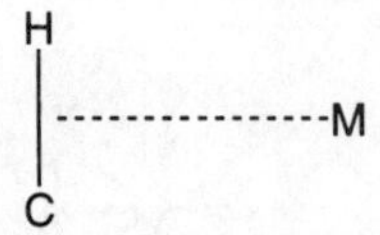

163

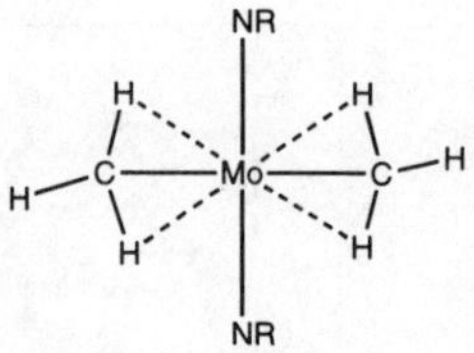

164

molecule is involved in a hydrogen bond to a Cl ligand of a neighbouring molecule, leading to the formation of infinite and possibly even cooperative chains $Cl-Ir-H\cdots Cl-Ir-H\cdots Cl-Ir-H$. The $H-H$ bond length is 1.11 Å, which is elongated compared to the 0.74 Å value in free H_2 but still well within what is considered a covalent bond between the two H atoms.

3.5.3 Agostic interactions – $M\cdots(H-C)$

The agostic interaction is the three-centre two-electron interaction between an electron-deficient metal atom and a $C-H$ σ bond, **163**. Agostic interactions and hydrogen bonds are fundamentally different. Hydrogen bonds $X-H\cdots A$ can be formally designated as three-centre four-electron interactions, and in an ideal geometry, the $X-H$ vector is directed at A. In contrast, an agostic interaction is of the three-centre two-electron type, in that an electron-deficient metal makes a close approach to an electron-rich $C-H$ bond, and the $M\cdots(H-C)$ geometry is T-shaped. The agostic interaction is therefore a manifestation of metal atom Lewis acidity (Crabtree 1993). Agostic interactions are somewhat stronger than most of the hydrogen bonds described in this work, with bond energies in the range −7 to −15 kcal/mol. Early transition metals such as Ti, Ta and Zr are typically involved, and also Ni (Youngs et al. 1993) and Fe. These $M\cdots(H-C)$ interactions attract much interest because of the possible connection with $C-H$ bond activation by metal atoms.

In the agostic interaction, the position of the H atom can be severely deformed so that neutron diffraction is of particular utility. A nice example is provided by the neutron crystal structure of $[Mo(NC_6H_3Pr^i_2-2,6)_2Me_2]$ with two highly distorted methyl ligands, **164** (Cole et al. 1998). This distortion has been ascribed to multiple $C-H\cdots Mo$ agostic interactions. The $C-H\cdots Mo$ angles are around 50° (for a methyl H atom not involved in the agostic interaction, the corresponding angle is around 43°) while the $H\cdots Mo$ distances are 2.585 and 2.598 Å, significantly shorter than the distances to the 'free' H atoms (2.755 and 2.76 Å).

Agostic interactions are normally formed intramolecularly, but intermolecular examples have also been reported, as in the ion-pair adducts

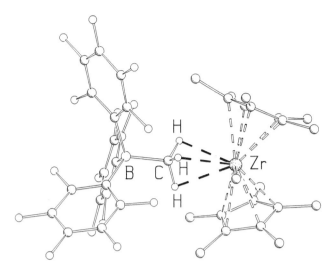

Fig. 3.131. Intermolecular agostic interaction in crystalline $[(Cp^*)_2ZrMe][(C_6F_5)_3BMe]$ (from Braga *et al.* 1997*a*).

between anionic Lewis acids and electron deficient and coordinatively unsaturated zirconocene compounds. The structure of $[(Cp^*)_2ZrMe][(C_6F_5)_3$ BMe] (Fig. 3.131) represents a prototype structure. These interactions are characterized by short $C \cdots Zr$ (2.64 Å) and $H \cdots Zr$ (2.43–2.54 Å) distances in the close approach of the B-atom-bound CH_3 group to the electron deficient Zr atom; the $C-H \cdots Zr$ angles are in the range 83–89° (Braga *et al.* 1996*b*).

Interestingly, the electron deficient Li atom can participate in an agostic interaction. As an example, the structure of $[LiBMe_4]$ determined by Rhine *et al.* (1975) is shown in Fig. 3.132. The structure consists of layers formed by four-coordinated Li^+ and three-coordinated BMe_4 units. These moieties are linked via short $C-H \cdots Li^+$ contacts with $C \cdots Li^+$ distances of 2.19 Å. In a CSD study on this kind of contact, Braga *et al.* (1996*b*) noted ranges of 1.89–2.20 Å for the $H \cdots Li^+$ separation, 2.16–2.45 Å for the $C \cdots Li^+$ separation, and 74–113° for the $C-H \cdots Li^+$ angle.

To summarize, agostic interactions are formed when an acidic metal atom accepts electron density from $C-H$ σ bonds. Both intramolecular and inter-molecular agostic interactions are specific to organometallic systems and have no counterpart in organic crystal chemistry.

3.6   Other varieties

In the previous sections and chapters, $A-H$ groups which have a partial positive charge on the H atom have been discussed. These groups form hydrogen

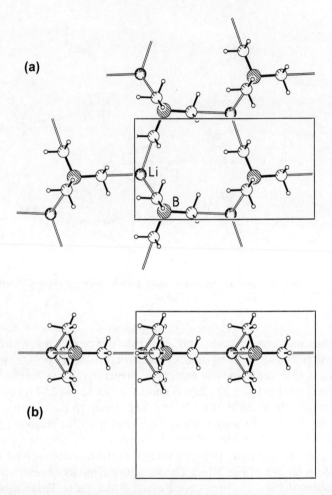

Fig. 3.132. C—H···Li⁺ interactions in [LiBMe₄] (Rhine *et al.* 1975) shown in two projections. All four methyl groups are involved in interactions with Li⁺ ions, leading to a layered structure. See Braga *et al.* 1996*b*.

bonds with acceptors that have a surplus negative charge. If the atom A has a *smaller* electronegativity than H, however, the polarity of the A—H bond is reversed and there could be a partial negative charge on the H atom (see also Section 2.2.11). Typical examples are the hydrides of B, Al and Ga and indeed of many transition metals. This leads to a possibility of new classes of inter-molecular interactions. The negative end of an $A^{\delta+}—H^{\delta-}$ group can serve as an *acceptor* in a hydrogen bond with a normal donor $X^{\delta-}—H^{\delta+}$, or it can form a directional interaction with a more general 'donor' X^+. These two situations lead respectively to what are now termed the *dihydrogen bond* and the *inverse hydrogen bond*.

3.6.1 *The dihydrogen bond – X — H ··· H — M*

In the *dihydrogen bond*, M can be any element with the M—H bond having a $M^{\delta+}-H^{\delta-}$ polarization. Typically, these are Group III elements including B, and transition metal atoms. The hydridic polarization $M^{\delta+}-H^{\delta-}$ is obtained depending on the mode of coordination of the H atom, the electronic state of the central atom and the nature of the other ligand(s) bound. These M—H groups, either Group III or transition metal, may interact attractively with conventional $X^{\delta-}-H^{\delta+}$ donors. Both cases have attracted much recent attention (Crabtree *et al.* 1996) and are now disussed in turn.

3.6.1.1 *Group III hydrides*

A systematic study by Crabtree and co-workers was prompted by the striking difference in melting points between H_3C-CH_3 (–181°C) and the isoelectronic H_3B-NH_3 (+104°C). This difference may be ascribed to N—H···H—B hydrogen bonding. Because the crystal structure of the latter compound was not available, the system was investigated theoretically and a database study performed on N—H···H—B interactions (Richardson *et al.* 1995). In the CSD analysis, 26 intermolecular contacts of this kind were found with $d(\text{H}\cdots\text{H})$ in the range 1.7–2.2 Å (for normalized H atom positions). These distances are much shorter than observed in normal X—H···H—Y van der Waals contacts. It was found that the N—H···H angles tend to linearity (range 117–171°, average 149°), whereas the H···H—B angles are often strongly bent (range 90–171°, average 120°), with the most densely populated decile interval being 100–120°. In most of the examples retrieved, the N—H donor is positively charged, say in ammonium and pyridinium ions. The B—H bonds are typically from boron cage anions, aminoboranes or aluminoboranes wherein the B atom is expected to bear at least a partial negative charge.

To complement the CSD study, *ab initio* calculations were carried out on the dimer of H_3B-NH_3. The global minimum structure is shown schematically in Fig. 3.133 together with the distribution of point charges. The dimer is linked by two dihydrogen bonds with $d(\text{H}\cdots\text{H})$ = 1.82 Å. Notably, the N—H···H angle is roughly linear while the H···H--B angle is bent, as seen in the database study. In the charge distribution, a more negative charge on B than on H(B) is obtained, explaining the systematic bending of the H···H—B angle. This way, the positive H atom of the donor comes closer to the centre of the charge of the acceptor. The calculated energy is –6.1 kcal/mol for each bond and this is comparable to conventional N—H···A hydrogen bonds. Energies in the kcal/mol range were calculated also by Alkorta *et al.* (1996) for a series of (hypothetical) dihydrogen bonded systems.

Two examples of N—H···H—B dihydrogen bonds in crystals will also be shown. 1(*e*),3(*e*),5(*e*)-Trimethylcyclotriborazane, **165**, has three axial N—H groups that all point in one direction and three axial B—H groups that point in the opposite direction. In the crystal, columns are formed in which adjacent

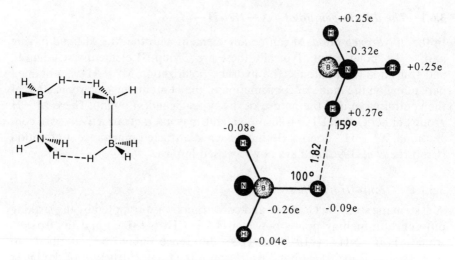

Fig. 3.133. Theoretical structure (PCI-80/B3LYP) of the H_3B-NH_3 dimer: *left*, schematic structure; *right*, one of the hydrogen bonds with Mulliken charge distribution (from Richardson *et al.* 1995).

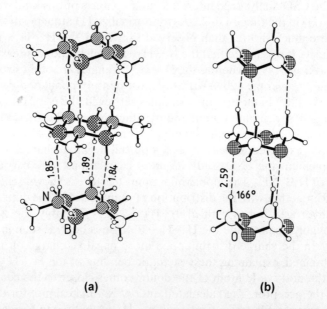

(a)   **(b)**

Fig. 3.134. (a) $N-H\cdots H-B$ dihydrogen bonds in $1(e),3(e),5(e)$-trimethylcyclotriborazane, **165** (based on Narula *et al.* 1986). (b) The related columnar structure of trioxane **166** (Busetti *et al.* 1969). Notice the similar hydrogen bond patterns.

molecules are linked by three short dihydrogen bonds with distances d between 1.84 and 1.89 Å, Fig. 3.134(a). Incidentally, trioxane **166** crystallizes in a similar columnar fashion, with $C-H\cdots O$ hydrogen bonds having the same function as the dihydrogen bonds in **165**, Fig. 3.134(b). Viewing these two struc-

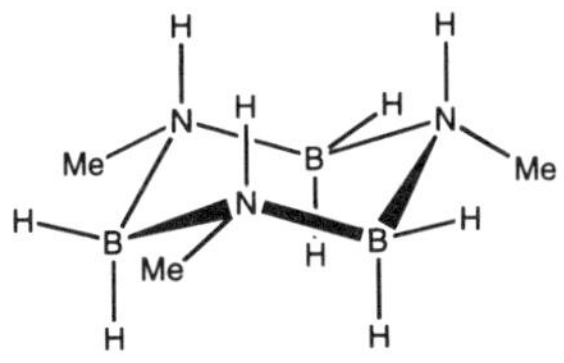

165

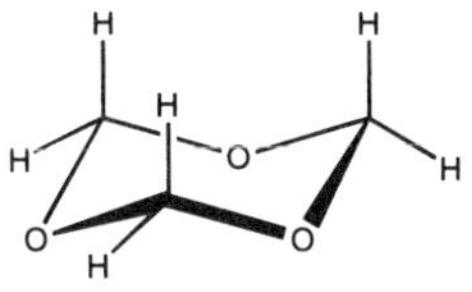

166

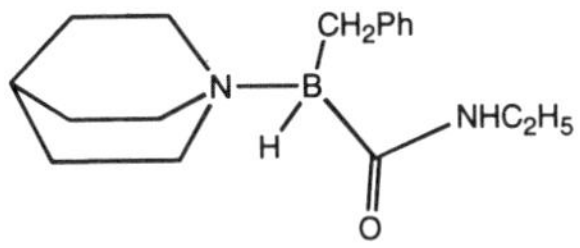

167

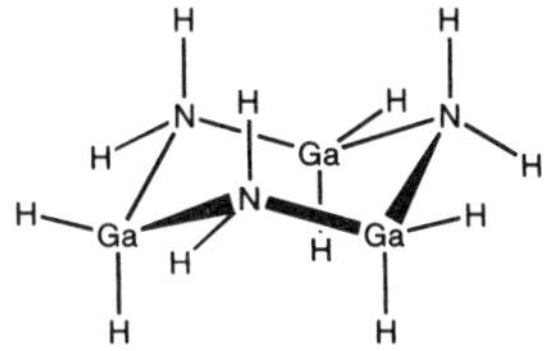

168

tures side by side indicates how in two molecules of similar shapes, the highly non-conventional acceptor H—B can take the role of the conventional O-acceptor without a change of the overall hydrogen bond architecture. Another revealing crystal structure is that of compound **167**. Here, molecular dimers are formed via N—H···H—B dihydrogen bonds with $d = 2.08$ Å, Fig. 3.135. Whereas one can argue that in compound **165**, the N—H donors have little choice other than to interact with the H—B moieties, this is not the case for **167**. The carbonyl group of this molecule would offer the opportunity to form conventional N—H···O=C hydrogen bonds, but in the crystal, dihydrogen bonds are formed and the carbonyl acceptor potential is satisfied by short C—H···O bonds from co-crystallized chloroform molecules.

An example of dihydrogen bonds with Ga—H$^{\delta-}$ acceptors was found by Campbell *et al.* (1998) in the crystal structure of cyclotrigallazane [H$_2$GaNH$_2$]$_3$, **168**. A close relation with the molecular structure of the borazane **165** is obvious, and also the intermolecular interactions are related. However, the molecules are now tilted with respect to the primary packing axis, and are connected by only two dihydrogen bonds with the shortest H···H distance being 1.97 Å, Fig. 3.136. The hydrogen bonds form an interconnected array H—Ga—N—H···H—Ga—N—H···H—Ga—N—H,   the   cooperative nature of which has been alluded to by the authors. The structure is furthermore related to that of 1,3,5-trigermanocyclohexane where, however, the corresponding C—H···H—Ge distance d is longer at 2.20 Å. This is not unexpected considering the relatively small electronegativity difference between C and Ge (Table 1.1). From theory, the authors estimate the N—H···H—Ga

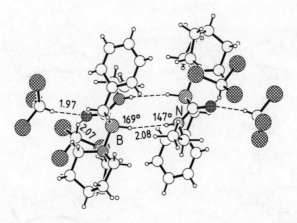

Fig. 3.135. Molecular dimer of **167** (Mills *et al.* 1989) linked by N—H···H—B dihydrogen bonds. Note also the chloroform molecules that are hydrogen bonded to the C=O groups.

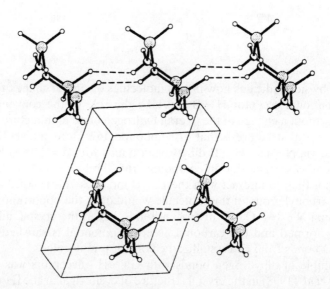

Fig. 3.136. N—H···H—Ga dihydrogen bonds in cyclotrigallazane **168** (from Campbell *et al.* 1998).

interaction to be worth around −3 kcal/mol in stabilizing energy, slightly less than the analogous N—H···H—Al interaction which was also calculated. The crystal structure of cyclotrigallazane is quite interesting but beyond the scope of the present work. Suffice to say, that if one considers the H···H dihydrogen bonds in this structure to represent H_2 in a state of incipient extrusion, the structure would collapse topotactically to that of cubic GaN, were the H_2

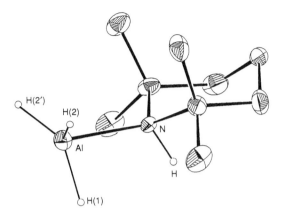

Fig. 3.137. H$_3$Al(2,2,6,6-tetramethylpiperidine), **169**. Because of the N−H$^{\delta+}\cdots{}^{\delta-}$H−Al interaction, the AlH$_3$ group is eclipsed with respect to N−H (from Atwood *et al.* 1994).

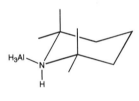

169

to be actually lost from the solid. The interested reader will further note that GaN adopts the wurtzite structure and not the cubic, or zinc blende, polymorph. Whether the latter could be produced via such solid state reactions is a matter for future research.

An novel intramolecular Al−H$^{\delta-}\cdots$H$^{\delta+}$−N interaction was found by Atwood *et al.* (1994) in the alane-amine **169** shown in Fig. 3.137. Due to the attraction of the oppositely charged H atoms on Al and N, the AlH$_3$ group is in an eclipsed orientation with respect to the N−H bond and not staggered as would be expected for say a methyl group. Though this interaction does not possess the geometry of a hydrogen bond, it is of relevance in the general context of non-conventional electrostatic interactions.

3.6.1.2 *Transition metal hydrides*

The polarity of the M−H bond can be tuned from M−H$^{\delta+}$ to M−H$^{\delta-}$ (Pearson 1985). In the case of M−H$^{\delta-}$, the system can interact attractively

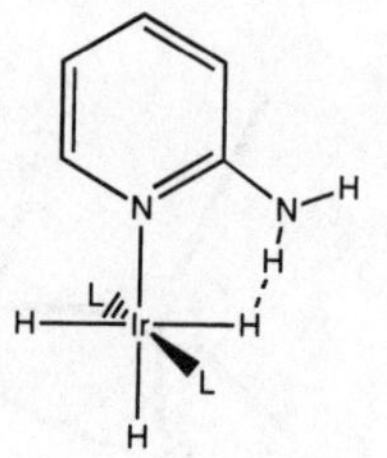

170

with $X-H$ groups of opposite polarity such as $O-H^{\delta+}$, $N-H^{\delta+}$ and $C-H^{\delta+}$, that is with 'normal' hydrogen bond donors. These dihydrogen bonds of transition metal hydrides are observed intra- and inter-molecularly. The intramolecular $X-H\cdots H-M$ interaction ($X = C, N, O, S$) can be weakly bonding in nature as demonstrated by spectroscopic (Fairhurst *et al.* 1995; Shubina *et al.* 1996) and diffraction (Wessel *et al.* 1995) experiments, and as discussed in theoretical studies (Liu and Hoffmann 1995). From solution NMR data, Crabtree *et al.* (1996) have provided an estimate of −5 kcal/mol for the $Ir-H\cdots H-N$ hydrogen bond in $[IrH_3(PPh_3)_2(2\text{-aminopyridine})]$, **170**, a compound specifically designed to study this unusual phenomenon. The authors conclude that this interaction is strong enough to be considered a full-fledged hydrogen bond, indeed comparable to a conventional $N-H\cdots OH_2$ hydrogen bond.

The complexes *cis*-$[IrH(OH)(PMe_3)_4][PF_6]$ and *cis*-$[IrH(SH)(PMe_3)_4][PF_6]$ represent interesting cases of contrast. X-Ray structural analyses show that the H atom bound to the metal-coordinated oxygen in the first complex establishes an intramolecular interaction with the hydride ligand ($Ir-H\cdots H-O$ 2.33 Å). This is shown in Fig. 3.138(a). In the isoelectronic thiol (Fig. 3.138(b)), however, the orientation of the SH group is such that the thiol H atom points outwards far away from the hydride ligand and directed towards the F atoms of the $[PF_6]^-$ anions (Stevens *et al.* 1990).

While the first examples of intramolecular dihydrogen bonds involving $Tr-H$ acceptors were obtained by serendipity, the first *inter*molecular interactions of this kind were constructed deliberately with rhenium polyhydrides. Two examples that have been studied with neutron diffraction are given in Fig. 3.139 (Patel *et al.* 1997). The shortest $H\cdots H$ distances are around 1.7 Å, and the bonding nature of the interaction has been verified with IR spectroscopy. For the example shown in Fig. 3.139(b), a 302 cm^{-1} red-shift of the $N-H$ stretching vibration was measured, similar to that found in a 'good' conventional hydrogen bond. Using the Iogansen equation (Section 1.3.2), a bond energy of −5.3 kcal/mol is estimated. Dihydrogen bonds formed by $C-H$ donors have been found by Xu *et al.* (1997) in some Ir and Ru complexes.

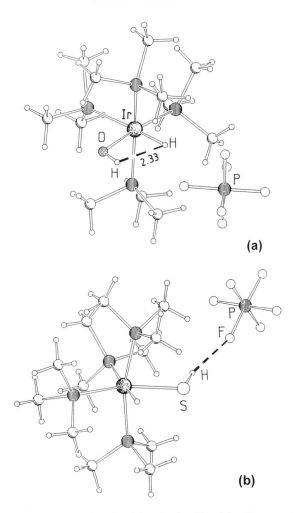

Fig. 3.138. Hydrogen bond interactions involving the O—H and S—H groups in crystalline (a) *cis*-[IrH(OH)(PMe₃)₄][PF₆] (Stevens *et al.* 1990) and (b) [IrH(SH)(PMe₃)₄][PF₆] (Milstein *et al.* 1986). In (a), an intramolecular dihydrogen bond O—H···H—Ir is formed, and in (b) a conventional intermolecular hydrogen bond S—H···F₆P⁻ (figures from Braga *et al.* 1998c).

The examples in Fig. 3.139 show dihydrogen bonds with bifurcated N—H donors. Morris and co-workers found an example with a bifurcated M—H acceptor. This case is intramolecular, and occurs in the X-ray crystal structure of the Ir-hydride shown in Fig. 3.140 (Park *et al.* 1994). Based on theoretical H atom positions, H···H separations of 1.9 Å are calculated.

Finally, we mention an interesting and quite extraordinary variation of the intramolecular dihydrogen bond, that is the interaction of a metal-bound

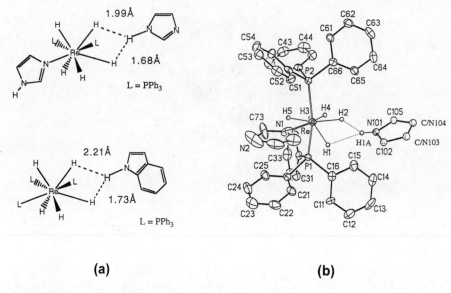

Fig. 3.139. Intermolecular dihydrogen bonds in Re-trihydrides as determined by low-temperature neutron diffraction. (a) Molecular structure and hydrogen bond geometry for two related crystal structures. (b) Crystal structure of the compound shown in (a), *top* (from Patel *et al.* 1997).

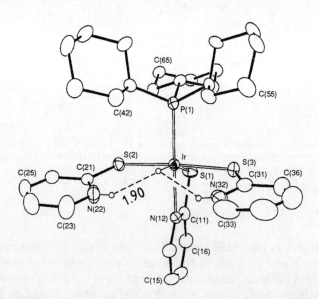

Fig. 3.140. Intramolecular N—H⋯H—Ir hydrogen bond with a bifurcated acceptor (adapted from Park *et al.* 1994).

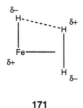

171

hydrogen molecule, η^2-H_2, with a hydrogen ligand. Such a hydrogen bond, **171**, occurs with an $H\cdots H$ separation of $1.86\,\text{Å}$ in the neutron crystal structure of $Fe(H)_2(\eta^2\text{-}H_2)(PEtPh_2)_3$ (van der Sluys *et al.* 1990), and also in a series of iridium trihydride complexes (Heinekey *et al.* 1990).

3.6.2 *The inverse hydrogen bond – $X—H^-\cdots A^+$*

If $X—H$ groups can have the polarity $X^{\delta+}—H^{\delta-}$, it follows that they should be able to form directional interactions of the kind $X^{\delta+}—H^{\delta-}\cdots A^{\delta+}$ which obey principles similar to those that hold for normal hydrogen bonds. Indeed such an arrangement would formally fulfil the Pimentel and McClellan definition of a hydrogen bond, in that it is bonding and that it involves the H atom (Section 1.1.1). Also Pauling's definition is formally satisfied because the H atom interacts with two other atoms, and forms a bond between them. In this sense, $X^{\delta+}—H^{\delta-}\cdots A^{\delta+}$ would be just another example of hydrogen bonding! Nevertheless, to highlight the reversed distribution of charge, Rozas *et al.* (1997*b*) have termed this interaction as the *inverse hydrogen bond*. This seems to be an apt description. The reader will note that the dihydrogen bonds $M—H\cdots H—X$ described in Section 3.6.1 could also be formally termed as a combination of a normal hydrogen bond $X—H^{\delta+}\cdots H^{\delta-}$ and an inverse hydrogen bond $M—H^{\delta-}\cdots H^{\delta+}$.

In theoretical terms, inverse hydrogen bonds have been treated by several authors mostly on hypothetical systems (see Rozas *et al.* 1997*b* and the references therein). Of some general interest is the so-called *lithium bond* $Li—H\cdots Li—H$ proposed by Pople and co-workers (Dill *et al.* 1977; DeFrees *et al.* 1987). An experimental example of an inverse hydrogen bond (though not so termed) was reported by Cotton *et al.* (1998). The diniobium compound $Nb_2(hpp)_4$ was prepared as an adduct with $NaEt_3BH$. The Na^+ cation is linked to the Et_3BH^- anion via a $B^-—H\cdots Na^+$ contact, **172**, in a situation that is the inverse of that found in the conventional hydrogen bonded adduct $N^+—H\cdots Cl^-$, **173**.

Interest in inverse hydrogen bonds is very recent, but a brief consideration of the chemical variety of the metallic elements coupled with chemical imagination would suggest many more varieties of this novel type of interaction. One will surely see more examples of this hydrogen bond in the future.

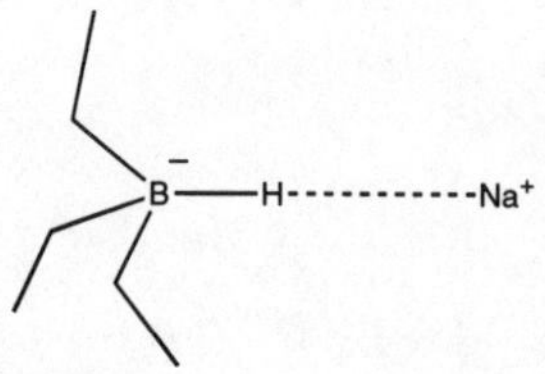

172 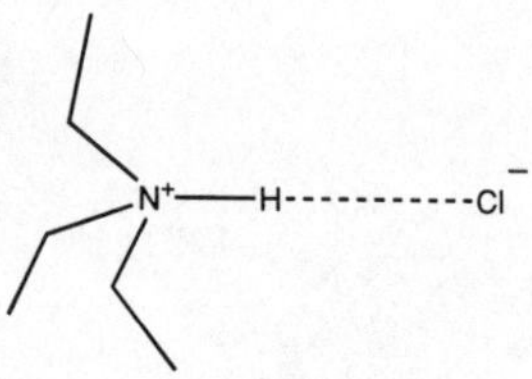   **173**

3.7 Summary

The study of non-conventional donors and acceptors has developed on the premise common to all hydrogen bonds, namely that they are largely electrostatic in nature. We have seen in this chapter that by taking the archetypical $C-H\cdots O$ hydrogen bond and then varying the acceptor, one obtains say the $C-H\cdots F$ and $C-H\cdots\pi$ bonds, or by varying the donor the $P-H\cdots O$ and $M-H\cdots O$ bonds. Modifications in both donor and acceptor moieties result in hydrogen bonds such as $O-H\cdots\pi$ and $X-H\cdots H-M$. In all these variations, the fundamental properties of the hydrogen bond are maintained. These variations are of further chemical interest because they could also involve changes in hardness–softness. In terms of strength, some of these novel types of hydrogen bond, especially $O-H\cdots\pi$ and $H\cdots H$, are more significant than the prototypical cases $C-H\cdots O/N$. So, it is largely their non-conventional nature rather than their weakness *per se* that makes them interesting from the viewpoint of the current work. The non-conventional agostic interaction is clearly of a different nature and it is again by no means weak. The widespread occurrence and diverse geometrical arrangements that characterize these hydrogen bonds of the non-$C-H\cdots O$ type hint that they will, in the future, find applications in supramolecular chemistry and crystal engineering. At the least, their role could be a supportive one in the construction of extended hydrogen bonded networks.

In Chapters 1, 2 and 3, we have defined the term *weak hydrogen bond*, justified its inclusion in the family of hydrogen bonds and have explored its range and variety. At the ridge of this work, one is well-poised to examine the role of this interaction type in supramolecular chemistry and structural biology. Indeed $C-H\cdots O/N$ and other weak hydrogen bonds have been employed in the construction of molecular assemblies and have been implicated in recognition schemes of several types. These matters form the subject of the next chapter.

4

The weak hydrogen bond in supramolecular chemistry

We have seen in Chapters 2 and 3 that the family of weak hydrogen bonds includes interactions that span a wide range of chemical variety. We have discussed several properties and attributes of these bonds. The effort that has been expended on the study of these interactions and the increased recent awareness of their importance have been largely motivated by the growth and development of the new subject of *supramolecular chemistry*. In this chapter, we review weak hydrogen bonds in the supramolecular context. More specifically, we discuss the relevance of these interactions in the analysis, design and synthesis of the structures of molecular assemblies, most notably crystals.

As its name suggests, supramolecular chemistry signifies chemistry beyond the molecule. It deals with implications of the fact that molecules can recognize one another via intermolecular interactions, typically in condensed media. The origins of the subject may be found in the work of Pedersen (1988), Cram (1988) and Lehn (1988) on molecular complexes and host–guest compounds, but the area has grown vastly in recent times (Lehn 1990; Atwood *et al.* 1996). Today, supramolecular chemistry connotes more an adherence to a new, distinct thought process in chemistry, one that asserts that the molecular structure is not the ultimate delimiter of the characteristic and representative properties of a chemical substance (Behr 1994; Lehn 1995). The idea of supramolecularity applies therefore not only to structure but extends also to properties and has grown around Lehn's analogy that 'supermolecules are to molecules and the intermolecular bond what molecules are to atoms and the covalent bond'.

This field has grown into two distinct branches – the study of supermolecules in solution (Hamilton 1996) and in the solid state, mainly in crystal structures (Desiraju 1995*b*, 1996*b*, 1997*b*). The concepts and principles of recognition and the nature of the interactions that mediate supramolecular construction are nearly the same in solution and in the solid state. Yet one must distinguish between these two situations. The most important distinction is not structural but lies in the lifetime of the interactions – supramolecular association in the crystal is time-stable whereas in solution it is not. This difference has led to a biological emphasis of solution-state supramolecular chemistry and to a materials emphasis of solid-state supramolecular chemistry. The former has been studied by physical and synthetic chemists interested in understanding biological processes, biosynthetic mimics and enzyme catalysis.

The latter has been developed by structural chemists and crystallographers to better understand non-covalent interactions for the design of novel materials and solid-state reactions. The present chapter will concentrate for the large part on solid-state supramolecular systems. Chapter 5 is biological in its emphasis.

4.1 The solid state – influence of weak hydrogen bonds on packing

Historically speaking, the solution and biological aspects of supramolecular chemistry were well advanced before solid-state and crystallographic concerns made their formal appearance. The nominal identification of a crystal as a supermolecule is usually attributed to Dunitz (1991) while Lehn (1995), more recently, has termed a crystal 'a very large supermolecule indeed'. One should note, however, that studies of crystalline molecular complexes of various types predate these developments by several decades. Foster's monograph (1969) on donor–acceptor complexes definitely conveys the supramolecular ethos, while Pfeiffer's study on such compounds (1927) seventy years ago might well be considered the source of the materials science stream of supramolecular chemistry, just as the biological stream began with Fischer's work (1894) thirty years earlier.

4.1.1 *The crystal as a supermolecule*

An organic crystal structure is an ideal paradigm of a supermolecule, a super-molecule *par excellence*, a nearly perfect periodic self-assemblage of millions upon millions of molecules, held together by short, medium and long-range interactions, to produce matter of macroscopic dimensions. Crystals are ordered supermolecular systems at a relatively high level of perfection. The high degree of order in a crystal structure is the result of complementary dispositions of shape features and functional groups in the interacting near-neighbour molecules. While electronic and steric complementarity are central to supramolecular chemistry, they are not altogether new in the world of crystallography. The early work of Kitaigorodskii (1973) on crystal packing firmly established ideas of shape-induced recognition between molecules. Accordingly, even for recognition between identical molecules, as is the case in most crystal structures, it is the dissimilar parts that come into close contact and not the similar surfaces – bumps fit into hollows just as a key fits into a lock (Dunitz 1995; Gavezzotti 1998). Conversely, identical parts of neigh-bouring molecules tend to avoid one another. Therefore space groups containing only rotation axes and mirror planes are found much less frequently when compared to those containing inversion centres, screw axes and glide planes (Nowacki 1942, 1943; Kitaigorodskii 1973; Brock and Dunitz 1994). Centrosymmetric close packing is preferred even for molecules that

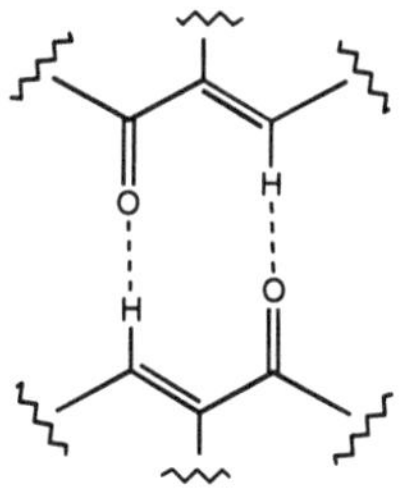

174

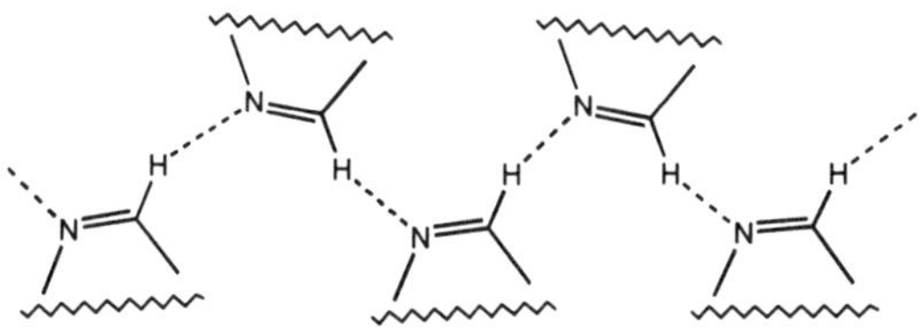

175

do not possess an inversion centre, and the four space groups $P\bar{1}$, $P2_1/c$, $C2/c$ and $Pbca$ account for more than half of all organic crystal structures. These close-packing arguments, based on the complementary recognition between molecules, are of an all-pervasive character, and packing coefficients in most single-component organic crystal structures lie in the range of 0.65 to 0.77.

The space group preferences of many heteroatom crystals parallel those derived on the basis of the Kitaigorodskii model, because the directional requirements of several common heteroatom contacts are in accord with the geometrical dictates of the same three symmetry elements that govern close packing: the inversion centre, the screw axis and the glide plane. This is generally not so well appreciated (Desiraju 1989a, p. 46) but several examples where patterns of weak hydrogen bonds are arranged about the common symmetry elements, have been given in Chapters 2 and 3. Two examples, the centrosymmetric **174** and the screw-symmetrical **175** are given here. These geometrical preferences of the common heteroatom interactions reinforce the close-packing tendencies with the result that there is a dominance of a small number of space groups which contain translational symmetry elements. Because the typical organic molecule has an irregular shape, low-symmetry crystal systems (triclinic, monoclinic, orthorhombic) are preferred over the high-symmetry ones.

The crystal structure of a molecule is a free energy minimum resulting from the overall optimization of the attractive and repulsive intermolecular inter-actions, which have varying strengths, directional preferences and distance-dependence properties. As a consequence, the understanding of the nature, strength and directionalities of intermolecular interactions is of fundamental importance in the design of solid-state supermolecules. Intermolecular inter-actions in organic compounds are of two types: isotropic, medium-range forces that determine close packing based on shape and size; and anisotropic long-range forces which are electrostatic and include hydrogen bonds and het-eroatom interactions. In organic salts, there are also isotropic long-range interactions between ions. The three-dimensional architecture in the crystal is the result then of the interplay between the conflicting or concurring demands of the isotropic van der Waals forces, whose magnitude is proportional to the surface size of the molecule, and the anisotropic (hydrogen bond) interactions, whose strengths are related to the chemical nature of the groups involved (Desiraju 1997a). The crystal structures of hydrocarbon molecules are largely dictated by close packing arguments while the structures of molecules con-taining heteroatoms and functional groups are dominated by hydrogen bonds and other anisotropic interactions.

In the context of this work, it is worthwhile to delineate further the role of weak hydrogen bonds in determining supramolecular structure. In terms of their effects on stronger hydrogen bonds, we have already categorized this role as innocuous, supportive or intrusive (Section 2.2.10.1). Without question, these interactions belong to the lower end in terms of an energy scale. Why then is their role not always innocuous or passive – or at best supportive?

4.1.2 Crystal structures wherein weak hydrogen bonds are important

The number of individual crystal structures in which weak hydrogen bonds have been reported to be important, has grown rapidly in recent years. As a consequence, these hydrogen bonds can no longer be overlooked. Several structures have already been discussed in Chapters 2 and 3, in particular in Section 2.2.10, and the reader is directed there for general comments. Here, we shall focus attention on crystal structures where the role of weak hydro-gen bonding in molecular recognition *per se* has been clearly identified, or where a larger group of compounds has been studied with the specific intent of examining the patterns of weak hydrogen bonds formed.

4.1.2.1 Quinonoid compounds

Quinones are of classical importance in the study of $C-H\cdots O$ hydrogen bonds (see Section 2.2.6). In a variation of the simpler 1,4-benzoquinone layer pattern, Keegstra *et al.* (1994) found that 2-methoxy-1,4-benzoquinone is net-worked in a planar arrangement with extensive involvement of short $C-H\cdots O$ bonds in the d range 2.38–2.59 Å (Fig. 4.1). In the bifurcated

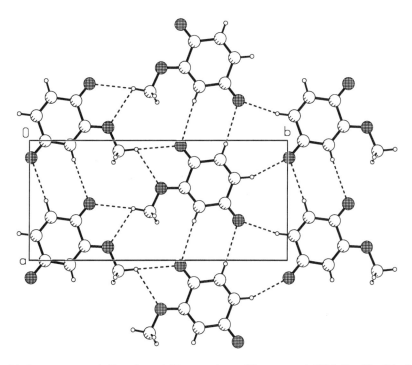

Fig. 4.1. Layer structure in 2-methoxy-1,4-benzoquinone (Keegstra *et al.* 1994). C—H···O hydrogen bonds are shown as dashed lines. Notice the similarity to 1,4-benzoquinone (Fig. 2.6).

arrangements, the sums of the angles around the relevant H atoms are 344° and 351°. Bock *et al.* (1996) found a similar pattern in 2,5-dimethoxy-1,4-benzoquinone. The study was extended by the Utrecht group to other 2,5-di-*n*-alkoxy-1,4-benzoquinones, and it was found that in the di-*n*-propoxy derivative too, molecules are organized into a two-dimensional layered structure (Keegstra *et al.* 1996*a*). Compounds with the alkyl chain length ranging from C_1 to C_{10} were further studied and it was seen that odd–even effects are observed in DSC behaviour concomitant with alkoxy chain length. The importance of C—H···O hydrogen bonds and the interplay with other kinds of intermolecular interactions in these structures have been clearly described by these authors. Tetrasubstituted derivates were also studied. In tetramethoxy-1,4-benzoquinone too, C—H···O interactions are important in the layer structure. However, in tetradecyloxy-1,4-benzoquinone, close-packing of the side-chains dominates as might have been expected (Keegstra *et al.* 1996*b*).

4.1.2.2 *Nitro compounds*

Nitro aromatics in particular have also attracted interest. In an early study, Pedireddi *et al.* (1992) showed the importance of C—H···O hydrogen bonds in a family of substituted β-nitrostyrenes. 3,4-Methylenedioxy-β-nitrostyrene,

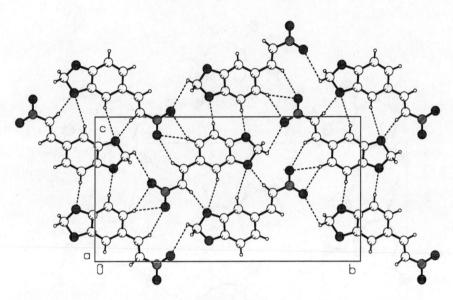

Fig. 4.2. Sheet structure of 3,4-methylenedioxy-β-nitrostyrene to show the C—H···O hydrogen bonds (from Pedireddi *et al.* 1992). *D* distances range from 3.15 to 3.68 Å.

176

176 (X, Y = CH$_2$O$_2$), is typical and its layer structure shows a profusion of short interactions with *D* in the range 3.15–3.82 Å (Fig. 4.2). In the packing of the 4-chloro derivative, **176** (X = Cl, Y = H), C—H···O hydrogen bonds are likewise important as are C—H···Cl interactions. However, in the 4-bromo and 4-iodo derivatives, polarization induced Br···O and I···O interactions are preferred to C—H···Br and C—H···I interactions. The nitro group now accepts both C—H groups and positively polarized halogen atoms. The Cl group is presumably too hard to participate in analogous Cl···O interactions (though these are known in other structures), or its presence may render the aromatic H atoms sufficiently acidic to favour C—H···Cl interactions. It is very difficult in cases such as these, where any of several possible weak interactions could occur, to state exactly why a particular interaction or packing arrangement is observed or not observed. One concludes that weak interactions such as C—H···X hydrogen bonds could play a significant role in determining crystal packing even though they might not contribute

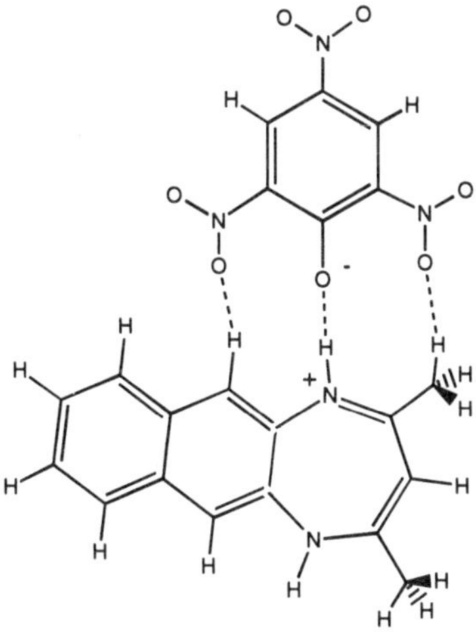

177

proportionately to the crystal stabilization energy. This apparent anomaly is crucial in the context of crystal engineering and is discussed further in Section 4.3.

An interesting study of molecular recognition and discrimination has been reported by T. Suzuki *et al.* (1992), who found that treatment of a mixture of 2,6- and 2,7-dimethylnaphthalene with 2,4,6-trinitrofluorenone resulted predominantly in the formation of the 1:1 donor–acceptor complex with 2,6-dimethylnaphthalene, with only a small amount of the 1:1 complex with the 2,7-isomer. These authors have ascribed the preferential formation of the complex with the 2,6-isomer to the presence of a larger number of C—H···O hydrogen bonds in its crystal structure. These interactions ($2.30 < d < 2.60$ Å; $145° < \theta < 175°$) are not only shorter but they are generally more linear than those found in the complex formed by the 2,7-isomer ($2.40 < d < 2.80$ Å; $130° < \theta < 160°$). In 2,4-dimethyl-1*H*-naphtho[2,3-*b*][1,4]diazepine hydropicrate, **177**, the diazepine molecule and the picrate moiety are linked with one strong N—H···O and two weak C—H···O bonds (Agafonov *et al.* 1994). The 1:1 molecular complex of 1,3,5-trinitrobenzene and dibenz[*a,c*]anthracene has been analysed by Carrell and Glusker (1997) in terms of a complete set of C—H···O hydrogen bonds that surrounds the molecules in a layered structure (Fig. 4.3). Using the CSD, Sharma and Desiraju (1994) surveyed several C—H···O hydrogen bond patterns in 2306 nitro compounds. It was concluded that these patterns, mostly formed by two-point recognition of nitro groups with alkyl and unsaturated groups, would be of utility in supramolecular assembly.

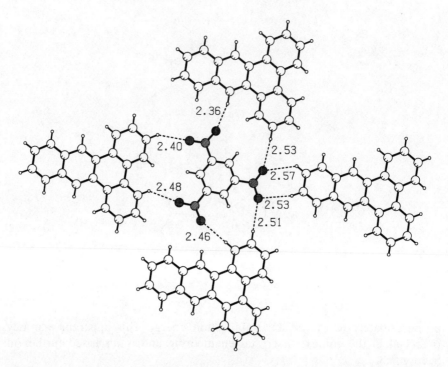

Fig. 4.3. Two-dimensional network of C—H···O hydrogen bonds in the 1:1 complex of trinitrobenzene and dibenzanthracene (from Carrell and Glusker 1997). Distances *d* are normalized.

4.1.2.3 Other compounds

Increasing interest in the role played by weak hydrogen bonds in crystal packing has led to several other systematic studies of the phenomenon. The crystal structures of a series of secondary arenedicarboxamides have been discussed by F. D. Lewis *et al.* (1996) in terms of an interplay between N—H···O, C—H···O and π···π interactions. A related situation is found in a benzimidazolin-2-one trimer studied by Elguero and co-workers (Diez-Barra *et al.* 1997). In both these cases, optimal use is made of all strong and weak donors and the carbonyl (amide) acceptors. Short C—H···O bonds (mean *D* value of 3.33 Å for six distinct interactions) are found in the structure of the 1:1 complex of phenylsulfonylacetamide with hexamethylphosphoramide, HMPA (Cragg-Hine *et al.* 1996). Davidson *et al.* (1996) have investigated complexes of HMPA with *ortho, meta* and *para*-carborane. They found remarkably short C—H···O interactions and that different patterns occur in the three isomers (Fig. 4.4). Very recently, the crystal packing of a substituted 1,2-dicarbadodecaborane has been explained in terms of short C—H···O hydrogen bonds formed by the carboranyl C—H groups and an acetyl group (Harakas *et al.* 1998). These authors have noted the attractiveness of the carborane skeleton for crystal engineering.

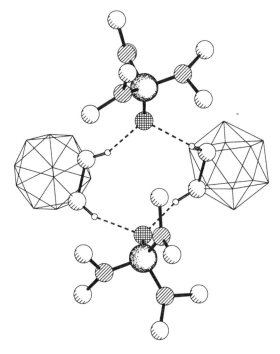

Fig. 4.4. C—H···O hydrogen bonds in the 1:1 adduct of *ortho*-carborane with hexamethylphosphoramide. Only H atoms involved in the hydrogen bonding are shown (redrawn after Davidson
et al. 1996).

An example of molecular recognition at the crystal/solvent interface involving C—H···O hydrogen bonding of methyl groups has been reported by Leiserowitz, Lahav and co-workers (Shimon *et al.* 1990). These authors studied the
effects of solvents and additives on crystal growth. On the surface of the polar
(*R,S*)-alanine crystals, the recognition sites for primary ammonium groups are
only on one particular face, whereas the opposite face recognizes carboxylate
groups. As the crystal grows, molecules are incorporated via NH_3^+ recognition
at one face, and via COO^- recognition at the other. By specifically blocking
recognition sites, crystal growth at particular faces can be inhibited. It was
found that methanol as an additive to the water solvent inhibits crystal growth
at the face involved in recognition of NH_3^+ but not at the other faces. This is
explained by occupation of the NH_3^+ sites by CH_3, associated with three
C—H···O bonds mimicking three N—H···O bonds formed by NH_3^+. So,
methanol molecules adhere strongly to this crystal face and prevent incorporation of alanine molecules. In support of this rationalization, it was shown
that CF_3OH does not inhibit crystal growth. This means that it is not an isosteric replacement of NH_3^+ by CH_3 that plays the decisive role but actually directional interactions.

Also of interest are crystal structures where there is an interplay between

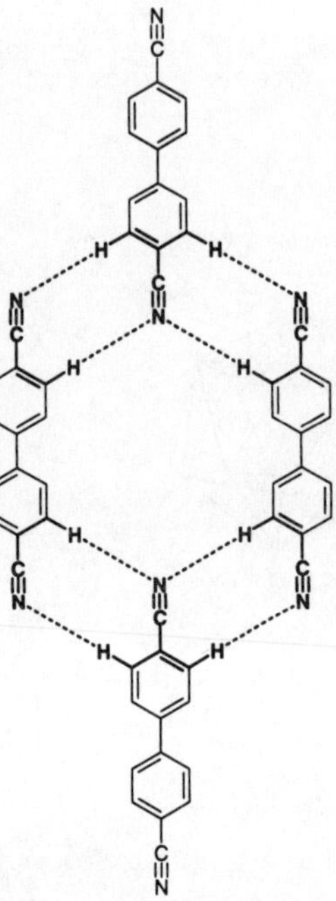

178

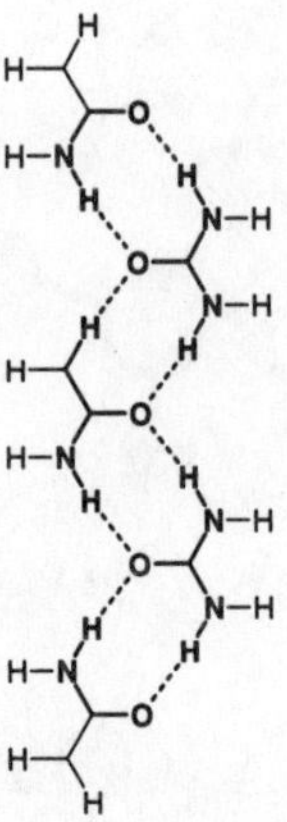

179

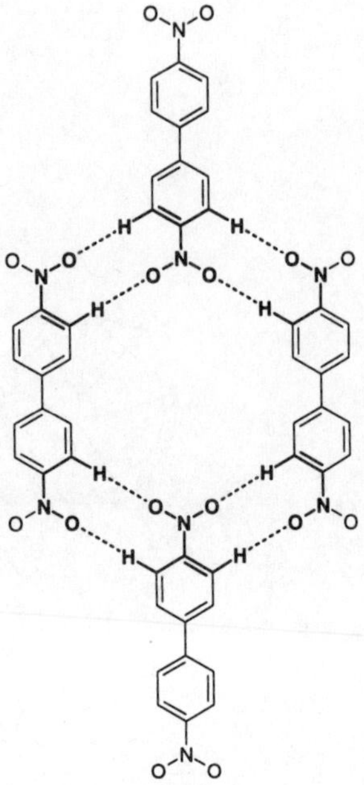

180

strong and weak hydrogen bonds (see Section 2.2.10). Molecular complexes offer much scope here. For instance, 4,4'-dicyanobiphenyl forms a 1:1 complex with urea wherein the aromatic host forms large hexagonal channels, **178**, via long $C-H\cdots N$ interactions ($D = 3.95, 3.84\,\text{Å}$), and the urea guest molecules are arranged in $N-H\cdots O$ ribbons, **179** (see also Fig. 3.81(a), Section 3.2.2), which fit completely and in a perpendicular manner within the host channels. A similar host framework, **180**, mediated via $C-H\cdots O$ interactions ($D = 3.84$, $3.97\,\text{Å}$) is found in the 1:1 complex formed by 4,4'-dinitrobiphenyl and urea. While these $C-H\cdots N$ and $C-H\cdots O$ hydrogen bonds are extremely weak, they must contribute in some measure to overall crystal stabilization because the observed urea ribbon pattern is quite uncommon in the absence of these particular hosts (Thaimattam *et al.* 1998). This lends credence to the idea that strong and weak hydrogen bonds must be considered jointly in the analysis and design of crystal structures, and further emphasizes the need to be able to understand the phenomenon of weak hydrogen bonding more fully.

Support for such ideas comes from the work of Braga *et al.* (1995, 1997*b,c* and the references cited therein) who have analysed a number of salts constituted with organometallic cations and organic anions. The cations and anions contain strong and/or weak hydrogen bonding functionalities and include moieties such as $[(\eta^5\text{-Cp})_2\text{Co}]^+$, $[(\eta^6\text{-C}_6\text{H}_6)_2\text{Cr}]^+$, tartrate and squarate. Further, some of the salts are hydrated and so contain additional donor and acceptor groups. It was noted that for a complete understanding of these structures, both $O-H\cdots O$ and $C-H\cdots O$ hydrogen bonding capabilities must be considered. The $C-H\cdots O$ bonds in these structures are at the shorter end of the usual distance range. This might be a consequence of the fact that they are both charge-assisted and parts of a cooperative network. It was observed by these authors that for a particular cation/anion combination, choosing a different relative stoichiometry often resulted in a significant change in the crystal structure obtained. These structures are too varied and still incompletely understood to be thought of in terms of deliberate design, but they do hint at future possibilities when organometallic compounds enter the mainstream of supramolecular chemistry and crystal engineering (Braga *et al.* 1998c).

4.2 Inclusion complexes

In this section, we discuss the the role of weak hydrogen bonds in a group of compounds that constitute a special category of supramolecular structures – the inclusion complexes. These structures are of fundamental interest, often of great practical importance, and also offer considerable aesthetic satisfaction. By inclusion complexes, we mean supramolecular compounds in which a guest molecule, or at least a part of it, is enclosed in a hollow host molecule. This is to some degree an arbitrary restriction. There are also host–guest complexes, in which the host cavity is constituted not by individual molecules but with finite or infinite assemblies of molecules. In a great number of lattice inclusion compounds, small organic molecules form host lattices in which guests are included more or less unspecifically (Vögtle 1991).

In the inclusion complexes discussed here, steric fit of the guest in the host is required, and also a matching of specific intermolecular interactions. These interactions can be strong (ionic, cation–π, $O/N-H\cdots O/N$ hydrogen bonds), but also weak ($C-H\cdots O$) and very weak ($C-H\cdots \pi$). Frequently, a blend of strong and weak interactions determines the inclusion characteristics. We restrict the following discussion to three classes of macrocyclic hosts which have fundamentally different inclusion properties: the crown ethers, oligoaryl hosts and cyclodextrins.

4.2.1 *Crown ethers*

In complexes of crown ethers, weak hydrogen bonds between the guest and the host occur frequently (reviewed by Goldberg 1984; Weber 1989). The best examined member of the family is 18-crown-6, that is *cyclo*-$(\text{CH}_2-\text{CH}_2-\text{O})_6$ (Pedersen 1988). In the uncomplexed crystal structure, and also in apolar solvents, 18-crown-6 adopts an elliptical shape which is stabilized by

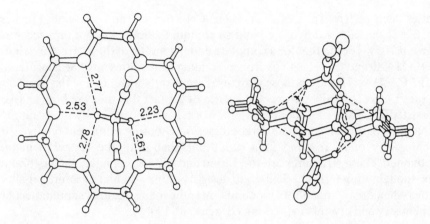

Fig. 4.5. 1:2 Complex of 18-crown-6 with malononitrile. In the left drawing, one of the guest molecules is omitted for clarity (from Steiner 1996a, after Kaufmann *et al.* 1977). Distances *d* are normalized.

intramolecular $C-H\cdots O$ hydrogen bonds (already shown in Section 2.2.9.1, Fig. 2.35(a)). When 18-crown-6 complexes a cation, it adopts a 'crown' shape with all O atoms turned inwards so that they can interact with the guest, while the CH_2 groups are turned outwards and form a lipophilic surface. Upon complexation with neutral molecules, hydrogen bonds are formed from the guest to the etheral O atoms of the host. A convincing example involving $C-H\cdots O$ hydrogen bonds is provided by the 1:2 complex of 18-crown-6 with the strong $C-H$ donor molecule malononitrile, $CH_2(CN)_2$, Fig. 4.5 (Kaufmann *et al.* 1977). The adduct is stabilized by a system of multifurcated $C-H\cdots O$ hydrogen bonds. Very similar host–guest interactions were later reported for the 1:2 complexes with the related guest molecules $CH_2Cl(CN)$ (Buchanan *et al.* 1992) and CH_2Cl_2 (Jones *et al.* 1994). For the 1:1 adduct 18-crown-6–malononitrile, the binding free energy in C_6D_6 has been determined by Reinhoudt and co-workers as $\Delta G° = -3.2$ kcal/mol, and binding energies with numerous derivatives of 18-crown-6 have also been reported (van Staveren *et al.* 1986).

To generalize the observations on 18-crown-6–malononitrile, Knöchel and co-workers studied a series of 18-crown-6 complexes with neutral molecules possessing the structural element XH_2 (X = O, N, C), and found good relationships within the series (Elbasyouny *et al.* 1983). In principle, OH_2, NH_2 and CH_2 units form the same kind of hydrogen bond motifs with the crown ether. Because of the possibility of forming multifurcated interactions, there are a number of different configurations. The two $X-H$ donors can form their shortest hydrogen bonds with opposite O atoms of the macrocycle, such as in Fig. 4.5, or with two O atoms separated by just one ether group. In 18-crown-6/cyanoacetic acid/water, **181**, different hydrogen bond configurations are formed within the same crystal structure. The complex with adiponitrile crys-

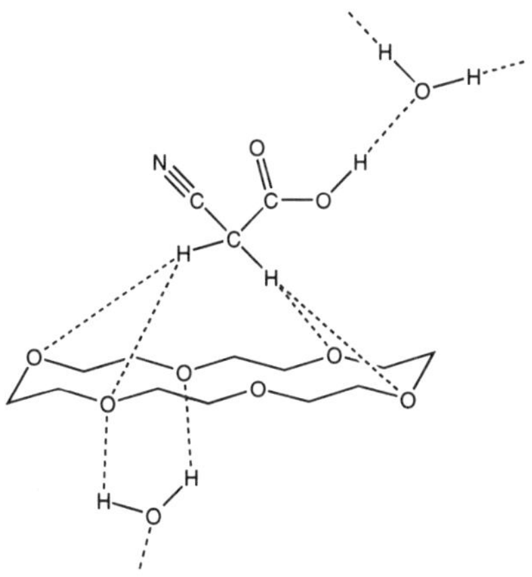

181 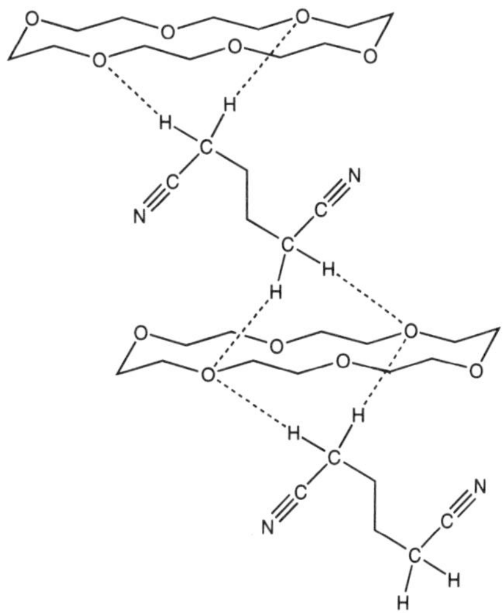   182

tallizes in chains where 18-crown-6 molecules alternate with the strong bifunctional CH_2 donor, **182**. On closer examination, it may be seen that complex
181 forms polymeric chains too, in which the cyanoacetic acid molecule on one
side of the crown ether forms an $O-H\cdots O$ hydrogen bond with the water
molecule on the other side of the next crown ether along the chain.

The above examples involve only relatively acidic $C-H$ donors. Much
weaker $C-H$ donors can also be engaged in crown ether complex stabilization. In particular for methyl groups interacting with 18-crown-6, a large
number of examples have been published (Table 4.1), beginning with a paper
of Goldberg (1975). The geometry of the methyl group allows for the formation of three $C-H\cdots O$ hydrogen bonds of good geometry with ether O atoms
situated on the same face of the crown, **183**. This is structurally analogous
to the favoured complexation mode for primary alkyl ammonium groups,
$R-NH_3^+$ ('perched' arrangement). The distances D are typically in a narrow
range 3.2–3.4 Å. There is a strong feeling that the relative strength of this
pattern lies in the *concerted* action of three weak but directional interactions.

There is a second mode of methyl interactions with 18-crown-6; if the $R-C$
axis of the $R-CH_3$ group is strongly tilted against the molecular axis of the
crown ether, only two of the $C-H$ groups form clear-cut hydrogen bonds,
whereas the third forms only a strongly bent interaction, **184**. Although the
third d separation can still be relatively short (2.6–2.8 Å), θ angles in the range
90–120° indicate only very weakly bonding nature, if at all. The geometrical
data in Table 4.1 show that no borderline can be drawn between geometries
183 and **184**, and indeed many intermediate cases do exist. Fine examples for

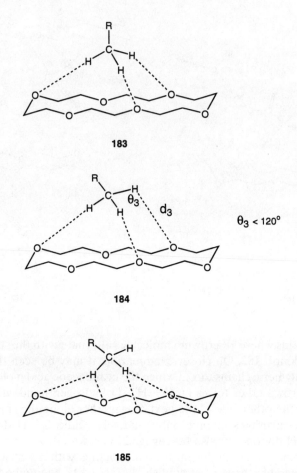

183

184

$\theta_3 < 120°$

185

both geometries have been reported by Caira and Mohamed (1993) for ternary complexes of 18-crown-6 with acetonitrile and the sulfonamides **187** and **188**, Fig. 4.6. Note in particular Fig. 4.6(b), which shows a mixed pattern of C(Me)—H···O, C(sp^2)—H···O and N—H···O hydrogen bonds. It is not unusual that multifurcated hydrogen bonds are formed in connection with the primary patterns **183** and **184**. This occurs particularly with more tilted geometries that allow the methyl group to dive deeply into the opening of the crown, **185** and Fig. 4.6(b).

Several crystal structures of substituted 18-crown-6 with neutral molecules have also been published. An interesting early example of isomer-dependent complexation involving C—H···O hydrogen bonds was reported by Damewood *et al.* (1988). The *cis-syn-cis* and *cis-anti-cis* isomers of dicyclohexane-18-crown-6 crystallize with malononitrile in different stoichiometries (1:1 and 1:2, respectively), and the different binding geometries are illustrated in Fig. 4.7.

Crown ethers can be used to build pseudorotaxanes. Stoddart and

Table 4.1 Complexes of 18-crown-6 stabilized by $C-H\cdots O$ hydrogen bonds donated by a methyl group

Guest molecule	d-range (Å)	D-range (Å)	θ-range (°)	Reference
(1)	2.31–2.88	3.08–3.39	169–90	Goldberg 1975
Me_2SO_2	2.27–2.53	3.31–3.36	164–128	Bandy et al. 1981
Me_2SO_4 (a)	2.29–2.66	3.22–3.44	159–112	Weber 1983
(b)	2.29–2.71	3.31–3.47	159–115	
CH_3-NO_2	2.23–2.49	3.24–3.32	154–133	Rogers and Green 1986
CH_3-NO_2, –150°C	2.22–2.63	3.14–3.31	174–108	Rogers and Richards 1987
CH_3-CN, triclinic (a)	2.20–2.63	3.20–3.32	159–113	Rogers et al. 1988
(b)	2.21–2.71	3.19–3.28	149–112	
CH_3-CN, monoclinic	2.24–2.63	3.25–3.35	154–118	Garrell et al. 1988
CH_3-CN, monoclinic	2.18–2.55	3.26–3.34	175–124	Weller et al. 1989
186	2.35–2.63	3.35–3.49	153–136	Chenevert et al. 1993
CH_3-CN and **187**	2.21–2.34	3.26–3.34	177–144	Caira and Mohamed 1993
CH_3-CN and **188**	2.29–2.76	3.25–3.37	179–107	Caira and Mohamed 1993
$PhSO_2-NMe-SO_2Me$ (a)	2.23–2.35	3.24–3.36	176–138	Henschel et al. 1996
(b)	2.26–2.45	3.20–3.32	151–136	

Distances d normalized. (1) Dimethyl acetylenedicarboxylate; (a), (b) refer to symmetry-independent molecules.

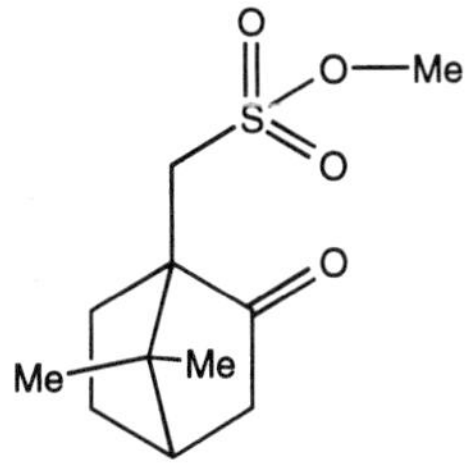

186

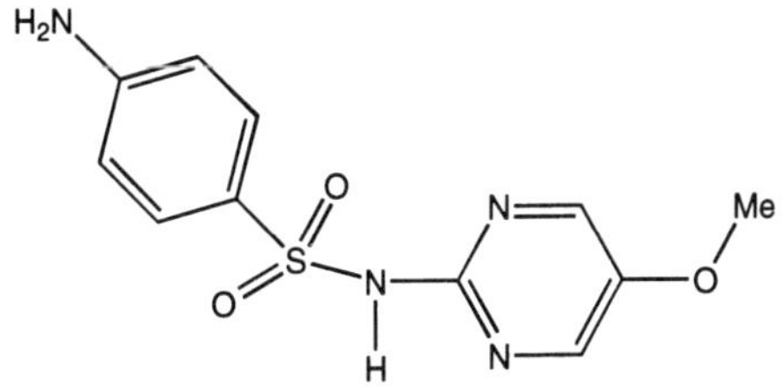

187

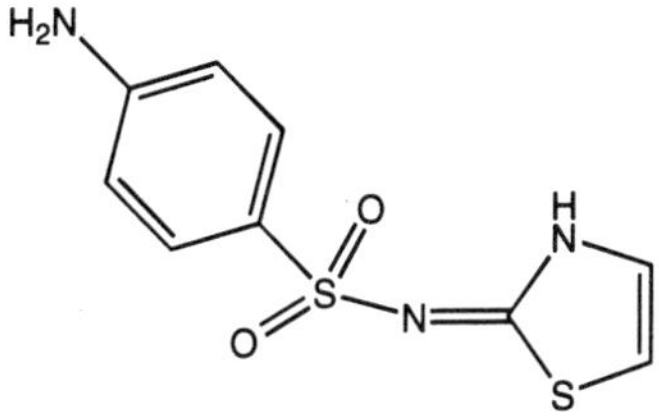

188

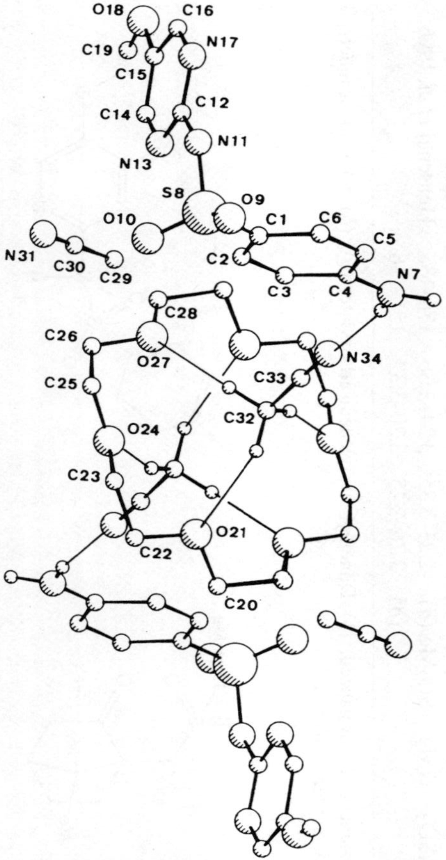

Fig. 4.6. Two ternary crown ether complexes with sulfonamides and acetonitrile: (a) 18-crown-6/compound **187**/MeCN (1/2/4); (b) 18-crown-6-compound **188**/MeCN (1:1) (from Caira and Mohamed 1993).

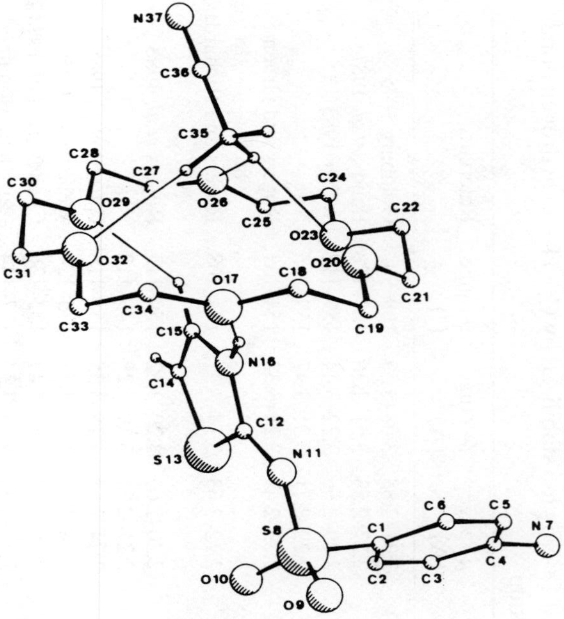

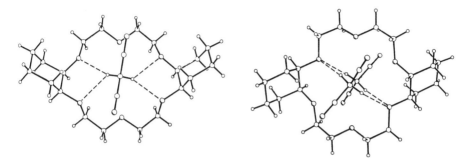

Fig. 4.7. Isomer-dependent complexation. Crystal structures of malononitrile complexed by the *cis-syn-cis* (*left*) and *cis-anti-cis* (*right*) isomers of dicyclohexano-18-crown-6 in (1:1) and (2:1) ratios, respectively (adapted from Damewood *et al.* 1988).

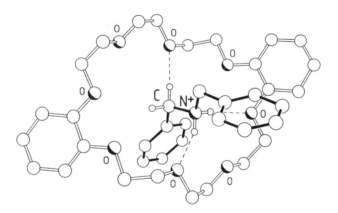

Fig. 4.8. A pseudorotaxane stabilized by N—H···O and C—H···O hydrogen bonds. Notice the *N,N*-dibenzylammonium cation encircled by a dibenzo-24-crown-8 macrocycle (from Ashton *et al.* 1996*a*).

co-workers have threaded elongated organic cations through substituted crown ethers, such as the *N,N*-dibenzylammonium cation through dibenzo-24-crown-8 (the counterion is PF_6^-). The complex is stabilized by two N$^+$—H···O and one C—H···O hydrogen bond, the latter with $D = 3.16$ Å, $\theta = 174°$ (Fig. 4.8, Ashton *et al.* 1996*a*). Concerted action of N$^+$—H···O and much weaker C—H···O hydrogen bonds is a recurrent feature of this series.

There are also examples of C—H···O hydrogen bonding in complexes of more heavily substituted crown ethers. The structure of 3,3'-(1,1'-bi-2-naphthol)-21-crown-5 monohydrate as reported by Goldberg (1978) is shown in Fig. 4.9. The single water molecule is too small to fill the relatively large macrocyclic cavity, so that part of the polyether chain adopts an irregular conformation which is associated with a transannular C—H···O hydrogen bond.

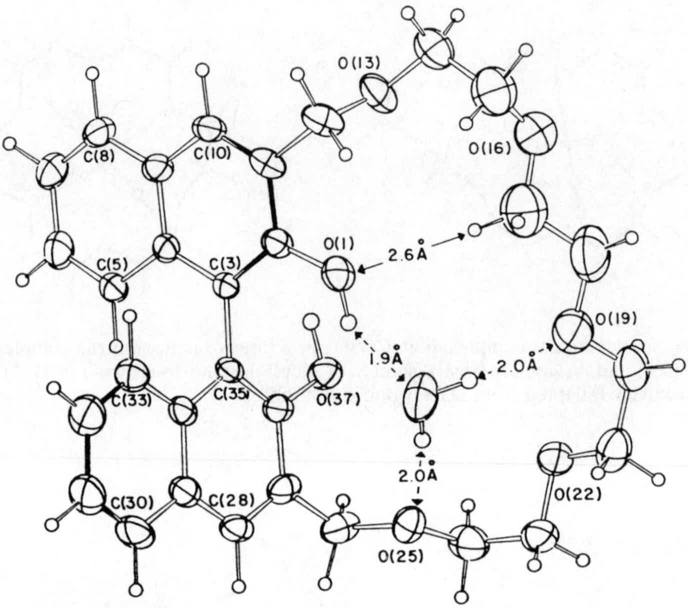

Fig. 4.9. Intramolecular C—H···O hydrogen bond in 3,3′-(1,1′-bi-2-naphthol)-21-crown-5 mono-hydrate (from Goldberg 1978).

A relationship to the behaviour of uncomplexed 18-crown-6 (Fig. 2.35(a)) is obvious. As a final example, the reader is referred to a series of hemispherands complexed with malononitrile. These molecules are composed of half a spherand and half a polyether moiety. Only one of four related crystal structures reported by Reinhoudt and co-workers is shown in Fig. 4.10. (Grooten-huis *et al.* 1987).

4.2.2 *Oligoaryl hosts*

With macrocyclic oligoaryl hosts, the molecular cavity is formed by phenyl groups which offer their electron-rich π-faces for host–guest interactions. Numerous molecular families of this kind, such as calixarenes, cyclophanes and carcerands, have been synthesized (Atwood *et al.* 1996). Normally, the aromatic rings are involved in cation binding, or in herringbone interactions with other aromatic rings. When complexing neutral molecules, the phenyl groups can serve as acceptors of X—H···π(Ph) hydrogen bonds, and because the cavities are formed by numerous phenyl rings, a number of such hydrogen bonds can be formed. Examples with pairs of Ph···H—X—H···Ph hydrogen bonds (X = O, C) in calix[4]arene complexes have been reported by Atwood and co-workers and have already been shown in Section 3.1.3 (Figs 3.17 and 3.26). An example with a macrocyclic oligophenylene encapsulating a chloroform molecule has been provided by Schlüter and co-workers, Fig. 4.11 (Hensel *et al.* 1997).

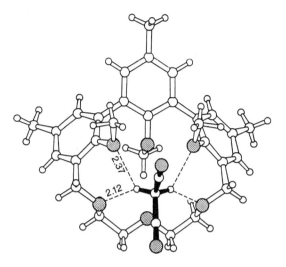

Fig. 4.10. A hemispherand complexing a malononitrile molecule (from Steiner 1996*a*, after Grootenhuis *et al.* 1987). Distances *d* are normalized.

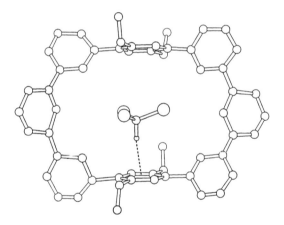

Fig. 4.11. A macrocyclic oligophenylene hosting a chloroform molecule. The $C-H\cdots Ph$ hydrogen bond is directed at the phenyl ring centroid with $D(\pi) = 3.53$ Å (from Hensel *et al.* 1997).

4.2.3 *Cyclodextrins (cycloamyloses)*

An important class of bio-organic host molecules are the cyclodextrins (CDs) or cycloamyloses (CAs) (reviewed by Harata 1998; Saenger *et al.* 1998; Saenger and Steiner 1998). These oligosaccharides are composed of $\alpha(1-4)$-linked D-glucose units, see Section 5.4. The best known of these are the cyclodextrins with six to eight glucose units, called α-CD, β-CD and γ-CD, respectively (or, alternatively, CA6, CA7 and CA8). The three hydroxyl groups per glucose are placed at the rims, making the molecules water-soluble. CDs are used in

aqueous solution, and their crystal structures are dominated by extended networks of cooperative O—H···O hydrogen bonds. On the other hand, the molecular cavity is lined by C—H groups and the glycosidic oxygen atoms which
link the glucose units, so that it is rather hydrophobic compared to the outer
surface. The CDs include a large variety of guest molecules with suitable sizes,
preferably those with apolar groups that are extracted by complexation from
the aqueous surrounding ('hydrophobic effect'). If a guest is too small to fill
the cavity volume, additional water molecules are included as space-fillers.

If polar molecules are included in cyclodextrin cavities, they have only
limited opportunities to satisfy their hydrogen bond potentials. As a resort,
weak host–guest hydrogen bonds of different kinds can be formed. A typical
example is β-CD–ethanol octahydrate, in which one ethanol and three water
molecules are enclosed in the molecular cavity (the other five water molecules
are placed in interstices). The crystal structure has been determined very accurately with low-temperature neutron diffraction, Fig. 4.12. Numerous
C—H···O hydrogen bonds are formed between the host and the guests, with
the shortest distances d around 2.4 Å (Steiner and Saenger 1992a). Note that
the C—H groups at the cavity surface are activated by adjacent O atoms.

An interesting case of C—H···O hydrogen bonding occurs in the complex

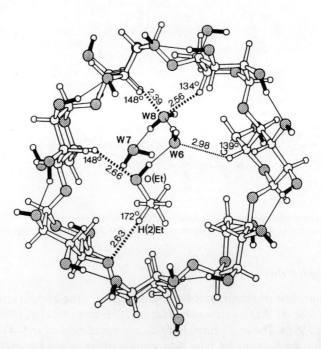

Fig. 4.12. Neutron structure of β-cyclodextrin–ethanol octahydrate at 15 K, with O—H···O
hydrogen bonds drawn as continuous lines, and C—H···O hydrogen bonds as dashed and dotted
lines (from Steiner and Saenger 1992a).

β-CD–1,5-pentanediol. In a systematic investigation of the inclusion geometry of short chain-like molecules in β-CD, it was found that the inclusion complexes with diethanolamine and 1,5-pentanediol crystallize isostructurally, and
that the guest molecules are included in practically the same orientations and
geometries, Fig. 4.13. The conventional N—H⋯O hydrogen bond formed by
the secondary amino group of the diethanolamine molecule is replaced in the

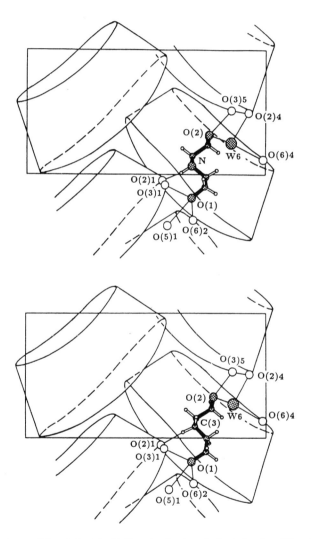

Fig. 4.13. Structures of β-cyclodextrin-diethanolamine hexahydrate (*top*) and β-cyclodextrin–1,5-
pentanediol hexahydrate (*bottom*), illustrating the isofunctional replacement of an N—H⋯O by
a C—H⋯O hydrogen bond. The β-cyclodextrin molecules are drawn schematically (after Steiner
et al. 1995*b*).

1,5-pentanediol complex by a C—H···O hydrogen bond donated to the same hydroxyl acceptor. The distances of the two hydrogen bonds are different ($d = 1.81$ and $2.49\,\text{Å}$, respectively), but the hydrogen bond pattern is conserved. This is interpreted as a case of isofunctional replacement of an N—H···O by a C—H···O hydrogen bond (Steiner *et al.* 1995*b*). It is also taken as the reason why the guest molecules are ordered in these examples, whereas chain-like guests without such stabilizing hydrogen bonds are typically disordered over several low-energy conformations.

If guest molecules with π-acceptor groups are complexed by cyclodextrins, the formation of C—H···π interactions with the cavity lining is possible. Consider for example the complex β-CD–but-2-yne-1,4-diol heptahydrate, shown in Fig. 4.14 (Steiner and Saenger 1995*a*). The arrangement of the molecular cluster in the CD cavity is stabilized by a complex system of O—H···O and C—H···O hydrogen bonds, and of C—H···π(C≡C) interactions. In quantum chemical calculations on this crystal structure, bond energies of around –1.0 kcal/mol were obtained for the C—H···O hydrogen bonds, and of –0.7 kcal/mol for the C—H···π interaction (Starikov *et al.* 1998). This assigns the polar host–guest interactions in cyclodextrin cavities to the weak end of the C—H···X hydrogen bond spectrum.

Hydrogen bonds in cyclodextrin cavities normally form very complex patterns of interconnected weak and moderate interactions. The weaker con-

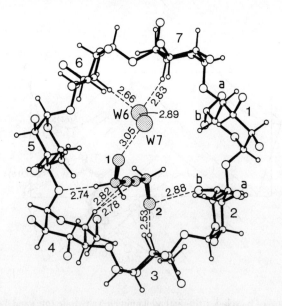

Fig. 4.14. β-Cyclodextrin–but-2-yne-1,4-diol heptahydrate stabilized by an interplay of O—H···O and C—H···O hydrogen bonds, and a C—H···π interaction (from Steiner and Saenger 1995*a*).

stituents mainly function so as to fill up hydrogen bond potentials that would otherwise be vacant. Each of the weak hydrogen bonds contributes only a small bond energy, and it is presumably their large number that makes them important in structure stabilization. It should be noted, however, that significant $C-H \cdots A$ hydrogen bonding is not observed in all cases where it would be possible in principle. An example is the inclusion complex of acetone in dimethyl-α-CD, in which the guest molecule is found oscillating more or less freely in the sterically available cavity volume (Steiner *et al.* 1996*e*).

In summary, we have delineated some of the roles played by weak hydrogen bonds in inclusion complexes with a selection of macrocyclic host molecules. The host molecules are chemically different and have different complexation properties. Crown ethers have a polar cavity and a lipophilic outer surface, and are primarily used for cation complexation in organic solvents. For cyclodextrins, the situation is reversed. They have a hydrophilic outer surface and a relatively apolar cavity, and are used primarily for complexation of partly apolar molecules in aqueous solution. Despite these differences, weak hydrogen bond effects play related roles in the stabilization of many inclusion complexes. Typically, weak hydrogen bonds are formed if 'unusual' guests are included. $C-H \cdots O$ hydrogen bonding allows crown ethers to complex almost non-polar neutral molecules instead of cations, and allows cyclodextrins to form stable complexes with alcohol and water molecules instead of the preferred bulky apolar guests. This means that weak hydrogen bonds can reverse the intrinsic affinities of these host species and in effect widen the range of guest molecules that can be included.

The characteristics described above are also relevant to the inclusion of guest molecules in cavities and recognition sites of biological macromolecules. As will be described in Chapter 5, partly hydrophobic and partly hydrophilic cavities occur very frequently in biological systems. Molecules enclosed in such cavities will normally form networks of strong and weak hydrogen bonds, all of which must be considered if the overall arrangement is to be rationalized fully.

4.3   Crystal engineering – promises and problems

Crystal engineering is the design of crystal structures for specific purposes and applications (Desiraju 1989*a*, 1997*a*; Gavezzotti 1996; Bürgi *et al.* 1998). Crystal structures are determined by intermolecular interactions and so the study and understanding of interactions, such as the weak hydrogen bond, are of considerable importance in the development of this new and rapidly growing subject (Aakeröy 1997; Zimmerman 1997; Anthony *et al.* 1998*b*). The description of a crystal as a supramolecular entity has had a direct impact on crystal engineering, for if a crystal is a supermolecule, crystal engineering is supramolecular synthesis in the solid state (Desiraju 1995a). Instead of building molecules with atoms and covalent bonds, one builds crystals (supermolecules) with molecules and non-covalent interactions according to a predetermined protocol.

The utilization of weak hydrogen bonds in crystal engineering strategies follows directly from their known involvement in crystal packing as has been discussed in Section 4.1, and also in the formation of bicomponent species as has been described in Section 4.2. Host–guest complexes are of importance in supramolecular chemistry because the very fact that two different molecular components have crystallized together is significant. When the interactions responsible for the co-crystallization are weak hydrogen bonds, one is more confident that such interactions can in turn be used for the *design* of new crystal structures.

4.3.1 *From molecular to crystal structure*

In the context of crystal engineering strategies a frequently asked question is 'Given the molecular structure of an organic substance, what is its crystal structure?' Though the purpose of this question appears to be obvious, the answer is not available at the present time at least in a general sense. The difficulties in answering this question stem from two counts. The first relates to the complementary nature of the recognition phenomenon (Pauling and Delbrück 1940; Dunitz 1996b; Desiraju 1997c). Because this complementarity is characteristic of both geometrical and chemical recognition, a functional group approach to crystal structure prediction is essentially futile. For instance, not all carboxylic acids have crystal structures wherein the supramolecular environment of the carboxylic group is identical or even similar. A particular carboxylic acid may adopt a particular crystal structure depending on other functional groups in the molecule. So while benzoic acid forms a centrosymmetric O—H···O dimer, **189**, pyrazine carboxylic acid forms a mixed dimer, **190** (Takusagawa *et al.* 1974), in which a weak C—H···O hydrogen bond is accomodated because of the preferential formation of a strong O—H···N hydrogen bond between the carboxyl group and the heterocyclic N atom. Still other carboxylic acids need not form a dimer at all – in 4-chlorocubane carboxylic acid, the rare *syn-anti* catemer, **191**, is favoured because of the assistance rendered to this motif by a good C—H···O bond formed by the relatively acidic cubyl C—H group (Kuduva *et al.* 1999). In yet another variation seen for instance in the non-centrosymmetric modification of 3,5-dinitrosalicylic acid, the carboxylic acid forms a hydrate wherein a complex array of strong and weak hydrogen bonds stabilize the packing (Desiraju 1991a). A crystal is truly a supramolecular entity and the molecules that constitute it interact with one another in their entirety during crystallization. The seemingly smallest of changes, small that is in molecular terms, can cause profound changes in the crystal structure adopted.

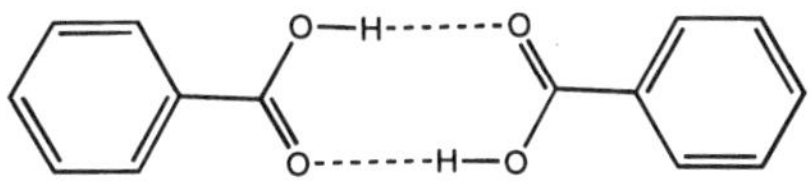

189

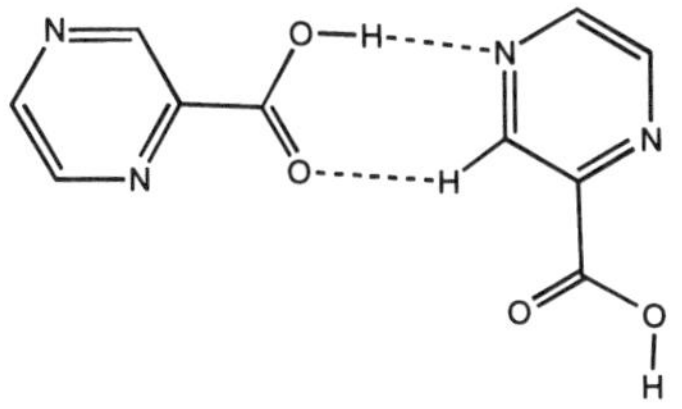

190

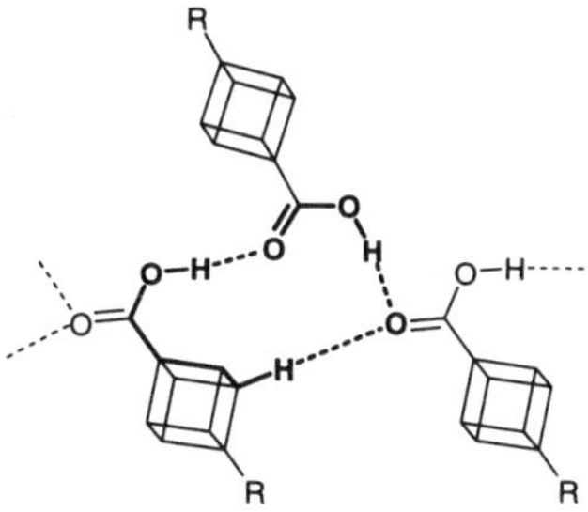

191

The second difficulty in predicting crystal structures lies in the fact that though the vast majority of organic structures tend towards close-packing, and indeed do achieve such a close-packing in their stable crystalline forms, the actual structure that is chosen from among the really large number of structures that might have been possible were close-packing alone of consequence, is often determined by weak long-range interactions which can critically direct the course of crystallization. This is a real problem. It means that weak interactions can exert an influence during crystallization that is out of all proportion to their final contribution to the overall crystal stabilization energy. Weakness of interaction does not generally lead to lack of specificity in recognition and polymorphism is not an across-the-board phenomenon. The roles played by some of the weaker interactions during crystallization could be of a specific nature.

Both these problems are more or less pervasive in all non-hydrocarbon crystals. For hydrocarbons, the interactions are exclusively of the van der Waals type and this means that crystal structures may be confidently predicted on the basis of the close-packing model, without interference from anisotropic and long-range interactions (Robertson 1951; Kitaigorodskii 1973; Williams 1974; Desiraju and Gavezzotti 1989). In heteroatom containing structures, however, the interactions are a complex mosaic of varying strengths, directionalities and distance-dependence characteristics. In such situations, direct and simple extrapolations from molecular to crystal structure are tricky and difficult in the best cases and impossible in the worst. The situation becomes hopeless if one is dealing with flexible molecules and if solvent inclusion is a possibility. We present now an example wherein some of the above-mentioned problems prevail, and which is at the same time relevant from the viewpoint of weak hydrogen bonding.

Ermer and Eling (1994) showed that mono- and poly-component systems that contain equal stoichiometries of hydroxy and primary amino groups, may be expected to crystallize with a predictable array of $O-H\cdots N$ and $N-H\cdots O$ hydrogen bonds. The OH group contains one hydrogen bond donor and two acceptor groups. The NH_2 group contains two donors and one acceptor, so that it complements the donor–acceptor capability of the OH group. Therefore, a system of equal stoichiometries of the two groups can be fully saturated as far as the (strong) hydrogen bond capability is concerned with all N and O atoms having tetrahedral connections. Indeed, the crystal structure predicted for these compounds is of the β-arsenic or ZnS type in which the atoms are tetrahedrally coordinated. A compound that nicely fits the Ermer–Eling model is 4-aminophenol and the expected tetrahedral network of $O-H\cdots N$ and $N-H\cdots O$ hydrogen bonds may be seen in Fig. 4.15. However, in the structures of the isomeric 2- and 3-aminophenols, the strong hydrogen bonded tetrahedral network is absent and instead $N-H\cdots\pi$ hydrogen bonds are observed. These crystal structures were redetermined using low-temperature neutron diffraction to adequately confirm the presence of $N-H\cdots\pi$ hydrogen bonds in both cases (Allen *et al.* 1997c). It was also seen that the 'normal' hydroxy–amino recognition seen in 4-aminophenol is not found in these structures. In 2-aminophenol, each $-OH$ group donates an $O-H\cdots N$ hydrogen bond and accepts $N-H\cdots O$ and $C-H\cdots O$ hydrogen bonds with different molecules. Each NH_2 group similarly donates and accepts a strong hydrogen bond with different molecules. The second H atom, however, participates in the $N-H\cdots\pi$ hydrogen bond. The hydrogen bonding array in 3-aminophenol is similar to that in 2-aminophenol. These features are illustrated in Fig. 4.16. While a tetrahedral environment around the O and N atoms is maintained in both these structures, the hydrogen bonds are not exclusively of the strong type.

The reason for these unexpected structures of 2- and 3-aminophenols compared with that of 4-aminophenol is understood in terms of the need to attain herringbone or T-shaped geometry of the phenyl rings in these two structures. Inspection of Fig. 4.17 indeed shows that the herringbone arrangement in the two cases is very similar. The driving force for the existence of the weaker $N-H\cdots\pi$ and $C-H\cdots O$ hydrogen bonds arises then from the need to adopt

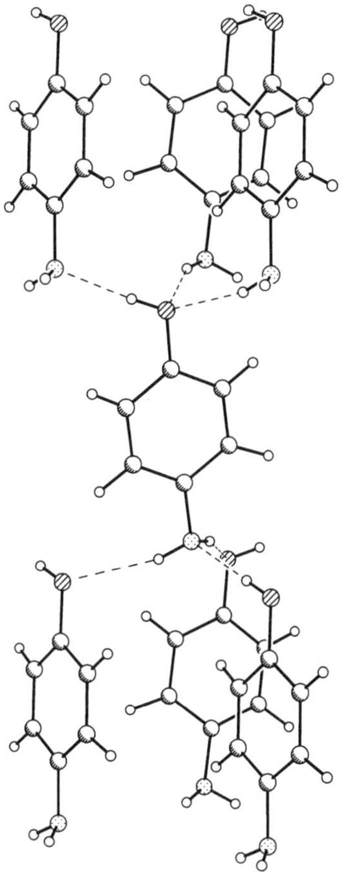

Fig. 4.15. Tetrahedral network formed by N—H···O and O—H···N hydrogen bonds in 4-aminophenol. Notice the favourable $\pi···\pi$ interactions (from Allen *et al.* 1997c).

a herringbone geometry of the phenyl rings, a geometry that renders the heteroatoms inaccessible for the formation of two conventional N—H···O hydrogen bonds. Interestingly, the herringbone geometry is also maintained in 4-aminophenol (Fig. 4.15) but here it is compatible with the β-arsenic network of strong hydrogen bonds. The herringbone geometry may be accordingly identified as the primary structural effect in these compounds. In summary therefore, simple molecules can have complex crystal structures with strong interactions being mediated by a blend of weaker interactions.

4.3.1.1 *Structural insulation*

The above example illustrates that straightforward connections between molecular and crystal structures are possible or difficult, depending on

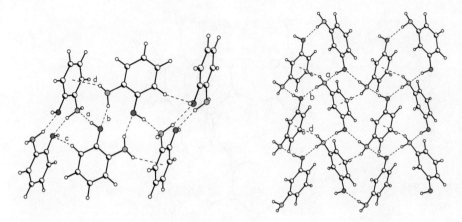

Fig. 4.16. Neutron structures of 2-aminophenol (*left*) and 3-aminophenol (*right*) to show various kinds of hydrogen bonding. 2-Aminophenol: O—H···N, a = 1.782; N—H···O, b = 2.141; C—H···O, c = 2.577; N—H···π, d = 2.309 Å. 3-Aminophenol: O—H···N, a = 1.758; N—H···O, b = 2.024; N—H···π, d = 2.409 Å. Note that the hydrogen bond patterns in these structures is different from that shown in Fig. 4.15 for 4-aminophenol (adapted from Allen *et al.* 1997*c*).

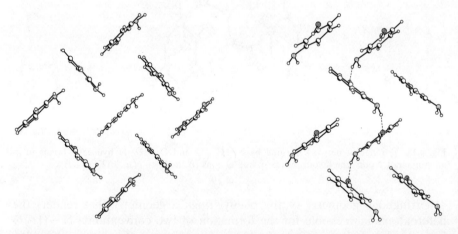

Fig. 4.17. Herringbone interactions in 2-aminophenol (*left*) and 3-aminophenol (*right*). Notice that the arrangement of aromatic rings is similar to that in the structure of benzene (from Allen *et al.* 1997*c*).

whether there is insulation or interference between the different sets of significant intermolecular interactions. Thus, in 4-aminophenol, the crystal structure retains a high degree of fidelity to other hydroxy–amino systems described by Ermer and Eling because the competing herringbone and hydrogen bonding interactions are effectively insulated, in other words the crystal structure can be formally partitioned into hydrogen bonding and herringbone domains. In 2- and 3-aminophenols, however, the structure-determining inter-actions interfere with one another and this leads to unexpected crystal struc-tures. Unfortunately, it is often difficult to anticipate from a casual inspection of the molecular structure if such insulation will or will not be present. This is a major problem in crystal engineering, indeed it is *the* problem, and at its heart lies our inability to accurately predict the crystal structure of an arbi-trary molecule from its molecular structure.

The failure of routine molecule-to-crystal extrapolation has immediate con-sequences for both computational and experimental approaches to crystal engineering, and these will now be detailed, with specific reference to weak hydrogen bonding. These consequences define the present challenges in the subject (Nangia and Desiraju 1998*a*).

4.3.2 *The computational approach*

If crystal structures were determined exclusively by isotropic interactions, their computational prediction would be routine. In this trivial situation, there would really be no need for a subject such as crystal engineering. So when Gavezzotti (1994) stated in his well-known review article that crystal struc-tures could not be predicted, he was effectively setting the stage for sophisti-cated computational efforts towards this very end. In our opinion his statement, which has sometimes been trivialized by others, is neither trite nor facetious. Rather, it is an explicit recognition of the fact that crystal structures are not readily derivable from molecular structures. Computational methods in crystal engineering attempt to model or to circumvent the critical but seem-ingly whimsical effects of directional interactions on the isotropic background, which in itself is largely non-discriminating in nature.

Distinctions can be made between the various computational methods to crystal structure prediction based on the sequence of operations by which mol-ecules are built up into three-dimensional crystals (Leusen *et al.* 1999). Three main routes have been popular. In the so-called 'static' method, clusters of molecules are generated based on known space group preferences and the directional requirements of the well-known intermolecular interactions. These clusters are then used to construct the three-dimensional crystal. Inherent in this classical approach, taken by Gavezzotti (1997) and Hofmann and Lengauer (1997), is that strong and relatively directional interactions such as $O-H\cdots O$ and $N-H\cdots O$ hydrogen bonding often determine the final packing. In this case, an effective way to arrive at the correct structure would be to simplify the computational problem by building the rudimentary molecular clusters first and to then derive the three-dimensional crystal from these units. The computational routine of the program *Promet3* follows such assumptions (Gavezzotti 1991*b*). This method runs into difficulties if the cluster–cluster interactions interfere with the cluster itself, that is if the chosen cluster is not an adequate module for the entire crystal. We are only restating here the problem of lack of sufficient structural insulation, in other words the presence of *structural interference*. For example, it would be problematic to correctly predict the crystal structure of 2-aminophenol by this method.

In the second method, one could begin with a one-dimensional periodic arrangement and build it up stepwise by increasing the dimensionality of the crystal. This 'Aufbau' type of crystal construction has been favoured by Perlstein *et al.* (1996) and in the sense that a non-zero-dimensional entity is used as a starting point, it might be expected to be of greater reliability. Finally, in the so-called 'dynamic' method, the three-dimensional packing is generated directly and the full symmetry is applied at all stages of the computation. This method is used in programs such as *UPACK* (*U*trecht crystal *pack*er) and the *Cerius²* (Polymorph Predictor) of Molecular Simulations Inc. The latter program was developed initially by Karfunkel and Gdanitz (1992) and subsequently refined by Leusen *et al.* (1999).

All three methods are dependent on the quality of the potentials used and in no case are the variable effects of weak hydrogen bonds handled satisfactorily. To the extent that atom charges are used, the electrostatic component of hydrogen bonding is taken into account, but none of these methods even begins to grapple with the long-range orienting effects of weak hydrogen bonds, effects that steer the crystallization pathway towards one or a small number of stable polymorphs. Typically, in the polymorph prediction sequence in *Cerius²*, a large number of structures are obtained within a 1.0 kcal/mol window. The correct (observed) structure is generally among these low energy choices but the program is still unable to discriminate between the choices. It is our opinion that such improvements will be possible if the effects of subtle interactions, such as the weak hydrogen bond, are taken into account more explicitly.

4.3.3 *The experimental approach – database research*

It may be seen from the above discussion that a stable crystal structure is the result of a trade-off between a large number of factors. Alternatively, one may say that the adoption of a particular crystal structure depends on a number of factors – enthalpic, entropic, kinetic – and is a multivariate phenomenon. Yet, it is a dictum of statistics that however diverse a group of observations, they may be correlated if a sufficiently large number of them is available for study. This is the basis for the experimental approach to crystal engineering, an approach which is based on statistical inference (Desiraju 1989a). Structural chemists and crystallographers have observed that some combinations of molecular functional groups are associated with only certain crystal patterns (Sarma and Desiraju 1986; Desiraju 1991a, 1996a). If a particular pattern is identified often enough, one may in favourable cases be able to correlate it with a particular molecular fragment(s), leading in turn to reliable strategies for crystal structure design. It is important to note that under such a condition, it may not be of crucial importance to have a full and rigorously accurate knowledge of all the intermolecular interactions involved. What is important is the *repetition* of particular patterns. In this way, the two above-mentioned problems associated with extrapolation from molecular to crystal structure are somewhat mitigated. By identifying crystal patterns that originate from sufficiently large (and hopefully representative) molecular fragments, the issue of complementarity is addressed. By scanning a sufficiently large number of crystal structures, the issue of long-range but weak interactions is approached, at least in part (Pedireddi *et al.* 1996).

The existence of crystallographic databases, especially the Cambridge Structural Database (CSD), that contain a large volume of high-quality crystal structure data (Allen *et al.* 1991; Allen 1998; see Section 1.3.1) have greatly facilitated such experimental approaches to structure design. It is not an exaggeration to state that the CSD is an essential tool for crystal engineering today. Various options for the analysis of intermolecular interactions (ISOSTAR, VISTA) and interactive and heuristic methods of illustrating crystal structures (PLUTO) reflect the growing ease with which one may attempt to handle and absorb large quantities of crystallographic information. The use of the CSD in ascertaining the nature of several types of weak hydrogen bonds has been adequately discussed in Chapters 2 and 3. By examining patterns based on these weak interactions, one can obtain an idea of the ruggedness of the interactions (Allen *et al.* 1998*b*).

4.3.4 *Crystal engineering in practice – supramolecular synthons*

The most robust and reliable of supramolecular patterns in crystals may be designated *supramolecular synthons*. This terminology recognizes that crystal engineering is a type of synthetic activity, and is based on the definition of a *synthon* in molecular chemistry by Corey (1967) who introduced this formalism in organic synthesis to logically trace the chemical thought process from starting material to the target substance. Corey defined synthons as 'structural units within molecules which can be formed and/or assembled by known or conceivable synthetic operations'. A synthon is usually smaller and less complex than the target molecule, and yet contains a substantial part of the vital bond connectivity and stereochemical information required to synthesize the goal substance. The analysis of a complex target molecule into simpler synthons is performed then through a series of rational bond disconnections, this exercise being termed *retrosynthesis* (Corey and Cheng 1989). Recognizing that crystal engineering is the solid-state supramolecular equivalent of organic synthesis, supramolecular synthons are 'structural units within supermolecules which can be formed and/or assembled by known or conceivable intermolecular interactions' (Desiraju 1995*a*). By analogy again with organic synthesis, the analysis of the complex interplay of close packing, hydrogen bonding and other interactions in a crystal structure (and by implication

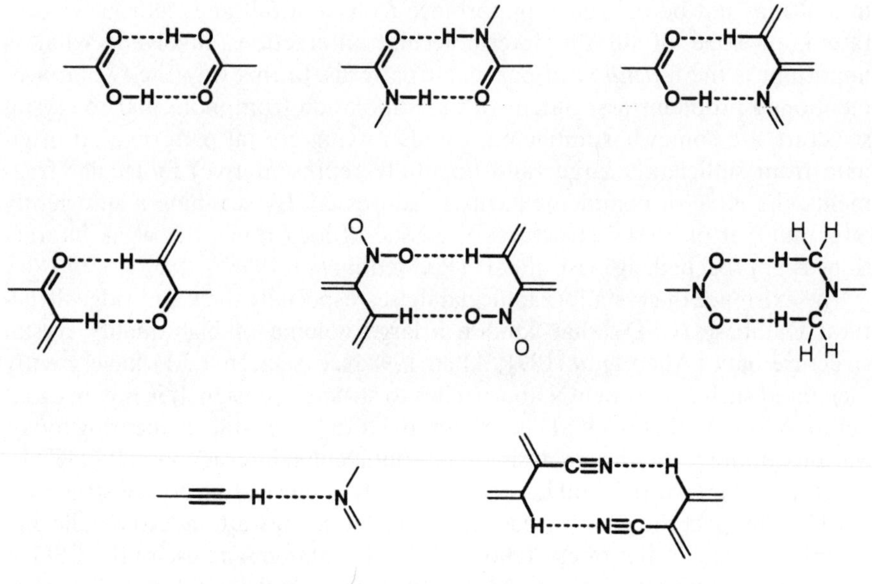

Fig. 4.18. Representative supramolecular synthons constructed with strong and weak hydrogen bonds.

during crystallization) may be termed *supramolecular retrosynthesis*. In that analysis and synthesis are carried out in opposite senses, the term 'retrosynthesis' aptly describes the procedure for the logical analysis of a structure, be it molecular or supramolecular (Nangia and Desiraju 1998*a*,*b*). Figure 4.18 shows some representative synthons constituted with strong and weak hydrogen bonds.

Supramolecular synthons are spatial arrangements of intermolecular interactions between complementary functional groups, and constitute the core of the retrosynthetic strategy for supramolecular structures. The synthon approach is advantageous in that it offers a considerable simplification in the understanding of crystal structures. The emphasis in crystal engineering may therefore be increasingly diverted from the constituent molecules to the topological features and geometrical connectivities of non-bonded interactions between molecules. With crystal structures defined as networks with the molecules being the nodes and the supramolecular synthons being the node connections, retrosynthesis may be performed on network structures to yield appropriate molecular structures. The advantage of such an approach in crystal engineering is that:

(1) supermolecule ⇒ molecule connections are easily established;

(2) comparisons between seemingly different crystal structures are facilitated;

(3)   the interference between supramolecular synthons can be strategically minimized; and

(4)   more than one combination of molecular and supramolecular synthons is seen to lead to similar crystal structures.

Operationally, such supramolecular retrosynthesis is carried out most conveniently with the CSD (Desiraju 1995*a*). To summarize then, supramolecular synthons are of significance in crystal engineering because they relate molecular and supramolecular structure. They are the smallest structural units which contain all the information inherent in the recognition events through which molecules assemble into supermolecules.

4.3.4.1   *Supramolecular retrosynthesis illustrated*

In the context of the present work, it is pertinent to note that supramolecular construction is feasible with weak hydrogen bonds and is not confined to just the strong ones (Aakeröy and Seddon 1993*a*; Mascal 1994; Subramanian and Zaworotko 1994; Burrows *et al.* 1995; Desiraju 1996*a*). This is so because of the electrostatic and therefore long-range character of all hydrogen bonds. A few generalizations are useful:

1.   H atom acidity, bond cooperativity and, sometimes, a rigid molecular geometry are important for the establishment of $C-H\cdots O$ based synthons.

2.   Molecules which have the possibility of forming $C-H\cdots O$ bonds and no other directional interaction are preferable choices for supramolecular construction. The absence of $O-H\cdots O$ and $N-H\cdots O$ forming functional groups is an advantage.

3   The number of hydrogen bond donors and acceptors should be matched so that bifurcation is avoided. This is difficult in donor-rich $C-H\cdots O$ systems.

4.   Molecular symmetry may be transformed into supramolecular symmetry using $C-H\cdots O$ directionality.

5.   Multiple matching of recognition sites is distinctly advantageous.

6.   All possible interactions of a molecule must be considered, because no single type of interaction controls crystal packing entirely.

Some of these ideas are illustrated by the $1:2$ molecular complex of 2,5-dibenzylidenecyclopentanone, **192**, and 1,3,5-trinitrobenzene, **193** (Fig. 4.19). The $N-H\cdots O$ and $N-H\cdots N$ based synthon **194** is well known. By replacing the strong hydrogen bonds in **194** with $C-H\cdots O$ bonds, the new target, **195**, was obtained. The choice of **193** as a precursor was then quite deliberate. The H atoms in this molecule are very acidic and the nitro group is a good acceptor in these cooperative situations. The choice of **192** as the second molecular component of the supermolecule was made by matching complementary groups. There are seven $C-H\cdots O$ bonds with $2.30 < d < 2.84\,\text{Å}$, 3.18

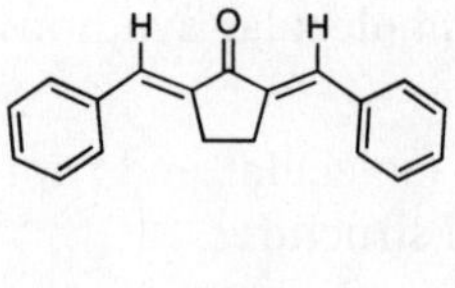

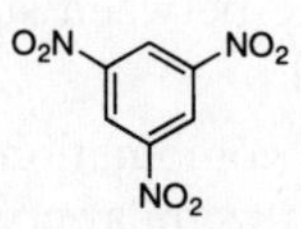

192   **193**

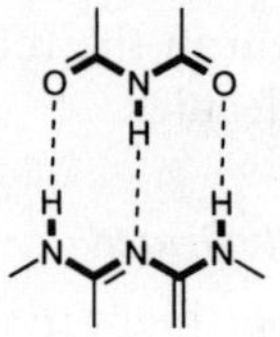

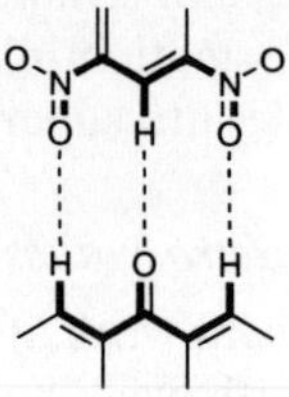

194   **195**

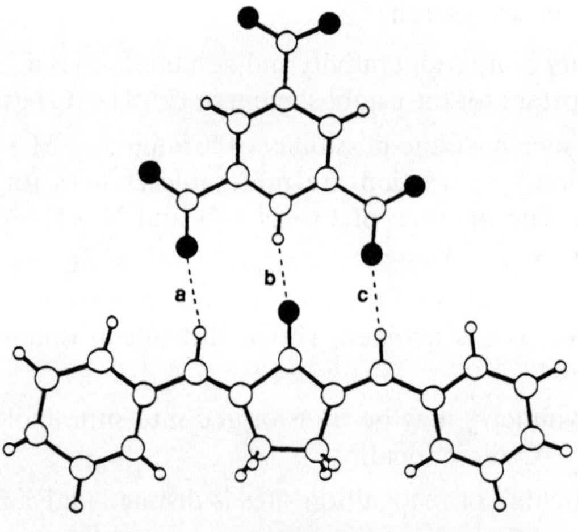

	C···O (Å)	H···O (Å)	C—H···O (°)
a	3.44	2.48	157
b	3.18	2.30	160
c	3.31	2.39	156

Fig. 4.19. Recognition pattern between dibenzylidenecyclopentanone, **192**, and trinitrobenzene, **193**, in their 1:2 complex. The C—H···O bonds are indicated. Only one of the two symmetry independent molecules of **193** is shown (from Biradha *et al.* 1993).

< D < 3.60 Å and 122° < θ < 160° (Biradha *et al.* 1993). A number of other molecular complexes based on picryl chloride and trinitrobenzene with other substituted dibenzylidene ketones behave similarly and have been described in detail by Nangia, Desiraju and co-workers (Biradha *et al.* 1997).

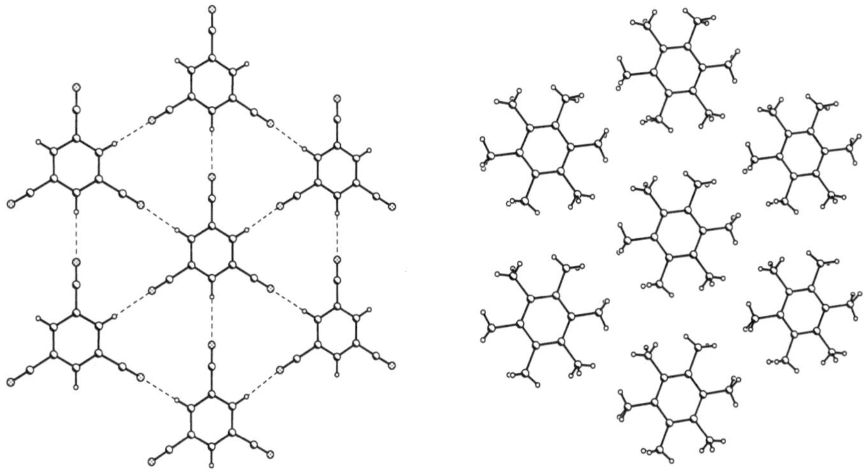

Fig. 4.20. Layer structures of 1,3,5-tricyanobenzene, **196** (*left*), and hexamethylbenzene, **197** (*right*), molecules in their 1:1 molecular complex. The C—H⋯N bonds (*D* = 3.47, 3.52 Å) are shown as dotted lines. The molecules of **197** are situated on inversion centres (Reddy *et al.* 1993).

Another example is provided by the 1:1 complex of 1,3,5-tricyanobenzene, **196**, and hexamethylbenzene, **197** (Reddy *et al.* 1993). This complex was obtained by retrosynthetic analysis of the trigonal network, **198**, leading to synthon **199** which in turn leads to the molecular precursor **196**. Such an analysis follows from the known linearity of C—H⋯N hydrogen bonds. Pure **196** contains a slightly modified form of network **198** (Reddy *et al.* 1995), while recrystallization of a 1:1 mixture of **196** and **197** leads to the desired network structures shown in Fig. 4.20. Conceivably, the presence of **197** leads to a certain amount of structural insulation which is conducive to the formation of the lamellar structure. The layered arrangement of **197** is also shown in Fig. 4.20, while the entire crystal structure is shown in Fig. 4.21 providing an idea of the donor–acceptor interactions between the two components of the molecular complex. This type of lamellar insulation is enhanced in the 1:1 molecular complex of **193** with trimethylisocyanurate, **200**, the design of which was accomplished retrosynthetically by Thalladi *et al.* (1995) via synthon **201** (Fig. 4.22). The distinct layers in both complexes have threefold symmetry, but a translational offset of layers in the former complex leads to a loss of global threefold symmetry, and the space group is *C2/c*. In the latter complex, however, the symmetry axes of adjacent layers coincide, the alternating layers are rotated exactly by 60° and threefold symmetry is fully retained, the space group being *P6̄* (Fig. 4.23).

It is still unclear when this kind of carry-over of molecular symmetry into the crystal is effective, contradicting, as it seems to, Kitaigorodskii's laws of close-packing (1973). However, when it occurs, it allows for the construction of elaborate supramolecular structures. The 1:2 complex of 1,3,5,7-

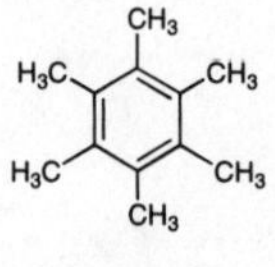

197

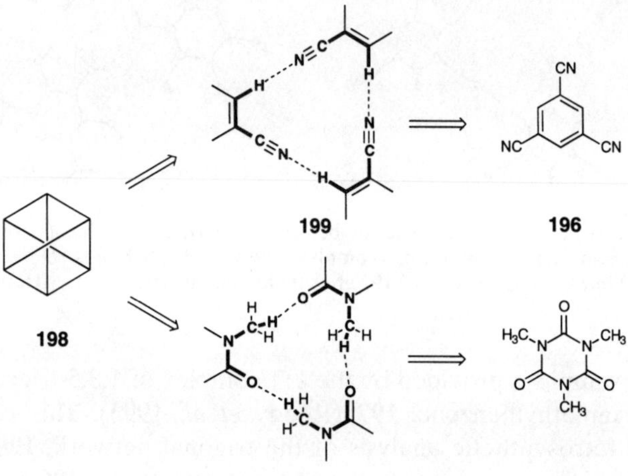

199   **196**

198

201   **200**

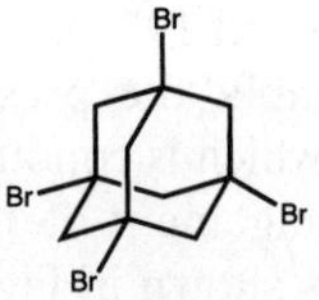

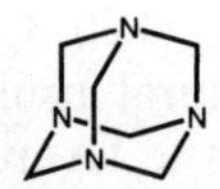

202   **203**

tetrabromoadamantane, **202**, and hexamethylenetetramine, **203** (Fig. 4.24), is face-centred cubic with molecules of **202** at the cell corners and molecules of **203** tetrahedrally situated in alternating octants (as in cubic ZnS). These molecules are held by four sets each of three equivalent C—H···N hydrogen bonds ($D = 3.77\,\text{Å}$, $\theta = 151°$) forming a large adamantoid cavity containing the second (disordered) molecule of **203** (Reddy *et al.* 1994). Such organic zeolites allow for the study of host–guest and other properties.

A crystal then may be thought of as a retrosynthetic target, and retrosynthetic thinking is accordingly of great importance in crystal engineering because it could be the basis of new and interesting approaches to structural

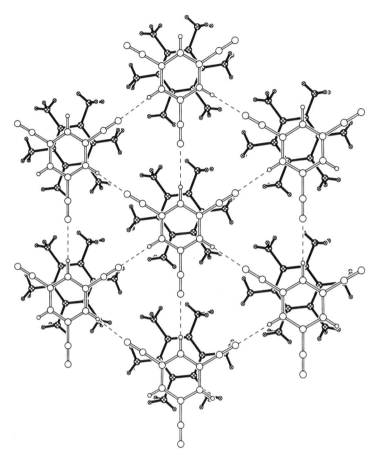

Fig. 4.21. Overlap of layers of **196** and **197** in their 1:1 molecular complex; see Fig. 4.20. Note that though the individual layers approximate to trigonal symmetry, this is lost in the third dimension because of the lateral translation between layers. This translation optimizes donor–acceptor interactions. The C—H···N bonds are shown as dotted lines (Reddy *et al.* 1993).

design. The ribbon based structure of terephthalic acid is well known, and consists of a linear array of phenyl rings and carboxyl dimer synthons, **4**. By replacing alternating synthons with the nitro-dimethylamino synthon, **19**, encountered already in *N,N*-dimethylnitroamine (Section 2.2.7 and Fig. 4.18), the crystal structure of the 1:1 complex of 4-nitrobenzoic acid and 4-(*N,N*-dimethylamino)benzoic acid is readily realized (Sharma *et al.* 1992). The ribbon arrangement in this structure is shown in Fig. 4.25.

4.3.4.2 *Engineering of structures and properties*

Crystal engineering today is properly concerned not only with the design of specific structures but also with the design of specific properties (Desiraju

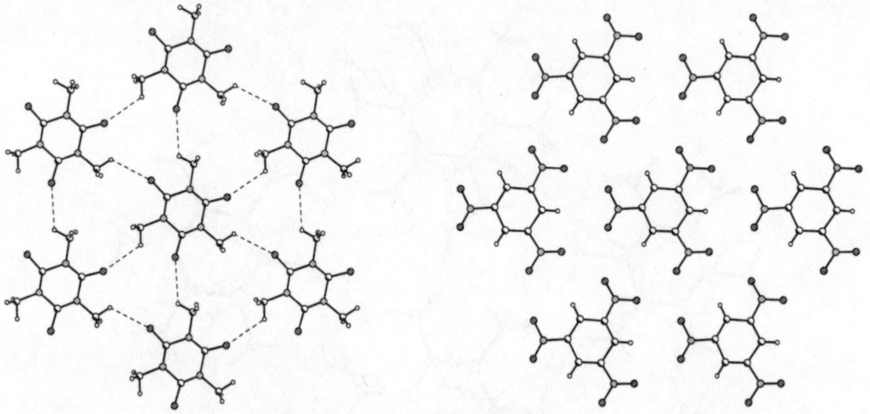

Fig. 4.22. Layer structures of trimethylisocyanurate, **200** (*left*), and 1,3,5-trinitrobenzene, **193** (*right*), molecules in their 1:1 molecular complex. The C—H⋯O bonds ($D = 3.41$ Å) are shown as dotted lines. The nearest O atoms in the layers of **193** are at van der Waals separation (Thalladi *et al.* 1995).

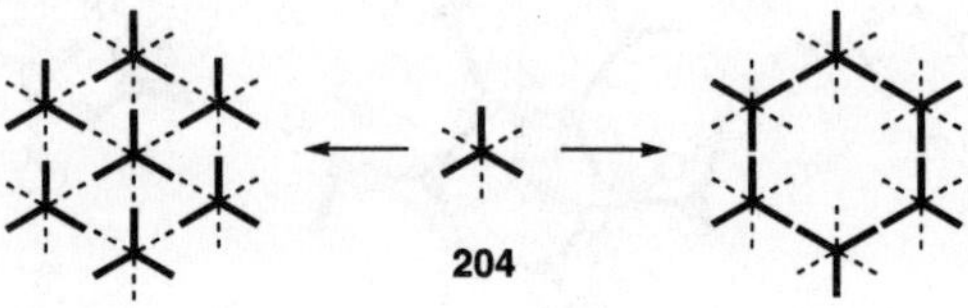

205   **206**

1997*b*). Very recently, a new class of SHG-active substances, namely octupolar molecules, has been proposed and been shown to display significant NLO behaviour at the molecular level (Zyss and Ledoux 1994). At the crystalline or supramolecular level, it has been shown that octupolar SHG is characteristic of trigonal networks (Ledoux *et al.* 1990). Inasmuch as two-dimensional systems are concerned, the crystal engineering problem amounts to steering the structure of an appropriately substituted trigonal molecule **204** towards the trigonal, non-centrosymmetric network structure **205** characterized by specific interactions between unlike groups in the molecular skeleton. The hexagonal, centrosymmetric network **206** characterized by close approaches between like groups has to be avoided. The reader will appreciate of course that the majority of trigonal molecules **204** adopt neither structure **205** nor **206** but some trivial close-packed arrangement, and this only renders the engineering problem more challenging.

The crystal structure of the complex between **193** and **200** therefore appeared to be a suitable starting point in the crystal engineering of an octupolar non-linear optical crystal. However, single component crystals are pre-

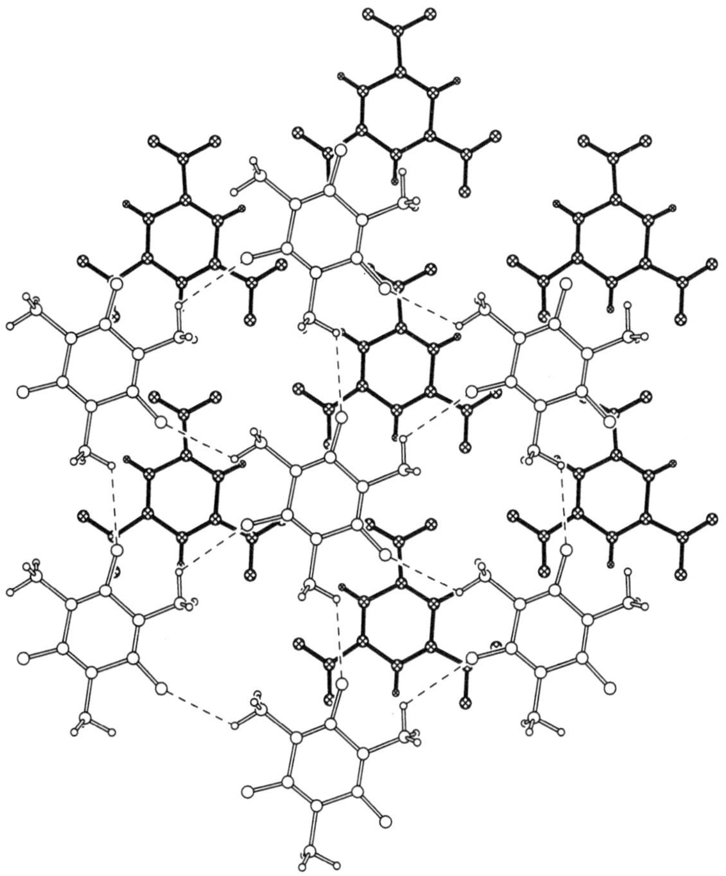

Fig. 4.23. Overlap of layers of **193** and **200** in their 1:1 molecular complex; see Figure 4.22. The trigonal symmetry of the individual layers is retained in the crystal and the threefold axes pass through the centres of the molecules as well as through the centre of the void space created as a result of the layer overlap. There is no significant ring–ring interaction between molecules of **193** and **200** in adjacent layers (Thalladi *et al.* 1995).

ferred to molecular complexes for NLO applications because of issues connected with material purification, crystal growth and optical characterization in both solution and the solid state. Therefore we turned our attention to the symmetrical isocyanurates, all of which have alternating C—H···O donors and acceptors in the molecular structure. Such an alternation is an essential prerequisite for the formation of network **205**.

Keeping such considerations in mind, tribenzyl isocyanurate, **207**, with its $C(sp^2)$–H groups was examined. The crystal structure of **207**, derived retrosynthetically from network **205** (Fig. 4.26), shows that the desired non-centrosymmetric trigonal structure has been obtained (Thalladi *et al.* 1997).

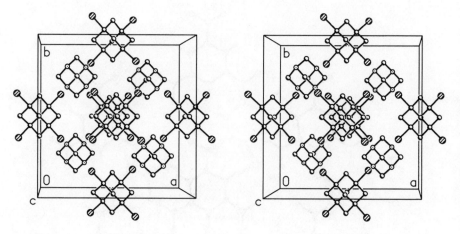

Fig. 4.24. Stereoview of the 1:2 molecular complex between 1,3,5,7-tetrabromoadamantane, **202**, and hexamethylenetetramine, **203**. There are two symmetry-independent molecules of **203**. One of them forms a super-adamantoid C—H···N mediated cage with alternating molecules of **202**, while the other, which is at the centre of the unit cell, is included within this cage (Reddy *et al.* 1994).

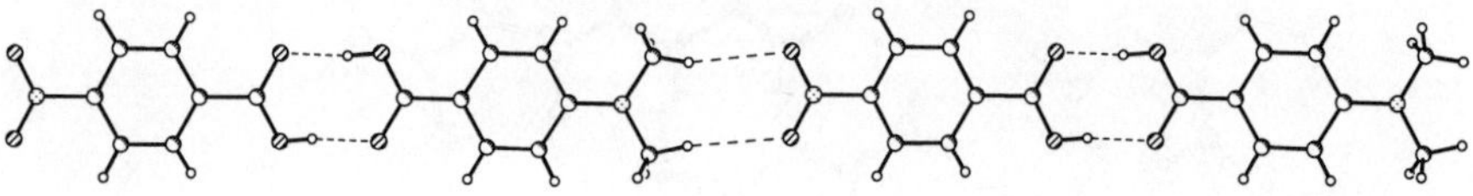

Fig. 4.25. O—H···O and C—H···O hydrogen bonds in the crystal structure of the 1:1 molecular complex of of 4-nitrobenzoic acid and 4-(*N,N*-dimethylamino)benzoic acid. Notice the lack of structural interference between strong and weak hydrogen bonds (Sharma *et al.* 1992).

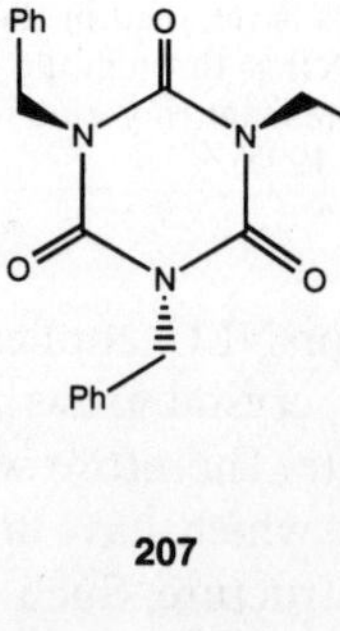

207

The molecules are far from planar. With respect to the central heterocyclic ring, two benzyl groups point in one direction whilst the third points in the other leading to an overall 'chair'-shape (Fig. 4.27). The layer structure in **207** is therefore corrugated and this increase in dimensionality could well assist in

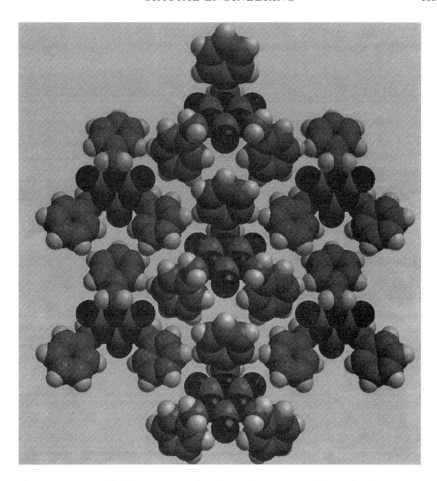

Fig. 4.26. Corrugated chiral layer structure in tribenzyl isocyanurate, **207**. Notice the interlocking of phenyl rings and the C—H···O hydrogen bonds. This is one of the first examples of a crystal where octupolar non-linear optical properties have been demonstrated (Thalladi *et al.* 1997).

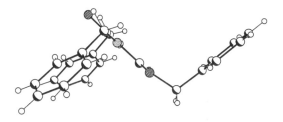

Fig. 4.27. Side view of a molecule of **207** in its crystal structure (Thalladi *et al.* 1997). Notice the step-shape of the molecule that leads to the layer corrugation seen in Fig. 4.26.

the $C-H\cdots\pi$ stacking of layers that results in overall three-dimensional non-centrosymmetry, confirmed by a weak SHG powder signal a tenth that of urea at $1.064\,\mu m$.

Control of structural dimensionality is of importance in crystals designed for NLO purposes. In Section 4.1.2, reference was made to the layered structure of 3,4-methylenedioxy-β-nitrostyrene. Related structures with corrugated layers have been obtained for 2,4-dinitro-3',4'-methylenedioxystilbene and 2,4-dinitro-3',4',5'-trimethoxystilbene by Sarma et al. (1994). In the latter compound the extent of corrugation is greater, and this is possibly the reason for adoption of the non-centrosymmetric space group $P2_12_12_1$ and with it a SHG signal of two-tenths that of urea. Structural control in the third dimension is always a challenge, and corrugation may just be sufficient to achieve parallel registry of two-dimensional motifs in some cases.

4.3.4.3 Other examples

In this section, we enumerate several examples where weak hydrogen bonds have been implicated in families of structures that have been engineered. Pyridine N-oxides form infinite chains and closed dimers constituted with $C-H\cdots O$ bonds and in a recent study (Bodige et al. 1997) it has been demonstrated that 2,2'-dithio-bis(pyridine N-oxide) supports the formation of network structures in its complexes with tetracyanobenzene and pyromellitic dianhydride. Karle et al. (1996) have shown that in imidazolium hydrogen squarate, the two-dimensional ionic sheet of $N-H\cdots O$ hydrogen bonds is further stabilized by $C-H\cdots O$ bonds. Weak $C-H\cdots O$ bonds have been identified in a group of styrylcoumarins and related systems by Venkatesan and co-workers (Moorthy and Venkatesan 1994; Moorthy et al. 1994). In an early study, Sarma and Desiraju (1987a) pointed out the importance of O atom substituents and consequently of $C-H\cdots O$ bonds in the adoption of $4\,\text{Å}$ crystal structures. Some of these are found to undergo topochemical solid state photochemical or Diels–Alder reactions (Sarma and Desiraju 1987b; Desiraju and Kishan 1989). Bishop and co-workers have developed an interesting family of helical tubulate inclusion compounds (Ung et al. 1995).

Co-crystallization of neutral molecules is always an interesting subject in the crystal engineering context, and in a recent intriguing study Dyer et al. (1998) have reported that trans-[Pd{PPh$_2$(C$_{16}$H$_{15}$)}$_2$Cl$_2$] and [Pd{PPh$_2$(C$_{16}$H$_{15}$)}Cl$_2$]$_2$, (where C$_{16}$H$_{15}$ is diphenyl[2.2]paracyclophanyl) co-crystallize with CH$_2$Cl$_2$. The solvent molecule is very important for the formation of the adduct because it forms two types of weak hydrogen bond with each of the two Pd-containing molecules. One of its acidic $C-H$ groups acts as a bifurcated donor with respect to a terminal and a bridging Cl group of the dinuclear species, while one of its two Cl atoms receives a hydrogen bond from an aromatic $C-H$ group of a cyclophanyl moiety in the mononuclear species. The solvent links, in effect, the two distinct molecular species and thus

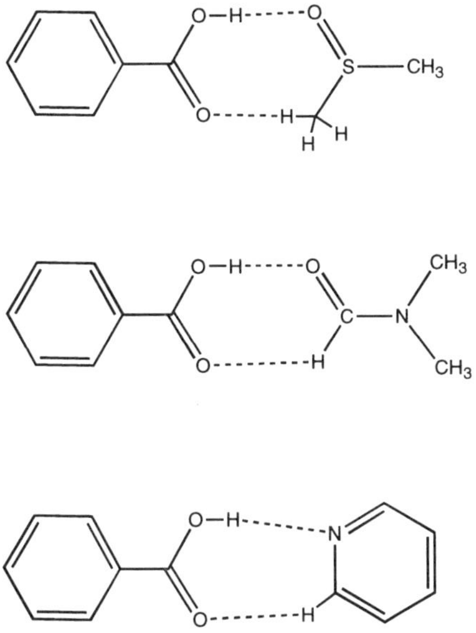

Fig. 4.28. Some supramolecular synthons formed by carboxylic acids with common organic solvents.

appears to actively aid in the formation of the mixed crystal. Indeed, barring a few very non-polar solvents, the inclusion of solvent molecules in a crystal structure is rarely an innocent event – solvents like pyridine, DMF, DMSO, CHCl$_3$, CH$_2$Cl$_2$, ethyl acetate and even acetone are often characterized by specific solute–solvent recognition motifs that involve weak hydrogen bonds, whether the solute contains groups capable of strong hydrogen bonding or not. The group of Weber has carried out a very wide range of studies on O—H···O and C—H···O mediated solvent binding (Csöregh et al. 1986, 1990; Weber et al. 1986, 1987, 1988, 1991; Weber 1987; Gallardo et al. 1995; Helmle et al. 1997). The patterns formed with DMSO, DMF and pyridine are given in Fig. 4.28 (see also **190**), but the full implications of this work are still to be understood properly. When progress is made along these directions, there is every reason to expect that the phenomenon of co-crystallization with solvent may actually be used as a design element in crystal engineering.

A combination of C—H···O, C—H···N and π···π interactions have been identified recently by Thalladi et al. (1998b) in a family of 2,4,6-triaryloxy-1,3,5-triazines that show potential for octupolar NLO activity. The distinction between non-centrosymmetric and centrosymmetric variants of this structure type is found to depend on a subtle interplay of these interactions. Several arrangements of C—H···O bonds have been found in 2-amino-5-nitropyridinium dichloroacetate and the 1:1 complex of 2-amino-5-nitropyridine with chloroacetic acid, both structures having been designed

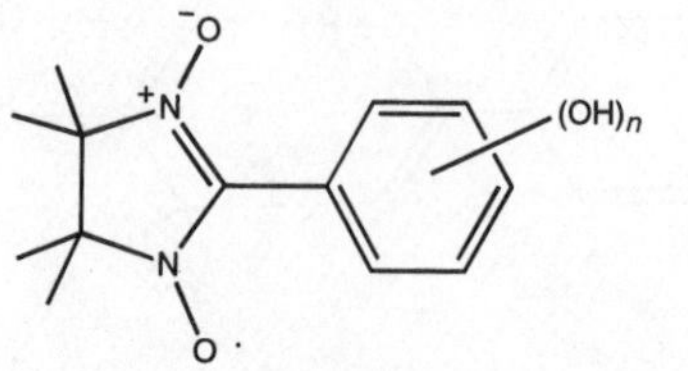

208

for quadratic NLO activity (Le Fur *et al.* 1996). The related dihydrogenophosphate acid salt is centrosymmetric but it was noted by the authors (Zaccaro *et al.* 1996) that the $N-H\cdots O$ and $C-H\cdots O$ bonds between the organic and inorganic sub-networks are 'of the same order of magnitude'. The same group have also reported (Le Fur *et al.* 1995) that in the non-centrosymmetric 1-ethyl-2,6-dimethyl-4(1*H*)-pyridinone trihydrate, a layer structure is mediated by several strong and weak hydrogen bonds. In the molecular metal area, the role of $C-H\cdots O$ bonds in determining packing preferences has been described in detail for a family of Re-based BEDT–TTF derivatives (Pénicaud *et al.* 1993).

The formation of hydrogen bonds in the crystal structures of open-shell molecules has implications for the extension of ferromagnetic interactions into multi-dimensions. Veciana and co-workers (Hernàndez-Gasio *et al.* 1994; Cirujeda *et al.* 1995) have studied substituted α-phenyl nitronyl nitroxide radicals, **208**. Though the unpaired electron is distributed mainly on the two NO groups and the α-carbon atom between them, a small and significant spin density is present on the rest of the nucleii too. It may be noted that while the O atoms of the NO groups support large positive spin densities, the H atoms of the methyl groups and the *ortho* and *para* H atoms of the phenyl rings support small negative spin densities. In this way, both the $O-H\cdots O-N$ and $C-H\cdots O-N$ hydrogen bonds in the crystal structures of these phenolic nitroxides can extend the ferromagnetic interactions into the network structure leading to bulk ferromagnetism. The NO group is certainly a good hydrogen bond acceptor, even better than the $C=O$ group. Romero *et al.* (1996) have utilized this $C-H\cdots O$ bond forming ability in their design of a molecule for magnetic properties, that contains both the nitroxide and the ethynyl group. This is a nitronyl nitroxide pyridine-based radical which, as expected, shows a pattern of $C-H\cdots O$ bonds ($d = 2.21\,\text{Å}$) between the ethynyl and nitroxide groups.

4.4 Recognition in solution and related phenomena

We have seen in Section 2.2.12 that hydrogen bonds of the $C-H\cdots O$ and $C-H\cdots N$ type exist in solution and have been invoked to explain chemical

209   **210**   **212**

Fig. 4.29. Template-directed synthesis of the tetracationic cyclophane, **212** (from Fyfe and Stoddart 1997).

phenomena in solution. Because of kinetic dictates, such weak interactions are of a transient nature under these conditions. Nevertheless, if the supramolecular assembly involves many of them at a time, their viability is expected to improve. Interactions wherein the polarization component dominates over the electrostatic one (like the $C(sp^3)-H\cdots\pi$) are further expected to persist in solvents of high dielectric constant. In this section, we briefly discuss the role played by weak hydrogen bonds in recognition phenomena in solution.

4.4.1 *Supramolecular assistance to molecular synthesis*

In their important review article, Fyfe and Stoddart (1997) distinguish two facets of modern synthetic supramolecular chemistry. These are *supramolecular synthesis*, exemplified in the solid state by crystal engineering, and *supramolecular assistance to molecular synthesis*. In the former, the target is a supramolecular species and is assembled with intermolecular interactions using premeditated molecular modules. In the latter, the target is a discrete molecular entity held together wholly by covalent and mechanical bonds, but it is only achieved because of the templating effects of various intermolecular interactions during the synthesis. These interactions can, in effect, be used to direct chemical reactions for the formation of specific compounds. The cases cited in Section 2.2.12 are simple examples of this concept.

For a more elaborate illustration, consider the cyclobis(paraquat-4,4′-biphenylene), **212** (Asakawa *et al.* 1996). This organic molecular 'square' is produced in minute quantities when the dication **209** is treated with the dibromide **210** in the absence of any templating reagent (Fig. 4.29). However, when the two compounds are reacted in the presence of the ferrocene-based template **211**, the yield of the desired **212** increases by a factor of 16. The templating action of **211** may be ascribed to the $C-H\cdots O$ hydrogen bonds between the polyether O atoms in the templating agent and the acidic H atoms, α to the quarternary N atoms, in the product. These bonds serve to align **209** and **210**

in a manner suitable for the formation of the desired product **212**. Also implicated in this mechanism are the $\pi\cdots\pi$ stacking interactions between the π-rich
Cp rings in **211** and the π-deficient heterocyclic rings in **212**. Studies on related
compounds show fine effects that corroborate the function of the $C-H\cdots O$
bonds. For example, the templating effect is better when the $C-H\cdots O$ bond
forming groups are located in better accessible sites (Amabilino 1998). Other
studies (Gunter *et al.* 1994; Ashton *et al.* 1996*b*; Gillard *et al.* 1997) deal with
related approaches.

4.4.2 *Drug design and biological recognition*

Structure-based drug design begins with macromolecular and small-molecule
crystallographic data (Böhm 1992; Boyd 1995). The use of macromolecular
data constitutes what may be termed a direct approach, while one approaches
the problem indirectly with small-molecule data. In the former, the
three–dimensional structure of the active site (receptor) is considered along
with that of a protein-drug complex of known activity. The goal of structure-
based drug-design is to predict, even approximately, the binding characteristics of modified compounds. The indirect approach uses small-molecule crystal
structure data, and consists of deriving the stereochemical requirements of the
unknown binding sites of the macromolecule by complementarity to the pharmacophore, common to a family of active molecules.

In both methods, the search for a lead molecule starts from the spatial coordinates of the binding sites, that is the topology of the pharmacophore or of
its complementary partner, the 'antipharmacophore', by computational or
experimental methods. The binding of a drug to its receptor is effected by van
der Waals, electrostatic and $\pi\cdots\pi$ stacking interactions and also by strong and
weak hydrogen bonding (Böhm and Klebe 1996). Once the relative spatial dispositions of the pertinent functional groups are known, the directionality of
intermolecular interactions of the receptor–substrate have to be determined.
A knowledge of the nature of these intermolecular interactions once again
assumes great importance, as there is not much difference between the crystallization of molecules and the docking of a drug molecule on a receptor site.
The reader will recognize both these as examples of supramolecular assembly.

In this context, let us examine the role played by weak hydrogen bonds in
a selected case of drug–receptor recognition. Engh *et al.* (1996) have described
the structures of three inhibitors complexed with the enzyme thrombin, which
is a serine protease similar to trypsin. Structure-based drug design has been
explored for this system to develop new antithrombotic drugs. Two of the
inhibitors are based on 4-aminopyridine (4AP) and one is based on naph-
thamidine. It was noted that the protonated 4AP donates a $C-H\cdots O$ hydrogen bond to a water molecule which, in turn, is in contact with the π face of a
tyrosine side-chain (Fig. 4.30). This water molecule disrupts a conventional
hydrogen bond that is usually formed between Trp215NH and 227PheCO. The

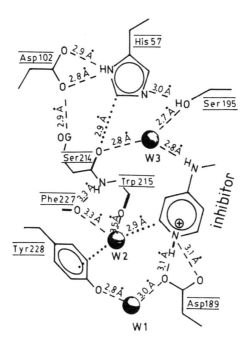

Fig. 4.30. Binding mode of the pyridinium group of a branched inhibitor to thrombin. Strong hydrogen bonds are shown as dashed lines and weak ones as dotted lines. Notice the $C-H\cdots O$ hydrogen bond between His57 and Ser214, and the $C-H\cdots O_W$ and $O_W\cdots$Ph contacts formed by the water molecule W_2 (redrawn from Engh *et al.* 1996).

binding of the 4AP in this manner also induces a rotation of the peptide plane of Ser214–Trp215. The observation that an inhibitor is both directly involved in such contacts and indirectly alters the hydrogen bond network and to some degree the local conformation of the protein, highlights the complexities of receptor–drug interactions and the difficulties that are inherent in their study. Furthermore, the role played by the water molecule illustrates that receptor–drug binding does not involve just the two molecules as is assumed simplistically but also – and in hardly forseeable ways – interactions with solvent molecules. Hydrogen bonds of the $C-H\cdots O$ and $O-H\cdots\pi$ type are often seen in proteins and this subject is reviewed in Chapter 5. Examples of drug-receptor recognition involving weak hydrogen bonds other than $C-H\cdots O$ will be discussed in Section 5.2.4.

The thiazolidine ring in the penams, a sub-family within the β-lactam class of antibiotics, exists in two conformations with axially and equatorially disposed carboxy groups. The axial conformation is shown in Scheme **213**. Since the overall shape of the antibiotic molecule is different in the two conformations, the biological activity associated with each of these conformers is also expected to be different. In a recent study of 24 penams with the axial carboxy

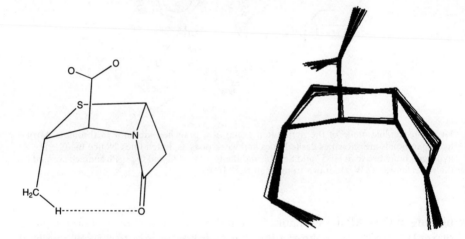

Fig. 4.31. Superposition plot to show that an intramolecular C—H···O hydrogen bond favours puckering of the thiazolidine ring in penams to the 3α-CO$_2$ axial conformation (from Nangia and Desiraju 1998c).

conformation, Nangia and Desiraju (1999) have noted that an intramolecular C—H···O bond between the C2-β-CH$_3$ and the lactam C=O group could influence the conformation of the thiazolidine ring. Of these 24 compounds, it was seen that 19 have *gem*-dimethyl groups at C2 and of these, 14 contain a total of 17 C—H···O hydrogen bonds in the *d* range 2.6–3.0 Å. Extra assistance is also provided by an intramolecular N—H···S bond and the authors estimated that these two weak intramolecular interactions are together worth slightly less than 3 kcal/mol. A superposition plot of the relevant compounds shows that the C—H···O bond plays a stabilizing role in maintaining the axial carboxy conformation in these compounds (Fig. 4.31).

Weak hydrogen bonding has also been invoked, in part, in the molecular theory of sweet taste. In general, sweetness is a property associated with

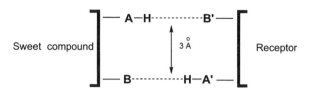

Fig. 4.32. A model for the sweetness receptor. Hydrogen bonds are shown as dashed lines (adapted from Shallenberger and Acree 1967).

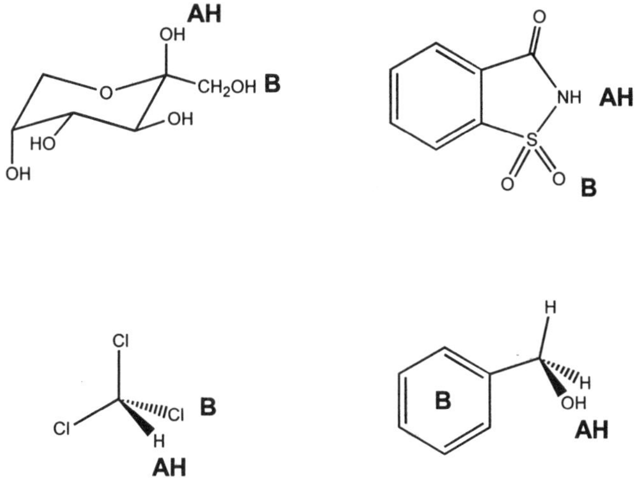

Fig. 4.33. Hydrogen bonding sites, AH and B in β-D-fructose, saccharin, chloroform and benzyl alcohol (following Shallenberger 1982).

hydrogen bonding and Shallenberger and Acree (1967) compared the molecular structures of various sweet-tasting molecules, and postulated that the saporous unit of these compounds is a bifunctional group consisting of an acidic (AH) and basic (B) moiety with an A—H proton to B distance of around 3 Å. Presumably, this glycophore or taste couple is matched by a complementary couple in the taste bud receptor site as shown in Fig. 4.32.

Using this concept, AH and B residues in various natural and artificial sweetening agents have been proposed. These are detailed in Fig. 4.33 for β-D-fructose, saccharin, chloroform and benzyl alcohol. From this it is clear that weak donors (C—H) and acceptors (π ring) may substitute for strong donors and acceptors in sweet compounds (Shallenberger *et al.* 1969). These concepts have been further elaborated taking into account also the hydrophobic nature of the sweetener (Shallenberger 1982), but in the context of the present work, it suffices to state that both weak and strong hydrogen bonds may be considered in these mechanisms. This is clearly an area where further work is required. For instance, it is known that there are physiological similarities between sweet and bitter principles. Accordingly, how similar or dissimilar are the modes of recognition of these compounds?

An area where the involvement of weak hydrogen bonding is probably implicated but is still to be firmly established is in the binding of fluoro-substituted drugs especially fluorinated steroids. Compounds labelled with radioactive ^{18}F may used in imaging techniques for the detection of cancers and in this context, much work has been published by the Katzenellenbogen group in Illinois (Van Brocklin *et al.* 1994; Choe *et al.* 1995 and the references cited therein). Generally, fluoro-substitution is used in such applications because of the isosteric relationship of F and H groups, leading supposedly to similar binding of the fluoro and non-fluoro analogues at the receptor site (Goldman 1969). It has been found that the fluoro analogue sometimes binds much better but sometimes not so well as the H analogue (Choe and Katzenellenbogen 1995). It is suggested here that enhanced binding could arise from the formation of $C-H\cdots F$ interactions at the receptor–steroid interface. In general too, several drug molecules are fluoro-substituted, and the study of weak hydrogen bonding interactions involving F is surely expected to lead to new insights into the mode of binding and modes of action of these important compounds.

While the significance of hydrogen bonding in physiological phenomena has been amply demonstrated for strong hydrogen bonds, the involvement and implication of weak hydrogen bonds in such mechanisms still remains to be systematically shown. However, the early results in this regard are promising and hint that present models could be further refined once the weaker inter-actions are taken into account. To conclude, these discussions pertaining to solution-based or *in-vivo* phenomena with all the attendant subtleties of kinetics and dynamics brings us now to the issue of weak hydrogen bonds in biological structures, the subject of the next chapter.

5

The weak hydrogen bond in biological structures

The structure and function of biological molecules is to a large degree determined by hydrogen bonding. This is the case for proteins, nucleic acids, carbohydrates, membranes and also the aqueous medium in which these components are held. The three-dimensional architecture of proteins and nucleic acids is stabilized by hydrogen bonds, biological recognition operates mainly by this mechanism, and the molecular mobility required for biological processes is directly connected with rapid formation and breaking of hydrogen bonds. Conventional hydrogen bonding in biological structures has been comprehensively reviewed by Jeffrey and Saenger (1991) and is touched upon here only when necessary.

5.1 Introduction

Weak hydrogen bonds in biological structures were observed as early as the 1960s. Sutor (1963) noted the existence of $C-H \cdots O$ interactions in purine and pyrimidine bases, while they were recognized in nucleosides by Shefter and Trueblood (1965) and by Sundaralingam (1966). Ramachandran and co-workers observed these hydrogen bonds in collagen (1965) and polyglycine II (1966). An $O-H \cdots Ph$ hydrogen bond in a cyclic peptide was reported by McPhail and Sim (1965). Despite this promising start, systematic studies of weak hydrogen bonding in the biological world are rather recent and have not yet been extended to all relevant systems. The characteristics of weak hydrogen bonds that have been described in Chapters 2 to 3 for organic structures are valid for biological structures too. This includes geometrical properties, the effects of donor acidity and acceptor basicity, and typical energies. However, one cannot extrapolate from this to the functional properties of weak hydrogen bonds. These can be very different in different systems and must therefore be studied separately in each case.

Typical biological molecules carry many groups that can potentially form weak hydrogen bonds. Various kinds of $C-H$ donors and π-acceptors are situated on the surfaces and also in the interior of biological macromolecules, almost necessarily leading to the formation of $C-H \cdots A$ and $X-H \cdots \pi$ interactions. The functional groups at the surfaces may be involved in weak hydrogen bonds that operate in water–biomolecule interactions, and also in recognition processes and structural stabilization of the molecular peripheries. Hydrogen bonding groups in the interior could play important roles in determining and stabilizing molecular conformations. Before these functions are discussed in any detail, it is necessary to highlight two fundamental differences between organic (or organometallic) and biological crystal structures.

5.1.1 *Biological structures are not time-stable*

Crystal structures of biological molecules are very different from organic ones in this important respect. The vast majority of organic crystal structures are time-stable. Barring disordered or spontaneously reactive crystals, organic molecules within crystals maintain their location, orientation and conformation, their tautomeric state and also their arrangement of intermolecular interactions indefinitely and ideally infinitely. Many degrees of freedom, such as rotation of flexible moieties, are frozen in by specific intermolecular interactions. Co-crystallized solvent molecules are commonly trapped in well-defined positions and orientations. Hydrogen bonds, even the weak ones, are formed by all symmetry-related molecules in any given crystal in exactly the same way. In other words, an organic crystal is, at least in the bulk, a low-entropy system; one might even call it 'thermodynamically dead'. Consequently, structure stabilizing roles can be analysed for interactions with dissociation energies of around the kT level at room temperature (0.6 kcal/mol).

Crystals of biological macromolecules are very different. The molecules are 'living' in the crystal lattice and often maintain their biological activity. The crystals are composed of about 30–80 per cent water, which fills large channels and cavities between the macromolecules. The diffusion of water, ions and small molecules within these channels may occur on a similar time scale as in solution. Structural fluctuations of the molecules, which are inherently coupled with biological activity, can also occur in the crystal as in solution. Biological crystals are high-entropy systems which mimic solutions in many ways. As a consequence, hydrogen bonds are not time-stable, often very short-lived and their structure-stabilizing roles must be viewed from a somewhat different perspective than one would for organic crystals. It is not a long lifetime that counts, but rather the specific kind (i.e. the non-random nature) of the intermolecular interactions.

In organic crystals, there is normally a single and well-defined array of intermolecular interactions. The crystallization process constitutes a definite decision in favour of this array. Alternative networks are most often mutually exclusive and can only be realized in different polymorphic modifications. In the fluctuating biological structures, however, alternative networks are realized at the same time in different unit cells of the crystal and continuously transform into one another. Though fascinating, this property can make interpretation of hydrogen bond patterns very difficult.

5.1.2 The crystallographic resolution problem

A further difference between typical organic and biological crystal structures is the accuracy to which they can be determined. Organic structures are routinely determined at atomic resolution with X-ray diffraction, with the H atom positions being directly obtained in difference Fourier calculations and refined isotropically. Uncertainties of geometrical hydrogen bond parameters are typically $\sigma \leq 0.01$ Å for distances D and about 0.01–0.03 Å for distances d (if H atom positions are normalized). In precision studies using low-temperature neutron diffraction, uncertainties of D and d can be reduced to around 0.001 Å or even better.

In routine protein structure determination, the crystallographic resolution is typically in the range 2.0–3.0 Å. The data-to-parameter ratio is poor, so that heavy restraints on covalent geometries and also on non-covalent interactions must be used in refinement. To avoid 'bad contacts' that would otherwise occur in large numbers, non-covalent contacts are restrained by parameters that can be selected by the user. Typical statistical uncertainties $\sigma(D)$ of non-covalent interatomic distances are then 0.2 Å, and furthermore these distances are severely affected by bias due to the restraints. The consequences of distance uncertainties $\sigma(D) \approx 0.2$ Å are illustrated for some examples in Table 5.1 (assuming a Gaussian resolution function). If a true $O \cdots O$ separation is 2.8 Å, it will be found with a probability $p = 67$ per cent within $D \pm 1\sigma = 2.6$–3.0 Å, with $p = 95$ per cent within 2.4–3.2 Å and with $p = 99.7$ per cent within 2.2–3.4 Å. These intervals are wide but only for the $D \pm 3\sigma$ interval does it lead to a problem at the short end because of a conflict with distance restraints. This means that 'good' conventional hydrogen bonds can be reasonably recognized. For $C-H \cdots O$ interactions, the situation differs in an unpleasant way. $C-H \cdots O$ bonds of acidic donors may have true distances D of 3.2 Å. With $\sigma(D) = 0.2$ Å, many of the observed contacts would have very short D values

Table 5.1   Effect of refinement uncertainty on non-covalent interatomic distances, shown for a standard uncertainty of $\sigma(D) = 0.2$ Å. For some true distances, the intervals are given in which they are found in a crystal structure with given probabilities p (assuming a Gaussian resolution function)

True D (Å)	Example	$D \pm 1\sigma$ ($p = 67\%$) (Å)	$D \pm 2\sigma$ ($p = 95\%$) (Å)	$D \pm 3\sigma$ ($p = 99.7\%$) (Å)
2.8	Good $O-H \cdots O$	2.6–3.0	2.4–3.2	2.2[a]–3.4
3.0	Good $N-H \cdots O$	2.8–3.2	2.6–3.4	2.4–3.6[b]
3.2	Short $C-H \cdots O$	3.0–3.4	2.8[a]–3.6[b]	2.6[a]–3.8[b]
3.5	'Normal' $C-H \cdots O$	3.3–3.7[b]	3.1–3.9[b]	2.9[a]–4.1[b]
4.0	No hydrogen bond	3.8–4.2	3.6[b]–4.4	3.4[b]–4.6

[a] Conflict with typical refinement restraints. [b] Severe interpretation problems.

< 3.0 or even < 2.8 Å, which are disallowed by the refinement restraints. In effect, the restraints push the true D distribution towards longer distances, making individual hydrogen bonds more difficult to recognize. Moderate $C-H \cdots O$ interactions ($D \approx 3.5$ Å) are comfortably recognized as hydrogen bonds at atomic resolution but an uncertainty $\sigma(D) = 0.2$ Å creates a range of observed values that conflicts with the restraints at the short end and goes out of the hydrogen bond range at the long end. Finally, long contacts with a true D of 4.0 Å may be observed as being short enough that they might be mistaken for hydrogen bonds. In all these situations, the interpretation of $C-H \cdots O$ geometries is very problematic. Calculation of theoretical H bond positions and allowing for the θ angles does help but it does not cure the main malady.

There is only one good way to overcome these problems and that is to use crystal structures determined at atomic resolution and refined without restraints. The statistical studies of Derewenda *et al.* (1995) and Fabiola *et al.* (1997) used protein structures with relatively good resolutions around 1.5 Å but still biased by refinement restraints. Many protein structures are currently being published at resolutions of 1.0 Å and better but unfortunately, weak hydrogen bond effects have not (yet) been at the focus of these studies. With the increasing awareness of the importance of the weaker hydrogen bond types, however, one can but hope to see substantial improvements in methodology and analysis in the coming years.

5.2 Peptides and proteins

We begin this discussion of biological molecules with the largest sub-group, namely peptides and proteins. A few general comments about these structures now follow, mainly with the aim to introduce the reader to the relevant hydrogen bonding moieties.

5.2.1 The building blocks – amino acids

Proteins are built from 20 kinds of α-amino acids, the structures of which are shown in Fig. 5.1 together with the standard atom labelling and their average occurrence in proteins. The variations in the side-chains lead to a great diversity of possible hydrogen bond types, ranging from the very strong $N^+-H \cdots O^-$ salt bridges to the very weak $C-H \cdots O$ interactions of methyl groups. The $C-H$ groups of these amino acids span a wide range of donor strengths. The methyl groups of the apolar residues are poor hydrogen bond donors, the $C_\beta-H$ groups are somewhat stronger and the $C-H$ bonds of the aromatic rings even stronger. The C_β of valine, isoleucine and threonine carry only one H atom, so that these $C_\beta-H$ groups should be stronger donors than the $C_\beta H_2$ groups of the other standard amino acids. The very important $C_\alpha-H$

groups are significantly activated by the neighbouring N atom. Because there is no additional activation by a C_β atom in glycine, its $C-H$ groups are weaker donors than the $C_\alpha-H$ of the other amino acids. $C_{\delta 1}-H$ of the tryptophan indole ring is adjacent to an $>N-H$ group, while lysine and arginine have CH_2 groups adjacent to charged N atoms. These donors are therefore activated. The strongest $C-H$ donors in proteins are $C_{\delta 2}-H$ and $C_{\epsilon 1}-H$ of histidine, in particular if the imidazole ring is charged. His$C_{\epsilon 1}-H$, which is adjacent to two N atoms, is even more activated than His$C_{\delta 2}-H$. The $S-H$ group of cysteine is also a weak donor but is not within the scope of this work. Apart from the weak $C-H$ donors, some amino acids also contain weak acceptors of hydrogen bonds. In proteins, these are phenylalanine, tyrosine, tryptophan and histidine.

Some amino acids may occur in proteins in different protonation states, depending on the pH and the local environment. This is of relevance for the side-chain of His which may be cationic, or neutral in the $N_{\delta 1}-H$ or $N_{\epsilon 2}-H$ tautomers, and the side-chains of Asp and Glu which are normally found as carboxylates but may also occur as free acids (Deacon *et al.* 1997). These variations naturally influence the activation of the neighbouring $C-H$ groups.

Apart from the 20 standard amino acids, certain proteins contain modified amino acids, and natural peptides may contain further variants. Furthermore, there are natural monomeric amino acids which have very different hydrogen bonding groups, like $-C\equiv N$, $-C=N^+=N^-$, $-N^+(Me)_3$ and organic halogen (Voet and Voet 1995). All this considerably adds to the variety of weak hydrogen bond types (Chapter 3), but can be regarded as exotica which are outside the scope of this work.

Amino acids polymerize by formation of the peptide linkage $O=C-N-H$, Fig. 5.2(b). The backbone of these polymers carries a dense sequence of strong $N-H$ donors and $C=O$ acceptors and the conformations of the molecules are largely determined by hydrogen bonds between these groups. In addition, conventional hydrogen bonds of different strengths are formed by the the side-chains. In consequence, typical protein and peptide structures are dominated by complex systems of $N-H\cdots O$ and $O-H\cdots O$ hydrogen bonds, many of which have partly ionic character, $N^+-H\cdots O$ and $N^+-H\cdots O^-$. The apolar side-chains often form large aggregates and, frequently, hydrogen bonded and apolar domains can be clearly distinguished. Conventional hydrogen bonding in proteins has been thoroughly described in the classical reviews of Baker and Hubbard (1984) and Jeffrey and Saenger (1991). Herringbone interactions between aromatic side-chains have been reviewed by Burley and Petsko (1988) and by Hunter *et al.* (1991).

Weak hydrogen bonding in peptides and proteins has attracted interest only recently and in any case after the analogous effects in DNA fragments. In this section, we first discuss the ubiquitous $C-H\cdots O$ hydrogen bonds and then the $X-H\cdots Ph$ hydrogen bonds which occur more rarely. Finally some

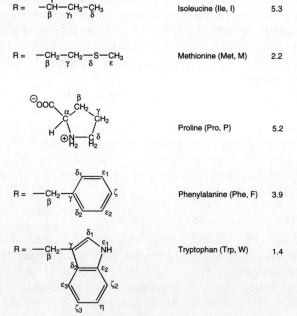

Fig. 5.1. The 20 standard amino acids and their average occurrences in proteins (after Schulz and Schirmer 1979). Only the L-isomers occur in natural proteins.

Polar side-chains

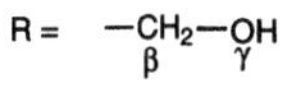 R = $-CH_2-OH$ (β, γ)	Serine (Ser, S)	6.8
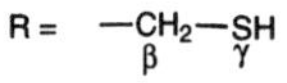 R = $-CH_2-SH$ (β, γ)	Cysteine (Cys, C)	1.9
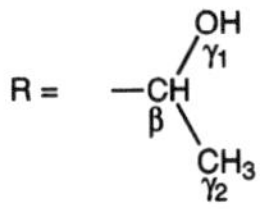 R = $-CH$ (β) with OH (γ_1) and CH_3 (γ_2)	Threonine (Thr, T)	5.9
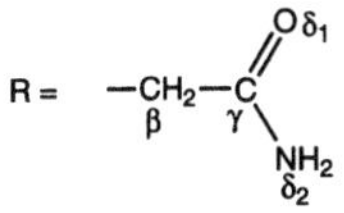 R = $-CH_2-C$ (β, γ) with O (δ_1) and NH_2 (δ_2)	Asparagine (Asn, N)	4.3
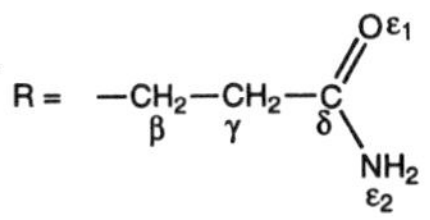 R = $-CH_2-CH_2-C$ (β, γ, δ) with O (ε_1) and NH_2 (ε_2)	Glutamine (Gln, Q)	4.3
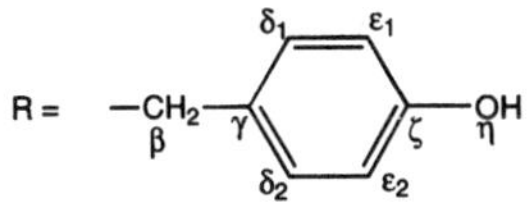 R = $-CH_2-$ (β, γ) ring with δ_1, ε_1, δ_2, ε_2, ζ, $-OH$ (η)	Tyrosine (Tyr, Y)	3.2

Charged side-chains

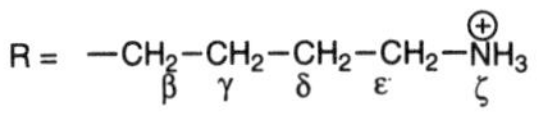 R = $-CH_2-CH_2-CH_2-CH_2-\overset{\oplus}{N}H_3$ (β, γ, δ, ε, ζ)	Lysine (Lys, K)	5.9
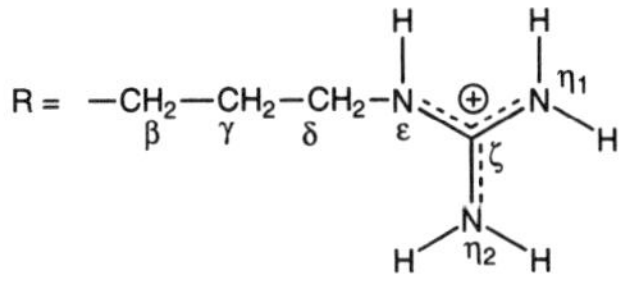 R = $-CH_2-CH_2-CH_2-N$ (β, γ, δ, ε) with guanidinium group $\overset{\oplus}{C}$ (ζ), N (η_1), N (η_2)	Arginine (Arg, R)	5.1
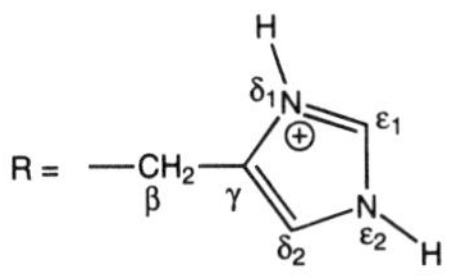 R = $-CH_2-$ (β, γ) imidazolium ring with $\delta_1 N$, ε_1, δ_2, ε_2 N-H	Histidine (His, H)	2.3
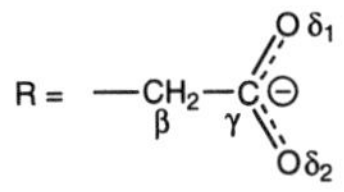 R = $-CH_2-C$ (β, γ) with O (δ_1) and $O^\ominus$ (δ_2)	Aspartic acid (Asp, D)	5.3
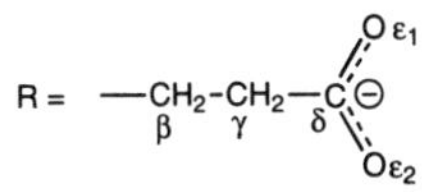 R = $-CH_2-CH_2-C$ (β, γ, δ) with O (ε_1) and $O^\ominus$ (ε_2)	Glutamic acid (Glu, E)	6.3

Fig. 5.1. *(cont.)*

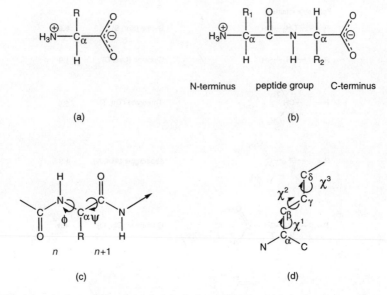

Fig. 5.2. (a) α-Amino acid in the zwitterionic form. (b) Dipeptide built of two α-amino acids. (c) Definition of the torsion angles ϕ and ψ which characterize the conformation of the molecular backbone. (d) Definition of the torsion angles χ which characterize the side-chain conformation.

reported functions of these interactions in protein–ligand recognition and in enzymatic activity are described.

5.2.2   C—H···O hydrogen bonds

In the first systematic description of C—H···O hydrogen bonding in amino acids, Jeffrey and Maluszynska (1982) analysed the intermolecular interactions in 32 amino acid neutron crystal structures. They found that C_α—H frequently donates C—H···O hydrogen bonds with typical distances d around 2.4 Å and with a lower limit of around 2.15 Å, indicating some significance. Compared to C_α—H, the C—H groups of the side-chains are involved in hydrogen bonding infrequently.

The first comprehensive statistical study on C—H···O interactions in proteins was published by Derewenda et al. (1995), who based their analysis on 13 well-refined crystal structures with crystallographic resolutions in the range 1.0–2.0 Å. These authors were aware that despite the good average resolution, some bias from refinement restraints was unavoidable in their data set. The distributions of non-covalent interatomic distances were determined for the atom pairs C···C, C···O and N···O. These are very similar in all 13 proteins, so that their sum can be used to smooth the statistical fluctuations (Fig. 5.3). The N···O distances show a pronounced maximum in the range 2.75–3.25 Å.

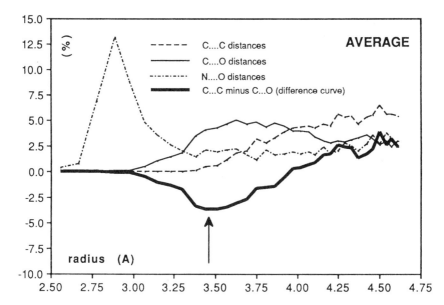

Fig. 5.3. Distribution of non-covalent interatomic distances (C···C, C···O and N···O) in 13 protein crystal structures in the 1.0–2.0 Å resolution range, and the C···C minus C···O difference distribution (from Derewenda *et al.* 1995).

These correspond to the conventional N−H···O hydrogen bonds that are dominant in protein structures. At longer distances, this distribution falls to nearly zero and rises only slowly with further increase in D. The distribution of C···C distances is close to zero for $D < 3.5$ and increases gradually with increasing distances. This indicates that C···C interactions are of the van der Waals type and that they are quite repulsive at short distances. The distribution of C···O distances lies between these extremes with the shortest contacts occurring at about 3.0 Å and a broad but clearly recognizable maximum in the range 3.5–4.0 Å. To illustrate the different behaviour of the C···C and the C···O distance distributions, their difference distribution is also shown. This has a minimum at about 3.5 Å. The shape of the C···O distribution is taken to mean that it represents a mixed population of van der Waals contacts and weak hydrogen bonds.

The shorter (C)H···O contacts ($d < 2.7$ Å) contributing to Fig. 5.3 may be inspected more closely and found for the most part to belong to one of four categories. By far the most common are C_α−H···O=C interactions in β-sheets. The second class are C−H···O=C contacts in α-helices with some preference for C_β−H donors. The third class is composed of contacts to buried polar side-chains, and the fourth consists of interactions with buried water molecules. The first three classes will be discussed in the following sections. Their role in protein hydration will be discussed in a wider context in Section 5.5.3.

Table 5.2 Mean geometry of $N-H\cdots O=C$ and $C_\alpha-H\cdots O=C$ hydrogen bonds in β-sheets of 11 high resolution protein crystal structures (Fabiola *et al.* 1997)

	Antiparallel β-sheet	Parallel β-sheet
Number of structures	9	3
Number of C/N$-$H$\cdots$O bonds	49	27
$N-H\cdots O=C$ bond geometry		
H$\cdots$O (Å)	1.94	2.00
N$\cdots$O (Å)	2.89	2.95
N$-$H$\cdots$O (°)	158	162
$C_\alpha-H\cdots O=C$ bond geometry		
H$\cdots$O (Å)	2.37	2.36
$C_\alpha\cdots$O (Å)	3.27	3.31
$C_\alpha-$H$\cdots$O (°)	141	146

5.2.2.1 *Secondary structure elements*

Parallel and antiparallel β-sheets. The most frequent $C-H\cdots O$ hydrogen bonds in proteins occur in parallel and antiparallel β-sheets (Derewenda *et al.* 1995). Prompted by earlier observations in peptides, Pattabhi and co-workers analysed the geometry of $C_\alpha-H\cdots O=C$ interactions in β-sheets in detail from a set of 11 protein structures in the very high resolution range 0.83–1.26 Å (Fabiola *et al.* 1997). Relatively restrictive cut-off criteria were used: $D < 3.5$ Å and $\theta > 130°$. On the basis of this analysis, the hydrogen bond schemes in antiparallel and parallel β-sheets is shown in Fig. 5.4, with mean geometries as given in Table 5.2. In both arrangements, each carbonyl O atom accepts a pair of hydrogen bonds, one from a peptide $N-H$ and one from the $C_\alpha-H$ group of the preceding residue, forming a chelated arrangement **214**. The hydrogen bond geometries are similar in antiparallel and parallel β-sheets, and the mean distances indicate a significant strength of the $C_\alpha-H\cdots O=C$ component (D around 3.30 Å, d around 2.37 Å). In Fig. 5.4, the pattern is drawn in an idealized way, whereas in real proteins, it is much less regular in its details and the geometries vary considerably from residue to residue. For the $C_\alpha-H\cdots O=C$ interactions, the distances d vary between 2.01 and 2.82 Å and the distances D from 2.91 Å to the cut-off value of 3.50 Å. The distributions of the parameters D, d and θ are shown in Fig. 5.5. The $C_\alpha-H\cdots O=C$ interaction satisfies the carbonyl acceptor directionality as effectively as the $N-H\cdots O=C$ hydrogen bond, Fig. 5.6. In the set of β-sheets studied, glycine does not form $C_\alpha-H\cdots O=C$ bonds and is always oriented in an unsuitable way. Very similar geometrical characteristics were found by Derewenda *et al.* (1995) in their work.

In β-sheets, side-chain $C-H$ groups may also form hydrogen bonds to the

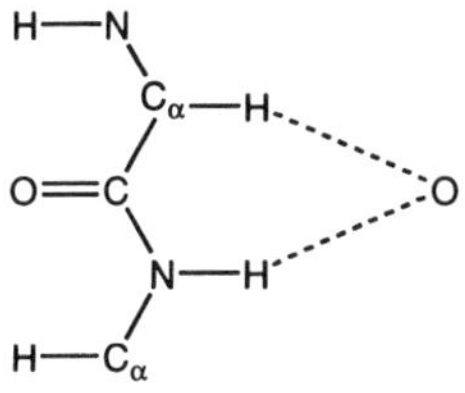

214 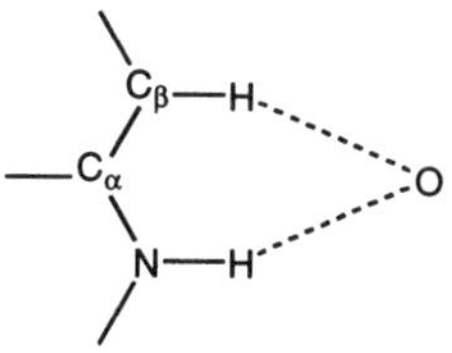   **215** 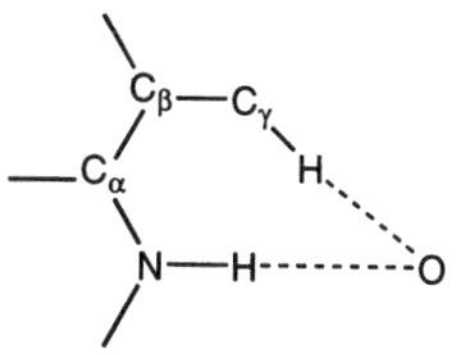   **216**

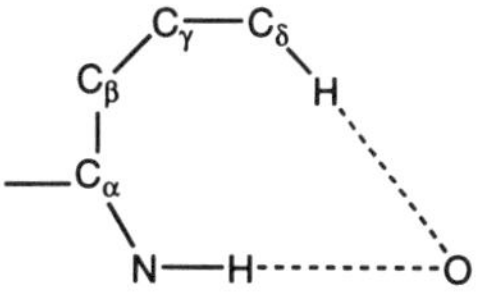

217 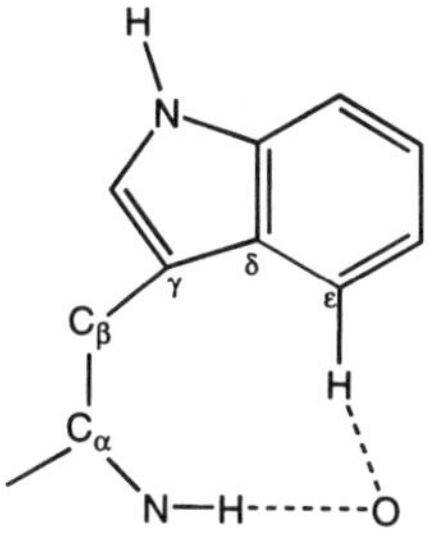   **218**

carbonyl groups. These interactions are rarer than the hydrogen bonds from C_α—H donors and the average geometry is less favourable. According to Fabiola *et al.* (1997), the distances *d* peak in the interval 2.5–2.6 Å and the angles θ in the interval 130–140°. Therefore, they are considered to be less important when compared to the systematic C_α—H$\cdots$O=C interactions.

β-Sheet structures also occur in small peptides. It is noteworthy that a parallel β-sheet structure involving C_α—H$\cdots$O=C hydrogen bonds was clearly identified by Chatterjee and Parthasarathy in crystalline in *N*-formyl-L-Met-L-Val as early as 1984, Fig. 5.7(a). A related parallel β-sheet is present in L-His-L-Ser trihydrate (Padiyar 1998). A miniature antiparallel β-sheet is formed by the dipeptide L-His-Gly and is shown in Fig. 5.7(b). In this case, the acceptor is not a peptide carbonyl group but an O atom of the C-terminal carboxylate (Steiner 1997c).

In the chelating motif **214**, the role of the C_α—H$\cdots$O=C interaction is not dominant but supportive to the stronger N—H$\cdots$O=C hydrogen bonds. β-Sheets are formed not because of the C_α—H$\cdots$O=C but because of the N—H$\cdots$O=C hydrogen bonds, and the former only provide some extra stability. When judging the supporting function of the C_α—H$\cdots$O=C interactions, it suffices to state that they occur systematically, in large numbers and in accord with the dominant N—H$\cdots$O=C arrangement. Though each of the individual C—H$\cdots$O=C bonds should contribute only about −1 kcal/mol of bond energy, their large numbers and their concerted action gives them a clearly recognizable relevance.

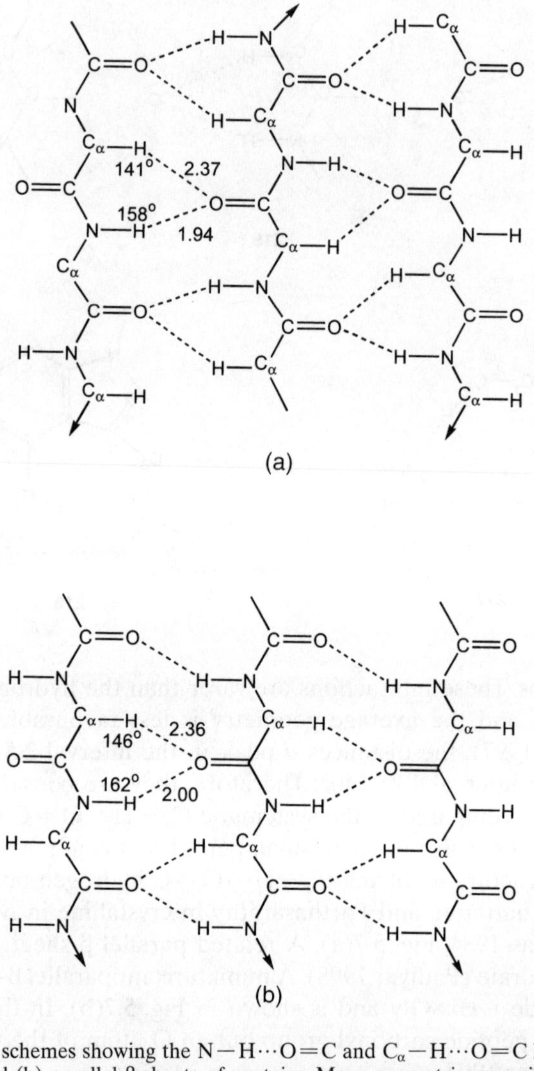

(a)

(b)

Fig. 5.4. Idealized schemes showing the $N-H\cdots O=C$ and $C_\alpha-H\cdots O=C$ hydrogen bonds in (a) antiparallel and (b) parallel β-sheets of proteins. Mean geometries are given as determined by Fabiola *et al.* (1997).

α-Helix. The second main secondary structure element of proteins is the α-helix. In the α-helix, the $C_\alpha-H$ bonds are oriented away from the helix axis and there are no systematic $C_\alpha-H\cdots O=C$ hydrogen bonds. Some $C-H\cdots O=C$ interactions are donated by side-chains, mainly by the weakly activated $C_\beta-H$ groups. These interactions are of the kind $C_\beta(i + 3)-H\cdots O=C(i)$, supportive to the standard hydrogen bond $N(i+4)-H\cdots O=C(i)$, Fig. 5.8. Although the distances d and D can be short,

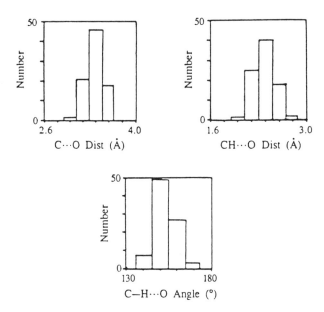

Fig. 5.5. Distribution of the parameters d, D and θ of the $C_\alpha-H\cdots O=C$ hydrogen bonds in parallel and antiparallel β-sheets (Fabiola *et al.* 1997).

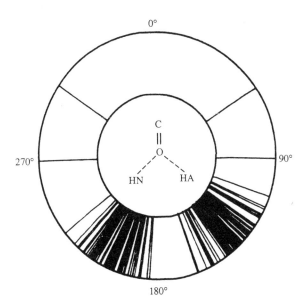

Fig. 5.6. Acceptor directionality of $C_\alpha-H\cdots H\cdots O=C$ and $N-H\cdots O=C$ hydrogen bonds in parallel and antiparallel β-sheets. Polar scatterplot of $H\cdots O=C$ angles (from Fabiola *et al.* 1997).

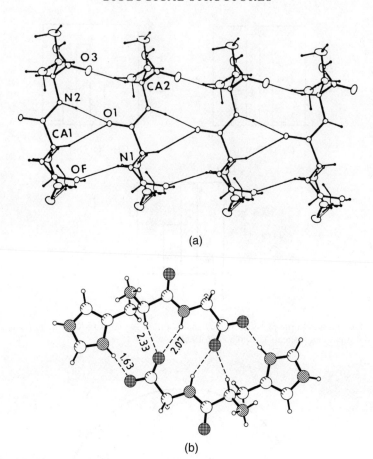

(a)

(b)

Fig. 5.7. Intermolecular β-sheet structures of small peptides containing C_α—H···O=C hydrogen bonds. (a) A parallel β-sheet formed by N-formyl-L-Met-L-Val trihydrate (from Chatterjee and Parthasarathy 1984). (b) Miniature antiparallel β-sheet formed by L-His-Gly chloride (Steiner 1997c).

the angles θ are far from linear (Derewenda *et al.* 1995), so that these interactions are certainly very weak.

Proline cannot donate N—H···O hydrogen bonds. If inserted in an α-helix, the regular pattern of $N(i)$—H···O=C($i-4$) hydrogen bonds is disrupted, leading to a kink in the helix. Chakrabarti and Chakrabarti (1998) have noted that in this situation, the activated proline $C_\delta H_2$ group is often involved in C—H···O interactions with carbonyl acceptors at positions ($i-3$), ($i-4$) or ($i-5$) depending on the local conformation. Frequently the two C_δH donors form a pair of interactions $C_\delta(i)$—H_1···O=C($i-3$), $C_\delta(i)$—H_2···O=C($i-4$). These compensate to some degree for the loss of the strong N—H···O hydro-

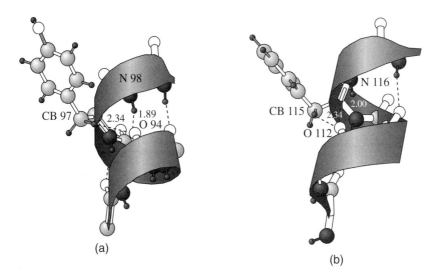

Fig. 5.8. C—H···O=C interactions in α-helices: (a) in *Saccharomyces cerevisiae* cytochrome *c*, PDB entry 1YCC; (b) in myoglobin, PDB entry 1MBA (from Derewenda *et al.* 1995).

gen bond. Still, the helix-breaking role of proline is undisputed and so this compensation can only be partial.

Collagen triple helix. Systematic interchain C_α—H···O=C hydrogen bonds in the collagen triple helix were recognized by Ramachandran and Sasisekharan (1965) based on low-resolution fibre diffraction data. In a 30-residue peptide model of the collagen triple-helix, Bella and Berman (1996) observed the hydrogen bond pattern shown in Fig. 5.9(a), which consists of two distinct repetitive hydrogen bond motifs. One is a pair of N—H···O=C and C_α—H···O=C hydrogen bonds analogous to those observed in β-sheets, **214**. In the other motif, a glycine residue donates a pair of hydrogen bonds to carbonyl acceptors of two adjacent residues. The mean geometries of these motifs are shown in Fig. 5.9(b) together with data obtained for a refined fibre diffraction model of collagen. The modes of approach to the C=O acceptors are shown in Fig. 5.9(c) for the different kinds of hydrogen bonds.

Polyglycine II helix. Shortly after their study on collagen, Ramachandran *et al.* (1966) reported the occurrence of systematic C_α—H···O=C hydrogen bonds in polyglycine II, which adopts another kind of helical structure. These fibre diffraction studies are, to our knowledge, the first experimental publications on weak hydrogen bonds in macromolecules. In polyglycine II, there are interstrand N—H···O=C and C_α—H···O=C hydrogen bonds with $D = 2.73$ and 3.20 Å, respectively, as shown already in Fig. 2.4 of Section 2.1.

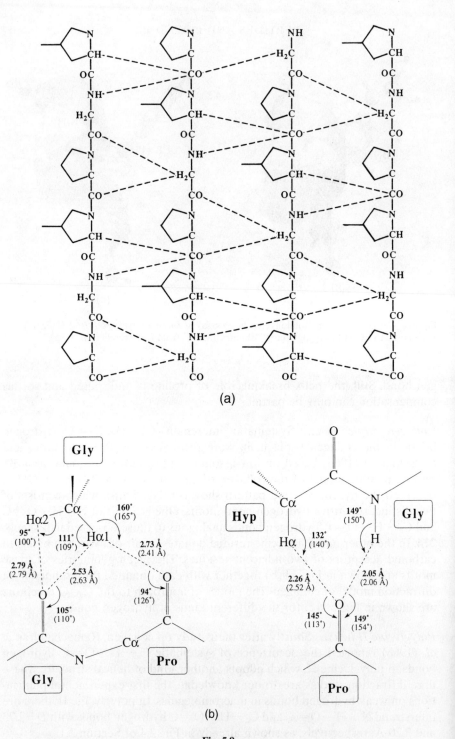

(a)

(b)

Fig. 5.9.

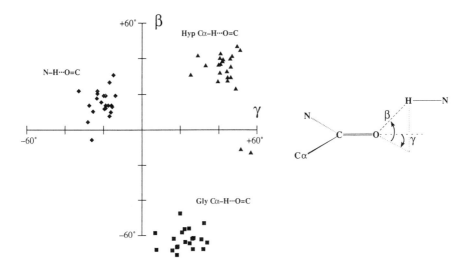

Fig. 5.9. N—H···O=C and C_α—H···O=C hydrogen bonds in collagen triple helices. (a) In the model peptide $(Pro-Hyp-Gly)_4-Pro-Hyp-Ala-(Pro-Hyp-Gly)_5$ (Hyp = γ-hydroxy-proline). The ideal collagen sequence is $(Pro-Hyp-Gly)_n$. (b) The two local hydrogen bond motifs in (a) with their mean geometries. Numbers in parentheses are valid for the model peptide, bold numbers are valid for a refined fibre diffraction model of collagen. (c) Polar scatterplot showing the acceptor directionality (from Bella and Berman 1996).

Chelating hydrogen bonds donated by $N(n)$—H and $C_\alpha(n - 1)$—H. The hydrogen bond pattern **214** involving N—H and C_α—H of the preceding residue occurs in β-sheets and the collagen helix with main-chain carbonyl acceptors, but it is also formed with side-chain acceptors and water molecules (see below). This remarkable recurrence suggests a closer examination of the circumstances under which it can be formed. The N—H and the C_α—H bonds point in the same direction when the angle ψ is around 120° and for glycine (which has two C_α—H bonds) also for ψ around −120° (Fig. 5.2(c)). According to the Ramachandran plot, these are nicely allowed values but they are adopted only in certain secondary structure elements. According to the mean ϕ and ψ angles given in Table 5.3, motif **214** can be formed only by the main-chain donors in parallel and antiparallel β-sheets, the collagen helix and the polyglycine and polyproline helices. Actually, suitable ψ-angles do occur also in the irregular loop regions which make up a major part of protein structures.

5.2.2.2   Side-chain interactions

C—H···O hydrogen bonds from amino acid side-chains are less investigated than hydrogen bonds from C_α—H. This could be because it is less likely that one would find systematic recurrences in patterns as compared to the hydrogen bonds in the secondary structure elements. The lack of repetitivity makes

Table 5.3 Average ϕ and ψ angles in secondary structure elements of proteins (Voet and Voet 1995)

Secondary structure	ϕ (°)	ψ (°)
α-Helix	−57	−47
Parallel β-sheet	−119	113
Antiparallel β-sheet	−139	135
3_{10}-Helix	−49	−26
π-Helix	−59	−70
Left-handed α-helix	57	−47
2.2_7 Ribbon	−78	59
Polyglycine II helix and poly- L-proline II helix	−79	150
Collagen	−51	153

interpretation uncomfortable and it is almost impossible to distinguish between functional and casual interactions. The situation is further complicated because there are so many different kinds of side-chain C—H groups. The residues which carry positive charges (His, Arg, Lys) often form short C—H···O interactions but they also form very strong N^+—H···O^- salt bridges, which by their dominance make the interpretation of adjacent weak hydrogen bonds problematic.

The characteristics of C—H···O contacts have been systematically investigated only for a few amino acid residues. In an early and interesting approach, Thomas *et al.* (1982) analysed the interatomic environments of 170 phenylalanine aromatic rings in 28 protein crystal structures. An edge-on approach of O atoms to the phenyl rings is preferred to an approach onto the faces. From the population densities of edge-on and face-on contacts, the free energy difference between these contact geometries was estimated. The obtained value of about −1 kcal/mol is interpreted as due to an attractive electrostatic interaction. This analysis was repeated and confirmed by Gould *et al.* (1985) for amino acid and peptide structures in the CSD. The shortest approaches of O atoms to the H atoms at the edges of phenyl rings are around $d = 2.45$ Å. The approaches to the hydrophobic side-chains of leucine, isoleucine and valine were also analysed in that study and it was found that typical distances D are longer when compared to phenylalanine. The edge-on preference of phenyl–O contacts was reconfirmed for proteins by Singh and Thornton (1990).

Greater interest has been devoted to the side-chain of histidine, probably because it is the strongest of all amino acid C—H donors and because of its functional role in catalytic processes (see Section 5.2.5). In peptides, it is frequently observed that all X—H groups of the imidazole ring form relatively short hydrogen bonds, in particular if the residue is positively charged. An example in the cyclic dipeptide *cyclo*-(L-His-L-Asp) trihydrate is shown in Fig. 5.10(a). A related situation in a protein has been found in the human plas-

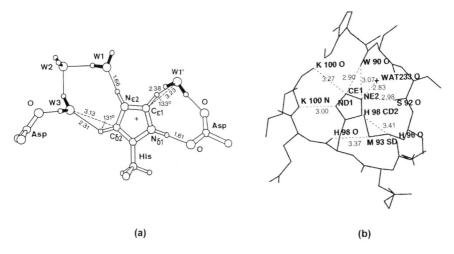

(a)   (b)

Fig. 5.10. Imidazole rings of histidine side-chains, in which both C—H groups donate short hydrogen bonds. (a) In *cyclo*-(L-His-L-Asp) (from Steiner 1995*d* based on Ramani *et al.* 1978). (b) In the human plasminogen kringle 4 (from Stec *et al.* 1997).

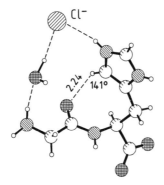

Fig. 5.11. Intramolecular C_δ—H···O=C hydrogen bond in Gly-L-His chloride dihydrate, stabilizing a non-standard side-chain conformation (Steiner 1996*c*).

minogen kringle 4 by Stec *et al.* (1997), Fig. 5.10(b). In the hydrated chloride salt of the dipeptide Gly-L-His, an unusual conformation of the histidine side-chain has been found and is associated with a short intramolecular C_δ—H···O=C hydrogen bond to the peptide carbonyl group, Fig. 5.11. This interaction stabilizes the non-standard conformation with $\chi^2 = 165°$; normally, χ^2 is around ±90° without an intramolecular side-chain to main-chain hydrogen bond (Steiner 1996*c*).

At a more general level, Derewenda *et al.* (1995) have pointed out that side-chain C—H···O hydrogen bonds can serve to satisfy acceptor potentials of buried polar groups. Two examples are shown in Fig. 5.12. In protein structures, polar groups are often found embedded in a more or less apolar envi-

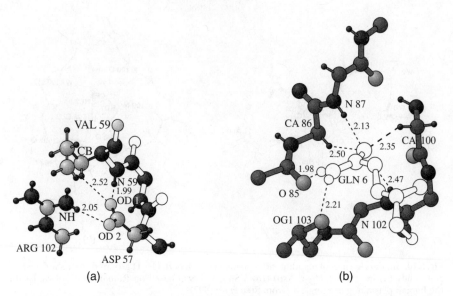

Fig. 5.12. C—H···O=C hydrogen bonds involving buried side-chain carbonyl groups. The acceptor potentials are only partially satisfied by N—H···O=C hydrogen bonds, leaving space for
C—H···O interactions. (a) In therimtase, PDB entry 1THM. (b) In human myeloma Fab fragment, PDB entry 2FB4 (from Derewenda *et al.* 1995).

ronment in which they cannot form a sufficient number of strong hydrogen
bonds. It would be misleading to consider these buried polar groups as completely devoid of hydrogen bonding – weak directional interactions mainly of
the C—H···O type are often formed with the surrounding groups. Buried
amide groups can also donate N—H···Ph hydrogen bonds (see below). A different mechanism to satisfy hydrogen bond potentials at least partially is by
the formation of very long and distorted O/N—H···O hydrogen bonds. On
the whole, polar side-chains can form a continuum of strong to weak to very
weak hydrogen bonds and it would be misleading to suggest a limit that separates satisfied from unsatisfied hydrogen bond potentials (McDonald and
Thornton 1994).

As mentioned earlier, interpretation of side-chain C—H···O hydrogen
bonds is made difficult by the relative lack of recurring patterns. However,
exceptions do exist and in a series of such motifs, a pair of hydrogen bonds is
donated by a peptide N—H and a C—H group of the same residue (**215–218**).
With the donor C_β—H, **215**, there is a structural resemblance to motif **214** in
which C_α—H is the donor. Functionally, however, **215** is certainly weaker
because of the weaker activation of C_β—H compared to C_α—H. An example
may be seen in Fig. 5.12(a) for a valine C_β—H group. With suitable side-chain
conformations, C_γ—H and also C_δ—H can be oriented in a way that allows
pairs of hydrogen bonds together with the peptide N—H, **216** and **217** (Koellner *et al.* 1999). The related pair of hydrogen bonds involving C_ε—H is found

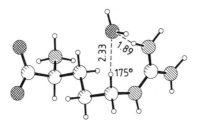

Fig. 5.13. Pair of C_δ—H$\cdots$O and N_η—H$\cdots$O hydrogen bonds donated by an arginine side-chain. In the neutron structure of L-Arg phosphate monohydrate (Espinosa *et al.* 1996).

only with tryptophan residues, **218**. Examples for **217** with tyrosine and tryptophan C_δ—H are shown in Section 5.5.3.1 with water acceptors (Fig. 5.72). As a final example of a recurrent pattern, chelating pairs of N_η—H$\cdots$O and C_δ—H$\cdots$O hydrogen bonds are frequently formed by the arginine side-chain (Fig. 5.13). In all these patterns composed of strong and weak hydrogen bonds, the weak constituent has a supportive role. If a pattern is composed of a moderately strong bond with an energy of say -4 kcal/mol and a weak one of say -1 kcal/mol, the supportive contribution is clearly too large to be neglected.

5.2.3 X—H$\cdots\pi$ hydrogen bonds

The second class of weak hydrogen bonds in proteins are formed to π-acceptors. Examples are known with all strong donor types that are present in proteins: main-chain and side-chain N—H, side-chain O—H, water molecules and O/N—H groups of substrate molecules. The acceptors are the side-chains of Phe, Tyr, Trp and occasionally His residues and also π systems of substrates. The general properties of O/N—H$\cdots\pi$ hydrogen bonds have been described in detail in Section 3.1, but it is pertinent to recall the most important characteristics.

(a)   The energy range of O/N—H$\cdots\pi$ hydrogen bonds is about 2–4 kcal/ mol for uncharged systems, with more examples found rather at the low end of this scale. These energies are smaller than in conventional O/N—H$\cdots$O hydrogen bonds but still of the same order of magnitude and clearly above those of typical C—H$\cdots$O interactions.

(b)   The directionality of X—H$\cdots\pi$ hydrogen bonds is extremely flexible.

(c)   The entire face of aromatic rings can serve as the acceptor, including the individual ring atoms.

Combination of properties (b) and (c) makes phenyl groups a 'target that is very easy to hit', that is, it can be approached by a donor from a very large volume in space with only little differences in energy.

Unambiguous N/O—H···π hydrogen bonds occur relatively rarely in proteins and many of the smaller proteins do not contain even one of these interactions. However, if they are formed, they often play an important functional role. In particular, there are a considerable number of examples where N/O—H···π hydrogen bonds are specifically involved in protein–ligand recognition. This sharply contrasts the roles of C—H···O interactions in proteins – every protein contains a very large number of C—H···O hydrogen bonds and for the larger proteins they occur in the thousands. Most of these interactions are weak to very weak and their functions are normally supportive at best. In this sense, C—H···O and N/O—H···π hydrogen bonds have opposite characteristics of abundance and individual importance. Proteins abound with the weakly polar Ph···Ph interactions and C—H···Ph interactions of other C—H groups but we do not consider these as hydrogen bonds.

X—H···π hydrogen bonding to phenyl rings in proteins attracted considerable attention from about the mid-1980s onwards, and this was triggered by articles from two prominent groups. In a combined X-ray and neutron diffraction study at 1.0 Å resolution, Wlodawer, Huber and co-workers found two unambiguous aromatic hydrogen bonds in bovine pancreatic trypsin inhibitor, BPTI (Wlodawer *et al.* 1984). The relevant section of BPTI is shown in Fig. 5.14. The side-chain of Tyr35 accepts two aromatic hydrogen bonds, one from

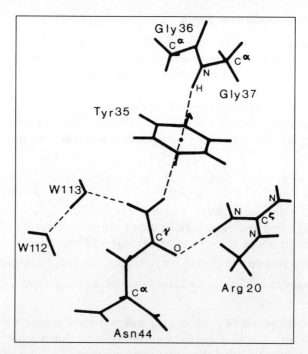

Fig. 5.14. Aromatic hydrogen bonds in bovine pancreatic trypsine inhibitor (adapted from Tüchsen and Woodward 1987).

the side-chain of Asn44 and the other from the peptide N—H of Gly37. Both interactions have distances d(M) of around 2.6 Å. The hydrogen bond nature of these contacts was confirmed by Tüchsen and Woodward (1987) using ^{1}H NMR spectroscopy, and based on solution ^{1}H NMR data, Wagner *et al.* (1987) showed that they are not confined to the crystal. The second pioneering work on aromatic hydrogen bonding in proteins is the finding of Perutz *et al.* (1986) that this interaction occurs in hemoglobin–drug binding (see below). The subsequent theoretical study of Levitt and Perutz (1988), where a typical N—H···Ph bond energy of –3 kcal/mol was calculated (for uncharged N—H), has already been discussed (Section 3.1.3.1). These relatively early contributions already emphasize the three main configurations of aromatic hydrogen bonds in proteins, that is side-chain to side-chain, main-chain to side-chain and protein–ligand. They also point at a functional importance, variable geometry and persistence in solution.

5.2.3.1   *Side-chain to side-chain interactions*

Side-chain to side-chain N—H···Ph interactions have been the subject of an early statistical study of Burley and Petsko (1986). In 33 protein crystal structures with resolutions of 2 Å or higher, they inspected the intermolecular surrounding of the aromatic groups of Phe, Tyr and Trp. They found that the amino groups of the Lys, Arg, Asn, Gln and His side-chains are preferentially located within 6 Å of the ring centroid, peaking sharply between 3 and 4 Å, whereas longer distances between 6 and 10 Å are much rarer (Fig. 5.15). The

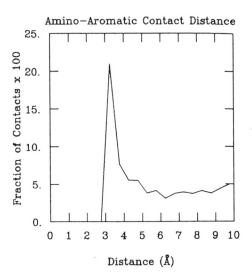

Fig. 5.15. Frequency of amino–aromatic contacts shorter than 10 Å in proteins displayed as a percentage of all atom–aromatic contacts (from Burley and Petsko 1986). Only some of the contacts in the sharp peak region represent hydrogen bonds, the others are from stacked arrangements.

amino groups are located preferably above the ring face and avoid the ring edge. This behaviour is clearly different from a distribution expected from random close packing of the side-chains and the directionality is opposite to that of the Ph$\cdots$O interactions which prefer the ring edge and avoid facial approach.

In a closer analysis of amino–aromatic interactions, Singh and Thornton (1990) found a similar distance distribution, but only a fraction of the contacts may be actually considered to be hydrogen bonds. The majority represent stacked arrangements, which also have short distances from the N atom to the ring centre. The N$\cdots$M distance is not a complete enough criterion for an aromatic hydrogen bond and the direction of the N$-$H vector has also to be taken into account.

Earlier studies concentrated on N$-$H donors and the Ph acceptors of Phe, Tyr and Trp side-chains. Hydroxyl and water donors were studied somewhat later (see Sections 5.2.4 and 5.2.5). An example with the heterocycle of a His residue acting as a π-acceptor was only recently reported by Gilliland and co-workers, who observed the interaction between two His side-chains of the mode (His)$N_\varepsilon$$-H\cdots\pi$(His) in human carboxyhemoglobin (Vásquez $et\ al.$ 1998).

Preference of stacked over hydrogen bonded arrangements.   The question of stacked versus hydrogen bonded amino–aromatic interactions was taken up again by Thornton and co-workers in an analysis of 55 non-homologous protein structures with resolutions of 2 Å and better (Mitchell $et\ al.$ 1994). Only sp^2-hybridized N atoms were considered because only here can the theoretical H atom positions be reliably calculated. 'N-above-ring' interactions are defined as those where an N atom lies within 20° of the perpendicular to the ring (ω, as defined in Fig. 3.4, Section 3.1.3, is <20°) and have at least one contact D(C) or D(M) < 3.8 Å. Interactions with interplanar angles <30° were defined as 'stacked', while 'hydrogen bonds' must have at least one angle θ(C) > 120°. Contacts that could not be clearly classified as stacked or as hydrogen bonds were called 'others'. Based on these definitions, the 115 N-above-ring interactions in the structural sample were divided into 30 hydrogen bonds, 76 stacked arrangements and 9 others (Table 5.4). This means that the stacked arrangements outnumber the aromatic hydrogen bonds by around 2.5:1. Because the analysis is based on 55 protein structures, it also means that clear-cut aromatic hydrogen bonds are rare. With the above cut-off definitions, they occur in only about half of all protein structures.

Structural data of small organic molecules show that aromatic hydrogen bonds can be formed with deviations from perpendicularity much greater than $\omega = 20°$ (Section 3.1.3). For example, the hydrogen bond in Fig. 3.12(b) has a value of ω(O) = 35.3°. The shallow potential energy surface is one of the main characteristics of the aromatic hydrogen bond. This means that the cut-off criteria used by Mitchell $et\ al.$ (1994) are unnecessarily restrictive in ω. The

Table 5.4 Numbers of 'N-above-ring' interactions in 55 protein structures (Mitchell *et al.* 1994). The main chain to Ph hydrogen bonds are restricted to the case $n > i + 2$

	H-bonded	Stacked	Others
Side-chain N / Phe	8	14	2
Main-chain N / Phe	5	19	1
Side-chain N / Tyr	6	15	3
Main-chain N / Tyr	11	28	3
Total	30	76	9

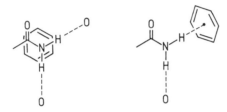

Fig. 5.16. Stacked and hydrogen bonded arrangements of amino and aromatic groups, shown for an asparagine side-chain. Both arrangements are associated with short N—Ph separations $< 4 \text{ Å}$.

cut-offs in D (3.8 Å) and in θ (120°) are better justified. With a relaxed ω cut-off (such $\omega < 35°$ or 40°), a larger number of contacts would be considered as aromatic hydrogen bonds.

To interpret the preference of stacked over clearly hydrogen bonded amino–aromatic interactions, one must take into account the conventional hydrogen bonds formed by the amino group as is illustrated for an asparagine side-chain in Fig. 5.16. In a stacked amino–aromatic contact, the amino group can form two strong N—H···O hydrogen bonds. In addition, with an appropriate offset that places positive charges on the asparagine side-chain over negative charges on the aromatic group, there can be a favourable electrostatic stacking interaction. For the N—H···Ph hydrogen bonded arrangement, only one additional N—H···O hydrogen bond can be formed and the stacking interaction is not possible. The energy sum of all interactions formed by the amino group is clearly larger for the stacked configuration, so that it is favoured in a competitive situation. If there is a local lack of strong acceptors, however, formation of an N—H···Ph hydrogen bond is more favourable than the presence of a dangling N—H donor.

Functionally important X—H···Ph hydrogen bonds between side-chains. The function of a particular side-chain hydrogen bond in a protein can often

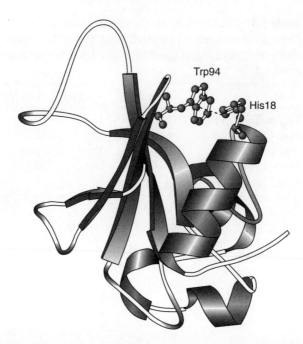

Fig. 5.17. Aromatic hydrogen bond between the side-chains of His18 and Trp94 of barnase. Ribbon diagram showing the major structural elements and the two relevant residues (adapted from Loewenthal *et al.* 1992).

not be quantified. Even strong hydrogen bonds often appear to be casual and it is not unusual that point mutations deleting one hydrogen bond have no recognizable effect on protein function or stability. In other instances, important roles of particular side-chain hydrogen bonds have been pinned down with considerable finesse. We now discuss such examples involving aromatic hydrogen bonds.

In the small enzyme barnase, the X-ray crystal structure suggests the presence of an aromatic hydrogen bond between His18 and Trp94 (Fig. 5.17). The distance between the ring centres is 4.0 Å and the interplanar angle is 43°. The histidine residue is at the C-terminus of an α-helix and acts as a C-cap, with the N_ε being hydrogen bonded with the backbone carbonyl three residues behind in the chain, as shown in Scheme **219**. The contact with the face of Trp94 is via N_δ. Loewenthal *et al.* (1992) studied the function of this aromatic hydrogen bond in great detail. In a series of mutations, Trp94 was replaced by other residues and the pK_a of His18 was determined by titration experiments and compared to that in denaturated enzyme. It was found that the pK_a of His18 is the highest in the native enzyme and falls in the mutation series as 'denaturated wild type' < Leu94 < Phe94 < Tyr94 < Trp94 (pK_a = 6.52, 7.13, 7.37, 7.64, 7.75, respectively, with uncertainties around 0.02 units). This means that the basic strength of the imidazole ring is modulated by the local environment and that the protonated form of histidine is stabilized by the aromatic hydrogen bond compared to direct contact with water. The strength of the hydrogen bond decreases in the series His-Trp > His-Tyr > His-Phe. In double mutation and unfolding experiments, it was found that the free energy of the interaction between the Trp94 and His18 side-chains relative to solvation with water amounts to −1.4 kcal/mol for the protonated and −0.4 kcal/mol for the unprotonated form of histidine.

In the isolated α-helical C-peptide, that is residues 1 to 13 of ribonuclease A, there is an aromatic hydrogen bond between the side-chains of His12 and Phe8. The difference of four positions along the chain corresponds to about one helix turn and is ideal for side-chain interactions. Shoemaker et al. (1990) showed by pH dependent solution experiments that the interaction is specific to the protonated form of histidine, **220**. Armstrong et al. (1993) further investigated the $(i, i+4)$ Phe–His interaction in α-helices by determining the degree of helicity in solution for a series of synthetic peptides, with circular dichroism. It was found that $(i, i+4)$ spaced pairs of Phe and His residues stabilize α-helices independent of the position within the helix and that this stabilization is much larger with protonated than with uncharged histidine side-chains. Phe, His pairs with $(i, i+5)$ spacing, **221**, do not show such helix stabilization.

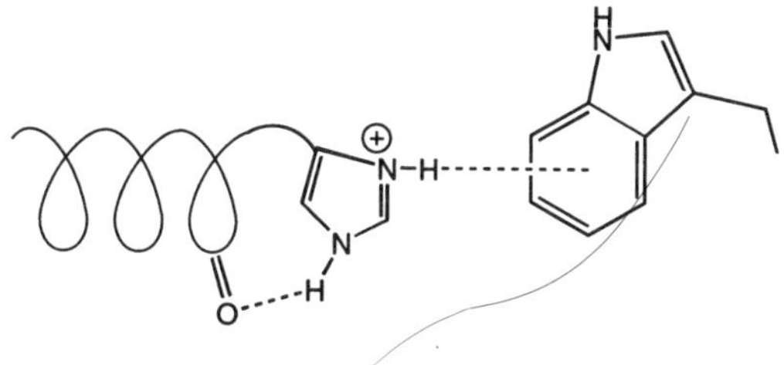

α-Helix   His 18⁺   Trp 94

219

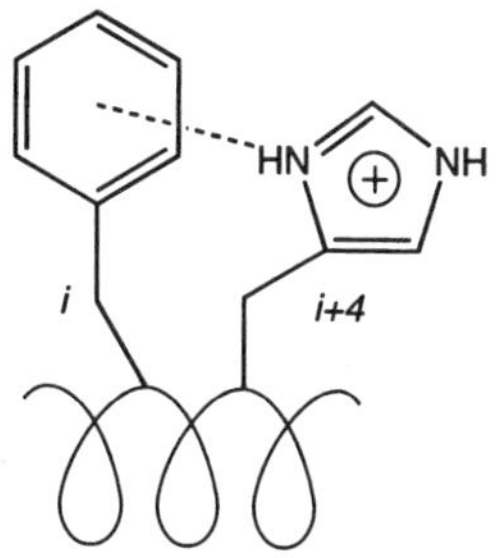

220

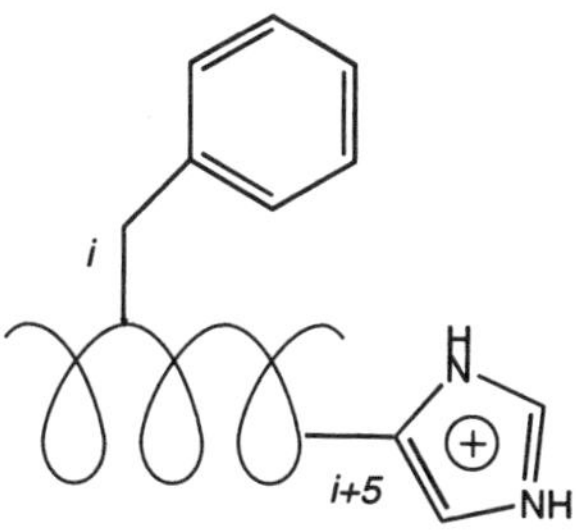

221

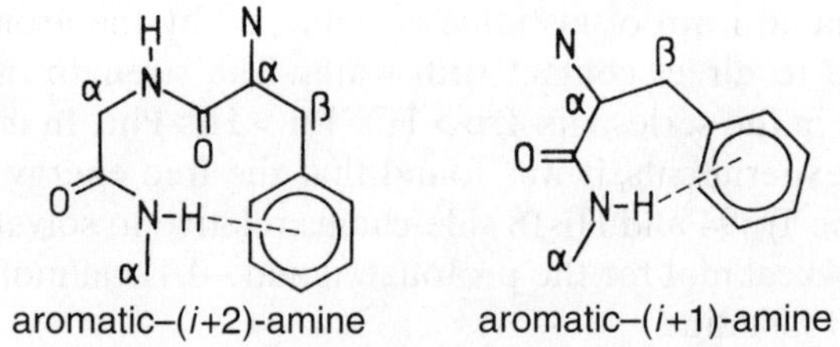

aromatic–(*i*+2)-amine   aromatic–(*i*+1)-amine

Fig. 5.18. Structure of aromatic–(*i* + 2) amine and aromatic–(*i* + 1) amine hydrogen bonds.

Replacement of a properly positioned Phe side-chain by a cyclohexyl group completely removes the effect. Solution experiments of this kind are very important and complement crystallography, because they can reveal the stabilizing function of aromatic hydrogen bonds, unlike structure analysis that provides only geometries and can hardly distinguish between functional and casual interactions.

5.2.3.2 *Main-chain to side-chain interactions*

The statistical survey of Mitchell *et al.* (1994) reveals a number of aromatic hydrogen bonds where the donor is a peptide $N-H$ group, Table 5.4. In the analysis, these interactions were only considered when the donor and acceptor are not close to each other in the polypeptide chain. If donor and acceptor *are* close to each other, a sterically different situation arises in which only geometrically constrained hydrogen bonds can be formed. In crystal structures, hydrogen bonds are observed that involve aromatic side-chains and peptide $N-H$ donors of the next and of the next but one residue (Fig. 5.18). In α-helices and β-sheets the peptide $N-H$ donors are engaged in $N-H\cdots O=C$ hydrogen bonds, so that main-chain to side-chain aromatic hydrogen bonding can occur only in the less structured parts of proteins.

In a database study of 297 representative protein crystal structures, Worth and Wade (1995) analysed in detail the interactions of Phe and Tyr side-chains with the peptide $N-H$ group of the second next residue. This is called the 'aromatic–(*i* + 2) amine interaction'. For all Phe and Tyr side-chains in the data sample, the geometries with respect to the (*i* + 2) amine were inspected. Aromatic–(*i* + 2) amine interactions of both the stacked and the hydrogen bonded kind can be formed; examples of both kinds are shown in Fig. 5.19 for BPTI (the hydrogen bonded example has already been shown in a different environment in Fig. 5.14). Out of 4490 side-chains, only 115 are involved in an aromatic–(*i* + 2) amine interaction, corresponding to a fraction of 2.6 per cent. This means that the interaction as such is rare. However, there are some pronounced sequence dependencies, Table 5.5. Tyrosine forms interactions with

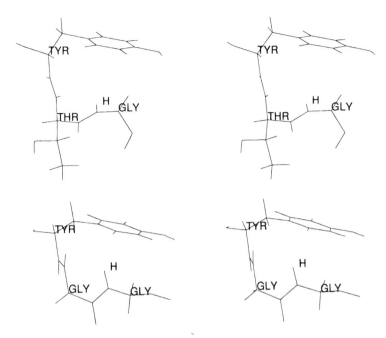

Fig. 5.19. Stereoview of two tripeptides in the crystal structure of bovine pancreatic trypsin inhibitor (BPTI), which have an aromatic–$(i + 2)$ amine interaction: *top'* BPTI$_{10-12}$ with a stacked interaction; *bottom'*: BPTI$_{35-37}$ with an aromatic hydrogen bond (from Worth and Wade 1995).

Table 5.5 Aromatic–$(i + 2)$ amine interactions in 297 proteins (Worth and Wade 1995). All tripeptide sequences starting with Phe or Tyr are analysed

Sequence[a]	Found	Interactions	%
Ar–X–X'	4490	115	2.6
Phe–X–X'	2324	41	1.8
Tyr–X–X'	2166	74	3.4
Ar–X–Gly	410	42	10.2
Ar–Gly–X'	384	4	1.0

[a] Ar=Phe, Tyr; X=any residue; X'=any residue except Pro.

the $(i + 2)$ amine almost twice as frequently as phenylalanine (fractions of 3.4 and 1.8 per cent, respectively), and glycine donates far more frequently than any other residue (10.2 per cent of all tripeptides Ar—X—Gly). On the other hand, the interaction is only formed rarely if glycine is inserted between the two residues involved.

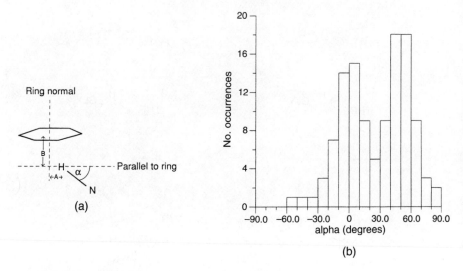

Fig. 5.20. (a) Definition of the angle α that can be used to distinguish stacked and hydrogen bonded aromatic–amine interactions. (b) Bimodal distribution of the angle α for 115 aromatic–$(i + 2)$ amine interactions in proteins (from Worth and Wade 1995).

The hydrogen bonded and stacked arrangements are distinguished by the angle α as defined in Fig. 5.20(a). For stacked interactions, α is around 0° and for aromatic hydrogen bonds, it is ideally 90°. The α distribution for the 115 interactions under study is shown in Fig. 5.20(b). The distribution shows two peaks of similar areas, one centred at 0° for ideally stacked geometries and one centred around 50° representing distorted aromatic hydrogen bonds. Ideal geometries with $\alpha = 90°$ cannot be formed because of the steric constraints by the covalent chain between donor and acceptor. There are also intermediate cases which can be classified as neither stacked nor hydrogen bonded. Assuming that about half of the aromatic–$(i + 2)$ amine interactions represent hydrogen bonds, this would mean that slightly more than 1 per cent of the aromatic side-chains in proteins are involved in this kind of interaction. Because most proteins contain far less than 100 phenylalanine and tyrosine residues, the average number of aromatic–$(i + 2)$ amine hydrogen bonds per protein is less than one. In the same data set, 286 aromatic–$(i + 1)$ amine interactions are found (involving 6.4 per cent of all Phe and Tyr side-chains), which are typically stacked.

To complement their statistical survey, Worth and Wade performed extensive computations on the aromatic–$(i + 2)$ amine interaction. For the strongly simplified model of free mutual orientation of donor and acceptor in vacuum, the potential energy surface shown in Fig. 3.6 is obtained (Section 3.1.3). In the absence of other potential acceptors, the hydrogen bonded arrangement is clearly favoured over the stacked one. If solvent interactions are included

Table 5.6 Geometries of aromatic–$(i + 1)$ and $(i + 2)$ amine hydrogen bonds in peptides (Steiner 1998d)

	Tyr-Tyr-Leu	Tyr-Tyr-Phe	L-Phe-Gly-Gly-D-Phe
Reference	(a)	(b)	(c)
Type	$i + 1$	$i + 1$	$i + 2$
Acceptor	Tyr	Tyr	Phe
d(C) range (Å)	2.41–3.73	2.47–4.16	2.63–3.29
d(C) spread (Å)	1.32	1.75	0.66
D(C) range (Å)	3.14–4.75	3.25–5.14	3.59–4.18
D(C) spread (Å)	1.61	1.89	0.59
d(M) (Å)	2.82	3.13	2.65
D(M) (Å)	3.78	4.07	3.74
θ(M) (°)	156	153	142

(a) Steiner 1998d; (b) Steiner et $al.$ 1998b; (c) Fujii et $al.$ 1987.

in the calculation, the situation changes and the stacked arrangement becomes favoured because it allows formation of an additional 'external' hydrogen bond. This parallels the observations of Mitchell et $al.$ (1994) on aromatic–amine interactions between side-chains. On the whole, it may be concluded that the aromatic–$(i + 2)$ amine interaction in proteins is of some importance but not as a driving force for folding. The importance lies in its corrective role, allowing an otherwhise vacant hydrogen bond donor capacity to be fulfilled. This provides a mechanism that allows a protein to miss out strong hydrogen bonds without a large enthalpic disadvantage, because they can be substituted by a weaker but still significantly bonding interaction.

Aromatic–$(i + 1)$ and $(i + 2)$ amine hydrogen bonds also occur occasionally in peptides where their geometry can be determined very accurately. Examples for both kinds of hydrogen bonds are shown in Figs 5.21 and 5.22, respectively, and geometrical data are given in Table 5.6. In the monohydrated tripeptide Tyr-Tyr-Leu, the two tyrosine side-chains are oriented in the same direction so that the N—H donor of the peptide unit linking the two residues is shielded from intermolecular contacts. Instead, the N—H group forms a short contact with one of the aromatic groups (Fig. 5.21(a); Steiner 1998d). The hydrogen bond geometry is best seen in projections onto and perpendicular to the aromatic plane, Fig. 5.21(b). Owing to the steric constraints within the arrangement, the hydrogen bond is necessarily off-centred. The hydrogen bond represents a six-membered intramolecular ring which is topologically analogous to the six-membered rings that are frequently formed by intramolecular O—H$\cdots$Ph hydrogen bonds (Fig. 3.21 in Section 3.1.3.4). Despite this distorted geometry, the interaction is still well inside the bonding region of the potential energy surface obtained by Worth and Wade (Fig. 3.6 in Section 3.1.3).

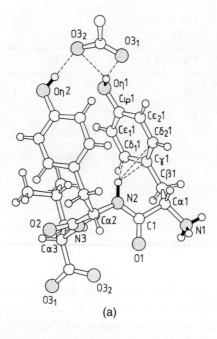

(a)

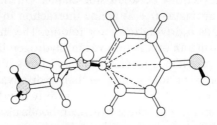

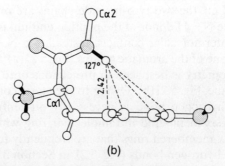

(b)

Fig. 5.21. Aromatic–$(i + 1)$ amine hydrogen bond in L-Tyr-L-Tyr-L-Leu monohydrate. (a) Overall view showing the molecular conformation. (b) Projection onto and parallel to the aromatic acceptor (from Steiner 1998d).

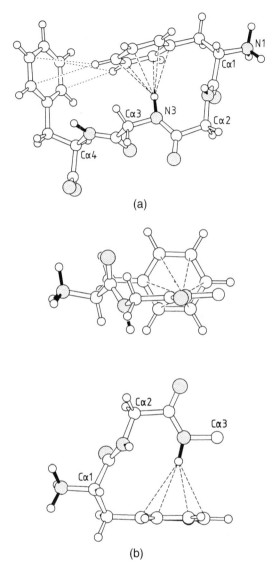

(a)

(b)

Fig. 5.22. Aromatic–$(i + 2)$ amine hydrogen bond in L-Phe-Gly-Gly-D-Phe trihydrate. (a) Overall view. (b) Projection onto and perpendicular to the aromatic acceptor (from Steiner 1998d, using the structure of Fujii *et al.* 1987).

An example for an aromatic–$(i + 2)$ amine hydrogen bond is found in the crystal structure of the trihydrated tetrapeptide L-Phe-Gly-Gly-D-Phe (Fig. 5.22; Fujii *et al.* 1987). In addition to the aromatic hydrogen bond, the peptide conformation is also stabilized by a Ph–Ph herringbone interaction. The aromatic hydrogen bond can adjust to a much better geometry than the hydrogen bond of the $(i + 1)$ kind.

Aromatic main-chain to side-chain hydrogen bonds in a series of synthetic peptides involving non-standard residues have been characterized by Toniolo and co-workers (Crisma *et al.* 1997); an example has already been shown in a different context in Fig. 3.25(a) (Section 3.1.3.4). Finally, it should be mentioned that aromatic $(i + 2)$ amine hydrogen bonds are also formed by small peptides in solution. This has been shown in circular dichroism and ^{1}H NMR measurements on BPTI fragments (Kemmink and Creighton 1993).

5.2.4 Protein–ligand interactions

The role of weak hydrogen bonds in protein folding is probably supportive and/or corrective with the key roles played by the stronger interactions. These are hydrogen bonds between the polar groups and hydrophobic interactions between the apolar groups. In molecular recognition processes involving proteins, on the other hand, several examples are known where weak hydrogen bonds play key roles and the recognition patterns cannot be rationalized without taking them into account.

Classical examples for weak polar interactions in protein–drug binding have been provided by Perutz *et al.* (1986) in a series of hemoglobin–drug complexes. Figure 5.23 shows the structure of the drug bezafibrate and its intermolecular contacts as it is bonded in the central cavity of hemoglobin. There are a number of $C-H\cdots O$ hydrogen bonds, an $N-H\cdots Ph$ hydrogen bond from an asparagine side-chain (shortest $N\cdots C$ distance 3.1Å) and also $C-H\cdots Ph$ interactions of weakly polar $C-H$ groups. In complexes with chlorinated drugs, there are systematic $C-H\cdots Cl$ interactions. The authors used relatively cautious terms to characterize the intermolecular interactions, but it was clearly concluded that the drug binding is determined by a concert of forces which are very different in strengths. In a later review, the $N-H\cdots Ph$ interactions were explicitly termed hydrogen bonds (Perutz 1993).

The example of ligand binding of the serine proteinase thrombin has already been mentioned in the context of drug design, Section 4.4.2. In this work, Engh *et al.* (1996) determined the structures of a series of thrombin–drug complexes and found intricate patterns are composed of $N-H\cdots O$, $O-H\cdots O$, $C-H\cdots O$ and $O-H\cdots Ph$ hydrogen bonds. In the example shown in Fig. 4.30, a water molecule forms short contacts to the aromatic face of a tyrosine side-chain of the enzyme and to the edge of a protonated pyridine ring of the ligand. All these interactions have to be taken into account if the mode of drug binding is to be understood.

Two further examples of protein complexes involving aromatic hydrogen bonds are shown in Fig. 5.24. The phosphotyrosine binding in the SH2 domain of the v-*src* oncogene involves a complex network of strong and weak hydrogen bonds, Fig. 5.24(a). In particular, there are $N^+-H\cdots O^-$ and $O/N-H\cdots O^-$

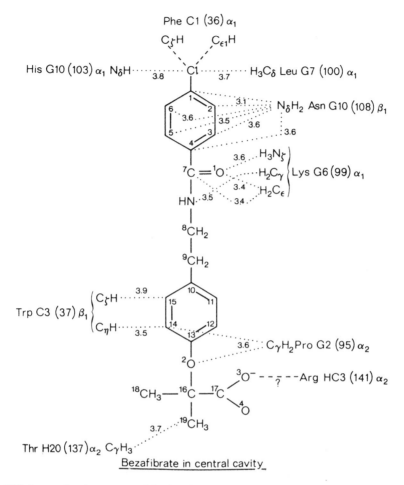

Phe C1 (36) α_1

Bezafibrate in central cavity

Fig. 5.23. Intermolecular contacts of the drug bezafibrate as bonded in the central cavity of hemoglobin (from Perutz *et al.* 1986).

hydrogen bonds to the phosphate group and a pair of $N^+ - H \cdots Ph$ hydrogen bonds to the aromatic ring. The aromatic ring is effectively sandwiched between two charged hydrogen bond donors and this would be a very unlikely arrangement from a conventional viewpoint (Waksman *et al.* 1992). In the complex of the tetracycline repressor with tetracycline, an even more complicated system of directional and non-directional intermolecular interactions is formed, involving Mg^{2+} ion coordination, $N - H \cdots O$ hydrogen bonds involving charged and uncharged groups, $C - H \cdots Ph$ interactions, van der Waals contacts and also an $O - H \cdots Ph$ hydrogen bond (Fig. 5.24(b); Hinrichs *et al.* 1994). The latter interaction seems to be of importance because the amino acid involved, Phe 86, is conserved in all known tetracycline repressor sequences.

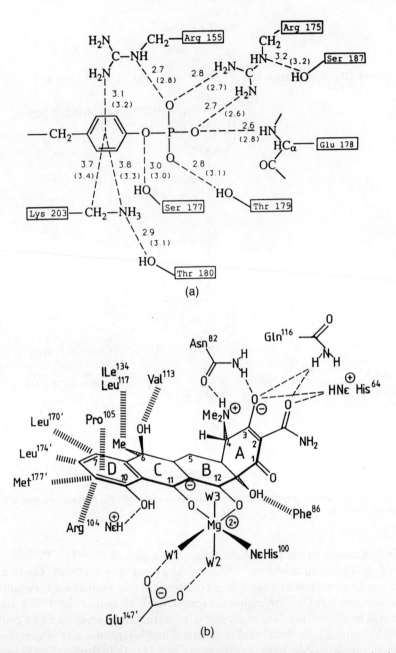

Fig. 5.24. Protein–ligand binding involving aromatic hydrogen bonds. (a) Phosphotyrosine in the binding site of the SH2 domain of v-src. The aromatic ring is sandwiched between two charged amino donors (from Waksman *et al.* 1992). (b) Binding of tetracycline to the tetracycline repressor (from Hinrichs *et al.* 1994).

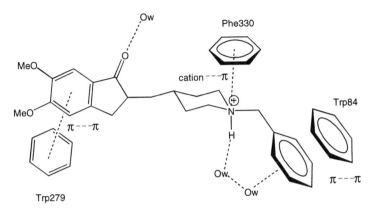

Fig. 5.25. Binding mode of the anti-Alzheimer drug E2020 in the active site gorge of acetyl-cholinesterase from *Torpedo californica*, dominated by π-stacking and cation–π interactions. There is also a water contact suggestive of an aromatic hydrogen bond (Kryger *et al.* 1998).

Molecular recognition in the active site gorge of acetylcholinesterase (AChE) is to an unusually large degree governed by non-conventional inter-molecular interactions. The substrate is a cationic molecule with a quarternary ammonium group. Unexpectedly, the charged group is not recognized and bonded by anions but by cation–π interactions involving aromatic amino acid side-chains (Sussman *et al.* 1991, also see Section 3.1.3.6). The binding mode of the AChE inhibitor E2020 has been described in detail by Kryger *et al.* (1998), Fig. 5.25. There are no direct hydrogen bonds between the enzyme and the inhibitor but a number of phenyl stacking interactions and a cation–π interaction of the E2020 tertiary ammonium group. In addition, there are O—H···O and O—H···Ph hydrogen bonds with water molecules which are also included in the active site gorge. A comparison of a number of inhibitor complexes shows binding dominated by non-conventional interac-tions as a common feature. Water molecules are always present in the active site region.

An interesting example of C—H···O hydrogen bonding in an engineered protein cavity has been reported by Goodin and co-workers. In cytochrome *c* peroxidase, an internal cavity was created by a Trp→Gly mutation, in which the large side-chain of tryptophan is deleted. This cavity binds a number of cationic heterocyclic compounds. In the native enzyme, the Trp N_ε—H donor forms a conventional hydrogen bond to an aspartate O atom (Fig. 5.26(a)). If the mutated enzyme binds imidazolium cations in the cavity, analogous N^-—H···O^-(Asp) hydrogen bonds are formed, which apparently determine the cation binding specificity (Fig. 5.26(b); Fitzgerald *et al.* 1994). Interestingly, the cations 2,3,4-trimethylthiazole and 3,4,5-trimethylthiazole bind to the cavity with essentially the same geometries as imidazolium ions and with C—H···O^- hydrogen bonds replacing the N^+—H···O^- salt bridges (Fig.

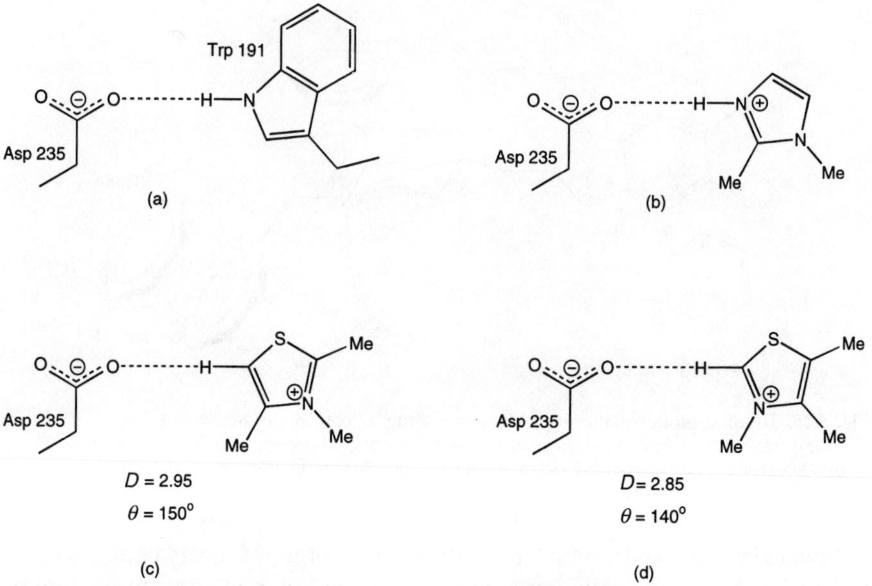

Fig. 5.26. Ligand binding in an engineered cavity of cytochrome c peroxidase. The cavity is obtained by mutation Trp191Gly, in which the Trp side-chain is deleted. (a) $N_{\varepsilon}-H\cdots O$ hydrogen bond in the native enzyme. (b) $N-H\cdots O$ hydrogen bond formed by 1,2-dimethylimidazole in the mutant cavity (Fitzgerald *et al.* 1994). (c, d) $C-H\cdots O$ hydrogen bonds of 2,3,4-trimethylthiazole and 3,4,5-trimethylthiazole in the mutant cavity (Musah *et al.* 1997).

5.26(c,d); Musah *et al.* 1997). This is a very persuasive example for isofunctional substitution of strong hydrogen bonds by $C-H\cdots O$ interactions.

A most important case of protein–ligand binding is the site-specific recognition of DNA by DNA-binding proteins. In this context, Zhurkin and co-workers have analysed the contact geometries in 43 crystal structures of protein–DNA complexes (Mandel-Gutfreund *et al.* 1998). A large number of short $C-H\cdots O$ contacts in the DNA major groove are found involving the $C(5)-H$ group of cytosine and the methyl group of thymine. The distances of these contacts are comparable with other typical $C-H\cdots O$ hydrogen bonds and it was concluded that these interactions contribute to the specificity of the recognition.

In an important paper, Berman and co-workers have shown that an aromatic hydrogen bond can substitute for a conventional hydrogen bond in sequence-specific protein–DNA recognition (Parkinson *et al.* 1996). *Escherichia coli* catabolite activator protein (CAP) binds to DNA specifically at a 22 base pair site. Upon binding, a hydrogen bond is formed between the carboxylate side-chain of CAP Glu181 and a cytosine N^4-H donor of a particular G-C base pair. This interaction cannot be formed with the alternative base pairs C-G, T-A and A-T and thereby contributes to the specificity of the

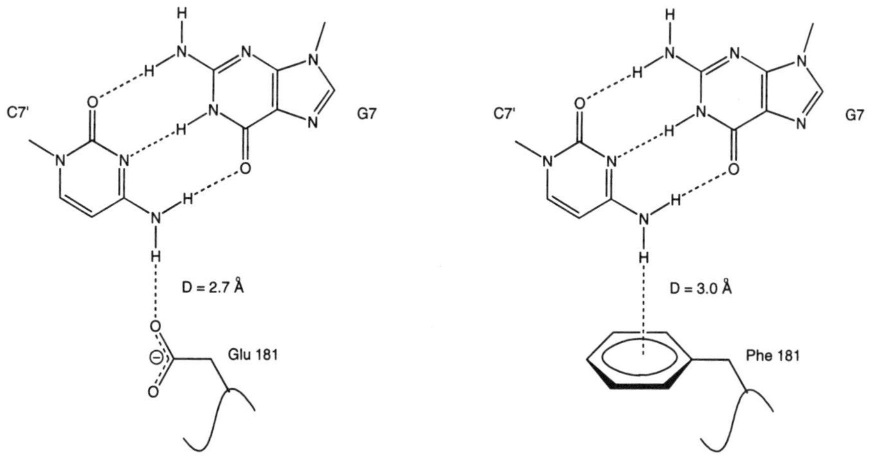

Fig. 5.27. Recognition of a 22 base pair DNA site by the CAP protein. Shown is a hydrogen bond responsible for the specificity towards a particular G-C pair. (a) Native CAP. (b) Glu181Phe mutant of CAP, where the N—H···O⁻ is replaced by an isofunctional N—H···Ph hydrogen bond (after Parkinson *et al.* 1996).

DNA fragment recognition. Truncation of the Glu 181 side-chain reduces the binding affinity and eliminates the specificity for the G-C pair. If the carboxylate side-chain is replaced by the aromatic side-chains of Phe, Tyr and Trp, the binding affinity is not reduced and the specificity for G-C recognition is maintained. In the crystal structures of CAP and the Glu181Phe mutant complexed with the DNA fragment that is recognized, it is actually observed that the N—H···O⁻ hydrogen bond formed by native CAP is replaced by an isofunctional N—H···Ph hydrogen bond in the Phe181 mutant (Fig. 5.27). The isofunctional nature is proven by good binding affinity and specificity of the mutant *in vivo*.

In summary, the importance of weak hydrogen bonds in protein–ligand recognition has been impressively demonstrated. N/O—H···Ph and C—H···O hydrogen bonds are important in a variety of diverse events such as enzyme–substrate binding and site-specific protein–DNA recognition. In some of these cases, the relevance of weak hydrogen bonds has been shown not only by structural studies but also by mutation experiments in which particular hydrogen bonds are modified or deleted.

5.2.5 *Enzymatic activity*

The roles of weak hydrogen bonds in enzymatic activity have not been rigorously studied. Nevertheless, two exemplary cases have been reported, one involving C—H···O and the other involving O—H···Ph hydrogen bonding.

A supportive but functional role of $(His)C_\varepsilon - H \cdots O = C$ hydrogen bonds in the enzymatic action of serine hydrolases has been suggested by Derewenda *et al.* (1994). In 22 crystal structures of serine hydrolases in the resolution range 1.2–3.1 Å (typically around 2.0 Å), it was observed that the active site histidine residue almost invariably forms a short $C_\varepsilon - H \cdots O$ contact to a backbone carbonyl group. In 17 of the structures (77 per cent), the contact has a distance d shorter than 2.6 Å with a mean value of 2.3 Å, and in the five other structures there is such a contact with d in the range 2.8–3.1 Å. Depending on the specific fold of the enzyme, the acceptor carbonyl group may be from diverse parts of the protein primary sequence. Examples for six serine hydrolases are shown in Fig. 5.28 and an independent example as described by Engh *et al.* (1996) has been shown in Fig. 4.30. For comparison, only 15 per cent of the non-active site histidine residues form a short $C_\varepsilon - H \cdots O$ interaction. The $(His)C_\varepsilon - H \cdots O = C$ hydrogen bond is clearly a conserved structural feature of all serine hydrolases and this must have been developed and maintained in evolution. This evolutionary effort suggests some function in the enzymatic process, which is performed by the catalytic triad Asp/His/Ser. A simple function might be that the hydrogen bond helps stabilize the orientation of the histidine group. As a more sophisticated function, it was proposed that the $C_\varepsilon - H \cdots O = C$ hydrogen bond affects the charge distribution within the imidazolium group so as to enhance the electronegative character of N_ε necessary for proton abstraction, **222**.

An influence of a second-sphere $O - H \cdots Ph$ hydrogen bond on a catalytic process has been described for glutathione (GSH) S-transferase by Armstrong and co-workers (Liu *et al.* 1993; Xiao *et al.* 1996). This class of enzymes catalyses the general reaction $GSH + R - X \rightarrow GSR + XH$. One of the key points in the catalytic mechanism is that the pK_a of GSH in the binary complex with the enzyme is almost three pK_a units lower than that of GSH in aqueous solution. In the crystal structure of a particular GSH S-transferase in complex with GSH, the hydroxyl group of a threonine side-chain forms an aromatic hydrogen bond with the face of an active site tyrosine residue, which in turn is hydrogen bonded with the glutathione thiolate ion, **223**. It is assumed that the aromatic hydrogen bond strengthens by cooperativity the functionally important hydrogen bond to the thiolate ion. This is supported by mutations that delete the aromatic hydrogen bond such as Thr→Val and Thr→Ala. Both these mutations increase the apparent pK_a of the enzyme-bound GSH by about 0.7 units. This means that the stability of the thiolate ion is reduced upon deletion of the second-sphere aromatic hydrogen bond. The functional role of this hydrogen bond was confirmed by Dietze *et al.* (1996), who incorporated an $O - H \cdots Ph$ hydrogen bond into a related enzyme that is lacking in it. In native rat A1-1 glutathione S-transferase, no aromatic hydrogen bond is formed to the active site tyrosine. Opposite the aromatic face, there is a phenylalanine side-chain, forming a $Ph \cdots Ph$ herringbone interaction. By the mutation Phe→Tyr, an aromatic hydrogen bond is created, **224**, which is of the same

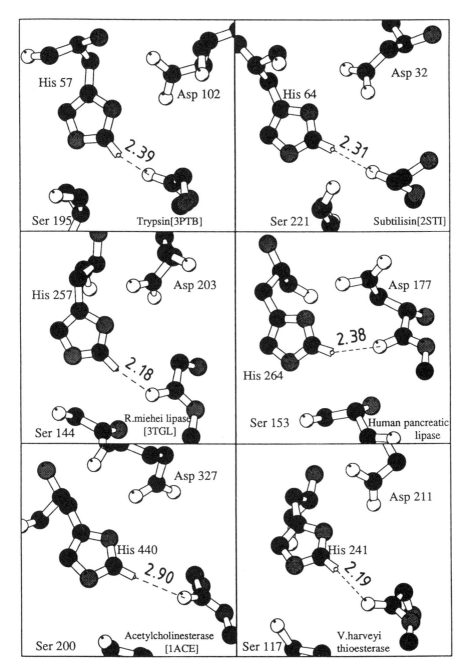

Fig. 5.28. Six examples of $(His)C_\epsilon\!-\!H\cdots O\!=\!C$ hydrogen bonds donated by the active site histidine side-chain of serine hydrolases (from Derewenda *et al*. 1994).

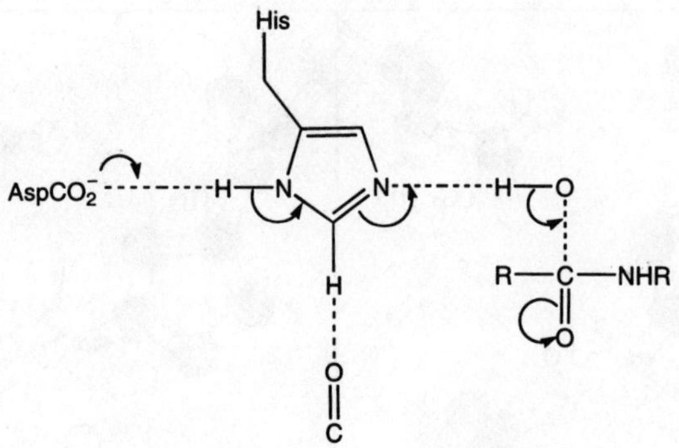

222

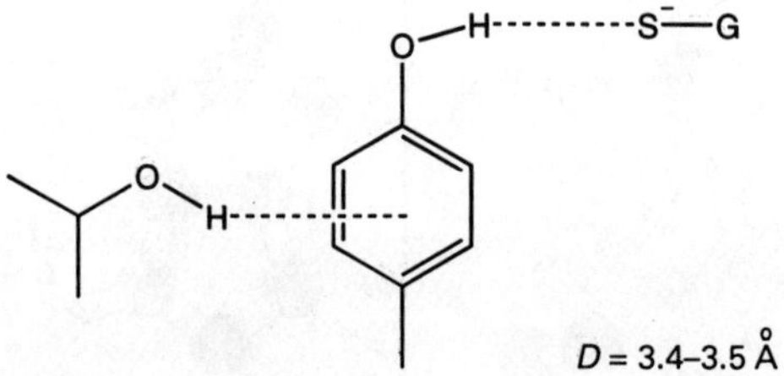

$D = 3.4–3.5$ Å

223

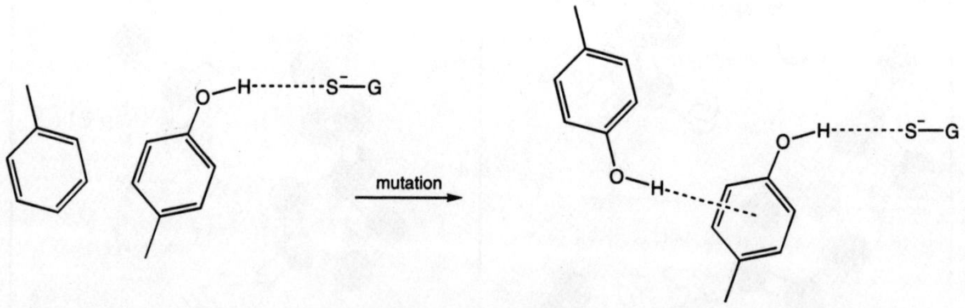

224

kind as in **223**. This mutation lowers the pK_a of the enzyme-bound GSH by 0.5 units, that is it stabilizes the ion significantly. This is the reverse effect to removal of the aromatic hydrogen bond from **223**.

Although the experimental evidence discussed here is only limited, it pro-

Table 5.7 Nomenclature for bases, nucleosides and nucleotides (abbreviations are given in parentheses)

Base	Nucleoside	Nucleotide
RNA constituents		
Uracil	Uridine (U)	Uridylic acid (pU or 5′-UMP)
Cytosine	Cytidine (C)	Cytidylic acid (pC or 5′-CMP)
Adenine	Adenosine (A)	Adenylic acid (pA or 5′-AMP)
Guanine	Guanosine (G)	Guanylic acid (pG or 5′-GMP)
DNA constituents		
Thymine	Deoxythymidine (dT)	Deoxythymidylic acid (pdT or 5′-dTMP)
Cytosine	Deoxycytidine (dC)	Deoxycytidylic acid (pdC or 5′-dCMP)
Adenine	Deoxyadenosine (dA)	Deoxyadenylic acid (pdA or 5′-dAMP)
Guanine	Deoxyguanosine (dG)	Deoxyguanylic acid (pdG or 5′-dGMP)

vides evidence that weak hydrogen bonding can play a role in enzymatic reactions. As might be expected, weak hydrogen bonds are not the driving force of enzymatic reactions but they are capable of modulating the bonding situation in catalytic sites to a significant degree. These fine-tuning mechanisms can sufficiently influence the catalytic properties, to be built into the catalytic site architectures by natural evolution.

5.3 Nucleic acids

The overall properties of nucleic acid structure have been reviewed repeatedly, including a comprehensive book by Saenger (1984). Nonetheless, their chemical structure and some conformational characteristics are described here briefly in the context of weak hydrogen bonding. The polymeric chains of deoxyribonucleic acid (DNA) and ribonucleic acid (RNA) are built of nucleotides, which consist of three distinct moieties: a purine or pyrimidine base, a ribose or 2′-deoxyribose and a monophosphate group. Polymerization occurs by formation of phosphodiester linkages between nucleotides. The structure of the bases and the standard atomic numbering scheme of purine and pyrimidine are shown in Fig. 5.29, and the structure of the nucleotides and their polymers is shown in Fig. 5.30. The nomenclature of the main nucleic acid constituents is given in Table 5.7.

The conformation of nucleotides is determined by the sugar pucker and the orientations of the base and the phosphate, respectively, with respect to the ribose. For the base, there are two distinct modes of orientation, called *anti* and *syn*, as shown in Fig. 5.31(a). In the slightly preferred *anti* conformation, the bulky part of the base is oriented away from the sugar moiety, thereby avoiding steric conflicts. The position of the O(5′) atom with respect to the

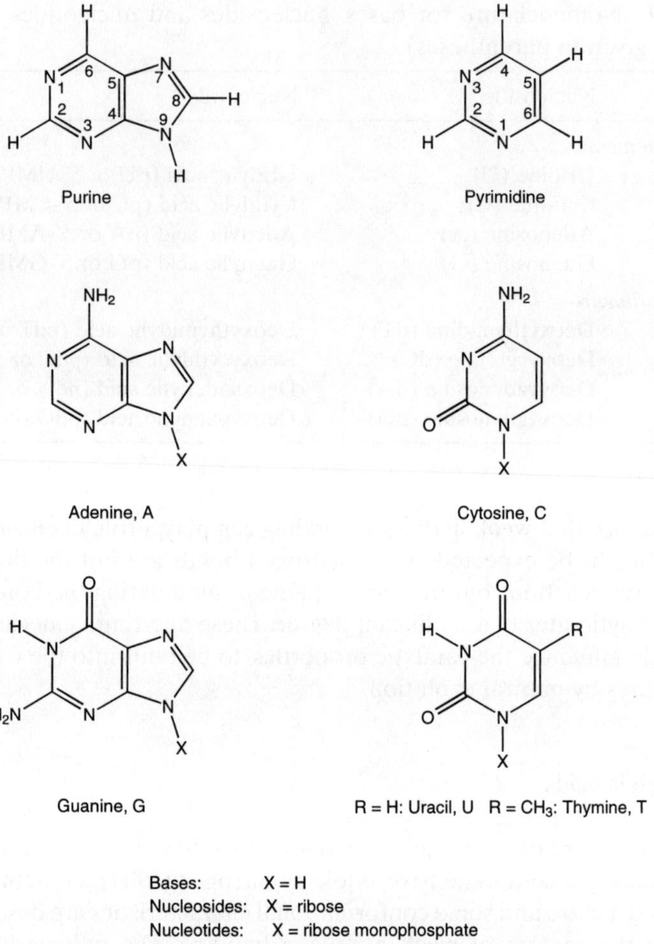

Fig. 5.29. Constitution and atomic numbering scheme of purine and pyrimidine, the DNA and RNA bases adenine, guanine, cytosine, uracil and thymine, and definition of the terms *nucleoside* and *nucleotide*.

ribose is allowed in three staggered conformations (Fig. 5.31(b)). The preferred +*sc* conformation positions O(5′) 'over' the ribose and brings it close to the base. Of the many possible ribose puckering modes, only the two twist conformations C(3′)-*endo* and C(2′)-*endo* are generally observed (Fig. 5.31(c)).

Polymeric DNA and RNA differ in two main points. The first is in the occurrence of the bases; uracil occurs only in RNA, whereas thymine (5-methyluracil) occurs in DNA. As for the second difference, the RNA ribose has a hydroxyl group whereas the DNA ribose has none. This apparently small

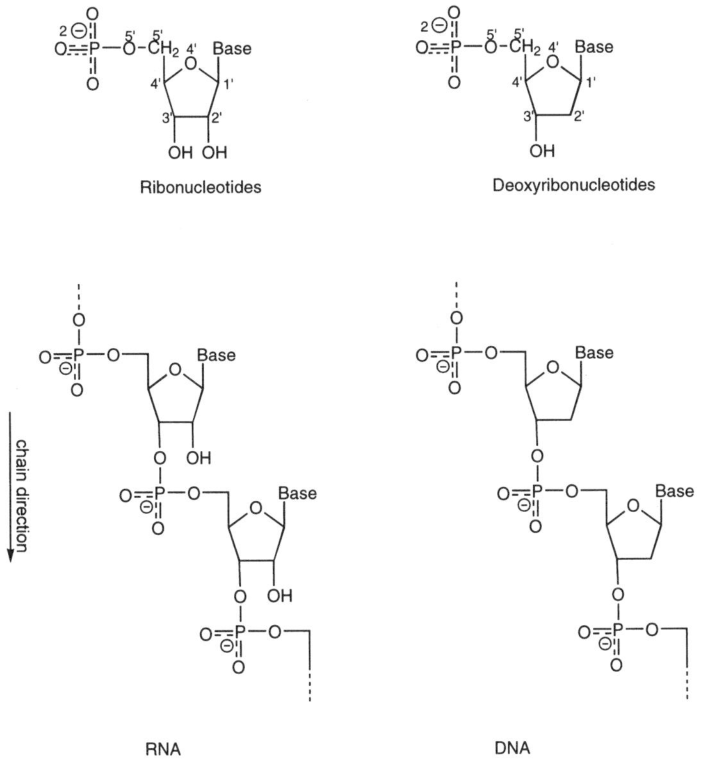

Fig. 5.30. Constitution and atomic numbering scheme of ribonucleotides and deoxyribonucleotides, and of RNA and DNA.

difference in the ribose unit leads to very large differences in the structures of DNA and RNA: whereas DNA is limited to helical structures, RNA can adopt a large number of helical and folded conformations. Furthermore, RNA polymers can contain a number of modified bases with different hydrogen bond properties, only a small selection of which are shown in Fig. 5.32.

Nucleic acids contain a variety of strong and weak hydrogen bonding functional groups. There are strong $N-H$ and $O-H$ donors, highly activated $C-H$ donors of the bases, moderately activated $C-H$ donors of the ribose and the very weakly activated methyl donor of thymine. As acceptors, there are the very strong phosphate groups, the N_ε atom and $C=O$ acceptors of the bases, and the ribose O atoms. Nucleic acid structure is therefore dominated by strong hydrogen bonds and base-stacking interactions. However, numerous $C-H \cdots O$ hydrogen bonds of different strengths and functions also do occur (Berger and Egli 1997; Wahl and Sundaralingam 1997). As will be shown in Section 5.5, weak hydrogen bonds with water molecules are also of importance.

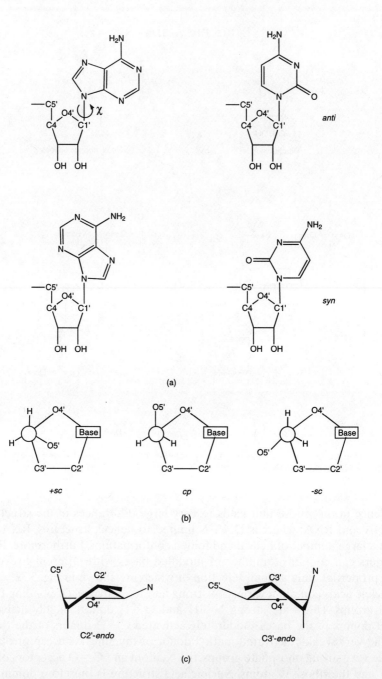

Fig. 5.31. Parameters defining nucleotide conformations. (a) The two main orientations of the base with respect to the ribose, as defined by the torsion angle $\chi = O(4') - C(1') - N(9/1) - C(4/2)$. (b) The three main orientations of the $C(5') - O(5')$ bond with respect to the ribose, as defined by the torsion angle $\gamma = O(5') - C(5') - C(4') - C(3')$. (c) The two preferred modes of ribose puckering: the twist conformations C(2')-*endo* and C(3')-*endo*. Many other ribose puckering modes are possible and do occur in certain nucleosides and nucleotides (following Saenger 1984).

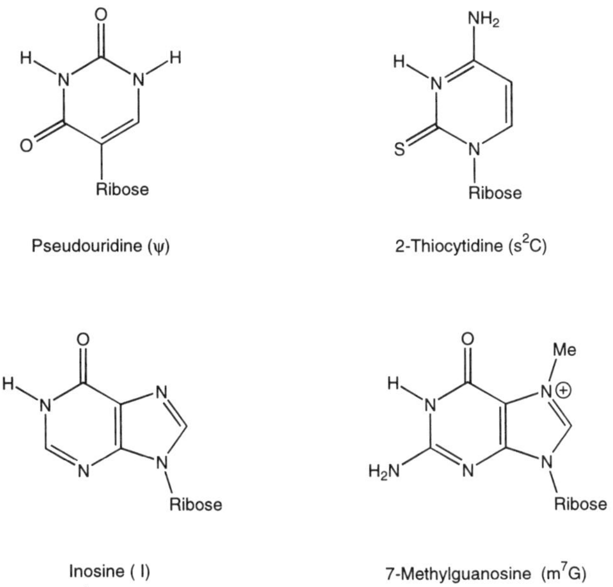

Pseudouridine (ψ)

2-Thiocytidine (s²C)

Inosine (I)

7-Methylguanosine (m⁷G)

Fig. 5.32. Small selection of the modified bases that occur in RNA but not in DNA. Base modifications may also include addition of larger groups which contain 'new' hydrogen bonding functional groups like COOH and $C(sp^3)OH$. A more complete list of modified bases has been compiled by Saenger (1984).

5.3.1 *Nucleic acid constituents*

The potential of purine and pyrimidine bases to donate $C-H\cdots O$ hydrogen bonds was recognized by Sutor (1962, 1963). The heterocyclic $C-H$ groups are activated by neighbouring N atoms, in particular purine $C(2)-H$ and $C(8)-H$ and pyrimidine $C(2)-H$, all of which neighbour two N atoms. These $C-H$ groups can donate hydrogen bonds to O atom acceptors with distances d down to about 2.1 Å. If the bases are protonated, the hydrogen bonds become even shorter. Contacts with such geometries would be taken as hydrogen bonds even if conservative van der Waals cut-off criteria are used. We have discussed caffeine monohydrate (Fig. 2.3(a)), uracil (Fig. 2.3(b)), 1-methylthymine (Fig. 2.9(d)) and hypoxanthine nitrate monohydrate (Fig. 2.9(e)) in Chapter 2. All these structures were determined at high resolution, leaving no doubt about the hydrogen bond geometries. In some cases, the functions of $N-H$ and $C-H$ donors are identical; in uracil for instance, one of the $C=O$ groups accepts two $N-H\cdots O$ and the other one accepts two $C-H\cdots O$ hydrogen bonds.

$C-H\cdots O$ hydrogen bonding by the acidic $C-H$ groups of the bases has long been invoked for nucleosides and nucleotides. Shefter and Trueblood (1965) found that in the Ba salt of uridine-5′-phosphate, a short intramolecu-

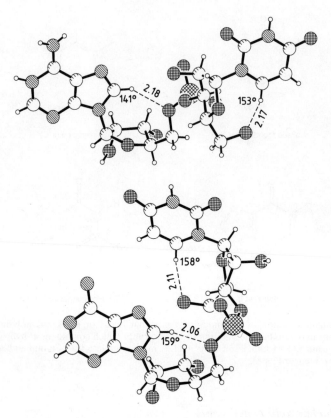

Fig. 5.33. Intranucleotide C(6/8)—H···O(5′) hydrogen bonding in the dinucleoside monosphosphate UpA hemihydrate There are two molecules of different overall conformation, which are drawn here in similar projections with respect to the adeninosine moiety (Sussman *et al.* 1972). Geometries are normalized.

lar contact with $D = 3.17\,\text{Å}$ is formed from the uridine C(6) to O(5′) of the sugar–phosphate link, and because there are related contacts in other nucleotides, they concluded that this interaction is a hydrogen bond and is of more general importance in nucleic acids. Analogous base-to-backbone hydrogen bonds are also formed by the purine C(8)—H group. Very appealing examples were discussed for the crystal structure of uridylyl-(3′,5′)-adenosine monosphosphate (UpA), independently determined by Sussman *et al.* (1972) and Rubin *et al.* (1972). In the asymmetric unit, there are two UpA molecules of different overall conformation but with very similar conformations of the nucleoside moieties. All these moieties exhibit short intramolecular hydrogen bonds of the same kind, C(8)—H···O(5′) for the adenine base and C(6)—H···O(5′) for the uracil base (Fig. 5.33). A modified nucleoside with an analogous hydrogen bond to a carboxylic acid acceptor in the position of O(5′) has already been shown in Fig. 2.35(b) (Takusagawa *et al.* 1979).

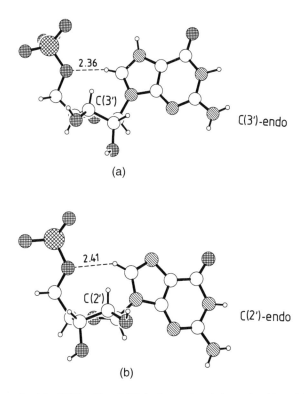

(a)

(b)

Fig. 5.34. Intranucleotide $C(6/8)-H\cdots O(5')$ hydrogen bonds in nucleotides with the ribose in $C(3')$-*endo* and $C(2')$-*endo* conformation. The sugar pucker has a major influence on the orientation of $O(3')$, at which polymerization occurs. Nucleotides in the $C(3')$-*endo* form are observed in A-type polynucleotide helices, and those in the $C(2')$-*endo* form in B-type helices. (a) Guanosine-5'-monophosphate trihydrate (Emerson and Sundaralingam 1980). (b) Disodium guanosine-5'-phosphate heptahydrate (Barnes and Hawkinson 1982). Geometries are normalized.

These short $C(6/8)-H\cdots O(5')$ contacts are certainly not forced interactions because they could easily be avoided by the flexible molecules.

It is worthwhile to look closer at the nucleotide conformations for which the intranucleotide $C(6/8)-H\cdots O(5')$ hydrogen bonds are formed. Obviously, the base must be oriented *trans* with respect to the ribose and the $O(5')$-orientation must be in the *+sc* range. The ribose can adopt either of the main conformations, $C(3')$-*endo* and $C(2')$-*endo*, as is shown for two examples of mononucleotides in Fig. 5.34. These two conformations are generally preferred for nucleotides in RNA and DNA, with $C(3')$-*endo* ribose pucker in A-type helices and $C(2')$-*endo* in B-type helices ('rigid nucleotide' concept of Yathindra and Sundaralingam 1973; Sundaralingam 1974). In consequence, the $C(6/8)-H\cdots O(5')$ hydrogen bonds are abundant in polymeric nucleic acids. Using semiempirical molecular orbital calculations, Amidon *et al.* (1974) estimated the bond energy of these hydrogen bonds to be around -2 kcal/mol.

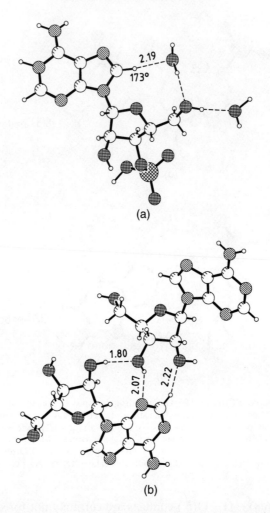

Fig. 5.35. Intermolecular C—H···O hydrogen bonding by adenine bases. (a) With a C(8)—H donor and a water acceptor in adenosine-3′-phosphate dihydrate (Sundaralingam 1966). (b) With a C(2)—H donor and a hydroxyl acceptor in adenosine (Lai and Marsh 1972). Geometries are normalized.

Although these calculations are somewhat outdated today, they fit into the frame of energy values obtained by modern calculations for C—H···O hydrogen bonds formed by relatively acidic C—H groups.

Acidic C—H groups of the bases can engage in hydrogen bonding in many other ways, depending on the molecular conformation and the local environment. Two examples of adenine bases forming short intermolecular C—H···O hydrogen bonds are shown in Fig. 5.35, one with a C(2)—H and one with a C(8)—H···O donor (Sundaralingam 1966, Lai and Marsh 1972, respectively).

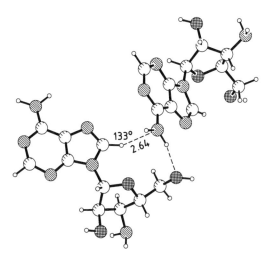

Fig. 5.36. Intermolecular hydrogen bond C(8)—H···N(6) in 3'-amino-3'-deoxyadenosine; discussed by Luisi *et al.* (1998) based on the structure of Sheldrick and Morr (1980). The geometry is normalized.

A more unusual intermolecular hydrogen bond C(8)—H···N(6) occurs in 3'-amino-3'-deoxyadenosine, Fig. 5.36 (discussed by Luisi *et al.* 1998 based on the structure of Sheldrick and Morr 1980). This example is of interest because the adenine N(6)H$_2$ group is often considered to be perfectly planar and incapable of accepting hydrogen bonds. Finally, an intramolecular C(5)—H···O=C hydrogen bond was found in the modified nucleoside N_4-acetylcytidine, Fig. 5.37 (Parthasarathy *et al.* 1974). This intramolecular C(5)—H···O interaction is of functional relevance; because the acetyl group is stabilized in an orientation distal to N(3), Watson–Crick base pairing is possible. All these examples in small nucleic acid constituents suggest that a large variety of C—H···O/N hydrogen bond configurations should be possible also in the polymeric nucleic acids.

5.3.2 *Polymeric DNA and RNA*

For oligomeric and polymeric nucleic acids, there are similar crystallographic resolution problems as for proteins. Hydrogen bonds are usually interpreted not on the basis of experimental evidence but from theoretical H atom positions, and the obtained geometries have large uncertainties. The discussion of weak hydrogen bonding effects is therefore relatively recent.

Several functional classes of weak hydrogen bonds can be distinguished in the stabilization of nucleic acids. The cases of base–backbone, base–base and backbone–backbone hydrogen bonds will be discussed in this section. Weak hydrogen bonds in the hydration of nucleic acids will be discussed in Section

5.5. Two studies on weak hydrogen bonding in protein–DNA recognition have already been mentioned in Section 5.2.4 (Fig. 5.27).

5.3.2.1   *Base–backbone C—H⋯O hydrogen bonds*

The shortest C—H⋯O hydrogen bonds in nucleic acids are the intranucleotide hydrogen bonds from purine C(8)—H and pyrimidine C(6)—H to O(5′). These stabilize the standard nucleotide conformations (Figs 5.33 and 5.34). Nucleotides adopt these overall conformations in most forms of RNA and DNA but considerable variations of the relevant torsion angles are allowed (Saenger 1984). The distance and angular characteristics of C(6/8)—H⋯O(5′) hydrogen bonds in RNA and A-DNA structures have been determined in a database analysis by Auffinger and Westhof (1997), Fig. 5.38. Because the nucleotides adopt the C(3′)-*endo* form in both these forms of nucleic acids, the pictures for RNA and A-DNA are very similar. The variation of individual hydrogen bond geometries is considerable but the clustering of the distances d appears to be in the region 2.1–2.6 Å. The angles θ are scattered between about 120° and 180°, with the clustering centred around

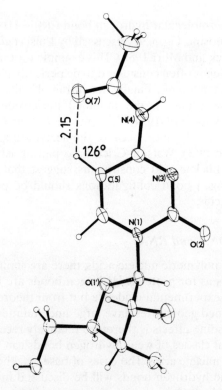

Fig. 5.37. Intramolecular C(5)—H⋯O hydrogen bond in the natural modified nucleoside N_4-acetylcytidine (adapted from Parthasarathy *et al.* 1974).

150°. As an example, the $C(6/8)$—H$\cdots$O(5′) hydrogen bonds formed in the Na salt of ApU are shown in Fig. 5.39. This self-complementary dinucleoside monophosphate forms a miniature double-helix of the A-RNA type (Seeman *et al.* 1976), and the base-pairing as well as the $C(6/8)$—H$\cdots$O(5′) hydrogen bonds are formed in the same way as in helical RNA. The occurrence of intranucleotide $C(6/8)$—H$\cdots$O(5′) hydrogen bonds in solution has been confirmed by deuterium exchange NMR experiments on the hybrid poly(rA)-poly(dT) by Benevides and Thomas (1988).

A more unusual kind of base-to-backbone C—H$\cdots$O hydrogen bond was reported to occur in d(CpG) steps of cytosine-rich left-handed Z-DNA (Egli and Gessner 1995). In these regions, the cytosine group adopts such a conformation that short intranucleoside contacts are formed between the cytosine $C(6)$—H and the ribose O(4′), as is shown in Scheme **225**. The geometry of this interaction, however, is rather unfavourable for a hydrogen bond, with $D \approx 2.70$, $d \approx 2.40$ Å and $\theta \approx 97°$.

In loop regions, other kinds of base-to-backbone C—H$\cdots$O hydrogen bonds are possible. Examples in the tRNA[Asp] anticodon hairpin have been described by Auffinger *et al.* (1996a). Figure 5.40(b) shows a base-to-base hydrogen bond $_{C36}C(5)$—H$\cdots$O(2)$_{U33}$ and Fig. 5.40(c) a base-to-ribose hydrogen bond $_{U35}C(5)$—H$\cdots$O(2′)$_{U33}$. In both cases, the donor is a pyrimidine $C(5)$—H group. The systematic intranucleotide hydrogen bonds $C(6)$—H$\cdots$O5′ are also drawn. To study the dynamic behaviour of the hydrogen bonding in tRNA[Asp], Auffinger and Westhof (1996) have performed 3 ns of multiple molecular dynamics simulations in a 17 nucleotide long hydrated fragment (Fig. 5.40(a)) neutralized with NH$_4^+$ counterions, starting with the experimental structure determined by Westhof *et al.* (1985). It was inferred that the conventional hydrogen bonds in canonical base pairs are typically

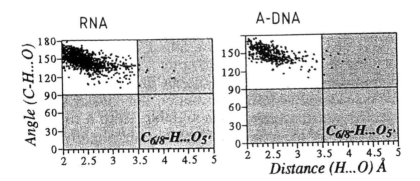

Fig. 5.38. Scatterplots of θ against d for purine $C(8)$—H$\cdots$O(5′) and pyrimidine $C(6)$—H$\cdots$O(5′) hydrogen bonds: *left*, in RNA, determined from the helical parts of 26 RNA and 12 RNA/protein crystal structures with resolutions in the range 1.4–3.0 Å; *right*, in A-DNA, determined from 54 A-DNA crystal structures. Based on theoretical H atom positions (from Auffinger and Westhof 1997).

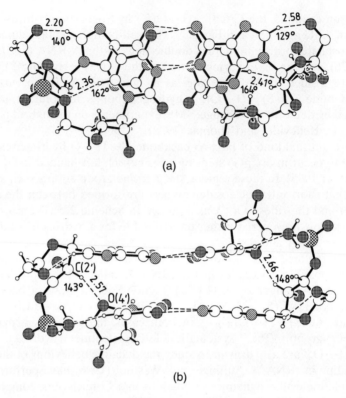

Fig. 5.39. Intramolecular C—H···O hydrogen bonding in a miniature A-RNA type double helix. Shown are intranuclotide and intranucleoside hydrogen bonds C(6/8)—H···O(5′) and intrastrand inter-ribose hydrogen bonds C(2′)—H···O(4′)(n + 1). The molecule is the self-complementary dinucleoside monophosphate ApU. (a) View perpendicular to the base pairs. (b) View into the plane of the base pairs (Seeman *et al.* 1976).

time-stable over the simulation period, whereas other conventional hydrogen bonds are more labile. Significantly, the two C(5)—H···O hydrogen bonds shown in Fig. 5.40 experience large distance fluctuations but are in themselves stable over long periods of time. In Fig. 5.41, three of the obtained trajectories are reproduced for comparison. The upper curve shows the distance fluctuations of the very stable central N—H···N hydrogen bond in a Watson–Crick C-G pair and the middle curve, on the same scale, the far larger fluctuations of the hydrogen bond $_{C36}$C(5)—H···O2$_{U33}$. On a different distance scale, the bottom curve shows an event of hydrogen bond breaking, as the interaction $_{U35}$C(5)—H···O2′$_{U33}$ is disrupted by insertion of a water molecule. Technical details on the computations are given in the legend of Fig. 5.41.

The hydrogen bonds discussed above are all formed within a single nucleic acid molecule. As an example of interduplex base-to-backbone C—H···O interactions, Berger and Egli (1997) reported systematic hydrogen bonds from

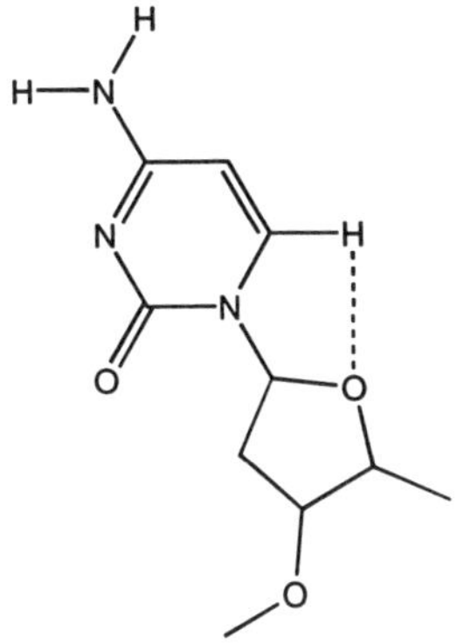

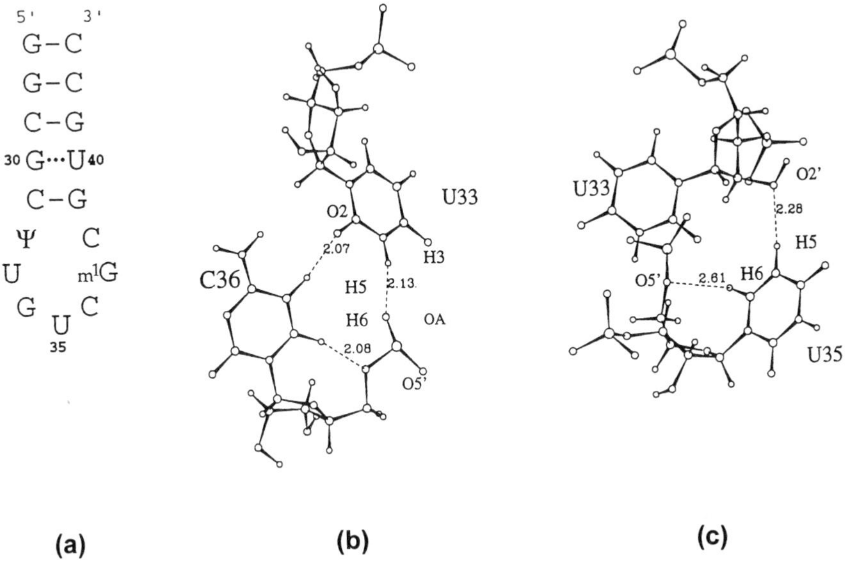

Fig. 5.40. C—H⋯O hydrogen bonds in the loop region of the tRNA^{Asp} anticodon hairpin. (a) Structure of the 17-nucleotide anticodon hairpin. The non-standard residue Ψ is defined in Fig. 5.32, and m^{1}G stands for 1-methylguanine. (b) C36 donating a base-to-base hydrogen bond C(5)—H⋯O(2) and an intranucleotide hydrogen bond C(6)—H⋯O(5'). The third hydrogen bond is of the type N—H⋯O—P. (c) U35 donating a base-to-ribose hydrogen bond C(5)—H⋯O(2') and an intranucleotide hydrogen bond C(6)—H⋯O(5') (from Auffinger *et al.* 1996a based on Westhof *et al.* 1985).

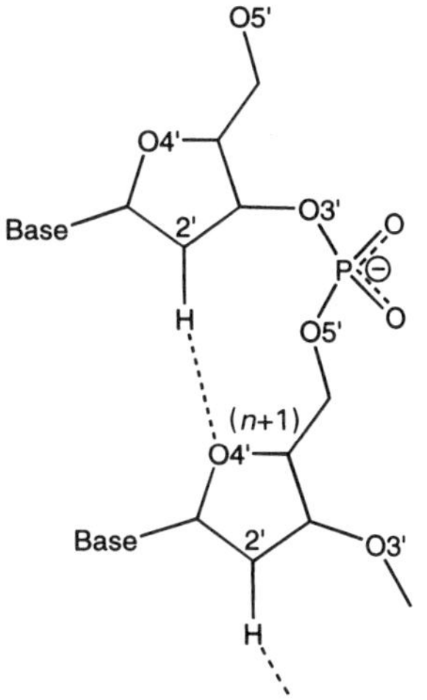

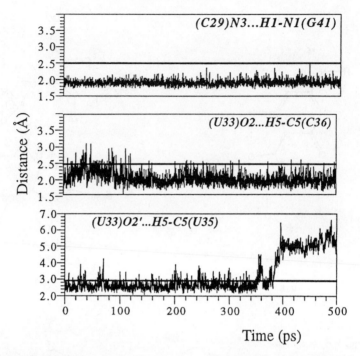

Fig. 5.41. Distance fluctuations of hydrogen bonds in MD simulations of the tRNA^{Asp} anticodon hairpin: *top*, the stable central $N-H\cdots N$ hydrogen bond in a Watson–Crick C-G pair; *middle*, the base-to-base hydrogen bond $_{C36}C(5)-H\cdots O2_{U33}$; *bottom*, the base-to-ribose hydrogen bond $_{C35}C(5)-H\cdots O2'_{U33}$, showing an event of hydrogen bond breaking; in other MD runs, this hydrogen bond was stable over the simulation period. Note the different distance scales. Computational procedure: series of six 500 ps simulations of the 17 nucleotide hairpin, neutralized with 16 NH$_4^+$ counterions and solvated by 1143 water molecules, 298 K, 2-fs time step, 9.0 Å truncation for Lennard–Jones interactions, Pearlman–Kim set of electrostatic point chages (from Auffinger and Westhof 1996).

guanine $C(8)-H$ to phosphate O atoms formed between coaxially packed Z-DNA fragments (Fig. 5.42). Notably, this interaction is formed between the strongest $C-H$ donor and the strongest acceptor present in nucleic acids. Hydrogen bonds of this kind can be formed only by bases in the *syn* and not in the *anti* conformation (Fig. 5.31(a)). When this DNA conformation is adopted in solution, the $C(8)-H$ group would be accessible to solvent interactions.

5.3.2.2   Base–base $C-H\cdots O$ hydrogen bonds

$C-H\cdots O$ hydrogen bonding between heterocyclic bases is well known. Unfortunately, for the most important base–base interactions in nucleic acids, that is Watson–Crick base pairing, the role and also the precise nature of the effect has not yet been clearly defined. The Watson–Crick C-G pair is connected by three conventional hydrogen bonds, whereas the T-A and U-A pairs are connected by only two, Fig. 5.43. Hunter and co-workers have suggested

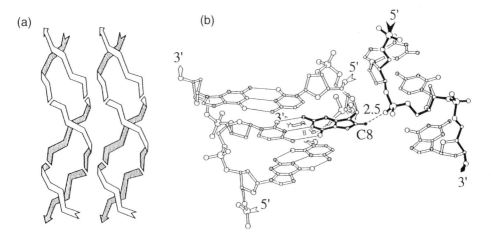

Fig. 5.42. Interduplex base-to-backbone interactions. (a) Side-by-side packed Z-DNA helices in d(CG)₃. (b) Base-to-backbone C(8)—H···O hydrogen bond at the interface between the duplexes (from Berger and Egli 1997).

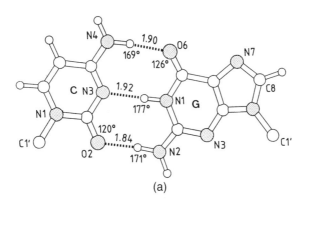

(a)

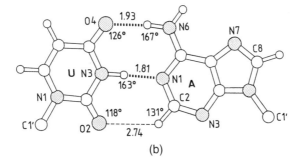

(b)

Fig. 5.43. Geometry of Watson–Crick nucleic acid base pairs as observed (a) for C-G in the crystal structure of sodium GpC nonahydrate (Rosenberg *et al.* 1976) and (b) for U-A in sodium ApU hexahydrate (Seeman *et al.* 1976). H atoms are in theoretical positions with bond lengths of N—H = 1.03 Å and C—H = 1.09 Å (from Starikov and Steiner 1997).

that in the T/U-A pairs, the contact $C(2)-H\cdots O(2)$ would complete a triply hydrogen bonded system related to that in the C-G pair (Leonard *et al.* 1995). Similar contacts occur in the Hoogsteen A-T pair and in the pair A(*anti*)-O8G(*syn*) as experimentally observed in a DNA dodecamer (Fig. 5.44), and also in pairs of the modified base 1,N^6-ethenoadenosine (Leonard *et al.* 1994). However, these contacts have very unfavourable geometries in terms of hydrogen bonding. The distances d are very long in the range 2.7 to 3.0 Å and the angles θ are strongly bent around 130°. Lone-pair directionality at the acceptor, though, is almost ideally obeyed. The geometries of these contacts are not voluntary but dictated by the interbase geometries that are determined by the conventional hydrogen bonds. Hunter and co-workers have been aware of these problems and have phrased their suggestion with care. It is notable that a possible bonding contribution of $C-H\cdots O$ interactions in non-Watson–Crick pairs was suggested by Ornstein and Fresco on computational grounds as far back as 1983 (see also Ornstein 1988).

Hunter's suggestion is considered very controversial. Jeffrey (1997) and Wahl and Sundaralingam (1997) have pointed out that the contact is forced by the two conventional hydrogen bonds, so that it is unclear whether it can be interpreted as a true hydrogen bond. On the other hand, the mere circumstance that a contact is dictated and not voluntary does not mean that it is destabilizing. A dictated contact can, in a fortuitous way, also be stabilizing (see Section 1.1.3 and Fig. 1.3). Based on results of *ab initio* calculations, Sponer *et al.* (1997) concluded that this kind of contact is not a hydrogen bond and the same is true for related $C(2)-H\cdots S(2)$ contacts in the pair A-^{2S}U (^{2S}U = 2-thiouracil).

To see if the $C(2)-H\cdots O(2)$ contact in A-U is stabilizing or destabilizing the base pair, *ab initio* calculations were performed on several molecular fragments (Starikov and Steiner 1997). For the fragment shown in Fig. 5.45, an intermolecular bond energy of -12.4 kcal/mol was obtained, which agrees well

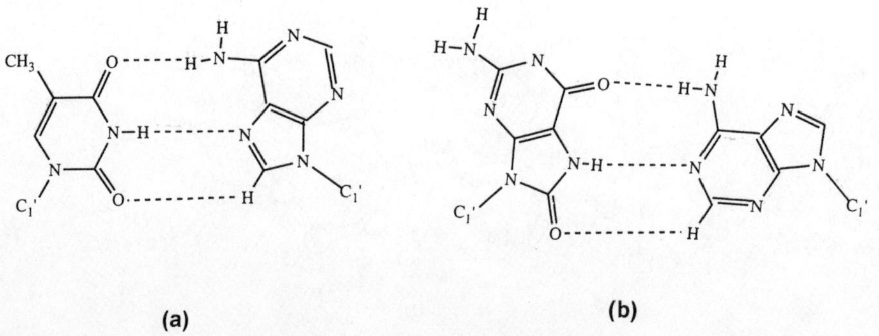

(a)   (b)

Fig. 5.44. Non-Watson–Crick base pairs with long and non-linear $C-H\cdots O$ contacts. (a) The A-T Hoogsteen pair with $d = 3.0$ Å and $D = 3.7$ Å. (b) The A(*anti*)-O8G(*syn*) base pair found in d[CGCAAATT(O8G)GCG] (from Leonard *et al.* 1995).

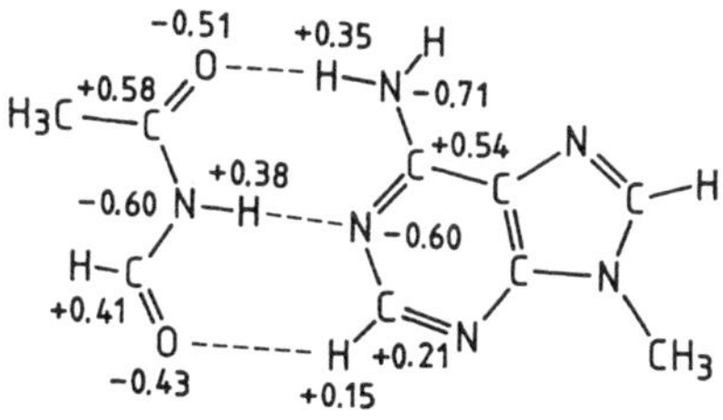

Fig. 5.45. Molecular fragments used for computations on the C(2)—H···O(2) interaction in the A-U base pair. Relevant calculated partial atomic charges are given (from Starikov and Steiner 1997).

with the experimental gas-phase value for the A-U pair, −13.0 kcal/mol (Yanson *et al.* 1979; Sukhodub 1987). After replacement of the C=O acceptor by a methyl group, the bond energy is reduced to −11.7 kcal/mol. The difference of −0.7 kcal/mol is a rough estimate of the C(2)—H···O(2) energy and amounts to about 6 per cent of the total dimer energy. The important result is not the numerical value, which is probably inaccurate, but the negative sign indicating that the interaction stabilizes the pair. The geometrical constraints imposed by the two conventional hydrogen bonds strongly distort the C—H···O interaction so that the energy contribution is far smaller than it would be in ideal geometry, but it is still not pushed into the destabilizing regime.

A look at the partial charges given in Fig. 5.45 shows that the computed charges on C(2)—H are positive for C as well as for H and have very similar values, with that on C being slightly larger. This means that the local dipole on C(2)—H is only small, whereas the total positive charge on the group, $(CH)^{\delta+}$, is relatively large. The interaction C(2)—H···O(2) can therefore, to an approximation, be considered as an attractive electrostatic interaction of $(CH)^{\delta+}$ with the negative end of the C=O dipole. One might argue whether the term 'hydrogen bond' is then appropriate (see also Section 2.2.11). Based on different *ab initio* calculations, which obtain a stabilizing nature but not the normal characteristics of C—H bond lengthening, the classification of this interaction as a hydrogen bond has been criticized by Hobza *et al.* (1998). In any case, its biological relevance lies in the question as to whether it stabilizes the pair or not and the present calculations favour the former function.

It is important to consider not only convergent hydrogen bonding of Watson–Crick base pairs but also their periphery. Watson–Crick pairs can be recognized by proteins and nucleic acids utilizing the peripheral hydrogen bonding functionalities of the pairs. Conventionally, this is attributed to N—H donors and N and O acceptors but some C—H donors are also available and these should be considered as well. For bases in the *anti* conformation with C(6/8)—H involved in intranucleotide hydrogen bonding, C(5)—H of cytosine and uracil, the methyl group at C(5) of thymine and to a limited degree C(2)—H of adenine are sterically accessible. In the absence of intranu-cleotide C(6/8)—H···O(5') hydrogen bonding, interactions with purine C(8)—H and pyrimidine C(6)—H are also possible and adenine C(2)—H can become fully available in mismatch pairs. Mandel-Gutfreund *et al.* (1998) have reported that cytosine C(5)—H and thymine C(5)—Me are in fact frequently involved in protein–DNA recognition patterns. For the other C—H groups mentioned, similar roles in DNA and RNA recognition processes can be readily anticipated.

In this context, the solution experiments of Marfurt and Leumann (1998) are of importance. The recognition characteristics of oligomer DNA duplexes by a third strand of DNA was systematically explored by these workers by variation of a single base in the third strand. Surprisingly, it is found that the Watson–Crick G-C pair is recognized by the synthetic nucleoside ⁷H [7-(2'-deoxy-β-ᴅ-ribofuranosyl)hypoxanthine] as well as by protonated methylcytidine, C⁺ (Fig. 5.46(a)). In the canonical triple C⁺·G-C, there are two conventional hydrogen bonds, whereas in the triple ⁷H·G-C, there is only one

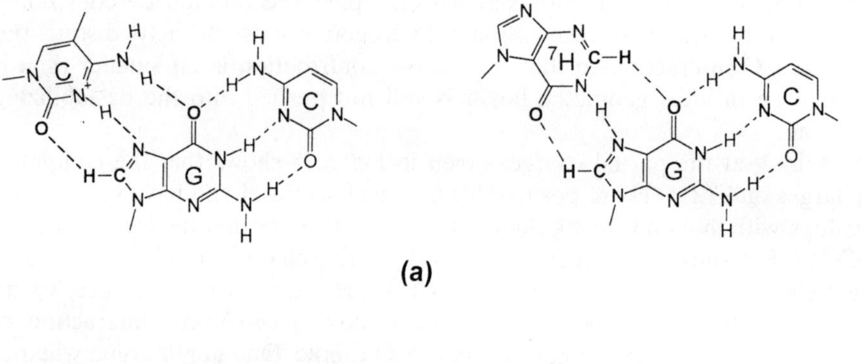

(a)

(b)

Fig. 5.46. C—H···O hydrogen bonds in the recognition of Watson–Crick base pairs by a third base, as occurring in recognition of duplex DNA by a third DNA strand. (a) Recognition of the G-C pair by C⁺ and by ⁷H, which are equally effective. Note that ⁷H forms only one conventional hydrogen bond with G. (b) Recognition of U by ⁷H. Upon deletion of the C—H···O hydrogen bond by fluorination, the binding enthalpy is reduced by about 1.1 kcal/mol (after Marfurt and Leumann 1998).

conventional hydrogen bond and two additional C—H···O interactions. Of
these, one is donated by C(8)—H of G and one by C(2)—H of ⁷H. In a different set of experiments, the binding of ⁷H to the Watson–Crick U-A pair is
investigated. The recognition operates by an N—H···O and a C—H···O
hydrogen bond formed between ⁷H and U (Fig. 5.46(b)). If the C—H···O
interaction is deleted by fluorination of the donor C(5)—H, the binding
enthalpy is reduced by about 1.1 kcal/mol. These and related experiments
show that the peripheral C—H groups of Watson–Crick pairs can play crucial
roles in molecular recognition. To complement their experiments, Marfurt and
Leumann (1998) calculated the heats of formation for the two base
pairs shown in Fig. 5.47. Using semi-empirical methods, they found that
the pair connected by two N—H···O hydrogen bonds is only 1.1 kcal/mol
more stable than the pair connected by an N—H···O and a C—H···O hydrogen bond. This means that in base–base recognition, a conventional hydrogen
bond can be quite easily replaced by a suitable C—H···O interaction.

In some non-Watson-Crick pairs, interbase C—H···O interactions are
formed which can be interpreted as hydrogen bonds with far less problems
than the one in the canonic pair A-U. In the crystal structure of the RNA
hexamer UUCGCG, Sundaralingam and co-workers observed a mismatch U-
U pair linked by an N—H···O and a C—H···O hydrogen bond, the so-called
'Calcutta pair'. This is shown in Fig. 5.48 (Wahl et al. 1996). The C—H···O
contact is obviously not a forced one and is cleanly interpreted as a hydrogen
bond. Notably, this pair of hydrogen bonds occurs in crystalline uracil itself in
an almost identical geometry (Fig. 2.3(b)). In loop regions of nucleic acids,
C—H···O hydrogen bonds can occur between bases that are not paired. An

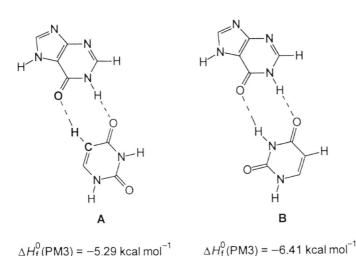

A

B

$\Delta H_f^0(PM3) = -5.29$ kcal mol^{-1} $\Delta H_f^0(PM3) = -6.41$ kcal mol^{-1}

Fig. 5.47. Calculated heats of formation of two ⁷H-dU base pairs (from Marfurt and Leumann 1998).

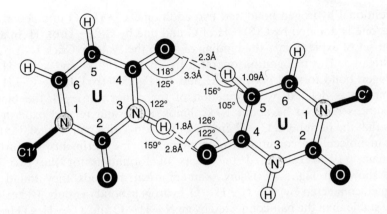

Fig. 5.48. The Calcutta U-U self base pair observed by Wahl *et al.* (1996) in the hexanucleotide r(UUCGCG) (from Wahl and Sundaralingam 1997).

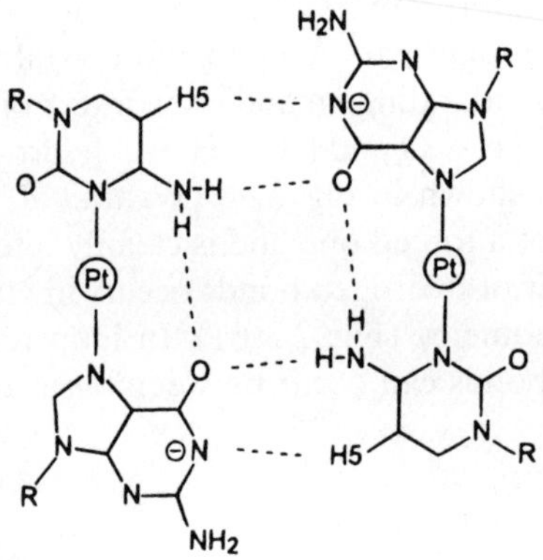

Fig. 5.49. Metallated G,C base quartet as proposed on the basis of solution ^{1}H NMR data (Metzger and Lippert 1996).

example from a tRNA discussed by Auffinger and Westhof has already been shown in Fig. 5.40.

More unusual is a metallated guanine-cytosine base quartet involving C—H···N hydrogen bonding from cytosine C(5)—H which has been characterized in solution by Metzger and Lippert (1996). In Fig. 5.49, the structure is shown as proposed from ^{1}H NMR data, which exhibits a large downfield shift for cytosine C(5)—H.

5.3.2.3   *Backbone–backbone C—H···O hydrogen bonds*

The ribose unit of the nucleic acid backbone has five C atoms which all can, at least in principle, donate C—H···O hydrogen bonds (Fig. 5.30). These C—H groups are all chemically and sterically different, and are clearly less

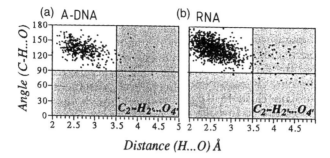

Fig. 5.50. Scatterplots of θ against d for intrastrand inter-ribose interactions $C(2')-H\cdots O(4')$ ($n + 1$) in A-forms of nucleic acids: (a) in 54 A-DNA fragments; (b) in helical regions of RNA from 26 RNA and 12 RNA/protein structures (from Auffinger and Westhof 1997).

activated than those of the bases. So, their hydrogen bonds are weaker than the base–backbone $C-H\cdots O$ interactions. Many repetitive hydrogen bond patterns have been found but this hydrogen bond type is not yet fully explored for all kinds and forms of nucleic acids.

Backbone–backbone $C-H\cdots O$ hydrogen bonds can be classified into interactions formed within one nucleic acid strand and those formed between different strands. Systematic intrastrand inter-ribose hydrogen bonds $C(2')-H\cdots O(4')(n + 1)$ in A-forms of nucleic acids, **226**, have been recognized by Auffinger and Westhof (1997). An example can be seen in the A-RNA type structure of ApU shown in Fig. 5.39. The geometric characteristics of these hydogen bonds are shown in Fig. 5.50 for A-DNA and RNA crystal structures. The average distance d is relatively long and the directionality is very fuzzy, pointing at the weakness of this interaction.

Systematic interstrand hydrogen bonds $C(1')-H\cdots O(4')$ and $C(4')-H\cdots O(4')$ have been found by Rich and co-workers within a four-stranded intercalated cytosine-rich DNA (Berger *et al.* 1996). In a zipper-like motif, adjacent ribose units of antiparallel strands are linked by mutual inter-actions of this kind (Fig. 5.51(a)).

Interstrand backbone–backbone $C-H\cdots O$ hydrogen bonds also occur between different duplexes and, more generally, between different nucleic acid entities. Then, they are specific not to a particular molecule but to a particu-lar crystal packing. An example of $C-H\cdots O$ hydrogen bonding between ribose $C(4')-H$ and $C(5')-H$ donors and $O(3')$ and phosphate O atom acceptors has been reported by Lippard and co-workers for a duplex DNA dodecamer binding cisplatin (Fig. 5.51(b); Takahara *et al.* 1996). Inter-ribose $C(2')-H\cdots O(4')$ hydrogen bonding was observed between the ends of symmetry-related RNA duplexes and this probably helps in aligning neighbouring duplexes into a quasi-infinite helix (Biswas and Sundaralingam 1997).

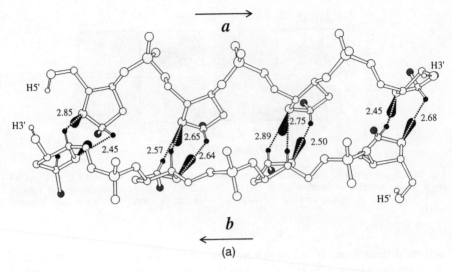

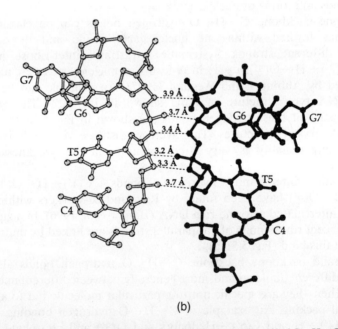

Fig. 5.51. Interstrand C—H···O hydrogen bonds. (a) Zipper-type motif of C—H···O(4′) hydrogen bonds in four-stranded intercalated d(CCCT). Distances d are given, the arrows indicate the strand polarity, the dark lobes indicate the O atom lone pair directions (from Berger *et al.* 1996). (b) Packing of a duplex DNA dodecamer binding the anticancer drug cisplatin. Hydrogen bonds are donated by ribose C(4′)—H and C(5′)—H groups, and accepted by O(3′) and phosphate O atoms (from Takahara *et al.* 1996).

5.4 Carbohydrates

Carbohydrates (or saccharides) are the most abundant biological molecules; cellulose is said alone to account for more than half of the carbon bound in the biosphere. The intermolecular interactions of carbohydrates are dominated by extensive and cooperative $O-H\cdots O$ hydrogen bond networks and $C-H\cdots O$ hydrogen bonds are also formed in large numbers. The structure and function of the latter is the subject of this discussion.

5.4.1 *Chemical constitution*

The basic units of carbohydrates are monosaccharides (or simple sugars). These molecules are well suited for hydrogen bonding, as can be seen from the structural formula of α-D-glucose, **227** or any other monosaccharide. Half the atoms of **227** can form strong hydrogen bonds (five OH groups and the ring O atom) and the rest are moderately activated $C-H$ groups. Such a molecular constitution leads to the formation of extended $O-H\cdots O$ hydrogen bond networks, which in general is a characteristic of all saccharides (Jeffrey and Saenger, 1991). In addition, the activated $C-H$ groups participate in many $C-H\cdots O$ interactions.

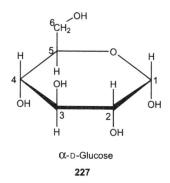

α-D-Glucose

227

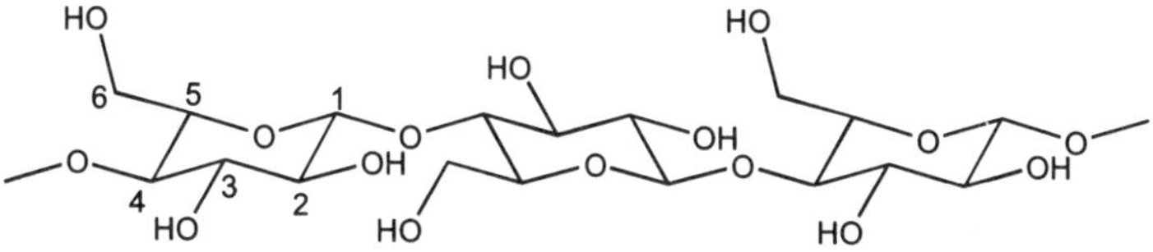

228

In modified sugars, hydrogen bonding groups are introduced or removed, altering the overall hydrogen bond properties. The simplest modification is a deletion of $-$OH groups leading to the deoxysugars, such as the $2'$-deoxyribose of DNA. In deoxysugars, the O/C ratio is smaller than in the native molecules and the average degree of C$-$H activation is lower. Some C$-$H groups may even become more or less unactivated. In the aminosugars, one or more $-$OH groups are replaced by amino or acetylamino groups. An important example is N-acetylglucosamine, which is the monomer building block of chitin. Other common substituents are carboxylic acid functionalities, which are often deprotonated in the organism so that the saccharide becomes an anion. All these alterations to the strong hydrogen bonding groups also modify the characteristics of weak hydrogen bonds occurring in the system.

Monosaccharides polymerize by formation of glycosidic bonds. This deletes one OH group per glucose and one becomes ethereal. In consequence, amylose carries only five O atoms per glucose, three of which are in OH groups. The number of C$-$H donors is unchanged upon polymerization and their degrees of activation are almost unaffected. Therefore, there will be about 2/5 less O$-$H$\cdots$O hydrogen bonds in polymeric than in monomeric saccharides (if unhydrated) and the competitive situation is changed in favour of the C$-$H$\cdots$O interactions. Actually, important functional roles of C$-$H$\cdots$O hydrogen bonds have been found for oligo- and polymeric saccharides, whereas they seem to be only innocuous to supportive in the simple sugars.

5.4.2 *Hydrogen bond geometry*

The carbohydrates are one of the primary model systems to explore O$-$H$\cdots$O hydrogen bonds and in this context, over 30 neutron diffraction studies have been performed (Jeffrey and Saenger 1991; Jeffrey 1997). Using this wealth of neutron crystal structures, the geometries of C$-$H$\cdots$O hydrogen bonds in carbohydrates have also been described (Steiner and Saenger 1992a). Based on 395 C$-$H bonds in 30 crystal structures, it was found that about 34 per cent of all C$-$H groups form intermolecular contacts to O atoms with $d < 2.7$ Å and $\theta > 90°$. This high fraction is certainly associated with the high density of acceptor atoms in the system. The shortest contact occurs in sucrose, with $d = 2.27$ Å and $\theta = 166°$, and the bulk of distances d are longer than 2.4 Å. This is clearly longer than the typical distances d observed with more activated C$-$H groups, but it still is clear that carbohydrates abound with C$-$H$\cdots$O hydrogen bonds.

The d–θ scatterplot of C$-$H$\cdots$O interactions in carbohydrates has already been shown in Section 2.2.3.1 for $d < 3.0$ Å (Fig. 2.17). The shortest contacts are relatively linear but this property is rapidly lost with increasing distance and from $d = 2.8$ Å onwards, contacts with all angles are observed. If the scatterplot is extended to $d < 5.0$ Å, Fig. 5.52, it becomes clear that there is a con-

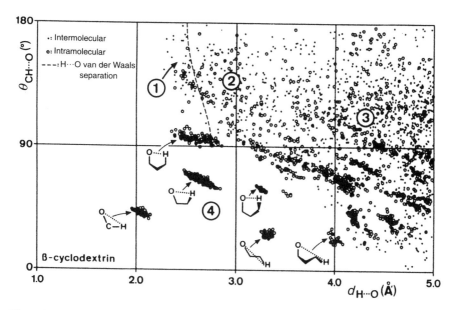

Fig. 5.52. Scatterplot of θ against d for *all* $C-H\cdots O$ contacts in 30 carbohydrate neutron crystal structures, no matter whether they represent hydrogen bonds or not. Circled numbers indicate regions of (1) $C-H\cdots O$ hydrogen bonds, (2) transition between hydrogen bonds and arrangements without bonding interactions, (3) long-distance arrangements without direct interaction, (4) specific types of intramolecular $C-H\cdots O$ contacts which are stereochemically dictated. The dashed line represents the $H\cdots O$ van der Waals separation. Note the continuous transition between regions (1), (2) and (3), and the arbitrary way in which the van der Waals separation cuts the distribution (from Steiner and Saenger 1992*a*).

tinuous transition from hydrogen bonds to the random scatter representing C—H/O geometries without direct interaction. Intramolecular $C-H\cdots O$ contacts are also included and show a number of distinct clusters that are demanded by the molecular conformation. The intramolecular contacts represented by the prominent cluster at $d \approx 2.4$–$2.8\,\text{Å}$ and $\theta \approx 95°$ are synaxial interactions in pyranoses and were later interpreted as slightly destabilizing (Steiner and Saenger 1998*a*).

   The pronounced smearing of the hydrogen bond directionality is certainly a result of the competitive situation, in which there are almost as many strong donors as weak ones. The $C-H\cdots O$ interactions, whatever their precise roles may be, are restricted to exist in a dense network of much stronger $O-H\cdots O$ hydrogen bonds and their directionality is too weak to compete successfully. Most of the intermolecular $C-H\cdots O$ geometries, though, are well within the bonding regime. Because there are so many of these distorted but weakly bonding $C-H\cdots O$ interactions, the sum of their enthalpic contributions will be considerable and their omission is misleading. In any case, the blurred character of Fig. 5.52 points at rather passive roles for $C-H\cdots O$ interactions in carbohydrates, even when they are relatively short. Exceptions are always possible and examples of functionally important $C-H\cdots O$ hydrogen bonds in carbohydrates will be discussed in the next section.

5.4.3 *Functionally important* $C-H\cdots O$ *hydrogen bonds*

$C-H\cdots O$ hydrogen bonds in carbohydrates can assume dominant functions if there is a local lack of strong $O-H\cdots O$ competitors. An important example is the hydrophobic internal cavities of cyclodextrins (cycloamyloses), which have already been discussed in the context of inclusion complexes (Section 4.2, Figs 4.12 to 4.14). These cavities lack $O-H$ donors and consequently, $C-H\cdots O$ hydrogen bonds often play significant roles in structure stabilization. Apart from weak host–guest interactions, cyclodextrins also form intramolecular $C-H\cdots O$ hydrogen bonds. In underivatized cyclodextrins, the orientation of neighbouring glucose units is systematically stabilized by the interglucose hydrogen bond pattern shown in Fig. 5.53(a). The dominant interaction is a bifurcated hydrogen bond between $O(3)_n$ and $O(2)_{n-1}$ with a minor component to $O(4)_n$. In addition, there is a weak interglucose hydrogen bond $C(6)_n-H\cdots O(5)_{n-1}$ with d in the range 2.39–2.91 Å, and with mean value 2.59 Å (Steiner and Saenger 1992a). This is a typical example of a supportive $C-H\cdots O$ hydrogen bond. If the $O(2)$ and $O(6)$ hydroxyl groups are methylated, the interglucose hydrogen bond pattern is fully conserved, Fig. 5.53(b) (Steiner and Saenger 1995b). If all hydroxyl groups are methylated, $O-H\cdots O$ hydrogen bonds can no longer be formed but the systematic $C-H\cdots O$ bonding persists, Fig. 5.53(c) (Caira *et al.* 1994; Steiner and Saenger 1996). In this new situation, the $C-H\cdots O$ interactions are no longer just supportive to a stronger hydrogen bond pattern but play a role of their own.

   If permethylated cyclodextrins are crystallized without guest molecules, they adopt shapes with largely reduced cavity volume (Caira *et al.* 1994; Steiner and Saenger 1998b). For permethyl-β-cyclodextrin, the reduction of the molecular cavity is achieved by chair inversion of one glucose residue, which leads to partial collapse of the annular structure. Most of the remain-

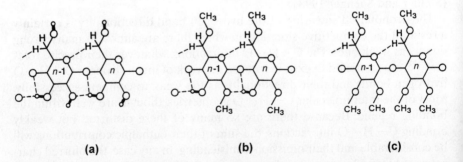

(a)   (b)   (c)

Fig. 5.53. Systematic intramolecular interglucose hydrogen bonds stabilizing the mutual orientation of the glucose residues in cyclodextrins: (a) in underivatized cyclodextrins; (b) in 2,6-dimethylated cyclodextrins; (c) in fully methylated cyclodextrins (adapted from Steiner 1996a).

ing volume is filled by methoxy groups which are stretched into the cavity or folded over it, Fig. 5.54. Notably, the methoxy groups are not folded in a random fashion but such that C—H···O hydrogen bonds of several kinds are formed. These involve terminal methyl groups and also some of the more activated C—H donors. Hence these interactions have a direct influence on the molecular conformation in the absence of stronger hydrogen bonds.

Cellulose, **228**, is one of the most important carbohydrates. Cellulose is the linear polymer of β-1,4-linked D-glucoses and because all the ring substituents are equatorial it is roughly ruler shaped, with the hydroxyl groups at the edges. The faces are formed by the axial ring H atoms and the O atoms O(4) and O(5) and are rather lipophilic in nature. Cellulose is polymorphic, but a feature common to all the polymorphs is O—H···O hydrogen bonding between the edges of the molecules and stacking of the faces. This leads to layered structures. In cellotetraose hemihydrate, a small molecule model for cellulose II, the molecules are stacked in such a way that systematic hydrogen bonds C(3)—H···O(4) and C(5)—H···O(4) are formed between the faces of molecules in adjacent layers (Fig. 5.55; Gessler *et al.* 1995). The geometries of these hydrogen bonds are close to ideal, with the parameters d, D and θ in the ranges 2.30–2.71 Å, 3.38–3.73 Å and 158–180°, respectively. The C—H···O(4) hydrogen bonds are presumably important in the fine-tuning of the stacking

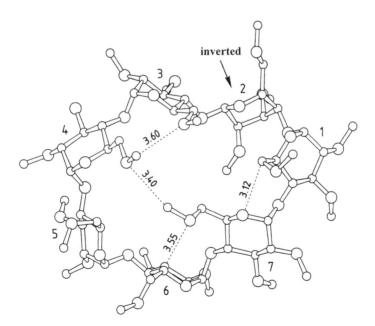

Fig. 5.54. Permethyl-β-cyclodextrin monohydrate with several short intramolecular C···O contacts. H-atoms are not drawn (from Steiner and Saenger 1998b, who repeated the structure of Caira *et al.* 1994).

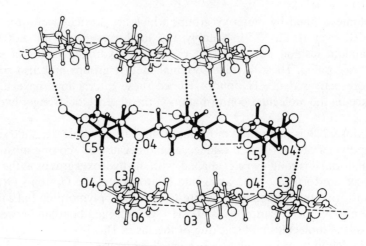

Fig. 5.55. Systematic $C(3)-H\cdots O(4)$ and $C(5)-H\cdots O(4)$ hydrogen bonds in cellotetraose hemihydrate, which is a structural model for cellulose II (adapted from Gessler *et al.* 1995).

arrangement and it can be assumed that related interactions are formed in polymeric cellulose II.

For cellulose I, no small molecule model is available and the published fibre diffraction crystal structures are inaccurate. Therefore, analysis of hydrogen bonds can only be tentative. Based on the fibre model of Gardner and Blackwell (1974), the hydrogen bond pattern shown in Fig. 5.56 may be derived. Within the cellulose layers, there are conventional hydrogen bonds involving the hydroxyl groups, and one of the $C(6)-H$ groups points at the $O(2)-H$ hydroxyl group of a neighbouring strand. The distance $d = 2.36\,\text{Å}$ is so strongly suggestive of $C-H\cdots O$ hydrogen bonding that it can be considered as real even if the structural model is only rough. Using the same fibre diffraction model, one can also anticipate tuning of the layer stacking by intermolecular $C-H\cdots O(4)$ hydrogen bonding.

5.5 Water molecules

Water is found in crystals in an enormous variety of chemical environments. This range extends from the gas hydrates to water of crystallization in inorganic salts, water molecules in organometallic, organic and bio-organic structures, and to the water that makes up a major part of typical protein and nucleic acid crystal structures. The eleven polymorphs of ice itself testify to the many factors, significant and subtle, that govern the structural chemistry of this unique molecule. The peculiar nature of water derives from the fact that the

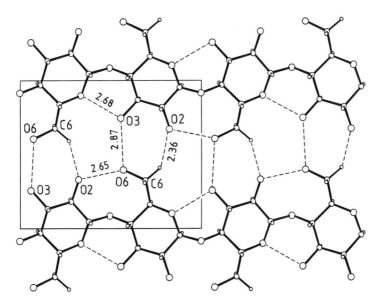

Fig. 5.56. O−H⋯O and C−H⋯O hydrogen bonding in a layer of cellulose I, derived from the fibre diffraction model of Gardner and Blackwell (1974). Because the model is refined using heavy restraints and hydroxyl hydrogen atoms have not been located, the interpretation must be taken as preliminary.

entire molecular surface is comprised of strong hydrogen bond donor and acceptor groups.

The relative importance of co-crystallized water molecules is not the same in different systems. In inorganic salts, water is normally an integral and indispensible part of ion coordination and is therefore of direct functional importance. In many organometallic and organic structures, however, water appears to act primarily as a polar space-filler, that is as a simple bridging functionality between hydrogen bonding groups. This is not a very sophisticated function and in consequence, water molecules attract less attention in organic structural chemistry (see, however, Desiraju 1991*b*). In biological structures, water is of an importance that can hardly be overemphasized (Jeffrey and Saenger 1991; Westhof 1993). The structures and functions of biological molecules are directly linked to the aqueous medium. The structure of nucleic acids depends critically on the composition of the surrounding solvent. Proteins contain internal water molecules that can be considered as integral parts of the molecular structure and the conformation of the surface residues is also determined by interactions with the solvent. In crystal structures, proteins and nucleic acids are heavily hydrated (solution mimics) and the structural properties of the biomolecule–solvent interface can be inferred from the crystalline state.

Unfortunately, water H atoms in typical biological structures cannot be located. Therefore, a detailed and reliable analysis of water interactions can only be performed using small and medium-sized organic crystal structures. In the following section, water coordination in organic crystal structures will be described with a focus on weak hydrogen bonding, and in turn this will serve as a basis for the discussion of these effects in biological systems. Inorganic hydrates and gas hydrates are, despite their general importance, not within the scope of this book. For an introduction into such systems, the interested reader is referred to Jeffrey (1997).

5.5.1 *Organic hydrates*

5.5.1.1 *Conventional hydrogen bonds*

Despite its simple structure, the water molecule adopts a multitude of hydrogen bond functions. Each of the two O_W—H groups can donate normal, bifurcated or trifurcated hydrogen bonds, so that the number of such contacts ranges between two and six. Formally, the O_W acceptor has two electron lone pairs that would in conventional terms tend to be approached by two donors. In the sterically undisturbed gas phase dimers, hydrogen bonds to water molecules are actually directed at the conventional lone-pairs (Legon and Millen 1987*b*), but the acceptor directionality is so weak that even slight steric disturbances completely smear it in the solid state. In statistical studies on water coordination geometries, no bimodal distribution of the preferred donor approach to O_W is observed (Jeffrey and Saenger 1991), and it rather seems that the whole surface area of O_W can act as a proton acceptor. This allows the water molecule to accept between one and five hydrogen bonds.

Although the water coordination function is in principle very versatile, there are certain preferred modes of association. When considering ion coordination and conventional hydrogen bonds in inorganic hydrates, Falk and Knop (1973) determined the coordination frequencies of O_W as shown in Fig. 5.57. The number of accepted interactions ranges between one and four but the majority of water molecules accept one (41 per cent) or two (54 per cent) ion contacts or hydrogen bonds. Jeffrey and Maluszynska (1990) surveyed the stereochemistry of 478 water molecules in hydrates of small biological molecules. When considering only conventional hydrogen bonds, they found strong preferences for the onefold (35 per cent) and twofold (60 per cent) acceptor functions. The preferred donor functions are twofold (60 per cent) and threefold (26 per cent). A total of 17 different hydrogen bond configurations were observed. The most frequent configuration is formally tetrahedral but this made up only 35 per cent of all water molecules in their sample.

5.5.1.2 C—$H \cdots O_W$ *hydrogen bonds*

The role of C—$H \cdots O_W$ hydrogen bonds in water coordination has been studied in a statistical analysis of 101 water molecules in 46 very accurate

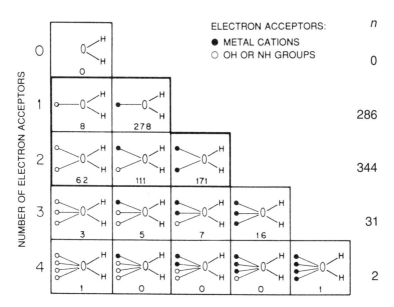

Fig. 5.57. Acceptor functions of water molecules in inorganic hydrates. Conventional hydrogen bonds and interactions with metal cations are considered (adapted from Falk and Knop 1973).

small-molecule neutron crystal structures (Steiner and Saenger 1993a). Over 50 per cent of these water molecules accept $C-H \cdots O_W$ interactions with $d < 3.0$ Å (Fig. 5.58(a)); that is a very significant fraction. The distribution of $C-H \cdots O_W$ hydrogen bond distances shown in Fig. 5.58(b) represents an average of many different types of $C-H$ donors, and accidentally, very acidic $C-H$ groups are not present in the sample. For acidic $C-H$ donors, much shorter hydrogen bonds to water molecules have been observed in X-ray studies (see Section 2.2.2, Table 2.3). Based on a moderate distance cut-off value of $d < 2.8$ Å, the water acceptor functions shown in Fig. 5.59 are obtained. It is obvious that $C-H$, $O/N-H$ and metal ions can coordinate to water molecules in virtually any combination, with the double acceptor function being clearly preferred. For reasons of completeness, the donor functions of the 101 water molecules in the sample are shown in Fig. 5.60. The donor and acceptor functions of Figs 5.59 and 5.60 can be freely combined to yield a vast variety of possible water coordination geometries.

In 39 cases in Fig. 5.59, $C-H$ donors are involved. In 21 of these, the donors complete a twofold water acceptor function and in the other 18 cases $C-H$ participates in higher coordination of water, with a fivefold acceptor function as the extreme. The common feature in all these structures is that the conventional water coordination is incomplete or strongly distorted, leaving 'free' acceptor potential that can be filled by one or more $C-H \cdots O_W$ interactions. Two examples of twofold water acceptor functions are shown in Fig. 5.61, one

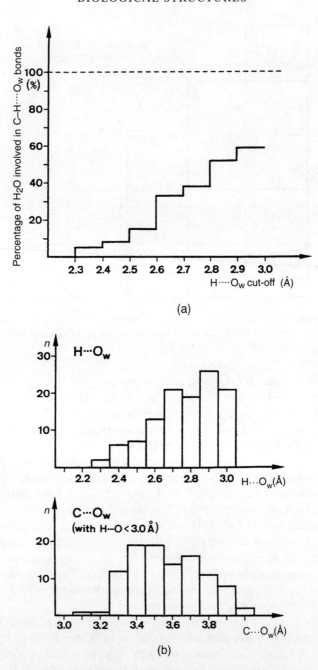

(a)

(b)

Fig. 5.58. C—H···O hydrogen bonding of water molecules in neutron crystal structures. (a) Percentage of water molecules accepting C—H···O_w contacts shorter than a given d. (b) Histograms of distances d and D (from Steiner and Saenger 1993a).

X–H can be O–H, N–H.   M⁺: metal ion

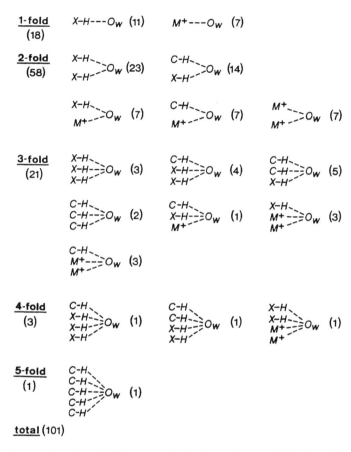

Fig. 5.59. Acceptor functions of 101 water molecules in neutron crystal structures. The numbers show the frequencies of occurrence. Cut-off distances were selected as $d < 2.8$ Å for C—H$\cdots$O$_W$, $d < 3.0$ Å for O/N—H$\cdots$O$_W$, and $D < 3.0$ Å for M$^+\cdots$O$_W$ interactions (from Steiner and Saenger 1993a).

with a C_α—H donor of asparagine and one with an O—CH$_2$—C donor of coenzyme vitamin B$_{12}$. Two examples with higher acceptor functions are shown in Fig. 5.62; in one, there are one O—H$\cdots$O$_W$ and two C—H$\cdots$O$_W$ hydrogen bonds and in the other the situation is reversed. Related arrangements are also observed if water molecules are coordinated to metal cations in such a way that a part of the O$_W$ acceptor remains sterically accessible.

It is of interest to examine the further consequences of *neglecting* C—H$\cdots$O$_W$ hydrogen bonds in the analysis of these structures. In individual cases, this simply leads to water configurations that appear to be very

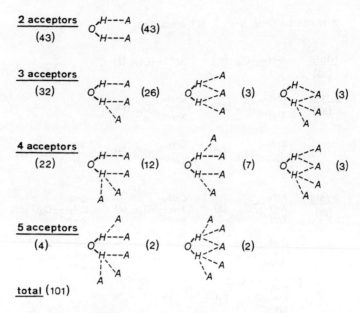

Fig. 5.60. Donor functions of 101 water molecules in neutron crystal structures. The numbers show the frequencies of occurrence. Distance cut-off $d < 3.0\,\text{Å}$ (from Steiner and Saenger 1993a).

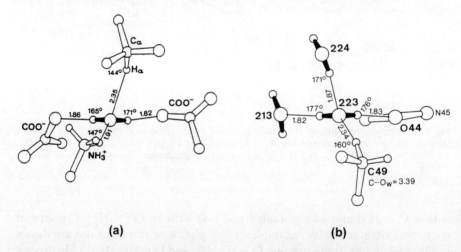

Fig. 5.61. Water molecules accepting one conventional and one $C-H\cdots O_W$ hydrogen bond. (a) In asparagine monohydrate at 15 K (structure from Weisinger-Lewin *et al.* 1989; figure from Steiner and Saenger 1993a). (b) In vitamin B_{12} coenzyme at 15 K (structure from Bouquiere *et al.* 1993; figure from Steiner and Saenger 1993b).

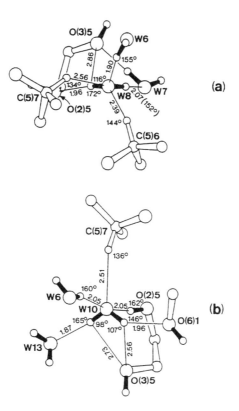

Fig. 5.62. Water molecules accepting three hydrogen bonds of which at least one is from a C—H group: (a) in β-cyclodextrin ethanol octahydrate at 15 K; (b) in β-cyclodextrin·11.6 D$_2$O at 120 K (Zabel *et al.* 1986; from Steiner and Saenger 1993*a*).

peculiar, with this peculiarity disappearing if the C—H···O$_W$ contacts are considered to be hydrogen bonds. The statistical consequences are summarized in Table 5.8. If, in the neutron sample of 101 water molecules, the C—H···O$_W$ interactions were neglected, 44 water molecules would have one and 46 would have two accepted hydrogen bonds or ion contacts. Three would accept no hydrogen bond or ion contact at all. If C—H···O$_W$ hydrogen bonds *are* considered, the three apparent 'non-acceptors' all become hydrogen bonded, the number of single acceptors falls to less than half what it was and the number of double acceptors increases significantly. The double acceptor function then is clearly the preferred one, outnumbering either the single or triple acceptors roughly by a factor of three. Acceptor functions greater than three are rare even with C—H···O$_W$ interactions included. These data suggest that the primary function of C—H···O$_W$ hydrogen bonds in water coordination is to 'fill up' the preferred twofold acceptor function – if this is not possible with conventional hydrogen bonding.

Table 5.8 Acceptor functions of the water molecules in Fig. 5.59, depending on whether $C-H\cdots O_W$ hydrogen bonds are neglected or not

	Number of accepted interactions					
	0	1	2	3	4	5
$C-H\cdots O_W$ neglected	3	44	46	7	1	0
$C-H\cdots O_W$ included	0	18	58	21	3	1

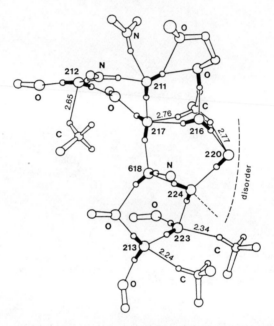

Fig. 5.63. Hydrogen bond network in the ordered pocket region of vitamin B_{12} coenzyme. Note the many $C-H\cdots O_W$ hydrogen bonds (from Steiner and Saenger 1993b, based on the structure determined by Bouquiere *et al.* 1993).

The validity of these principles in larger hydrated molecules was tested and verified with coenzyme vitamin B_{12}, for which Bouquiere *et al.* (1993) have provided an accurate low-temperature neutron diffraction model. In this crystal structure, there is a pocket region which contains an ordered water cluster whereas in an adjacent solvent channel, the water molecules are disordered even at 15 K. The water molecules in the partially polar and partially apolar pocket accept a large number of $C-H\cdots O_W$ hydrogen bonds from $C-H$ groups of the cavity lining, Fig. 5.63 (Steiner and Saenger 1993b). Note in particular W213 and W223, for which the $C-H\cdots O_W$ interactions complete almost ideal tetrahedral coordination geometries.

Table 5.9 Water molecules which accept no other than $C-H\cdots O_W$ hydrogen bonds as occurring in certain substance classes (Steiner 1995d)

Compound class	n H$_2$O analysed	n H$_2$O accepting only $C-H\cdots O_W$
Amino acids and peptides	222	12 = 5%
Nucleosides and nucleotides	133	0
Carbohydrates	124	0
Pyrimidines and purines	114	6 = 5%
Alkaloids	51	8 = 16%
Steroids	28	0

In Fig. 5.59, there are three water molecules that accept only $C-H\cdots O_W$ hydrogen bonds. To ascertain if this kind of water interaction occurs more frequently, a database analysis was performed on X-ray crystal structures of small biological molecules (Steiner 1995d). Out of 672 water molecules, 26 accept no conventional hydrogen bonds but only $C-H\cdots O_W$ interactions. Although this is only a small fraction, it is worthwhile to look at these water molecules more closely. The number of accepted $C-H\cdots O_W$ hydrogen bonds ranges from one to four and, as expected, the most frequent acceptor function is twofold (11 cases). This means that tetrahedral water coordination can be achieved *even in the complete absence of conventional hydrogen bond donors.* Two examples with onefold acceptor function of O_W have already been shown: hypoxanthinium nitrate monohydrate in Fig. 2.9(e) (Section 2.2.2) and a water molecule bonded to $C_\delta-H$ of a charged histidine side chain in Fig. 5.10(a). An example each of two-, three- and four-fold $C-H\cdots O$ acceptor functions of water molecules is shown in Fig. 5.64. Note that in all these examples, the water stabilization would be completely mysterious were the $C-H\cdots O_W$ hydrogen bonds to be neglected.

The occurrence of water molecules that accept only $C-H\cdots O_W$ hydrogen bonds is highly sample dependent, Table 5.9. About 5 per cent of the water molecules in hydrated amino acids, peptides, pyrimidines and purines are of this type and around 16 per cent of those in hydrated alkaloids. No examples are found in crystal structures of nucleosides, nucleotides, carbohydrates and steroids. The former classes of molecules share a common feature – a lack of conventional hydrogen bond donors often combined with the presence of acidic $C-H$ groups. This is most pronounced for the alkaloids, which often have only one strong donor (N^+-H) and many activated $C-H$ groups. Carbohydrates are at the other extreme, because they typically have many hydroxyl groups and their $C-H$ groups are weakly activated.

We conclude then that $C-H$ donors often participate in the coordination of water molecules similar to $O-H$ and $N-H$. Certainly, water molecules prefer to accept strong hydrogen bonds or to coordinate to metal cations.

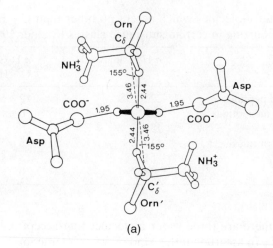

(a)

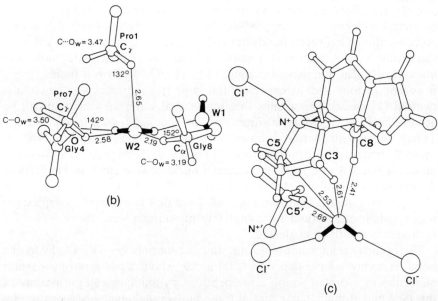

(b)   (c)

Fig. 5.64. Water molecules accepting only C—H···O$_W$ hydrogen bonds: (a) in L-ornithine L-aspartate hemihydrate (Salunke and Vijayan 1983); (b) in the uncharged peptide *cyclo*-(L-Pro-Gly)$_2$ trihydrate (Shoham *et al.* 1989); (c) in the alkaloid norsecurinine hydrochloride monohydrate. Note the chelation of the water molecule in the concave part of the alkaloid molecule (Joshi *et al.* 1986). Geometries are normalized (from Steiner 1995*d*).

However, if these are not available in sufficient numbers and in suitable enough configurations in a given local environment, a water molecule will resort to accepting the weaker C—H···O$_W$ hydrogen bonds rather than leaving its acceptor potential unsatisfied (Fig. 5.65). In these arrangements,

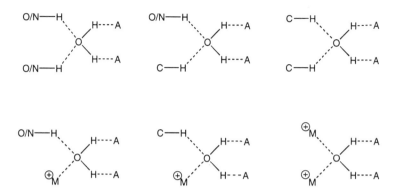

Fig. 5.65. The six possiblities to satisfy the double acceptor function of water molecules with O/N−H donors, C−H donors and metal cations. All these configurations are actually observed in crystal structures.

O/N−H⋯O_W and C−H⋯O_W hydrogen bonds have the same functions, with differences only in the strengths. Combined O/N−H⋯O_W and C−H⋯O_W hydrogen bonds occur very commonly but pure C−H⋯O_W hydrogen bonding is rare.

5.5.1.3 O_W−H⋯π hydrogen bonds

In small-molecule crystal structures, there are relatively few examples of water molecules donating hydrogen bonds to π-acceptors. In Chapter 3, examples have been shown with Ph acceptors of the tetraphenylborate anion (Fig. 3.16(b)) and of a calixarene (Fig. 3.17) and with the C≡C acceptor of a terminal alkyne (Fig. 3.40). A more systematic study has been performed on O_W−H⋯Ph hydrogen bonds in small hydrated peptides. Following the observation that a water molecule in Tyr-Tyr-Phe dihydrate donates a hydrogen bond to the phenylalanine side-chain, the occurrence of O_W−H⋯Ph hydrogen bonds in other peptides was studied in a database analysis (Steiner *et al.* 1998*b*). Five examples in well-refined crystal structures were found with geometries as listed in Table 5.10. The geometries are very variable, with some of the water O atoms residing almost exactly over the aromatic centroids, whereas others are more off-centred. The distances of O_W to the aromatic centroids, $D(M)$, are in the range 3.26–3.63 Å and the distances $d(M)$ are in the range 2.42–2.90 Å. This flexible geometry is typical for aromatic hydrogen bonds in general (Section 3.1.3).

In Fig. 5.66, four examples of O_W−H⋯Ph hydrogen bonds in peptides are shown. In Tyr-Tyr-Phe dihydrate, the two symmetry independent water molecules are hydrogen bonded to each other and form an interesting system of hydrogen bonds with their surroundings (Fig. 5.66(a)). Both water molecules have a fourfold coordination which is achieved by participation of non-

Table 5.10 Geometries of $O_W-H\cdots Ph$ hydrogen bonds in small hydrated peptides (Steiner *et al.* 1998*b*)

	YYF[a]	BIHXUL10[b]	JECYUL[c]	SOJPAJ[d]	TALVAD[e]
Acceptor type	Phe	Phe	Tyr	Tyr	Phe
$\omega(O_W)$ (°)	6.8	3.3	20.9	21.0	17.0
$D(M)$ (Å)	3.26	3.28	3.61	3.63	3.45
$D(C)$ range (Å)	3.40–3.68	3.49–3.63	3.38–4.31	3.41–4.31	3.33–4.07
$D(C)$ spread (Å)	0.28	0.14	0.93	0.90	0.74
$d(M)$ (Å)	2.42	2.48	2.87	2.90	2.47
$d(C)$ range (Å)	2.42–3.11	2.57–3.09	2.79–3.54	2.51–3.78	2.46–3.17
$d(C)$ spread (Å)	0.69	0.52	0.75	1.27	0.71
$\theta(M)$ (°)	143	138	132	132	171
$\theta(C)$ range (°)	118–172	112–165	106–158	115–152	141–159

[a] L-Tyr-L-Tyr-L-Phe dihydrate (Steiner *et al.* 1998*b*). [b] *Cyclo*-(L-Pro-L-Val-L-Phe-L-Phe-L-Ala-Gly) tetrahydrate, cycloamanide A tetrahydrate (Chiang *et al.* 1982). [c] L-Asp-L-Arg-L-Val-L-Tyr tetrahydrate (Feldman and Eggleston 1990). [d] L-Pro-L-Tyr monohydrate (Klein *et al.* 1991). [e] *Cyclo*-(L-Ser-L-Phe-L-Leu-L-Pro-L-Val-L-Asn-L-Leu) tetrahydrate, evolidine tetrahydrate (Eggleston *et al.* 1991). Distances *d* normalized.

conventional hydrogen bonds for W_1 by donating an aromatic hydrogen bond and for W_2 by accepting a $C_\alpha-H\cdots O_W$ interaction.

In Pro-Tyr monohydrate, Fig. 5.66(b), the water molecules form a simpler hydrogen bond pattern, donating an $O-H\cdots O^-$ and $O-H\cdots Ph$ hydrogen bond. The latter is not directed at the ring centre but at the midpoint of an aromatic $C-C$ bond, with a distance from H to the bond centre of 2.48 Å. The $O_\eta-H$ group of the accepting tyrosine side-chain donates a hydogen bond to the water molecule in the next unit cell, so that an infinite chain is formed: $O_W-H\cdots Ph-O_\eta-H\cdots O_W-H\cdots Ph-O_\eta-H$. One can speculate as to whether or not this chain is cooperative. Mutation experiments on glutathione S-transferase have strongly suggested the cooperative nature of a finite hydrogen bond chain $O-H\cdots Ph-O_\eta-H\cdots S^-$ occurring in that system (Section 5.2.5), so that cooperativity of the hydrogen bond chain in Pro-Tyr monohydrate may also be anticipated.

In the tetrahydrate of the cyclic hexapeptide cycloamanide A, an infinite chain of interconnected water molecules is formed (Fig. 5.66(c)). Within the chain, two of the water molecules play conventional roles whereas the third one donates a short and well centred $O_W-H\cdots Ph$ hydrogen bond to a phenylalanine side-chain. For this water chain, the intermolecular interactions are clearly dominated by conventional hydrogen bonds but the additional $O_W-H\cdots Ph$ interaction is required to stabilize the arrangement as a whole. The most complicated water structure of these examples occurs in the tetrahydrate of the cyclic heptapeptide evolidine (Fig. 5.66(d)). A relatively large interstitial cavity is filled with four water molecules. The cavity wall is formed by polar groups but also by 'apolar' groups like $C_\alpha-H$ and phenylalanine and

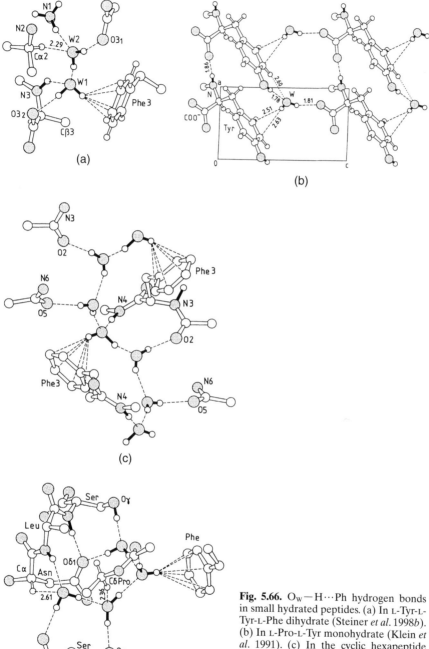

Fig. 5.66. O_W—H···Ph hydrogen bonds in small hydrated peptides. (a) In L-Tyr-L-Tyr-L-Phe dihydrate (Steiner *et al.* 1998*b*). (b) In L-Pro-L-Tyr monohydrate (Klein *et al.* 1991). (c) In the cyclic hexapeptide cycloamanide A tetrahydrate (Chiang *et al.* 1982). (d) In the cyclic heptapeptide evolidine tetrahydrate (Eggleston *et al.* 1991). Distances are normalized (from Steiner *et al.* 1998*b*).

proline side-chains. In this partly polar and apolar environment, the water cluster is stabilized by a highly complex system of conventional and non-conventional hydrogen bonds, including an $O_W-H\cdots Ph$ interaction. The water molecules do not avoid contact with the 'apolar' groups but engage in weak hydrogen bonds with them. This is clearly a more favourable outcome than unsatisfied hydrogen bond potentials.

The water interactions shown above are relatively rare examples found only in a handful of crystal structures. This suggests that they do not represent generally favourable configurations. On the other hand, they do provide an important mechanism as to how water molecules can find stable positions in 'unfriendly' environments. The general relevance of the crystal structures shown in Fig. 5.66 and the data in Table 5.10 lies in the occurrence of similar water interactions in proteins, especially when they occur in the active sites or substrate recognition sites. This is discussed in Section 5.5.3.

5.5.2 *Are there water molecules with vacant hydrogen bond potentials?*

In crystal structure publications, water molecules are occasionally reported which 'accept no hydrogen bonds' or of which one O_W-H group 'forms no hydrogen bond'. More frequently, it occurs that hydrogen bond tabulations contain such water molecules and authors are quiet about the matter in the main text. Many of these cases are examples where $C-H$ groups have not been considered as potential donors, or Ph groups as potential acceptors. It also occurs frequently that conventional hydrogen bonds are overlooked because restrictive distance cut-off criteria are used. Typical among these is the catastrophic criterion that the distance from H to O_W must be shorter than 2.4 or even 2.2 Å. In view of the data shown in the previous section, the question arises: are there *any* water molecules in crystals with completely unsatisfied hydrogen bond potentials? Or do all these cases disappear if weak hydrogen bonding effects are fully and properly taken into account?

In crystal structures of small biological molecules, water molecules have been examined which accept no conventional hydrogen bonds (Steiner 1995*d*). Twenty-six examples were found and they *all* accept $C-H\cdots O_W$ hydrogen bonds instead. In 23 of these examples, the shortest interaction has a distance of $d < 2.5$ Å, representing hydrogen bonds which are easy to recognize. The remaining three also accept at least one $C-H\cdots O_W$ interaction with $d < 2.7$ Å which is well inside the hydrogen bonding region. Indeed, no example of a reliably refined non-accepting water molecule in a crystal structure is known to either of us.

The situation is slightly different for water molecules with apparently unfulfilled donor functions. Some early examples certainly qualify for overlooked $O_W-H\cdots\pi$ hydrogen bonds. However, there *is* one case in a reliably refined low-temperature neutron crystal structure, where a water donor actually points only at $C-H$ groups. Figure 5.67 shows one of the symmetry-

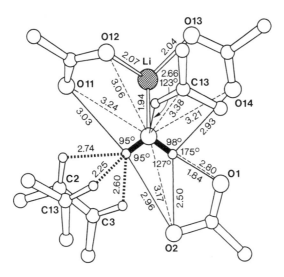

Fig. 5.67. The very peculiar O_W—H···H—C contacts of a water molecule in lithium hydrogen phthalate monohydrate. Section of the 15 K neutron crystal structure (Küppers *et al.* 1985). These interactions occur also at room temperature. Figure drawn for this book.

independent water molecules in the 15 K structure of lithium hydrogen phthalate monohydrate ($Z' = 2$; Küppers *et al.* 1985). One O_W—H donor is involved in a bifurcated conventional hydrogen bond and the other one points at three C—H groups, with the shortest H···H distance being only 2.25 Å. These must be destabilizing interactions. There are two very long and very bent O_W—H···O contacts with $d = 2.96$ and 3.03 Å and $\theta = 95$ and 95°, respectively, but they are subsidiary. The important hydrogen bonding potential head-on to O_W—H is clearly unsatisfied. This is a very exceptional case but crystal structures still carry surprises!

The translational mobility and the rotational freedom of the water molecule allows the formation of at least weak hydrogen bonds of both O_W—H donors and O_W acceptors in almost all crystal structures. The enthalpic penalty for unfulfilled hydrogen bond potential is efficiently avoided or kept low. Nevertheless, water molecules with one vacant donor potential *can* exist stably in crystals. Stable water positions with completely vacant acceptor potential have as yet not been observed.

5.5.3 *Macromolecular structures*

Hydration plays a crucial role in the structure and function of almost all biological macromolecules (exceptions are the anhydrous crystalline domains of cellulose and chitin). In this context, weak hydrogen bonding phenomena have been described for only two classes of molecules: proteins and nucleic acids.

Even for these, the developments are not very advanced at this time. Therefore, only a collection of individual results are presented in this section without critical review.

5.5.3.1 *Weak hydrogen bonds in protein hydration*

Since proteins carry so many $C-H$ donors and aromatic acceptors, weak hydrogen bonds in protein–water interactions may be expected. Unfortunately, water molecules are the least reliably refined parts of typical macromolecular structures, so that their hydrogen bonds are particularly difficult to detect and interpret. Hints that protein side-chains donate $C-H\cdots O_W$ hydrogen bonds can be found in a statistical analysis of protein hydration which is based on crystal structures with resolutions of 1.7 Å and better (Thanki *et al.* 1988). Though not discussed in the original article, some of the hydration patterns show numerous water molecules in positions that are suggestive of $C-H\cdots O_W$ hydrogen bonding. It is not surprising that this is most common for the histidine side-chain (Fig. 5.68(a)). For the side-chain of tryptophan,

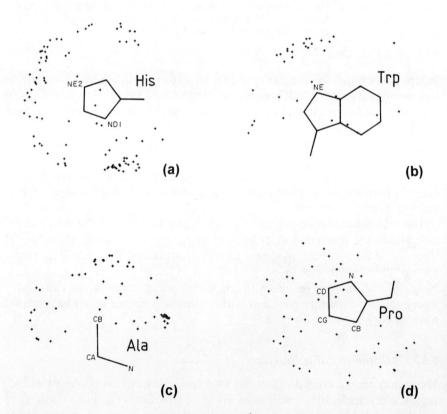

Fig. 5.68. Distribution of water molecules within 3.5 Å around the side-chains of (a) histidine, (b) tryptophan, (c) alanine, and (d) proline (from Thanki *et al.* 1988).

some water sites are found close to the most acidic of its C—H groups, that
is C_δ—H (Fig. 5.68(b)). Water molecules are also found close to the methyl
group of alanine and the CH_2 groups of proline but these are much weaker
interactions when compared to the ring C—H groups of histidine (Fig.
5.68(c,d), drawn for $D < 3.5$ Å). Some distance distributions of water contacts
to C atoms of apolar side-chains are shown in Fig. 5.69. These distributions
suggest that some fraction of the $C\cdots O_W$ interactions is associated with
$C—H\cdots O_W$ hydrogen bonds, but by no means all of them.

Equilibrium energies and distances of several $C—H\cdots O_W$ bonded dimers
have been calculated by Ornstein and Zheng (1997), Table 5.11 (MP2/A ab
initio computations). For comparison, values for $O/N—H\cdots O_W$ hydrogen
bonds with water and imidazole donors have also be computed. Equilibrium
energies range from −0.7 kcal/mol for the donor methane to −3.1 kcal/mol for
acetylene. Notably, the energy for the strong imidazole C_ε—H donor,
−3.0 kcal/mol, comes close to that for acetylene, and for the less activated imidazole C_δ—H it is still −2.0 kcal/mol. This means that $C—H\cdots O_W$ hydrogen
bonds formed by the strongest donors occuring in proteins have about half
the energy of hydrogen bonds from conventional donors. Hydrogen bonds
from the very weakly activated methyl groups amount to about 10 per cent in
energy terms of those donated by, say, water molecules.

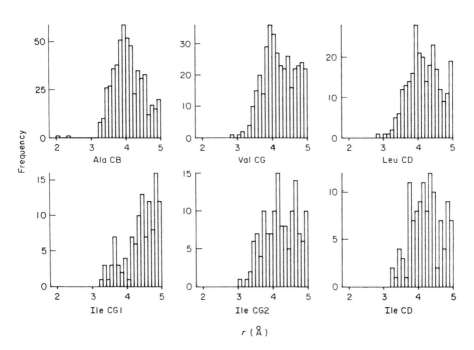

Fig. 5.69. Histograms of distances D of water contacts to apolar side chains of proteins (from
Thanki et al. 1988).

Table 5.11 Calculated equilibrium distances and energies of $X-H\cdots O_W$ ($X = C, O, N$) hydrogen bonds (MP2/A *ab initio* calculations by Ornstein and Zheng 1997)

Hydrogen bond	$d(H\cdots O_W)$ (Å)	$D(X\cdots O_W)$ (Å)	ΔE (kcal/mol)
$CH_4\cdots OH_2$	2.68	3.76	−0.7
$H_2C{=}CH_2\cdots OH_2$	2.54	3.62	−1.1
Benzene$\cdots OH_2$	2.47	3.55	−1.6
(Imidazole)$C_\delta-H\cdots OH_2$	2.45	3.52	−2.0
(Imidazole)$C_\varepsilon-H\cdots OH_2$	2.33	3.41	−3.0
$HC{\equiv}CH\cdots OH_2$	2.21	3.28	−3.1
$OH_2\cdots OH_2$	1.94	2.90	−5.4
(Imidazole)$N_\delta-H\cdots OH_2$	1.94	2.95	−6.8
(Imidazole)$N_\varepsilon-H\cdots OH_2$	1.91	2.86	−7.9

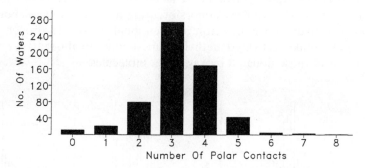

Fig. 5.70. Number of 'polar contacts' suggestive of conventional hydrogen bonds formed by buried water molecules in proteins, based on 75 high-resolution (< 2.5 Å) crystal structures (from Williams *et al.* 1994).

Internal water molecules of proteins. A more detailed view of $C-H\cdots O_W$ hydrogen bonding is possible for the internal (buried) water molecules in proteins, because they are normally more reliably refined than the surface water molecules. Internal water molecules are functionally important and because of this intrinsic interest, their conventional hydrogen bond coordination has been the subject of several statistical studies (Baker and Hubbard 1984; Hubbard *et al.* 1994; Williams *et al.* 1994). In the frequency distribution of 'polar contacts' formed by these water molecules, about 45 per cent are threefold and 18 per cent are only twofold or lower coordinated (Fig. 5.70; Williams *et al.* 1994). Assuming that many of the poorly hydrogen bonded water molecules might have higher coordination if $C-H\cdots O_W$ hydrogen bonding were to be considered, the very well-refined internal water cluster in the 1.7 Å structure of actinidin (Baker 1980) was analysed (Steiner and Saenger 1993*a*). A

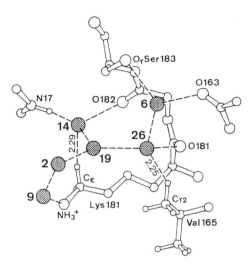

Fig. 5.71. Hydrogen bond coordination of internal water molecules in the enzyme actinidin. Two water molecules (W14 and W26) presumably accept C—H···O$_W$ hydrogen bonds, resulting in a tetrahedral coordination (from Steiner and Saenger 1993*a*, who used the coordinates of Baker 1980).

relevant section is drawn in Fig. 5.71, showing that two of the water molecules are actually involved in short C···O$_W$ contacts that complete a tetrahedral coordination geometry. The interpretation of these contacts as weak hydrogen bonds is supported by the observation that they occur in a very similar way in the related protein papain. Two further examples of internal water molecules accepting C—H···O$_W$ hydrogen bonds are described in the comprehensive article of Derewenda *et al.* (1995).

A single internal water molecule in an engineered internal cavity has been studied by Fersht and co-workers (Buckle *et al.* 1996). These authors performed cavity-creating mutations in the hydrophobic core of barnase in order to study the structural responses. In a mutation Ile → Ala, a small cavity with a volume of 31 Å^3 and which hosts a single ordered water molecule was created. This water molecule has only one contact with a conventional hydrogen bond partner (a backbone carbonyl, $D = 2.6$ Å), but many with C atoms of apolar side-chains forming the cavity wall (shortest with an IleC$_{\gamma 1}$, 2.9 Å; IleC$_{\gamma 2}$, 3.0 Å; IleC$_{\delta}$, 3.3 Å; AlaC$_{\beta}$, 3.4 Å). This can be interpreted as representing several C—H···O$_W$ hydrogen bonds, though the donors are very weak. The possibility that even 'hydrophobic' cavities are solvated has implications for free energies of unfolding of such mutants. Actually, the described Ile → Ala mutant is destabilized to a smaller degree than related mutants with unsolvated cavities and it may be concluded that this solvation has a stabilizing effect upon the protein folding.

Internal water molecules have been systematically investigated for a series of crystal structures of acetylcholinesterase (AChE) from *Torpedo californica*. Analysing four isomorphous structures of the native and complexed enzyme (resolution range 2.3–2.5 Å), it was found that the enzyme contains 38 internal water molecules which are fully conserved within the four structures (Koellner *et al.* 1999). Nineteen of the internal waters are isolated from all other water molecules and 19 are distributed over six larger cavities which contain between two and five molecules. The degree of conservation is so high that even $C \cdots O_W$ and $O_W \cdots Ph$ contacts occur in the same way in all four structures. This set of well-refined internal waters provides a wealth of structural information on weak protein–water hydrogen bonding, three examples of which are shown in Figs 5.72 to 5.74. An isolated water molecule forming three

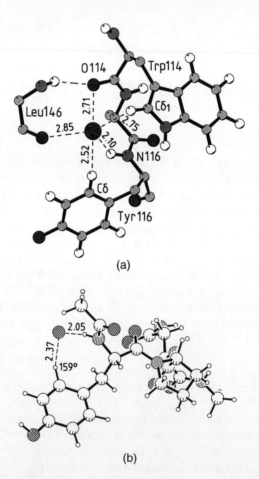

Fig. 5.72. (a) An isolated internal water molecule in acetylcholinesterase from *Torpedo californica* complexed with E2020 (Kryger *et al.* 1998), forming three conventional and two $C-H \cdots O_W$ hydrogen bonds. (b) *N*-acetyl-L-Tyr-L-Pro-*N*-methyl-L-asparaginamide (Montelione *et al.* 1984) containing the same pair of hydrogen bonds $N-H \cdots O_W$ and $C_\delta-H \cdots O_W$ donated by a tyrosine residue (Koellner *et al.* 1999).

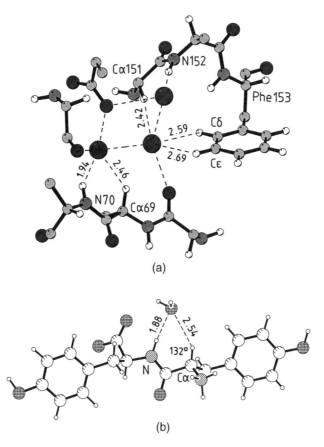

(a)

(b)

Fig. 5.73. (a) Internal cluster of three water molecules in acetylcholinesterase from *Torpedo californica* complexed with E2020 (Kryger *et al.* 1998), forming several conventional and $C-H\cdots O_W$ hydrogen bonds. Note the pair of hydrogen bonds $_{70}N-H\cdots O_W$ and $_{69}C_\delta-H\cdots O_W$, which form motif **214**. (b) L-Tyr-L-Tyr dihydrate (Cotrait *et al.* 1984) also containing motif **214** with a water acceptor (Koellner *et al.* 1999).

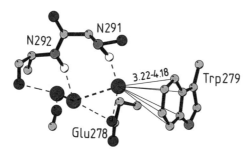

Fig. 5.74. Internal water molecule in acetylcholinesterase from *Torpedo californica* complexed with E2020 (Kryger *et al.* 1998) presumably forming an aromatic hydrogen bond $O_W-H\cdots Ph$ (Koellner *et al.* 1999).

conventional and two $C-H\cdots O_W$ hydrogen bonds is shown in Fig. 5.72(a). One of the $C-H\cdots O$ interactions is donated by a tryptophan $C_\delta-H$ group and the other one by a tyrosine $C_\delta-H$. The latter interaction forms, together with a conventional hydrogen bond from the peptide $N-H$ of the same residue, a chelating hydrogen bond motif. This motif has been mentioned as a recurring interaction pattern involving amino acid side-chains, **217**, in Section 5.2.2.2. For comparison, the same motif in a small peptide is shown in Fig. 5.72(b).

Figure 5.73(a) shows an internal water cluster formed by three molecules. Apart from a number of conventional hydrogen bonds, there are also clear $C-H\cdots O_W$ interactions. One water molecule is in short contact with the edge of a phenylalanine side-chain, interacting with two $C-H$ groups. The water molecule to the left accepts a pair of hydrogen bonds, one from a $C_\alpha-H$ and the other one from the peptide $N-H$ donor of the next residue. This pair of interactions constitutes pattern **214** that occurs frequently in main-chain to main-chain interactions (Section 5.2.2.1). A small molecule analogue for this motif is also shown (Fig. 5.73(b)). Finally, Fig. 5.74 shows an internal cluster of three water molecules, one of which is in short contact to the face of a phenyl ring. The six individual distances $D(C)$ are in the range 3.22–4.18 Å and this is acceptable for an $O_W-H\cdots Ph$ hydrogen bond (Table 5.10). Because it is conserved in four crystal structures, it can be interpreted as an aromatic hydrogen bond with some confidence even though the water H atoms are not seen.

The examples of the internal water molecules in AChE show that concerted occurrence of conventional and non-conventional hydrogen bonds is the normal situation in protein–water interactions. Actually, all six internal water clusters with $n \geq 2$ are involved in non-conventional hydrogen bonding (though not every water molecule within these clusters), and for all clusters these interactions help to rationalize the geometrical arrangement. It is expected that water molecules in surface clefts and pockets behave essentially in the same way. Water molecules having more intimate contact with the bulk solvent might behave differently in the sense that positions with poor conventional hydrogen bond capabilities are not well populated.

Water in active sites and substrate binding sites. In several instances, non-conventional hydrogen bonding has been reported for water molecules in active sites and substrate binding sites of enzymes. It is remarkable that in these environments, $O_W-H\cdots Ph$ hydrogen bonding is observed relatively frequently. The examples of aromatic hydrogen bonding in thrombin complexes (Engh *et al.* 1996; Fig. 4.30 of Section 4.4.2) and the complex of acetylcholinesterase with the drug E2020 (Kryger *et al.* 1998; Fig. 5.25) have already been discussed. The most accurately refined example of an $O_W-H\cdots Ph$ hydrogen bond in a protein is reported by Helliwell and co-workers in the 0.94 Å low-temperature crystal structure of native concanavalin A (Deacon *et al.* 1997). In the substrate-free saccharide-binding site of this metalloprotein, a

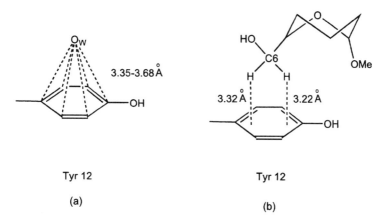

Tyr 12   Tyr 12

(a)   (b)

Fig. 5.75. Intermolecular interactions of the Tyr12 aromatic ring in the saccharide binding site of concanavalin A. (a) $O_W \cdots$ Ph contact in the hydrated substrate-free binding site, suggestive of an aromatic hydrogen bond. (b) Monosaccharide $C(6)H_2$ from a bound substrate contacting the aromatic ring, suggestive of weakly polar $C-H \cdots$ Ph interactions (Deacon *et al.* 1997).

water molecule is placed almost exactly over the centre of an aromatic ring. The distances $D(C)$ lie in the narrow range 3.35–3.68 Å, suggesting an aromatic hydrogen bond close to the ideal geometry (compare Table 5.10). In the difference electron density of this very high resolution structure, even a water hydrogen atom oriented towards the aromatic ring centre is visible. With bound substrate, an exocyclic CH_2 group of the saccharide is placed at about the same position as the water molecule and the $C-H$ groups are oriented towards the aromatic ring (Fig. 5.75). This suggests that upon substrate binding, an aromatic hydrogen bond formed in the solvated substrate binding site is replaced by a weak $C-H \cdots$ Ph interaction.

Summary. Though the experimental material is limited at present, there is little doubt that non-conventional hydrogen bonds occur frequently in protein–solvent interactions. $C-H \cdots O_W$ hydrogen bonds typically function in satisfying water acceptor potentials in partly or mainly hydrophobic surroundings. This allows water molecules to find stable positions even in sites that lack conventional hydrogen bond donors. The occurrence of $O_W-H \cdots$ Ph hydrogen bonds in proteins was challenged not so long ago (Flanagan *et al.* 1995), but today it is firmly established. Specific roles in protein functions can be readily conceived and their elucidation is a promising field of structural research for the future.

5.5.3.2 *Weak hydrogen bonds in nucleic acid hydration*

The structure of the polymeric nucleic acids strongly depends on hydration effects (Saenger 1984; Westhof 1993), and a full understanding of these phenomena is therefore of great importance. $C—H\cdots O_W$ hydrogen bonds have long been observed in nucleic acid constituents (Section 5.5.1), but crystallographic resolution problems impeded discussion of these effects in the polymeric forms. Detailed investigations in this field have been carried out only recently.

Groove hydration of helical nucleic acids is characterized by conserved water sites at well-defined positions. Timsit and co-workers analysed the hydration pattern around methylated CpG steps in three high-resolution crystal structures (1.7, 2.15 and 2.2 Å) of A-DNA decamers (Mayer-Jung *et al.* 1998). In the plane of unmodified cytosine bases, there are two major groove hydration sites as shown in Fig. 5.76(a). W_A hydrogen bonds to the N(4) amino group and W_B hydrogen bonds to a phosphate O atom and the cytosine $C(5)—H$ donor. The intranucleotide $C(6)—H\cdots O(5')$ hydrogen bond that is typical for A-forms of nucleic acids is also drawn. Upon 5-methylation of the cytosine base, this systematic hydration pattern is surprisingly well conserved (Fig. 5.76(b)). The water molecules are slightly displaced and do not hydrogen bond to each other any more but form similar hydrogen bonds with the nucleotide as before. W_A is still hydrogen bonded to N(4) but is now also within $C—H\cdots O_W$ hydrogen bonding distance to the m^5C methyl group ($D = 3.5$ Å). Similarily, W_B is still hydrogen bonded to the phosphate O atom and is in addition engaged in a contact to the m^5C methyl group ($D = 3.7$ Å). The intranucleotide hydrogen bond $C(6)—H\cdots O(5')$ is conserved and is now supported by a base–phosphate $C—H\cdots O$ interaction of the methyl group. These observations indicate that cytosine methylation does not prevent hydration of the base edge but only slightly alters the hydration pattern.

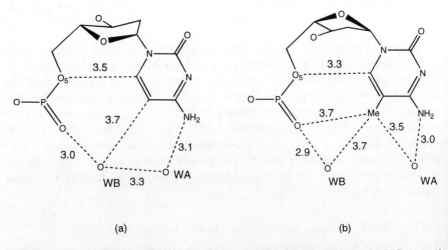

(a)   (b)

Fig. 5.76. Major groove hydration of (a) cytosine and (b) 5-methylated cytosine in CpG steps of A-DNA decamers (after Mayer-Jung *et al.* 1998).

Molecular dynamics simulations of tRNA.   Molecular dynamics (MD) simulations are an important computational method which is complementary
to structural investigations of biological systems. If carefully performed,
MD simulations yield information of a kind that cannot be obtained with
diffraction methods: typical lifetimes of hydrogen bonds, magnitudes of the
geometrical fluctuations during the lifetime, residence times of water molecules in particular sites and so on. For an introductory review on the method,
see the paper by Auffinger and Westhof (1998). A landmark MD study on
the anticodon hairpin of tRNA^{Asp} (Fig. 5.40(a)) by Auffinger *et al.* (1996*b*) has
provided many insights into the hydration of C—H groups in tRNA and is
described here in some detail (the technical procedure is described in
the legend of Fig. 5.41). The overall hydration behaviour of the C—H groups
is judged by three quantities: the radial distribution function of water O
and H atoms around H(C) as averaged over the whole simulation period (6
times 500 ps), the time averaged 'water hydrogen bonding percentage' of particular C—H groups, and the water molecule residence times in particular
hydrogen bonds. An arrangement is interpreted as a 'hydrogen bond' if
$d < 3.0$ Å and $\theta > 135°$.

The different tRNA C—H groups are not equally accessible to the solvent.
The solvent accessibility can be qualitatively seen in the 'hydrogen bond percentage' averaged over all 17 nucleotides (Table 5.12). The purine C(8)—H
and pyrimidine C(6)—H groups preferentially form intranucleotide hydrogen
bonds to O(5'), such as those shown in Fig. 5.40, and are not accessible to
solvent any longer. Similarly, the ribose C(2')—H groups form systematic
intramolecular contacts to O(4')$_{n+1}$, **226**, and have poor solvent accessibility
across the MD set. The other C—H groups are well accessible to solvent and
all have water O atoms closer than $d = 3.0$ Å in around 50 per cent of the simulation time. Despite similar accessibilities to solvent, these groups have quite

Table 5.12 'Hydrogen bonding percentages' in the
hydrated anticodon loop of tRNA^{Asp}, averaged over
all nucleotides, giving an idea of the average solvent
accessibilities. Because there is no adenosine in the
structure model, there is no value for C(2)—H. MD
simulations of Auffinger *et al.* (1996*b*)

Ribose	C, U and Ψ bases	G base
C(1')—H 47%	C(5)—H 48%	C(8)—H 9%
C(2')—H 3%	C(6)—H 1%	
C(3')—H 47%		
C(4')—H 52%		
C(5')—H$_1$ 48%		
C(5')—H$_2$ 43%		

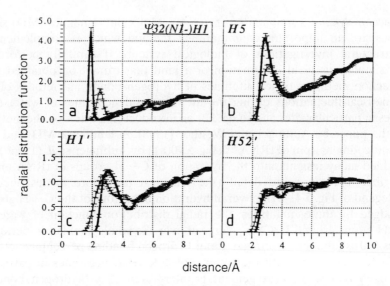

Fig. 5.77. Radial distribution functions (RDFs) of water O atoms around different kinds of H(C) atoms of tRNA in an MD simulation of the hydrated anticodon loop of tRNA^{Asp}. (a) RDF of a hydrated strong N$-$H donor. (b) Average over all pyrimidine C(5)$-$H donors. (c) Average over all ribose C(1')$-$H donors. (d) Average over all ribose C(5')$-$H$_2$ donors. The thick line represents H$\cdots$O$_W$ contacts, and the thin line represents H$\cdots$H$_W$ contacts (from Auffinger *et al.* 1996*b*).

different 'hydrophicity profiles' as judged from the time-averaged radial distribution functions (RDFs) of O$_W$ around H (Fig. 5.77). As an example for a strong hydrogen bond donor, the RDF of the Ψ32 N(1)$-$H group, which is fully accessible to solvent, is shown in Fig. 5.77(a). The profile shows a sharp peak close to 1.8 Å, representing an N$-$H$\cdots$O$_W$ hydrogen bond of stable geometry. The RDF averaged over all pyrimidine C(5)$-$H donors has a related appearance (Fig. 5.77(b)), the only difference being that the peak is broader and shifted to a longer distance, 2.6 Å. There is no adenine base in the anticodon loop of tRNA^{Asp}, but on a shorter simulation on the whole tRNA molecule, a similar but slightly broader profile is found also for adenine C(2)$-$H. This means that the acidic C$-$H groups of the bases, if exposed to solvent, form *bona fide* hydrogen bonds with water molecules. For the less acidic ribose C$-$H groups, much broader RDFs were calculated. The sharpest one is obtained for C(1')$-$H (Fig. 5.77(c)), which still has the shape expected for a hydrogen bond like interaction. Many of the water contacts, however, are associated also with short H$_C$$\cdots$H$_W$ contacts indicative of poor directionality of the interaction. For C(3')$-$H, C(4')$-$H, C(5')$-$H1 and C(5')$-$H2, the RDFs become increasingly featureless in this sequence, with the extreme case of C(5')$-$H2 shown in Fig. 5.77(d). Although C(5')$-$H2 frequently forms short contacts with

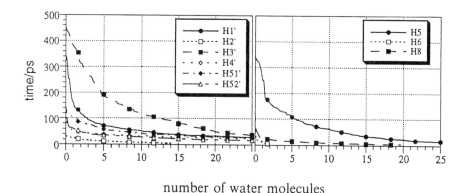

Fig. 5.78. Number of water molecules bound to the different types of C−H donors as a function of the contact time in an MD simulation of the hydrated anticodon loop of tRNA^{Asp} (from Auffinger *et al.* 1996*b*).

water molecules, it appears that the interaction is very poorly specific and directional.

Of further importance is the typical residence time of water molecules in particular hydration sites. In Fig. 5.78, the number of water molecules forming hydrogen bonds with particular C−H groups were plotted against the C−H⋯O$_W$ contact time. The plots show that certain water molecules remain hydrogen bonded to C−H groups for relatively long times > 200 ps, whereas others are more labile and participate only in diffuse hydration. The longest residence times are found at C(3′)−H and C(1′)−H of the sugar and C(5)−H of the bases. Residence times of this magnitude are characteristic of specific interactions, not of arbitrary 'hits' by diffusing molecules.

Auffinger *et al.* have also characterized the stereochemical environment of several long-lived C−H⋯O$_W$ hydrogen bonds, only one of which is shown here. In Fig. 5.79, a water molecule is shown bridging two phosphate O$_A$ atoms. It accepts a hydrogen bond from a water molecule that is not shown in the figure and completes its coordination sphere by two additional C−H⋯O$_W$ hydrogen bonds, one from a uracil C(5)−H donor and the other one from a ribose C(3′)−H. The trajectories in a 500 ps simulation are also shown. The distance fluctuations of the C−H⋯O$_W$ interactions are much larger than those of the conventional hydrogen bonds. There are events of reversible hydrogen bond breaking but the water position as such remains stable over the whole simulation period.

These MD simulations provide a whole set of postulates concerning weak hydrogen bonding in tRNA and on weak hydrogen bonding in general, going beyond results known from diffraction studies. The importance of C−H⋯O$_W$

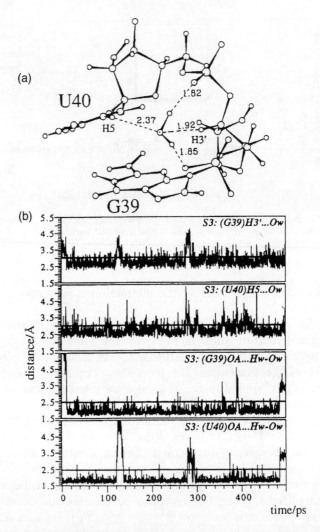

Fig. 5.79. Water molecule accepting long-lived $C-H\cdots O_W$ hydrogen bonds in an MD simulation on the hydrated anticodon loop of tRNA[Asp]. (a) Snapshot extracted from a simulation. (b) Time course of the $C-H\cdots O_W$ and $O_W-H\cdots O$ hydrogen bond distances. The horizontal lines drawn at particular distances are reference lines only (Auffinger *et al.* 1996*b*).

hydrogen bonds in nucleic acid hydration is emphasized and it is shown that these interactions must definitely be considered in order that one may understand the dynamical properties and tertiary folding of these biomacromolecules.

6

Conclusions

Conclusions of scientific books can easily dissipate into abstracts and we do not wish to present a summary here of all that has been said in the preceding chapters. The weak hydrogen bond was first identified in 1935, but only in the early nineties did it really permeate into the consciousness of chemists and biologists. In all fairness though, research activity in the subject progressed in a logical fashion during the intervening period. Looking back, it only seems natural that the study of this interaction was explored by spectroscopists before it could be tackled by crystallographers. For an interaction type as weak as this, spectroscopy alone could provide the unambiguous confirmation of its hydrogen bond status. And yet, as the systems studied became more complex, it was only crystallography and its sister-science, database research, that could provide the necessary quantification of the several fine structural effects that characterize this interaction type.

Indeed, there is little doubt that the study of weak hydrogen bonding is a study of subtleties. Weak hydrogen bonds could not have been studied before strong hydrogen bonds were understood properly, in the same way that chemists did not venture too deeply into the study of hydrogen bonding itself till the covalent paradigm was fully established. Yet, it is sobering to note in hindsight that the subtle and delicate effects associated with weak hydrogen bonding are not merely superfluous addenda to already well-established structural models, but rather are necessary for their proper appreciation and on occasion even comprehension. This is especially true in structural supramolecular chemistry where it has been perceived that a crystal structure is often not the result of hierarchic interaction preferences but a convolution of a large number of strong and weak interactions, each of which affect the rest intimately. Methods for codification of crystal structures must take this into account if they are to be accurate and useful. Of course, the goal of a subject like crystal engineering is to design systems where the interaction preferences *are* hierarchic, or in other words where the interaction interference is at a minimum. Take for instance, the recently discussed structure of 4-cyano-4'-ethynylbiphenyl, **229** (Fig. 6.1) (Langley *et al.* 1998). The $C{\equiv}C-H{\cdots}N{\equiv}C$ synthon is identical to that observed in cyanoacetylene forty years ago (Shallcross and Carpenter 1958; Fig. 2.1(b) in Section 2.1) and in effect one might term **229** the biphenylogous extension of cyanoacetylene. However, most crystal structures are not so predictable and the challenge posed by weak

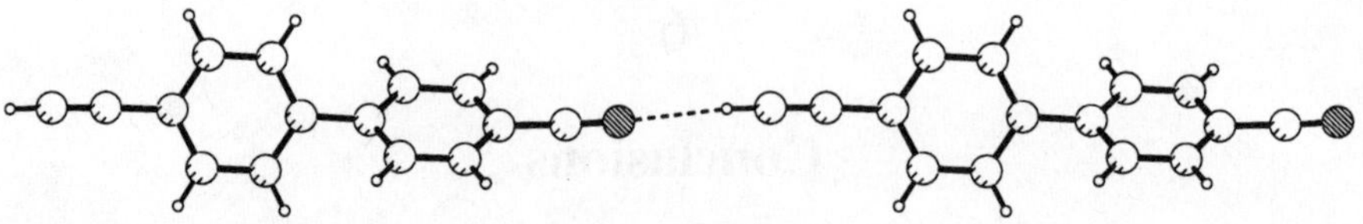

Fig. 6.1. Crystal structure of 4-cyano-4'-ethynylbiphenyl, **229**, to show the linear C≡C—H···N≡C synthon (d = 2.20 Å, θ = 176°). Note the similarity to cyanoacetylene, Fig. 2.1(b) (from Langley *et al.* 1998).

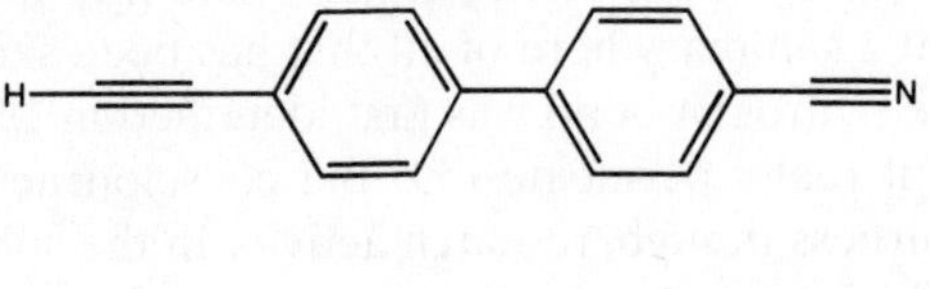

229

hydrogen bonding effects to the dogma of crystal engineering remains a real one.

The differences between strong and weak hydrogen bonds often confused and delayed progress in the field. We have mentioned the notorious van der Waals cut-off often in this text. Only after database research became popular, did it become clear that this criterion which was harmless for strong hydrogen bonds was misleading for weak hydrogen bonds, and in a general sense without scientific basis for *all* hydrogen bonds. In this sense, a completely new technique was required before meaningful progress could be made. Again, the dichotomy between refinement restraints and interaction length uncertainties in macromolecular structure refinements is a much more serious problem for weak hydrogen bonds than for stronger interactions. Methods, attitudes and even instincts that have developed over decades for strong hydrogen bonding may be inappropriate, insufficient or plain incorrect in the study of weak hydrogen bonding. At the same time, the commonalities between all types of hydrogen bonding interactions provide a unifying theme and indeed constitute the heart of this work. The electrostatic character of hydrogen bonds such as C—H···O, O—H···π or even C—H···π is just sufficient to observe fine structural effects. In this context, two examples mentioned earlier in this book may be further amplified.

Figures 6.2 and 6.3 show the nearly isostructural toluene and chlorobenzene solvates of 2,3,7,8-tetraphenyl-1,9,10-anthyridine, **60** (Madhavi *et al.* 1997; see also Section 3.1.3.5). It is almost axiomatic that the effects of the weakest hydrogen bonds are heavily influenced by other interactions in crystal structures, rendering their very study close to intractable. Yet this case is different. The anthyridine molecules form a pocket that is just sufficient for the inclusion of toluene or chlorobenzene molecules. In principle, the guest molecules

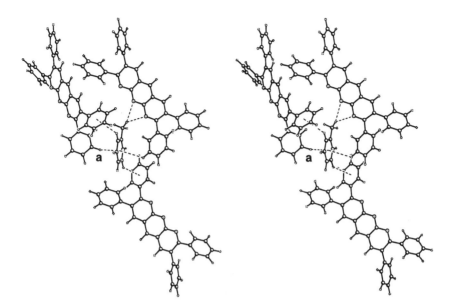

Fig. 6.2. Stereoview of the toluene solvate of tetraphenylanthyridine, **60**. The host molecules are inversion- and glide-related. Contact **a** is a C—H···π interaction between a C—H group of the heterocycle and the π-ring of the toluene. Notice the cooperative scheme of weak interactions (from Madhavi *et al.* 1997).

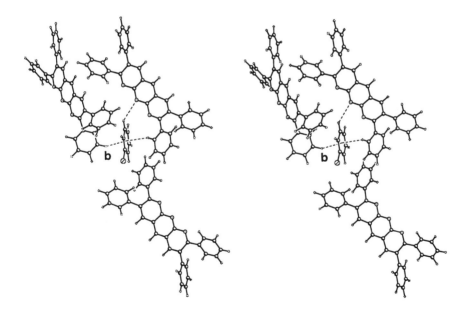

Fig. 6.3. Stereoview of the chlorobenzene solvate of heterocyle, **60**. The arrangement of host molecules is exactly as in Fig. 6.2. The Cl group is positioned in a direction opposite to that of the CH₃ group in Fig. 6.2. The C—H···π contact **b** is 0.07 Å longer than the corresponding one in Fig. 6.2. The pattern of weak interactions is less dense than in Fig. 6.2, concomitant with the reduced π-basicity of chlorobenzene relative to toluene (from Madhavi *et al.* 1997).

may locate themselves in one of two possible orientations with the substituent groups, CH_3 or Cl, pointing up or down in the direction of the figure; the guest molecule also has some lateral freedom between $C-H$ groups of the host that face each other across the pocket. In reality, the toluene guest is located such that the $C-H\cdots\pi$ contact (**a** in Fig. 6.2) is short (2.54 Å), and also such that its methyl group forms $C-H\cdots N$ interactions. The chlorobenzene molecule points in the other direction and the $C-H\cdots\pi$ contact (**b** in Fig. 6.3) is longer (2.61 Å). While the guest molecules would seem to have a free choice between these possibilities in terms of orientation and lateral positioning, neither guest is disordered and the orientations and locations are as might have been predicted from hydrogen bonding considerations. One comes away then with a distinct feeling that there is sufficient electrostatic character in these $C-H\cdots\pi$ interactions to pin down the guest position unambiguously. Shall we call these interactions hydrogen bonds? Throughout the book, we have not used any of the formal hydrogen bond definitions in a stringent way and have remained flexible. We note here that literal interpretations of these definitions can easily lead to man-made complications that have little chemical substance.

Take again the case of the isostructural steroidal lactones, **36** and **37** (see

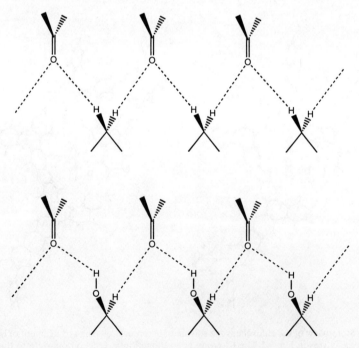

Fig. 6.4. Equivalent supramolecular synthons of 2-oxa-4-androstene-3,17-dione, **37** (*top*), and 6α-hydroxy-2-oxa-4-androstene-3,17-dione, **36** (*bottom*). Notice the interchange of weak and strong hydrogen bonds (from Anthony *et al.* 1998*a*).

also Section 2.2.10.3). The interaction mimicry between these isostructural compounds is shown in Fig. 6.4 (Anthony *et al.* 1998*a*). A $C-O-H\cdots O$ hydrogen bond in the former is replaced by a $C-H\cdots O$ interaction in the latter with little other change. This kind of isostructuralism has been noted elsewhere (Steiner *et al.* 1995*b*) but one could always maintain somewhat fixedly that such isostructurality arises from non-specific packing considerations. What makes this example different is that the two lactones form a substitutional solid solution. Unless there is a close degree of complementarity and pattern matching between the $O-H\cdots O$ and $C-H\cdots O$ interactions, such behaviour would be highly unlikely. So even in crystallographic examples involving fairly complex systems, one may still find clear and unambiguous evidence of the essential hydrogen bonding nature of these weak interactions. More examples that provide greater satisfaction will undoubtedly emerge from the wider variety of chemical systems now being studied.

As mentioned earlier, the range of biological systems studied thus far is not sufficient to provide a comprehensive account of the importance of weak hydrogen bonding in this class of compounds. However, even the developments to date yield enough that might be deemed persuasive evidence. Most notably, one may mention folding patterns in proteins, base-pair interactions in nucleic acids and coordination of water molecules. The role of computational and molecular dynamics studies in exploring the fluctuating patterns of hydrogen bonds in biomolecules will surely increase in the coming years.

Weak hydrogen bonds belong to a domain without fixed boundaries where rigid classifications or demarcations are unwise. We have sought to convey the delicate yet specific nature of these hydrogen bonds in a wide range of chemical and biological systems, in the confidence that one may yet appreciate the beauty and ever-changing character of these interactions that lie between the light and shade of structural science.

Appendix

Some bibliographic statistics

There are around 880 references cited in this book.

References per year

1950–59	1
1960–69	5
1970–79	7
1980–84	15
1985–89	21
1990–94	36
1995–98	97

The development of the relevant literature with time reflects some interesting points. The roots of the subject reach far back in time, but development has been slow and during some long periods even stagnant. A sharp increase of the literature directly related to weak hydrogen bonding occurs in the mid-1990s.

References to journals (≥2 per cent)

Journal of the American Chemical Society	105	11.9%
Chemical Communications	76	8.6%
Acta Crystallographica, Section B	60	6.8%
Acta Crystallographica, Section C	38	4.3%
Journal of the Chemical Society, Perkin Transactions 2	36	4.1%
Journal of Physical Chemistry	30	3.4%
Angewandte Chemie	25	2.8%
Journal of Molecular Biology	25	2.8%
Journal of Molecular Structure	22	2.5%
Journal of Chemical Physics	19	2.2%

All ACS journals: 209 (23.7%). All RSC journals: 146 (16.6%). All sections of *Acta Crystallographica*: 128 (14.6%). *Nature* (0.9%) and *Science* (1.5%) failed to enter the above list.

References

Pages on which each reference is cited are given in square brackets.

Aakeröy, C. B. (1997). Crystal engineering. Strategies and architecture. *Acta Crystallographica, Section B*, **53**, 569–86. [315]

Aakeröy, C. B. and Seddon, K. R. (1993*a*). The hydrogen bond and crystal engineering. *Chemical Society Reviews*, **22**, 397–407. [7, 325]

Aakeröy, C. B. and Seddon, K. R. (1993*b*). The crystal structures of pyridinium chloride revisited. Evidence for extensive C—H···Cl hydrogen-bond interactions. *Zeitschrift für Naturforschung*, **48b**, 1023–5. [249, 251]

Abdul-Sada, A. K., Al-Juaid, S., Greenway, A. M., Hitchcock, P. B., Howells, M. J., Seddon, K. R. and Welton, T. (1990). Upon the hydrogen-bonding ability of the H4 and H5 protons of the imidazolium cation. *Structural Chemistry*, **1**, 391–4. [248, 250]

Abola, E. E., Sussman, J. L., Prilusky, J. and Manning, N. O. (1997). Protein data bank archives of three-dimensional macromolecular structures. *Methods of Enzymology*, **277**, 556–71. [23]

Acharya, K. R., Gowda, D. S. S. and Post, M. (1979). The structure of chloramphenicol. *Acta Crystallographica, Section B*, **35**, 1360–3. [219]

Agafonov, V., Dubois, P., Moussa, F., Cense, J. M. and Toscani, S. (1994). Structure of 2,4-dimethyl-1*H*-naphtho[2,3-b][1,4]diazepine hydropicrate – Solid-state assembly via C—H···O hydrogen bonding. *Journal of the Chemical Society, Perkin Transactions* 2, 2007–10. [299]

Ahmed, N. A., Kitaigorodsky, A. I. and Sirota, M. I. (1972). Crystal structure of *p*-diethynylbenzene. *Acta Crystallographica, Section B*, **28**, 2875–7. [178, 181]

Albinati, A., Meile, S. V., Arnoldi, A. and Galli, R. (1985). Structure of ergosterol-biosynthesis inhibitors. I. Structure of 1-(4-fluorophenyl)-1-(3-pyridyl)-but-3-yn-1-ol, $C_{15}H_{12}FNO$. *Acta Crystallographica, Section C*, **41**, 97–9. [194]

Albinati, A., Bakhmutov, V. I., Caulton, K. G., Clot, E., Eckert, J., Eisenstein, O. *et al.* (1993). Reaction of H_2 with $IrHCl_2P_2$ (P = P^iPr_3 or P^tBu_2Ph): stereoelectronic control of the stability of molecular H_2 transition metal complexes. *Journal of the American Chemical Society*, **115**, 7300–12. [278, 279]

Albrecht, G. and Corey, R. B. (1939). The crystal structure analysis of glycine. *Journal of the American Chemical Society*, **61**, 1087–103. [2]

Aldrich, P. D., Legon, A. C. and Flygare, W. H. (1981). The rotational spectrum, structure, and molecular properties of the ethylene–HCl dimer. *Journal of Chemical Physics*, **75**, 2126–34. [127, 128, 129]

Aldrich, P. D., Kukolich, S. G. and Campbell, E. J. (1983). The structure and molecular properties of the acetylene–HCN complex as determined from the rotational spectra. *Journal of Chemical Physics*, **78**, 3521–30. [128, 129]

Al-Juaid, S. S., Al-Nasr, A. K. A., Eaborn, C. and Hitchcock, P. B. (1991). Hydrogen-bonding interaction between a hydroxy group and the π-electrons of an aryl group. Crystal structure of [*tris*(dimethylphenylsilyl)methyl]methylsilanol. *Journal of the Chemical Society, Chemical Communications*, 1482–4. [149]

Alkorta, I. and Elguero, J. (1996). Carbenes and silylenes as hydrogen bond acceptors. *Journal of Physical Chemistry*, **1996**, 19367–70. [244, 245]

Alkorta, I. and Maluendes, S. (1995). Theoretical study of C—H···O hydrogen bonds in H_2O–CH_3F, H_2O–CH_2F_2 and H_2O–CHF_3. *Journal of Physical Chemistry*, **99**, 6457–60. [74–5]

Alkorta, I., Elguero, J. and Foces-Foces, C. (1996). Dihydrogen bonds (A—H···H—B). *Chemical Communications*, 1633–4. [283]

Alkorta, I., Rozas, I. and Elguero, J. (1998). Non-conventional hydrogen bonds. *Chemical Society Reviews*, **27**, 163–79. [243]

Allen, F. H. (1986). A systematic pairwise comparison of geometric parameters obtained by X-ray and neutron diffraction. *Acta Crystallographica, Section B*, **42**, 515–22. [7]

Allen, F. H. (1998). The development, status and scientific impact of crystallographic databases. *Acta Crystallographica, Section A*, **54**, 758–71. [23, 323]

Allen, G. and Colclough, R. O. (1957). Hydrogen bonding of the thiol group in phosphinodithioic acids. *Journal of the Chemical Society*, 3912–15. [259]

Allen, F. H., Kennard, O. and Taylor, R. (1983). Systematic analysis of structural data as a research technique in organic chemistry. *Accounts of Chemical Research*, **16**, 146–53. [25, 49]

Allen, F. H., Bergerhoff, G. and Sievers, R. (1987). *Crystallographic databases*. International Union of Crystallography, Chester, UK. [23]

Allen, F. H., Davies, J. E., Galloy, J. J., Johnson, O., Kennard, O., Macrae, C. F. and Watson, D. G. (1991). The developments of versions 3 and 4 of the Cambridge Structural Database system. *Journal of Chemical Information and Computer Sciences*, **31**, 187–204. [23, 48, 323]

Allen, F. H., Lommerse, J. P. M., Hoy, V. J., Howard, J. A. K. and Desiraju, G. R. (1996a). The hydrogen bond C—H donor and π-acceptor characteristics of three-membered rings. *Acta Crystallographica, Section B*, **52**, 734–45. [58, 195]

Allen, F. H., Howard, J. A. K., Hoy, V. J., Desiraju, G. R., Reddy, D. S. and Wilson, C. C. (1996b). First neutron diffraction analysis of an O—H···π hydrogen bond. 2-Ethynyladamantan-2-ol. *Journal of the American Chemical Society*, **118**, 4081–4. [70, 83, 84, 166]

Allen, F. H., Bird, C. M., Rowland, R. S. and Raithby, P. R. (1997a). Resonance-induced hydrogen bonding at sulfur acceptors in $R_1R_2C = S$ and $R_1CS_2^-$ systems. *Acta Crystallographica, Section B*, **53**, 680–95. [25, 88, 227, 228, 229, 254]

Allen, F. H., Bird, C. M., Rowland, R. S. and Raithby, P. R. (1997b). Hydrogen-bond acceptor and donor properties of divalent sulfur (Y—S—Z and R—S—H). *Acta Crystallographica, Section B*, **53**, 696–701. [88, 227, 229, 230, 254]

Allen, F. H., Hoy, V. J., Howard, J. A. K., Thalladi, V. R., Desiraju, G. R., Wilson, C. C. and McIntyre, G. J. (1997c). Crystal engineering and correspondence between molecular and crystal structures. Are 2- and 3-aminophenols anomalous? *Journal of the American Chemical Society*, **119**, 3477–80. [318, 319, 320]

Allen, F. H., Baalham, C. A., Lommerse, J. P. M. and Raithby, P. R. (1998a). Carbonyl–carbonyl interactions can be competitive with hydrogen bonds. *Acta Crystallographica, Section B*, **54**, 320–9. [79]

Allen, F. H., Raithby, P. R., Shields, G. P. and Taylor, R. (1998b). Probabilities of formation of bimolecular cyclic hydrogen-bonded motifs in organic crystal structures: a systematic database analysis. *Chemical Communications*, 1043–4. [323]

Allerhand, A. and Schleyer, P. v. R. (1963). A survey of C—H groups as proton donors in hydrogen bonding. *Journal of the American Chemical Society*, **85**, 1715–23. [30–1, 120]

Alyea, E. O., Ferguson, G. and Kannan, S. (1998). Intermolecular hydrogen···metal interactions. The crystal structure of {cis-[PdCl$_2$(TPA)$_2$]}·H_2O, a water soluble palladium (II) tertiary phosphine complex. *Chemical Communications*, 345–6. [272, 275]

Amabilino, D. (1998). Personal communication. [338]

Amidon, G. L., Anik, S. and Rubin, J. (1974). An energy partitioning analysis of base–sugar intramolecular C—H⋯O hydrogen bonding in nucleosides and nucleotides. In *Structures and conformation of nucleic acids and protein–nucleic acid interactions* (ed. M. Sundaralingam and S. T. Rao), pp. 487–524. University Park Press, Baltimore. [391]

Anderson, D. J. and Muchmore, C. R. (1995). Rearrangements of 5-azidoisoxazoles. *Journal of Heterocyclic Chemistry*, **32**, 1189–96. [58]

Andreetti, G. D., Coghi, L., Nardelli, M. and Sgarabotto, P. (1971). Crystal and molecular structure of 1-(2-aminoethyl)biguanide-isothiocyanato copper(II) thiocyanate. *Journal of Crystal and Molecular Structure*, **1**, 147–54. [197]

Anthony, A., Jaskólski, M., Nangia, A. and Desiraju, G. R. (1998a). Isostructurality in crystalline oxa-androgens. A case of C—O—H⋯O and C—H⋯O interaction mimicry and solid solution formation. *Chemical Communications*, 2537–8. [104, 105, 106, 444, 445]

Anthony, A., Desiraju, G. R., Jetti, R. K. R., Kuduva, S. S., Madhavi, N. N. L., Nangia, A. *et al.* (1998b). Crystal engineering. Some further strategies. *Crystal Engineering*, **1**, 1–18. [315]

Arduengo, A. J., III, Gamper, S. F., Tamm, M., Calabrese, J. C, Davidson, F. and Craig, H. A. (1995). A *bis*(carbene)–proton complex. Structure of a C—H—C hydrogen bond. *Journal of the American Chemical Society*, **117**, 572–3. [243–4]

Arlinghaus, R. T. and Andrews, L. (1984). Infrared spectra of the PH_3, AsH_3, and SbH_3⋯HX hydrogen bonded complexes in solid argon. *Journal of Chemical Physics*, **81**, 4341–51. [239]

Armstrong, K. M., Fairman, R. and Baldwin, R. L. (1993). The (*i, i + 4*) Phe-His interaction studied in an alanine-based α-helix. *Journal of Molecular Biology*, **230**, 284–91. [369]

Asakawa, M., Dehaen, W., L'abbé, G., Menzer, S., Nouwen, J., Raymo, F. M. *et al.* (1996). Improved template-directed synthesis of cyclo*bis*(paraquat-*p*-phenylene). *Journal of Organic Chemistry*, **61**, 9591–5. [337]

Ashton, P. R., Chrystal, E. J. T., Glink, P. T., Menzer, S., Schiavo, C., Spencer, N. *et al.* (1996a). Pseudorotaxanes formed between secondary dialkylammonium salts and crown ethers. *Chemistry. A European Journal*, **2**, 709–28. [309]

Ashton, P. R., Menzer, S., Raymo, F. M., Shimizu, G. K. H., Stoddart, J. F. and Williams, D. J. (1996b). The template-directed synthesis of cyclo*bis*(paraquat-4,4′-biphenylene). *Chemical Communications*, 487–90. [338]

Atwood, J. L., Harnada, F., Robinson, K. D., Orr, G. W. and Vincent, R. L. (1991). X-Ray diffraction evidence for aromatic π-hydrogen bonding to water. *Nature*, **349**, 683–4. [144, 145, 161]

Atwood, J. L., Bott, S. G., Jones, C. and Raston, C. L. (1992). Aluminium-fused *bis-p-tert*-butylcalix[4] arene. A double cone with two π-arene⋯H interactions for included methylene chloride. *Journal of the Chemical Society, Chemical Communications*, 1349–51. [153–5]

Atwood, J. L., Koutsantonis, G. A., Lee, F. -C. and Raston, C. L. (1994). A thermally stable alane-secondary amine adduct; [H_3Al(2,2,6,6-tetramethypiperidine)]. *Journal of the Chemical Society, Chemical Communications*, 91–2. [287]

Atwood, J. L., Davies, J. E. D., MacNicol, D. D. and Vögtle, F. (ed.) (1996). *Comprehensive supramolecular chemistry*, Vol. 2. Pergamon, Oxford. [293, 310]

Aubry, A., Protas, J., Moreno-Gonzales, E. and Marraud, M. (1977). Structure cristalline du tétraphénylborate de tributylammonium monohydraté. Un exemple d'interaction H—π (*in French*). *Acta Crystallographica, Section B*, **33**, 2572–8. [144]

Auffinger, P. and Westhof, E. (1996). H-Bond stability in the tRNA^{Asp} anticodon hairpin. 3 ns of multiple molecular dynamics simulations. *Biophysical Journal*, **71**, 940–54. [80, 395–6, 398]

Auffinger, P. and Westhof, E. (1997). Rules governing the orientation of the 2'-hydroxyl group in RNA. *Journal of Molecular Biology*, **274**, 54–63. [394, 395, 405]

Auffinger, P. and Westhof, E. (1998). Simulations of the molecular dynamics of nucleic acids. *Current Opinion in Structural Biology*, **8**, 227–36. [437]

Auffinger, P., Louise-May, S. and Westhof, E. (1996a). Molecular dynamics simulations of the anticodon hairpin of tRNA^{Asp}. Structuring effects of C—H···O hydrogen bonds and of long-range hydration forces. *Journal of the American Chemical Society*, **118**, 1181–9. [395, 397]

Auffinger, P., Louise-May, S. and Westhof, E. (1996b). Hydration of C—H groups in tRNA. *Faraday Discussions*, **103**, 151–73. [437–40]

Aullón, G., Bellamy, D., Brammer, L., Bruton, E. and Orpen, A. G. (1998). Metal-bound chlorine often accepts hydrogen bonds. *Chemical Communications*, 653–4. [211, 215–17, 219]

Badger, R. M. and Bauer, S. H. (1937). Spectroscopic studies of the hydrogen bond, II. The shift in O—H vibrational frequency in formation of a hydrogen bond. *Journal of Chemical Physics*, **5**, 839–51. [26]

Baert, F., Fouret, R., Sliwa, H. and Sliwa, H. (1983). Structural study of two isomeric (2-phenyl-3-chromanyl)methanols. *Acta Crystallographica, Section B*, **39**, 444–50. [149, 150]

Baker, A. W. (1958). On the *cis–trans* equilibria in *o*-halophenols. *Journal of the American Chemical Society*, **80**, 3598–600. [202]

Baker, E. N. (1980). Structure of actinidin, after refinement at 1.7 Å resolution. *Journal of Molecular Biology*, **141**, 441–84. [430, 431]

Baker, E. N. and Hubbard, R. E. (1984). Hydrogen bonding in globular proteins. *Progress in Biophysics and Molecular Biology*, **44**, 97–179. [347, 430]

Baker, A. W. and Shulgin, A. T. (1958). Intramolecular hydrogen bonds to π-electrons and other weakly basic groups. *Journal of the American Chemical Society*, **80**, 5358–63. [125, 202]

Bakshi, P. K., Linden, A., Vincent, B. R., Roe, S. P., Adhikesavalu, D., Cameron, T. S. and Knop, O. (1994). Crystal chemistry of tetraradial species. Part 4. Hydrogen bonding to aromatic π-systems. Crystal structures of fifteen tetraphenylborates with organic ammonium cations. *Canadian Journal of Chemistry*, **72**, 1273–93. [137, 142, 143, 144]

Balegroune, F., Braunstein, P., Grandjean, D., Matt, D. and Nobel, D. (1988). Complexes with functional phosphines. 13. Reactivity of coordinated phosphino enolates with metallacycles toward chlorophosphines with diastereoselective formation of P—O and P—C bonds. Alkaline hydrolysis of coordinated phosphinites leading to oxodiarylphosphoranido ligands. Synthesis and molecular structure of *cis*-[Pd{Ph$_2$PCH=C(O)Ph} {CH$_2$C$_9$H$_6$N)}]. *Inorganic Chemistry*, **27**, 3320–5. [47]

Balocchi, F. A., Williams, J. H. and Klemperer, W. (1983). Molecular beam studies of C$_6$F$_6$, C$_6$F$_3$H$_3$ and C$_6$H$_6$ complexes of HF. The rotational spectrum of C$_6$H$_6$–HF. *Journal of Physical Chemistry*, **87**, 2079–84. [129–30]

Bandy, J. A., Truter, M. R. and Vögtle, F. (1981). The structure of the 1,4,7,10,13,16-hexaoxacyclooctadecane (18-crown-6) *bis*-(dimethyl sulfone) complex. *Acta Crystallographica, Section B*, **37**, 1568–71. [307]

Banks, R. E., DuBoisson, R. A., Pritchard, R. G. and Tipping, A. E. (1995). 2,3,4,5-Tetrafluoro-*N*-(4-fluorophenyl)-6-hydroxybenzamide. An example of a combined inter- and intra-molecular O···H···O bifurcated hydrogen bond. *Acta Crystallographica, Section C*, **51**, 1427–9. [206]

Barnes, C. L. and Hawkinson, S. W. (1982). Structure of disodium guanosine 5'-phosphate heptahydrate. *Acta Crystallographica, Section B*, **38**, 812–7. [391]

Barrett, D. M. Y., Kahwa, I. A., Mague, J. T. and McPherson, G. L. (1995). Preparations, crystal structures, and unusual proton NMR characteristics of some phthalimides. *Journal of Organic Chemistry*, **60**, 5946–53. [107]

Batail, P., LaPlaca, S. J., Mayerle, J. J. and Torrance, J. B. (1981). Structural characterization of the neutral-ionic phase transition in tetrathiafulvalene-chloranil. Evidence for C—H···O hydrogen bonding. *Journal of the American Chemical Society*, **103**, 951–3. [38]

Bats, J. W. (1976). The crystal structure of 2,5-dimercaptothiadiazole. *Acta Crystallographica, Section B*, **32**, 2866–70. [258]

Batsanov, A. S., Davidson, M. G., Howard, J. A. K, Lamb, S. and Lustig, C. (1996). Phosphonium ylides as hydrogen bond acceptors. Intermolecular C—H···C interactions in the crystal structure of triphenylphosphonium benzylide. *Chemical Communications*, 1791–2. [245, 246]

Battaglia, L.P., Corradi, A.B., Menabue, L., Pellacani, G. C., Prampolini, P. and Saladini, M. (1982). Ternary complexes of copper(II) with *N*-protected amino-acids and *N*-methylimidazole. Crystal and molecular structures of bis(*N*-acetyl-α-alaninato)bis(*N*-methylimidazole)copper(II) dihydrate. *Journal of the Chemical Society, Dalton Transactions*, 781–5. [274]

Bau, R., Teller, R. G., Kirtley, S. W. and Koetzle, T. F. (1979). Structures of transition metal hydride complexes. *Accounts of Chemical Research*, **12**, 176–83. [277]

Bednowitz, A. L. and Post, B. (1966). Direct determination of the crystal structure of β-fumaric acid. *Acta Crystallographica*, **21**, 566–71. [38]

Behr, J.-P. (ed.) (1994). *The lock-and-key principle*, Perspectives in supramolecular chemistry, Vol. 1. Wiley, Chichester. [293]

Behrens, P., van de Goor, G. and Freyhardt, C. C. (1995). Structure-determining C—H···O—Si hydrogen bonds in cobaltocenium fluoride nonasil. *Angewandte Chemie, International Edition in English*, **34**, 2680–2. [107]

Bella, J. and Berman, H. (1996). Crystallographic evidence for C^α—H···O=C hydrogen bonds in a collagen triple helix. *Journal of Molecular Biology*, **264**, 734–42. [357, 359]

Benevides, J. M. and Thomas, G. J. Jr. (1988). A solution structure for poly(rA)·poly(dT) with different furanose pucker and backbone geometry in rA and dT strands and intrastrand hydrogen bonding of adenine 8CH. *Biochemistry*, **27**, 3868–73. [395]

Benghiat, V. and Leiserowitz, L. (1972). Molecular packing modes. Part VIII. Crystal and molecular structures of but-3-ynoic acid. *Journal of the Chemical Society, Perkin Transactions 2*, 1772–8. [178]

Bennis, V. and Gallagher, J. F. (1998). N—H···π (pyrrole) intermolecular interactions in 1,4-*bis*(di-2-pyrrolylmethyl)benzene. *Acta Crystallographica, Section C*, **54**, 130–2. [191]

Berger, I. and Egli, M. (1997). Role of backbone oxygen atoms in the organization of nucleic acid tertiary structure. Zippers, networks, clamps, and C—H···O hydrogen bonds. *Chemistry. A European Journal*, **3**, 1400–4. [387, 396–8, 399]

Berger, I., Egli, M. and Rich, A. (1996). Inter-strand C—H···O hydrogen bond stabilizing four-stranded intercalated molecules. Stereoelectronic effects of $O_{4'}$ in cytosine-rich DNA. *Proceedings of the National Academy of Sciences of the USA*, **93**, 12116–21. [405, 406]

Bergerhoff, G., Hundt, R., Sievers, R. and Brown, I. D. (1983). The inorganic crystal structure database. *Journal of Chemical Information and Computer Sciences*, **23**, 66–9. [23]

Berkovitch-Yellin, Z. and Leiserowitz, L. (1982). Atom–atom potential analysis of the packing characteristics of carboxylic acids. A study based on experimental electron density distributions. *Journal of the American Chemical Society*, **104**, 4052–64. [38, 73]

Berkovitch-Yellin, Z. and Leiserowitz, L. (1984). The role played by C—H···O and C—H···N interactions in determining molecular packing and conformation. *Acta Crystallographica, Section B,* **40**, 159–65. [98, 104]

Berman, H. M., Olson, W. K., Beveridge, D. L., Westbrook, J., Gelbin, A., Demeny, T. *et al.* (1992). The nucleic acid database. A comprehensive relational database of three-dimensional structures of nucleic acids. *Biophysical Journal*, **63**, 751–9. [23]

Bernardinelli, G., Geoffroy, M. and Franzi, R. (1991). Triphenylmethanethiol. *Zeitschrift für Kristallographie*, **195**, 147–8. [264, 265]

Bernstein, J., Cohen, M. D. and Leiserowitz, L. (1974). The structural chemistry of quinones. In *The chemistry of the quinonoid compounds. Part I* (ed. S. Patai), pp. 37–110, Interscience, New York. [36, 98, 120]

Bernstein, J., Davis, R. E., Shimoni, L. and Chang, N. -L. (1995). Patterns in hydrogen bonding: functionality and graph set analysis in crystals. *Angewandte Chemie, International Edition in English*, **34**, 1555–73. [25]

Bertolasi, V., Gilli, P., Ferretti, V. and Gilli, G. (1996). Resonance-assisted O—H···O hydrogen bonding. Its role in the crystalline self-recognition of β-diketone enols and its structural and IR characterization. *Chemistry. A European Journal*, **2**, 925–34. [68, 262]

Bhattacharyya, P., Novosad, J., Phillips, J., Slawin, A. M. Z., Williams, D. J. and Woolins, J. D. (1995). *bis*(Bidentate) complexes of imino*bis*(diphenylphosphine chalcogenides). [M{N(XPPh$_2$)$_2$-X,X'}$_2$] (X = S or Se; M = Ni, Pd or Pt). *Journal of the Chemical Society, Dalton Transactions*, 1607–13. [234]

Biradha, K. and Zawarotko, M. J. (1998). A supramolecular analogue of cyclohexane sustained by aromatic C—H···π interactions. Complexes of 1,3,5-trihydroxybenzene with substituted pyridines. *Journal of the American Chemical Society*, **25**, 6431–2. [156]

Biradha, K., Sharma, C. V. K., Panneerselvam, K., Shimoni, L., Carrell, H. L., Zacharias, D. E. and Desiraju, G. R. (1993). Solid state supramolecular assembly via C—H···O hydrogen bonds. Crystal structures of the complexes of 1,3,5-trinitrobenzene with dibenzylideneacetone and 1,5-dibenzylidenecyclo-pentanone. *Journal of the Chemical Society, Chemical Communications*, 1473–5. [326]

Biradha, K., Nangia, A., Desiraju, G. R., Carrell, C. J. and Carrell, H. L. (1997). C—H···O hydrogen bonded multi-point recognition in molecular assemblies of dibenzylidene ketones and 1,3,5-trinitrobenzenes. *Journal of Materials Chemistry*, **7**, 1111–22. [326]

Biswas, R. and Sundaralingam, M. (1997). Crystal structure of r(GUGUGUA)cD with tandem G·U/U·G wobble pairs with strand slippage. *Journal of Molecular Biology*, **270**, 511–19. [405]

Blaschette, A., Nagel, K. H. and Jones, P. G. (1993). Polysulfonylamines, XLIV. Synthesis of [Na(15-crown-5)][N(SO$_2$CH$_3$)$_2$] and crystal structure at − 95 °C. *Zeitschrift für Naturforschung*, **48b**, 893–7. [107]

Bochmann, M., Bwembya, G. C., Whilton, N., Song, X., Hursthouse, M. B., Coles, S. J. and Karaulov, A. (1995). Synthesis of selenophosphinic and tellurophosphinic amides and amidato complexes. Crystal structures of $Bu_2^iP(Te)NH(C_6H_{11})$, $[Ti(\eta\text{-}C_5H_5)Cl_2$ $\{Bu_2^iP(Se)Npr^i\}]$ and $[TiCl_2\{Bu_2^iP(Se)N(C_6H_{11})\}_2]\cdot C_7H_8$. *Journal of the Chemical Society, Dalton Transactions*, 1887–92. [237, 238]

Bock, H., Dienelt, R., Schödel, H. and Havlas, Z. (1993). The $C-H\cdots O$ hydrogen bond adduct of two trinitromethanes to dioxane. *Journal of the Chemical Society, Chemical Communications*, 1792–3. [44, 47]

Bock, H., Nick, S., Seitz, W., Nather, C. and Bats, J. W. (1996). Structure of charge-perturbed or sterically overcrowded molecules, 80. Structural changes of *p*-benzoquinone by donor and acceptor substituents. *Zeitschrift für Naturforschung*, **51b**, 153–71. [297]

Bodige, S. G., Rogers, R. D. and Blackstock, S. C. (1997). Supramolecular networks via pyridine *N*-oxide $C-H\cdots O$ hydrogen bonding in the crystal structures of 2,2′-dithiobis(pyridine *N*-oxide) and its complexes with 1,2,4,5-tetracyanobenzene and pyromellitic dianhydride. *Chemical Communications*, 1669–70. [334]

Boenigk, D. and Mootz, D. (1988). The system pyridine–hydrogen fluoride at low temperatures. Formation and crystal structures of solid complexes with very strong NHF and FHF hydrogen bonding. *Journal of the American Chemical Society*, **110**, 2135–9. [209, 213]

Boese, R., Bläser, D. and Kuhn, N. (1994). Crystal structure of 1,3-diisopropyl-4,5-dimethylimidazolium chloride *tris*(trichlormethane), $C_{11}H_{21}ClN\cdot 3CHCl_3$. *Zeitschrift für Kristallographie*, **209**, 837–8. [248, 250]

Böhm, H.-J. (1992). LUDI. Rule-based automatic design of new substituents for enzyme inhibitor leads. *Journal of Computer-Aided Molecular Design*, **6**, 593–606. [338]

Böhm, H.-J. and Klebe, G. (1996). What can we learn from molecular recognition in protein–ligand complexes for the design of new drugs? *Angewandte Chemie, International Edition in English*, **35**, 2588–614. [338]

Bondi, A. (1964). Van der Waals volumes and radii. *Journal of Physical Chemistry*, **68**, 441–51. [20]

Bone, R. G. A., Murray, C. W., Amos, R. D. and Handy, N. C. (1989). Stationary points on the potential energy surface of $(C_2H_2)_3$. *Chemical Physics Letters*, **161**, 166–74. [176]

Boorman, P. M., Gao, X. and Parvez, M. (1992). X-Ray structural characterization of a thiolate salt displaying a very strong $S-H\cdots S$ hydrogen bond. *Journal of the Chemical Society, Chemical Communications*, 1656–8. [262]

Booth, C. A., Foxman, B. M. and Jaufmann, J. D. (1987). In *Crystallographically ordered polymers*, ACS Symposium Series, No. **337**, (ed. D. J. Sandman), pp. 95–105. American Chemical Society, Washington, DC. [174]

Borwick, S. J., Howard, J. A. K., Lehmann, C. W. and O'Hagan, D. (1997). 1-(4-bromophenyl)-2-fluoroethanone (2,4-dinitrophenyl)hydrazone containing a particularly short fluorine–hydrogen bond. *Acta Crystallographica, Section C*, **53**, 124–6. [205, 206]

Bouquiere, J. P., Finney, J. L., Lehmann, M. S., Lindley, P. F. and Savage, H. F. J. (1993). High-resolution neutron study of vitamin B_{12} coenzyme at 15 K. Structure analysis and comparison with the structure at 279 K. *Acta Crystallographica, Section B*, **49**, 79–89. [154, 418, 420]

Bowden, B. F., Coll, J. C., Ditzel, E., Mitchell, S. J. and Robinson, W. T. (1982). Studies of Australian soft corals. XXVIII. The structure determination of two A diterpenenes from the genus *xenia* (*alcyonacea*). *Australian Journal of Chemistry*, **35**, 997–1002. [186, 187]

Boyd, D. B. (1995). Computer-aided molecular design. In *Encyclopedia of Computer*

Science and Technology, Vol. 33, Suppl. 18 (ed. A. Kent and J. G. Williams), pp. 41–71. Marcel Dekker, New York. [338]

Boyd, D. R., Evans, T. A., Jennings, W. B., Malone, J. F., O'Sullivan, W. and Smith, A. (1996). Edge-to-face aromatic interactions in alkenes, nitrones and imines. *Chemical Communications*, 2269–70. [158]

Braden, D. A. and Gard, G. L. (1996). Crystal structure of (fluorosulfonyl)fluoroacetic acid. A novel four-center hydrogen bond system involving a C—H donor. *Inorganic Chemistry*, **35**, 1912–14. [45]

Braga, D., Grepioni, F., Biradha, K., Pedireddi, V. R. and Desiraju, G. R. (1995). Hydrogen bonding in organometallic crystals. 2. C—H···O hydrogen bonds in bridged and terminal first-row metal carbonyls. *Journal of the American Chemical Society*, **117**, 3156–66. [56, 57, 64, 67, 71, 72, 89, 271, 302]

Braga, D., Grepioni, F., Tedesco, E., Biradha, K. and Desiraju, G. R. (1996a). Hydrogen bonding in organometallic crystals. 4. M—H···O hydrogen-bonding interactions. *Organometallics*, **15**, 2692–9. [53, 277, 278, 279]

Braga, D., Grepioni, F., Biradha, K. and Desiraju. G. R. (1996b). Agostic interactions in organometallic compounds. Cambridge Structural Database studies. *Journal of the Chemical Society, Dalton Transactions*, 3925–30. [281]

Braga, D., Grepioni, F., Tedesco, E., Biradha. K. and Desiraju, G. R. (1997a). Hydrogen bonding in organometallic crystals. 6. X—H···M hydrogen bonds and M···(H—X) pseudo-agostic bonds. *Organometallics*, **16**, 1846–56. [273, 274, 276, 281]

Braga, D., Costa, A. L., Grepioni, F., Scaccianoce, L. and Tagliavini, E. (1997b). Organic-organometallic crystal synthesis. 1. Hosting paramagnetic [(η⁶-arene)₂Cr]⁺ (arene = benzene, toluene) in organic anion frameworks via O—H···O and C—H···O hydrogen bonds. *Organometallics*, **16**, 2070–9. [302]

Braga, D., Angeloni, A., Grepioni, F. and Tagliavini, E. (1997c). Organic-organometallic crystal synthesis. 2. Organic frameworks constructed around [(η⁵-C₅H₅)₂Co]⁺ via charge-assisted O—H···O and C—H···O hydrogen bonds. *Organometallics*, **16**, 5478–85. [302]

Braga, D., Grepioni, F., Tagliavini, E., Novoa, J. J. and Mota, F. (1998a). C—H···O hydrogen bonds in the mixed-valence salt [(η⁶-C₆H₆)₂Cr]⁺[CrO₃(OCH₃)]⁻ and the breakdown of the length/strength analogy. *New Journal of Chemistry*, 755–7. [111]

Braga, D., Grepioni, F. and Tedesco, E. (1998b). X—H···π (X = O, N, C) hydrogen bonds in organometallic crystals. *Organometallics*, **17**, 2669–72. [136–7, 138, 199, 271]

Braga, D., Grepioni, F. and Desiraju, G. R. (1998c). Crystal engineering and organometallic architecture. *Chemical Reviews*, **98**, 1375–405. [270, 271–2, 274, 289, 303]

Brammer, L. (1999). Direct and indirect roles of metal centres in hydrogen bonding. In *Implications of molecular and materials structure for new technologies* (ed. J. A. K. Howard and F. H. Allen), pp. 197–210. Kluwer, Dordrecht. [201, 271, 272]

Brammer, L. and Zhao, D. (1994). A hydrogen-bonded chain involving both N—H···N and N—H···Co hydrogen bonds. Low-temperature X-ray crystal structure of [(NMP)₃H₂][Co(CO)₄]₂ (NMP = N-methylpiperazine). *Organometallics*, **13**, 1545–7. [272, 274]

Brammer, L., Charnock, J. M., Goggin, P. L., Goodfellow, R. J., Koetzle, T. F. and Orpen, A. G. (1987). Hydrogen bonding by cisplatin derivatives.Evidence for the formation of N—H···Cl and N—H···Pt bonds in [NPr₄ⁿ]₂{[PtCl₄]·cis-[PtCl₂(NH₂Me)₂]}. *Journal of the Chemical Society, Chemical Communications*, 443–4. [219]

Brammer, L., Charnock, J. M., Goggin, P. L., Goodfellow, R. J., Koetzle, T. F. and Orpen, A.G. (1991). The role of transition metal atoms as hydrogen bond acceptors.

A neutron diffraction study of [NPr$_4^n$]$_2$[PtCl$_4$].*cis*-[PtCl$_2$(NH$_2$Me)$_2$] *Journal of the Chemical Society, Dalton Transactions*, 1789–98. [272]

Brammer, L., McCann, M.C., Bullock, R.M., McMullan, R.K. and Sherwood, P. (1992). Et$_3$NH$^+$Co(CO)$_4^-$: Hydrogen-bonded adduct or simple ion pair? Single-crystal neutron diffraction study at 15K. *Organometallics*, **11**, 2339–41. [272]

Brammer, L., Zhao, D., Ladipo, F. T. and Braddock-Wilking, J. (1995). Hydrogen bonds involving transition metal centres – a brief review. *Acta Crystallographica, Section B*, **51**, 632–40. [270, 275]

Brammer, L., Klooster, W. and Lemke, F. R. (1996). Neutron diffraction studies of [Cp(PMe$_3$)$_2$RuH] and [Cp(PMe$_3$)$_2$RuH$_2$]BF$_4$ at 20 K. *Organometallics*, **15**, 1721–7. [214, 215]

Brauer, D. J., Gol, F., Hietkamp, S. and Stelzer, O. (1986). Synthesis and structure of functionalized 1,2,3,5-diazadiphospholanes (*in German*). *Chemische Berichte*, **119**, 2767–76. [239, 241]

Brédas, J. L. and Street, G. B. (1988). Theoretical studies of the structure of the benzene–hydrogen fluoride complex. *Journal of the American Chemical Society*, **110**, 7001–5. [134]

Bricklebank, N., Godfrey, S. M., McAuliffe, C. A. and Pritchard, R. G. (1993). Triphenylphosphonium bromide. *Acta Crystallographica, Section C*, **49**, 1017–18. [267, 269]

Britton, D., Gleason, W. B. and Glick, M. (1981). Xanthocillin, 1,4-*bis*(4-hydroxyphenyl)-1,3-butadiene-2,3-diisonitrile, C$_{18}$H$_{12}$N$_2$O$_2$. *Crystal Structure Communications*, **10**, 1497–500. [243]

Brock, C. P. and Duncan, L. L. (1994). Anomalous space-group frequencies for monoalcohols C$_n$H$_m$OH. *Chemistry of Materials*, **6**, 1307–12. [232]

Brock, C. P. and Dunitz, J. D. (1994). Towards a grammar of crystal packing. *Chemistry of Materials*, **6**, 1118–27. [294]

Brodkin, J. S. and Foxman, B. M. (1996). Preparation, structure, and solid-state reactivity of lanthanum propynoates. *Chemistry of Materials*, **8**, 242–7. [41]

Brown, I. D. (1992). Chemical and steric constraints in organic solids. *Acta Crystallographica, Section B*, **48**, 553–72. [68]

Buchanan, G. W., Rodrigue, A., Bensimon, C. and Ratcliffe, C. I. (1992). The 18-crown-6·2 chloroacetonitrile complex. X-Ray crystal structure and solid phase motions of guest and host as studied by variable temperature ^{13}C CPMAS NMR spectroscopy. *Canadian Journal of Chemistry*, **70**, 1033–41. [304]

Buchholz, S., Harms, K., Marsch, M., Massa, W. and Boche, G. (1989*a*). Model of a solvent-shared ion pair with N—H···C hydrogen bonds bewteen amine and carbanion – crystal structure of [fluorenyllithium·2ethylenediamine]$_∞$. *Angewandte Chemie, International Edition in English*, **28**, 72–3. [145]

Buchholz, S., Harms, K., Marsch, M., Massa, W. and Boche, G. (1989*b*). Hydrogen bonds between an NH$_4^+$ ion and a carbanion – crystal structure of ammonium 1,2,4-tricyanocyclopentadienide. *Angewandte Chemie, International Edition in English*, **28**, 73–5. [145]

Buckle, A. M., Cramer, P. and Fersht, A. R. (1996). Structural and energetic responses to cavity-creating mutations in hydrophobic cores. Observation of a buried water molecule and the hydrophilic nature of such hydrophobic cavities. *Biochemistry*, **35**, 4298–305. [431]

Bürgi, H. -B. and Dunitz, J. D. (1983). From crystal statics to chemical dynamics. *Accounts of Chemical Research*, **16**, 153–61. [21, 68, 87]

Bürgi, H. -B., Hulliger, J. and Langley, P. J. (1998). Crystallisation of supramolecular materials. *Current Opinion in Solid State and Materials Science*, **3**, 425–30. [315]

Burks, J. E., van der Helm, D., Chang, C. Y. and Ciereszko, L. S. (1977). The crystal and molecular structure of briarein A, a diterpenoid from the gorgonian *briareum asbestinum*. *Acta Crystallographica, Section B*, **33**, 704–9. [218]

Burley, S. K. and Petsko, G. A. (1986). Amino–aromatic interactions in proteins. *FEBS Letters*, **203**, 139–43. [365–6]

Burley, S. K. and Petsko, G. A. (1988). Weakly polar interactions in proteins. *Advances in Protein Chemistry*, **39**, 125–89. [347]

Burrows, A. D., Chan, C. W., Chowdhry, M. M., McGrady, J. E. and Mingos, D. M. P. (1995). Multidimensional crystal engineering of bifunctional metal complexes containing complementary triple hydrogen bonds. *Chemical Society Reviews*, 329–39. [325]

Busetti, V., Del Pra, A. and Mammi, M. (1969). The structure of trioxane at low temperature. *Acta Crystallographica, Section B*, **25**, 1191–4. [284]

Caira, M. R. and Mohamed, R. (1993). Stabilizing role of included solvent in ternary complexation. Synthesis, structures and thermal analyses of three 18-crown-6/sulfonamide/acetonitrile inclusion compounds. *Acta Crystallographica, Section B*, **49**, 760–8. [306, 307, 308]

Caira, M. R., Griffith, V. J., Nassimbeni, L. R. and Oudtshoorn, B. V. (1994). Unusual $^1C_4$ conformation of a methylglucose residue in crystalline permethyl-β-cyclodextrin monohydrate. *Journal of the Chemical Society, Perkin Transactions 2*, 2071–2. [410, 411]

Calabrese, J. C., McPhail, A. T. and Sim, G. A. (1966). Hydrogen bonds. Part I. The ethynyl–H···O interaction in crystals of propargyl 2-bromo-3-nitrobenzoate. *Journal of the Chemical Society (B)*, 1235–8. [35–6, 41]

Calabrese, J. C., McPhail, A. T. and Sim, G. A. (1970). Hydrogen bonds. Part II. The ethynyl–H···O interaction in crystals of triethylprop-2-ynylammonium *p*-bromobenzenesulphonate, *Journal of the Chemical Society (B)*, 282–90. [35–6]

Calderazzo, F., Fachinetti, G., Marchetti, F. and Zanazzi, P. F. (1981). Preparation and crystal and molecular structure of two trialkyl amine adducts of $HCo(CO)_4$ showing a preferential $NR_3H^+ \cdots [(OC)_3Co(CO)]^-$ interaction. *Journal of the Chemical Society, Chemical Communications*, 181–3. [272, 274]

Cameron, A. F., Cheung, K. K., Ferguson, G. and Robertson, J. M. (1969). Laurencia natural products. Part I. Crystal structure and absolute stereochemistry of laurencin. *Journal of the Chemical Society (B)*, 559–64. [35]

Campbell, J. P., Hwang, J.-W. Young, Jr., V. G., Von Dreele, R. B., Cramer, C. J. and Gladfelter, W. L. (1998). Crystal engineering using the unconventional hydrogen bond. Synthesis, structure, and theoretical investigation of cyclotrigallazane. *Journal of the American Chemical Society,* **120**, 521–31. [285, 286]

Carrell, H. L. and Glusker, J. P. (1997). The molecular complex of dibenz[*a,c*]anthracene and trinitrobenzene. *Structural Chemistry*, **8**, 141–7. [299, 300]

Carroll, M. T. and Bader, R. F. W. (1988). An analysis of the hydrogen bond in BASE–HF complexes using the theory of atoms in molecules. *Molecular Physics*, **65**, 695–722. [89]

Cavaglioni, A. and Cini, R. (1997). The first crystal structure of a rhodium complex with the antileukaemic drug purine-6-thione; synthesis and molecular orbital investigation of new organorhodium(III) compounds. *Journal of the Chemical Society, Dalton Transactions*, 1149–58. [155]

Chakrabarti, P. and Chakrabarti, S. (1998). C—H···O hydrogen bond involving proline residues in α-helices. *Journal of Molecular Biology*, **284**, 867–73. [356]

Chaney, J. D., Goss, C., R., Folting, K., Santarsiero, B. D. and Hollingsworth, M. D.

(1996). Formyl C—H···O hydrogen bonding in crystalline bis-formamides? *Journal of the American Chemical Society*, **118**, 9432–3. [114]

Chao, I. and Chen, J. -C. (1996). Resolving the puzzling eclipsed conformation of the methyl group in a tricyclic orthoamide trihydrate. *Angewandte Chemie, International Edition in English*, **35**, 195–7. [97]

Chao, M. and Schempp, E. (1977). An X-ray and NQR study of 4-aminopyridine and related aromatic amines. *Acta Crystallographica, Section B*, **33**, 1557–64. [193]

Chardin, A., Laurence, C. and Berthelot, M. (1996). 2,6-di-*tert*-butylpyridine. A π-hydrogen-bond base. *Journal of Chemical Research (S)*, 332–3. [125, 126, 127]

Chatterjee, A. and Parthasarathy, R. (1984). Conformation and hydrogen bonding of *N*-formylmethionyl peptides. *International Journal of Peptide and Protein Research*, **24**, 447–52. [353, 356]

Chenevert, R., Gagnon, R. and Belanger-Gariepy, F. (1993). 18-Crown-6-*bis*(methyl-10-camphor sulphonate) (1/2). *Acta Crystallographica, Section C*, **49**, 1796–9. [307]

Chiang, C. C., Karle, I. L. and Wieland, T. (1982). Unusual intramolecular hydrogen bonding in cycloamanide A, cyclic (LPro-LVal-LPhe-LPhe-LAla-Gly). *International Journal of Peptide and Protein Research*, **20**, 414–20. [424, 425]

Choe, Y. S. and Katzenellenbogen, J. A. (1995). Synthesis of C-6 fluoroandrogens. Evaluation of ligands for tumor receptor imaging. *Steroids*, **60**, 414–22. [342]

Choe, Y. S., Lidström P. J., Chi, D. Y., Bonasera, T. A., Welch, M. J. and Katzenellenbogen, J. A. (1995). Synthesis of 11β-[^{18}F]fluoro-5α-dihydrotestosterone and 11β-[^{18}F]fluoro-19-nor-5α-dihydrotestosterone. Preparation via halofluorination-reduction, receptor binding, and tissue distribution. *Journal of Medicinal Chemistry*, **38**, 816–25. [342]

Churchill, M. R. and Bueno, C. (1983). Structural studies on polynuclear osmium carbonyl hydrides. 24. Crystal structures of the 60-electron cluster (μ-H)$_3$Os$_3$Co(CO)$_9$(η^5-C$_5$H$_5$). *Inorganic Chemistry*, **22**, 1510–15. [278]

Cirujeda, J., Mas, M., Molins, E., Panthou, F. L. d., Laugier, J., Park, J. G. *et al.* (1995). Control of the structural dimensionality in hydrogen-bonded self-assemblies of open-shell molecules. Extension of intermolecular ferromagnetic interactions in α-phenyl nitronyl nitroxide radicals into three dimensions. *Journal of the Chemical Society, Chemical Communications*, 709–10. [336]

Ciunik, Z. (1997). Intramolecular C—H···N hydrogen bond interactions in 1-(2-hydroxy-iminopyranosyl)pyrazoles. Results of crystallographic and semiempirical studies. *Journal of Molecular Structure*, **436–437**, 173–9. [97]

Ciunik, Z. and Jarosz, S. (1998). Hybride interactions (stacking + H-bonds) between molecules bearing benzyl groups. *Journal of Molecular Structure*, **442**, 115–19. [156–7]

Ciunik, Z., Berski, S., Latajka, Z. and Leszczynski, J. (1998). New aspects of weak C—H···π bonds: intermolecular interactions between alicyclic and aromatic rings in crystals of small compounds, peptides and proteins. *Journal of Molecular Structure*, **442**, 125–34. [157]

Cockroft, J. K. and Fitch, A. N. (1990). The solid phases of deuterium sulphide by powder neutron diffraction. *Zeitschrift für Kristallographie*, **193**, 1–19. [254, 259]

Cole, J. M., Gibson, V. C., Howard, J. A. K., McIntyre, G. J. and Walker, G. L. P. (1998). Multiple α-agostic interactions in a metal–methyl complex: the neutron structure of [Mo(NC$_6$H$_3$Pr$^i_2$-2,6)$_2$Me$_2$]. *Chemical Communications*, 1829–30. [280]

Coleman, P. C., Coucourakis, E. D. and Pretorius, J. A. (1980). Crystal structure of retrosine. *South African Journal of Chemistry*, **33**, 116–19. [186]

Cooke, S. A., Corlett, G. K., Evans, C. M. and Legon, A. C. (1997). The rotational spectrum of the benzene–hydrogen bromide complex. *Chemical Physics Letters*, **272**, 61–8. [129]

Cooke, S. A., Corlett, G. K. and Legon, A. C. (1998a). Is pyridinium hydrochloride a simple hydrogen-bonded complex $C_5H_5N \cdots HCl$ or an ion pair $C_5H_5NH^+ \cdots Cl^-$? An answer from its rotational spectrum. *Journal of the Chemical Society, Faraday Transactions*, **94**, 837–41. [128, 130]

Cooke, S. A., Corlett, G. K. and Legon, A. C. (1998b). Rotational spectrum of thiophene $\cdots HCl$. *Journal of the Chemical Society, Faraday Transactions*, **94**, 1565–70. [128, 130]

Coppens, P. (1964). Neutron diffraction study of 2-nitrobenzaldehyde and the $C-H \cdots O$ interaction. *Acta Crystallographica*, **17**, 573–8. [112]

Coppens, P. (1997). *X-ray charge densities and chemical bonding*. Oxford University Press, New York. [22]

Cordeiro, J. M. M. (1997). $C-H \cdots O$ and $N-H \cdots O$ hydrogen bonds in liquid amides investigated by Monte Carlo simulation. *International Journal of Quantum Chemistry*, **65**, 709–17. [80]

Corey, R. B. (1938). The crystal structure of diketopiperazine. *Journal of the American Chemical Society*, **60**, 1598–604. [2]

Corey, E. J. (1967). General methods for the construction of complex molecules. *Pure and Applied Chemistry*, **14**, 19–37. [323]

Corey, E. J. and Cheng, X.-M. (1989). *The Logic of Chemical Synthesis*, Wiley, New York. [323]

Corey, E. J. and Rohde, J. J. (1997). The application of the formyl $C-H \cdots O$ hydrogen bond postulate to the understanding of enantioselective reactions involving chiral boron Lewis acids and aldehydes. *Tetrahedron Letters*, **38**, 37–40. [119]

Corey, E. J., Rohde, J. J., Fischer, A. and Azimioara, M. D. (1997). A hypothesis for conformational restriction in complexes of formyl compounds with boron Lewis acids. Experimental evidence for formyl $CH \cdots P$ and $CH \cdots F$ hydrogen bonds. *Tetrahedron Letters*, **38**, 33–6. [114, 118–19]

Cotrait, M., Bideau, J. P., Beurskens, G., Bosman, W. P. and Beurskens, P. T. (1984). Structure and conformation of L-tyrosyl-L-tyrosine dihydrate, $C_{18}H_{20}N_2O_5 \cdot 2H_2O$. *Acta Crystallographica, Section C*, **40**, 1412–16. [433]

Cotton, F. A. and Luck, R. L. (1989). Strong interaction between an aliphatic carbon–hydrogen bond and a metal atom: the structure of (diethylbis(1-pyrazolyl)borato) allyldicarbonylmolybdenum (II). *Inorganic Chemistry*, **28**, 3210–13. [280]

Cotton, F. A., Daniels, L. M., Jordan, G. T. and Murillo, C. A. (1997). The crystal packing of *bis*(2,2'-dipyridylamido)cobalt(II), $Co(dpa)_2$, is stabilized by $C-H \cdots N$ bonds. Are there any real precedents? *Chemical Communications*, 1673–4. [49, 120]

Cotton, F. A., Malonie, J. H. and Murillo, C. A. (1998). Triply bonded Nb_2^{4+} tetragonal lantern compounds: some accompanied by novel $B-H \cdots Na^+$ interactions. *Journal of the American Chemical Society*, **120**, 6047–52. [291]

Cox, S. R., Hsu, L.-Y. and Williams, D. E. (1981). Nonbonded potential function models for crystalline oxohydrocarbons. *Acta Crystallographica, Section A*, **37**, 293–301. [73]

Crabtree, R. H. (1993). Transition metal complexation of σ bonds. *Angewandte Chemie, International Edition in Engish*, **32**, 789–805. [280]

Crabtree, R. H., Siegbahn, P. E. M., Eisenstein, O., Rheingold, A. L. and Koetzle, T. F. (1996). A new intermolecular interaction: unconventional hydrogen bonds with element-hydride bonds as proton acceptors. *Accounts of Chemical Research*, **29**, 348–54. [283, 288]

Cragg-Hine, I., Davidson, M. G., Edwards, A. J., Lamb, E., Raithby, P. R. and

Snaith, R. (1996). The isolation and structure of a highly stable, metallation-resistant and multiply hydrogen-bonded sulfonylamide-phosphine oxide adduct, $PhSO_2CH_2C(=O)NH_2 \cdot O = P(NMe_2)_3$, PSA·HMPA (PSA = phenylsulfonylacetamaide, HMPA = hexamethylphosphoramide). *Chemical Communications*, 153–4. [300]

Cram, D. J. (1988). The design of molecular hosts, guests and their complexes. *Angewandte Chemie, International Edition in English*, **27**, 1009–12. [293]

Crisma, M., Formaggio, F., Valle, G., Toniolo, C. Saviano, M., Iacovino, R. *et al.* (1997). Experimental evidence at atomic resolution for intramolecular $N-H \cdots \pi$ (phenyl) interactions in a family of amino acid derivatives. *Biopolymers*, **42**, 1–6. [152, 154, 376]

Csöregh, I., Sjogren, A., Czugler, M., Cserzö, A. and Weber, E. (1986). Versatility of the 1,1′-binaphthyl-2,2′-dicarboxylic acid host in solid-state inclusion: Crystal and molecular structures of the dimethylformamide (1:2), dimethyl sulphoxide (1:1), and bromobenzene (1:1) clathrates. *Journal of the Chemical Society, Perkin Transactions 2*, 507–13. [335]

Csöregh, I., Czugler, M., Ertan, A., Weber, E. and Ahrendt, J. (1990). Solid-state binding of dimethyl sulphoxide involving carboxylic host molecules. X-Ray crystal structures of four inclusion species. *Journal of Inclusion Phenomena and Molecular Recognition in Chemistry*, **8**, 275–87. [335]

Csöregh, I., Weber, E., Hens, T. and Czugler, M. (1996). Simultaneous electrophile–nucleophile $Cl \cdots \pi$ interactions stabilizing solid state inclusions. A new tool for supramolecular crystal engineering. *Journal of the Chemical Society, Perkin Transactions 2*, 2733–9. [149, 163]

Damewood, J. R. Jr., Urban, J. J., Williamson, T. C. and Rheingold, A. L. (1988). Isomer-dependent complexation of malononitrile by dicyclohexano-18-crown-6. *Journal of Organic Chemistry*, **53**, 167–71. [306, 309]

Dance, I. and Scudder, M. (1995). The sextuple phenyl embrace, a ubiquitous concerted supramolecular motif. *Chemical Communications*, 1039–40. [156]

Dance, I. and Scudder, M. (1996). Concerted supramolecular motifs: linear columns and zigzag chains of multiple phenyl embraces involving Ph_4P^+ cations in crystals. *Journal of the Chemical Society, Dalton Transactions*, 3755–69. [156]

Dance, I. and Scudder, M. (1998). Crystal supramolecularity: elaborate six-, eight- and twelve-fold converted phenyl embraces in compounds $[M(PPh_3)_3]^z$ and $[M(PPh_3)_4]^z$. *New Journal of Chemistry*, 481–92. [156]

Dastidar, P. and Goldberg, I. (1996). Zinc-*meso*-tetra-*p*-tolylporphyrin and its chlorotoluene channel-type clathrate with π–π and $C-H \cdots \pi$ interaction modes stabilizing the porphyrin host lattice. *Acta Crystallographica, Section C*, **52**, 1976–89. [156]

David, J. G. and Hallam, H. E. (1964). Infra-red solvent shifts and molecular interactions. Part 8. Acidity of thiophenols. *Transactions of the Faraday Society*, **60**, 2013–16. [254]

David, J. G. and Hallam, H. E. (1965). Hydrogen-bonding studies of thiophenols. *Spectrochimica Acta*, **21**, 841–50. [259]

Davidson, M. G. (1995). Protonation of an ylide leads to a unique $C-H \cdots O$ hydrogen-bonded dimer. The first synthesis, isolation and X-ray structural characterisation of a phosphonium aryloxide. *Journal of the Chemical Society, Chemical Communications*, 919–20. [107]

Davidson, M. G. and Lamb, S. (1997). The first structural characterisation of a phosphonium amide. Synthesis, isolation and molecular structure of $(Ph_3PEt)^+(NPh_2)^-$. *Polyhedron*, **16**, 4393–5. [120]

Davidson, M. G., Lambert, C., Lopez-Solera, I., Raithby, P. R. and Snaith, R. (1995).

Aggregation of metalated organics by hydrogen bonding. Synthesis and crystal structures of 2-aminophenoxy–aluminium and salen–aluminium ligand-separated ion pairs. *Inorganic Chemistry*, **34**, 3765–79. [249]

Davidson, M. G., Hibbert, T. G., Howard, J. A. K., MacKinnon, A. and Wade, K. (1996). Definitive crystal structures of *ortho*-, *meta*- and *para*-carboranes. Supramolecular structures directed solely by $C-H\cdots O$ hydrogen bonding to hmpa (hmpa = hexamethylphosphoramide). *Chemical Communications*, 2285–6. [300, 301]

Davies, T. and Staveley, L. A. K. (1957). The behaviour of the ammonium ion in the ammonium salt of tetraphenylboron by comparison of the heat capacities of the ammonium, rubidium, and potassium salts. *Transactions of the Faraday Society*, **53**, 19–30. [130, 142]

Davies, P. J., Veldman, N., Grove, D. M., Spek, A. L., Lutz, B. T. G. and van Koten, G. (1996). Organoplatinum building blocks for one-dimensional hydrogen-bonded polymeric structures. *Angewandte Chemie, International Edition in English*, **35**, 1959–61. [219, 220]

Dawoodi, Z. and Martin, M. J. (1981). Neutron diffraction structure analysis of $[(\mu\text{-}H)(\mu\text{-}NCHCF_3)\,Os_3(CO)_{10}]$ at 20 K. *Journal of Organometallic Chemistry*, **219**, 251–7. [277]

Deacon, A., Gleichmann, T., Kalb (Gilboa), A. J., Price, H., Raftery, J., Bradbrook, G. *et al.* (1997). The structure of concanavalin A and its bound solvent determined with small-molecule accuracy at 0.94 Å resolution. *Journal of the Chemical Society, Faraday Transactions*, **93**, 4305–12. [347, 434, 435]

De Alencastro, R. B. and Sandorfy, C. (1972). A low temperature infrared study of self-association in thiols. *Canadian Journal of Chemistry*, **50**, 3594–600. [259]

Deeg, A. and Mootz, D. (1993). Adducts arene/hydrogen chloride at low temperatures: contributions to formation and crystal structure (*in German*). *Zeitschrift für Naturforschung B*, **48**, 571–6. [139]

DeFrees, D. J., Raghavachari, K., Schlegel, H. B., Pople, J. A. and Schleyer, P. v. R. (1987). Binary association complexes of LiH, BeH_2, and BH_3. Relative isomer stabilities and barrier heights for their interconversion energy barriers in the dimerization reactions. *Journal of Physical Chemistry*, **91**, 1857–64. [291]

DeLaat, A. M. and Ault, B. S. (1987). Infrared matrix isolation study of hydrogen bonds involving $C-H$ bonds. Alkynes with oxygen bases. *Journal of the American Chemical Society*, **109**, 4232–6. [31]

Delaigue, X., Hosseini, M. W., Kyritsakas, N. de Cian, A. and Fischer, J. (1995). Synthesis and structural studies on *p-tert*-butyl-1,3-dihydroxy-2,4-disulfanylcalix[4]arene and its mercury complex. *Journal of the Chemical Society, Chemical Communications*, 609–10. [255, 256]

Del Bene, J. E. (1988). *Ab-initio* molecular orbital study of the structures and energies of neutral and charged complexes with H_2O and the hydrides AH_n (A = N, O, F, P, S and Cl). *Journal of Physical Chemistry*, **92**, 2874–80. [12]

Del Bene, J. E., Frisch, M. J. and Pople, J. A. (1985). Molecular orbital study of the complexes $(AH_n)_2H^+$ formed from NH_3, OH_2, FH, PH_3, SH_2 and ClH. *Journal of Physical Chemistry*, **89**, 3669–74. [12]

Derewenda, Z. S., Derewenda, U. and Kobos, P. M. (1994). $(His)C^\varepsilon-H\cdots O=C<$ hydrogen bond in the active sites of serine hydrolases. *Journal of Molecular Biology*, **241**, 83–93. [382, 383]

Derewenda, Z. S., Lee, L. and Derewenda, U. (1995). The occurrence of $C-H\cdots O$ hydrogen bonds in proteins. *Journal of Molecular Biology*, **252**, 248–62. [346, 350–1, 352, 356, 357, 361, 362, 431]

Desiraju, G. R. (1989*a*). *Crystal engineering. The design of organic solids.* Elsevier, Amsterdam. [21, 48, 120, 295, 315, 322]

Desiraju, G. R. (1989*b*). Distance dependence of C—H···O interactions in some chloroalkyl compounds. *Journal of the Chemical Society, Chemical Communications*, 179–80. [50, 51]

Desiraju, G. R. (1990). Strength and linearity of C—H···O bonds in molecular crystals. A database study of some terminal alkynes. *Journal of the Chemical Society, Chemical Communications*, 454–5. [51]

Desiraju, G. R. (1991*a*). The C—H···O hydrogen bond in crystals. What is it? *Accounts of Chemical Research*, **24**, 270–6. [39, 98, 120, 316, 322]

Desiraju, G. R. (1991*b*). Hydration in organic crystals: prediction from molecular structure. *Journal of the Chemical Society, Chemical Communications*, 426–8. [413]

Desiraju, G. R. (1992). C—H···O hydrogen bonding and the deliberate design of organic crystal structures. *Molecular Crystals Liquid Crystals*, **211**, 63–74. [120]

Desiraju, G. R. (1995*a*). Supramolecular synthons in crystal engineering – a new organic synthesis. *Angewandte Chemie, International Edition in English*, **34**, 2311–27. [25, 316, 323, 325]

Desiraju, G. R. (ed.) (1995*b*). *The crystal as a supramolecular entity*, Perspectives in supramolecular chemistry, Vol. 2. Wiley, Chichester. [293]

Desiraju, G. R. (1996*a*). The C—H···O hydrogen bond. Structural implications and supramolecular design. *Accounts of Chemical Research*, **29**, 441–9. [20, 98, 120, 322, 325]

Desiraju, G. R. (1996*b*). The supramolecular concept as a bridge between organic, inorganic and organometallic crystal chemistry. *Journal of Molecular Structure*, **374**, 191–8. [293]

Desiraju, G. R. (1997*a*). Designer crystals: intermolecular interactions, network structures and supramolecular synthons. *Chemical Communications*, 1475–82. [18, 296, 315]

Desiraju, G. R. (1997*b*). Crystal engineering: solid state supramolecular synthesis. *Current Opinion in Solid State and Materials Science*, **2**, 451–4. [293, 329–30]

Desiraju, G. R. (1997*c*). Crystal gazing: structure prediction and polymorphism. *Science*, **278**, 404–5. [316]

Desiraju, G. R. and Gavezzotti, A. (1989). From molecular to crystal structure. Polynuclear aromatic hydrocarbons. *Journal of the Chemical Society, Chemical Communications*, 621–3. [156, 318]

Desiraju, G. R. and Kishan, K. V. R. (1989). Crystal chemistry of some alkoxyphenyl-propiolic acids. The role of oxygen and hydrogen atoms in determining stack structures of planar aromatic compounds. *Journal of the American Chemical Society*, **111**, 4838–43. [334]

Desiraju, G. R. and Murty, B. N. (1987). Correlation between crystallographic and spectroscopic properties for C—H···O bonds in terminal acetylenes. *Chemical Physics Letters*, **139**, 360–1. [41, 42]

Desiraju, G. R. and Sharma, C. V. K. M. (1991). C—H···O hydrogen bonding and topochemistry in crystalline 3,5-dinitrocinnamic acid and its 1:1 donor–acceptor complex with 2,5-dimethoxycinnamic acid. *Journal of the Chemical Society, Chemical Communications*, 1239–41. [99]

Desiraju, G. R., Murty, B. N. and Kishan, K. V. R. (1990). Unexpected hydrogen bonding in the crystal structure of (4-chlorophenyl)propiolic acid. Role of C—H···O hydrogen bonds in determining O—H···O networks. *Chemistry of Materials*, **2**, 447–9. [100]

Desiraju, G. R., Kashino, S., Coombs, M. M. and Glusker, J. P. (1993). C—H···O packing motifs in some cyclopenta[a]phenanthrenes. *Acta Crystallographica, Section B*, **49**, 88–92. [68, 94, 98]

DesMarteau, D. D., Xu, Z. Q. and Witz, M. (1992). N-fluorobis[(perfluoroalkyl)sulfonyl]imides. Reactions with some olefins via α-fluoro carbocationic intermediates. *Journal of Organic Chemistry,* **57**, 629–35. [203]

Deumal, M., Cirujeda, J., Veciana, J., Kinoshita, M., Hosokoshi, Y. and Novoa, J. J. (1997). Theoretical analysis of the crystal packing of nitronyl nitroxide radicals: the packing of the α-2-hydro nitronyl nitroxide radical. *Chemical Physics Letters*, **265**, 190–9. [75]

Dietze, E. C. Ibarra, C., Dabrowski, M. J., Bird, A. and Atkins, W. M. (1996). Rational modulation of the catalytic activity of A1–1 glutathione S-transferase. Evidence for incorporation of an on-face (π···HO—Ar) hydrogen bond at tyrosine-9 *Biochemistry*, **35**, 11938–44. [382]

Diez-Barra, E., Dotor, J., de la Hoz, A., Foces-Foces, C., Enjalbal, C., Aubagnac, J. L. *et al.* (1997). N,N'-Linked 1,2-ethanediyl-poly(benzimidazolin-2-ones) and the X-ray crystal structure of a benzimidazolin-2-one trimer. *Tetrahedron*, **53**, 7689–704. [300]

Dill, J. D., Schleyer, P. v. R., Binkley, J. S. and Pople, J. A. (1977). Molecular orbital theory of the electronic structure of molecules. 34. Structure and energies of small compounds containing lithium or beryllium. Ionic, multicenter and coordination bonding. *Journal of the American Chemical Society*, **99**, 6159–73. [291]

Donohue, J. (1968). Selected topics in hydrogen bonding. In *Structural Chemistry and Molecular Biology* (ed. A. Rich and N. Davidson), pp. 443–65, Freeman, San Francisco. [35, 49, 115, 120]

Donohue, J. (1983). Hydrogen bonding. Some glimpses into the distant past. In *Crystallography in North America* (ed. D. McLachlan and J. P. Glusker), pp. 314–15. American Crystallographic Association. [35, 120]

Dougherty, D. A. (1996). Cation–π interactions in chemistry and biology. A new view of benzene, Phe, Tyr, and Trp. *Science,* **271**, 163–8. [159]

Dougill, M. W. and Jeffrey, G. A. (1953). The structure of dimethyl oxalate. *Acta Crystallographica*, **6**, 831–7. [31]

Drew, M. G. B. and Willey, G. R. (1986). Hydrogen bonding in phenylhydrazone derivatives of benzophenone. Crystal and molecular structures of benzophenone (2-nitrophenyl)hydrazone, containing an intramolecular NO_2···NH···π bifurcated hydrogen bond, and benzophenone (4-nitrophenyl)hydrazone, containing an intramolecular NH···π hydrogen bond. Comments on hydrogen bond vagaries in various types of phenylhydrazone. *Journal of the Chemical Society, Perkin Transactions 2*, 215–20. [147–8]

Driess, M., Monsé, C., Boese, R. and Bläser, D. (1998). Synthesis and crystal structure analysis of tetraphosphanylsilane and identification of tetraphosphanylgermane. *Angewandte Chemie, International Edition in English*, **37**, 2257–9. [268–9, 270]

Dulmage, W. J. and Lipscomb, W. N. (1951). The crystal structure of hydrogen cyanide, HCN. *Acta Crystallographica*, **4**, 330–4. [29, 30, 31]

Dunitz, J. D. (1991). Phase transitions in molecular crystals from a chemical viewpoint. *Pure and Applied Chemistry*, **63**, 177–85. [294]

Dunitz, J. D. (1995). Thoughts on crystals as supermolecules. In *The crystal as a supramolecular entity*, Perspectives in supramolecular chemistry, Vol. 2. (ed. G. R. Desiraju), pp. 1–30, Wiley, Chichester. [294]

Dunitz, J. D. (1996a). Personal communication. [9]

Dunitz, J. D. (1996b). Weak intermolecular interactions in solids and liquids. *Molecular Crystals and Liquid Crystals*, **279**, 209–18. [316]

Dunitz, J. D. and Taylor, R. (1997). Organic fluorine hardly ever accepts hydrogen bonds. *Chemistry. A European Journal*, **3**, 89–98. [88, 202–3, 205]

Dyer, P. W., Dyson, P. J., James, S. L., Suman, P., Davies, J. E. and Martin, C. M. (1998). A remarkable example of co-crystallisation: the crystal structure of the mononuclear and dinuclear diphenyl[2.2]paracyclophanylphosphine palladium(II) chloride complexes *trans*-[Pd{PPh$_2$(C$_{16}$H$_{15}$)}$_2$Cl$_2$]·[Pd{PPh$_2$(C$_{16}$H$_{15}$)}Cl$_2$]$_2$·0.6CH$_2$Cl$_2$. *Chemical Communications*, 1375–6. [334]

Dykstra, C. E. (1988). Intermolecular electrical interaction: a key ingredient in hydrogen bonding. *Accounts of Chemical Research*, **21**, 355–61. [17]

Eaton, P. E., Galoppini, E. and Gilardi, R. (1994). Alkynylcubanes as precursors of rigid-rod molecules and alkynylcyclooctatetraenes. *Journal of the American Chemical Society*, **116**, 7588–96 [181]

Eggleston, D. S., Baures, P. W., Peishoff, C. E. and Kopple, K. D. (1991). Conformation of cyclic heptapeptides. Crystal structure and computational studies of evolidine. *Journal of the American Chemical Society*, **113**, 4410–16. [424, 425]

Egli, M. and Gessner, R. V. (1995). Stereoelectronic effects on deoxyribose O4′ on DNA conformation. *Proceedings of the National Academy of Sciences of the USA*, **92**, 180–4. [395]

Elaiwi, A., Hitchcock, P. B., Seddon, K. R., Srinivasan, N., Tan, Y.-M., Welton, T. and Zora, J. A. (1995). Hydrogen bonding in imidazolium salts and its implications for ambient-temperature halogenoaluminate(III) ionic liquids. *Journal of the Chemical Society, Dalton Transactions*, 3467–72. [249]

Elbasyouny, A., Brügge, H. J., von Deuten, K., Dickel, M., Knöchel, A., Koch, K. U. *et al.* (1983). Host–guest complexes of 18-crown-6 with neutral molecules possessing the structure element XH$_2$ (X = O, N, C). *Journal of the American Chemical Society*, **105**, 6568–77. [304]

Emerson, J. and Sundaralingam, M. (1980). Zwitterionic character of guanosine 5′-monophosphate (5′-GMP). Redetermination of the structure of 5′-GMP trihdrate. *Acta Crystallographica, Section B*, **36**, 1510–13. [391]

Engh, R. A., Brandstetter, H., Sucher, G., Eichinger, A., Baumann, I., Bode, W. *et al.* (1996). Enzyme flexibility, solvent and 'weak' interactions characterize thrombin–ligand interactions: implications for drug design. *Structure*, **4**, 1353–62. [338, 339, 376, 382, 434]

Erickson, J. A. and McLoughlin, J. I. (1995). Hydrogen bond donor properties of the difluoromethyl group. *Journal of Organic Chemistry*, **60**, 1626–31. [58]

Eriksson, A. and Hermansson, K. (1983). Analysis of the thermal parameters of the water molecule in crystalline hydrates studied by neutron diffraction. *Acta Crystallographica, Section B*, **39**, 703–11. [70]

Ermer, O. and Eling, A. (1994). Molecular recognition among alcohols and amines: super-tetrahedral crystal architectures of linear diphenol–diamine complexes and aminophenols. *Journal of the Chemical Society, Perkin Transactions 2*, 925–44. [318]

Eskudero, Kh. M., Akimov, V. M., Antipin, M. Yu., Lindeman, S. V. and Struchkov, Yu. T. (1992). The structure of thiourea perchlorate. *Russian Journal of Inorganic Chemistry*, **37**, 383–5 (translated from *Zhurnal Neorganishkoi Khimii*, **37**, 767–70). [257]

Espinosa, E., Lecomte, C., Molins, E., Veintemillas, S., Cousson, A. and Paulus, W. (1996). Electron density study of a new non-linear optical material, L-arginine phosphate monohydate (LAP). Comparison between $X - X$ and $X - (X + N)$ refinements. *Acta Crystallographica, Section B*, **52**, 519–34. [363]

Etter, M. C. (1990). Encoding and decoding hydrogen-bond patterns of organic compounds. *Accounts of Chemical Research*, **23**, 120–6. [25, 161]

Etter, M. C. and Panunto, T. W. (1988). 1,3-*bis*(*m*-nitrophenyl)urea. An exceptionally good complexing agent for proton acceptors. *Journal of the American Chemical Society*, **110**, 5896–7. [91–2]

Fabiola, G. F., Krishnaswamy, S., Nagarajan, V. and Pattabhi, V. (1997). C—H···O hydrogen bonds in β-sheets. *Acta Crystallographica, Section D*, **53**, 316–20. [346, 352, 353, 354, 355]

Fairhurst, S. A., Henderson, R. A., Hughes, D. L., Ibrahim, S. K. and Pickett, C. J. (1995). An intramolecular W—H···O=C hydrogen bond? Electrosynthesis and X-ray crystallographic structure of [WH$_3$(η^1- OCOMe)(Ph$_2$PCH$_2$CH$_2$PPh$_2$)$_2$]. *Journal of the Chemical Society, Chemical Communications*, 1569–70. [278, 288]

Falk, M. and Knop, O. (1973). Water in stoichiometric hydrates. In *Water. A comprehensive treatise* (ed. F. Franks), pp. 55–113. Plenum, New York. [414, 415]

Fan, M.-F., Lin, Z., McGrady, J. E. and Mingos, D. M. P. (1996). Novel intermolecular C—H···π interactions. *Ab initio* and density functional theory study. *Journal of the Chemical Society, Perkin Transactions 2*, 563–8. [164, 170, 171]

Feil, D. and Loong, W. S. (1968). The crystal structure of thiourea nitrate. *Acta Crystallographica, Section B*, **24**, 1334–9. [257]

Feldman, S. H. and Eggleston, D. S. (1990). Structure of the tetrahydrate of the *N*-terminal tetrapeptide from angiotensin II. *Acta Crystallographica, Section C*, **46**, 678–82. [420]

Ferguson, G. and Islam, K. M. S. (1966). C—H···O hydrogen bonding. Part II. The crystal and molecular structure of *o*-chlorobenzoacetylene. *Journal of the Chemical Society (B)*, 593–600. [35, 36]

Ferguson, G. and Tyrrell, J. (1965). C—H···O hydrogen bonding. *Journal of the Chemical Society, Chemical Communications*, 195–7. [35]

Ferguson, G., Gallagher, J. F., Glidewell, C. and Zakaria, C. M. (1994). O—H···π(arene) intermolecular hydrogen bonding in the structure of 1,1,2-triphenylethanol. *Acta Crystallographica, Section C*, **50**, 70–3. [164]

Ferguson, G., Glidewell, C., Royles, B. J. L. and Smith, D. M. (1996). 1,1-Ferrocenediylbis(2-phenylethanedione). A three-dimensional network generated by short C—H···O hydrogen bonds. *Acta Crystallographica, Section C*, **52**, 2465–8. [108]

Ferstanding, L. L. (1962). Carbon as a hydrogen bonding base and carbon–hydrogen–carbon bonding. *Journal of the American Chemical Society*, **84**, 3553–7. [243, 245]

Feyereisen, M. W., Feller, D. and Dixon, D. A. (1996). Hydrogen bond energy of the water dimer. *Journal of Physical Chemistry*, **100**, 2993–7. [12]

Fischer, E. (1894). Einfluss der Configuration auf die Wirkung der Enzyme. *Chemische Berichte*, **27**, 2985–93. [294]

Fitzgerald, M. M., Churchill, M. J., McRee, D. E. and Goodin, D. B. (1994). Small molecule binding to an artificially created cavity at the active site of cytochrome *c* peroxidase. *Biochemistry*, **33**, 3807–18. [379, 380]

Flanagan, K., Walshaw, J., Price, S. L. and Goodfellow, J. M. (1995). Solvent interactions with π ring systems in proteins. *Protein Engineering*, **8**, 109–16. [435]

Fluck, E., Savara, J., Neumüller, B. Riffel, H. and Thurn, H. (1986). Preparation and structures of methylphosphonium chloride (*in German*). *Zeitschrift für Anorganische und Allgemeine Chemie*, **536**, 129–36. [267, 269]

Foster, R. (1969). *Organic charge-transfer complexes*. Academic, New York. [294]

Fraser, G. T., Suenram, R. D., Lovas, F. J., Pine, A. S., Hougen, J. T., Lafferty, W. J. and Muenter, J. S. (1988). Infrared and microwave investigations of interconversion tunneling in the acetylene dimer. *Journal of Chemical Physics*, **89**, 6028–45. [127, 128, 129]

Fraser, G. T., Lovas, F. J., Suenram, R. D., Gillies, J. Z. and Gillies, C. W. (1992). Microwave and infrared spectra of $C_2H_4 \cdots HCCH$. Barrier to two-fold internal rotation of C_2H_4. *Journal of Chemical Physics*, **163**, 91–101. [127]

Freyhardt, C. C. and Wiebcke, M. (1994). Bronsted conjugate acid–base species $B(OH)_3/[BO(OH)_2]^-$ coexist in the crystalline solid $(NEt_4)_2[BO(OH)_2]_2 \cdot B(OH)_3 \cdot 5H_2O$. *Journal of the Chemical Society, Chemical Communications*, 1675–6. [58]

Frisch, M. J., Pople, J. A. and Del Bene, J. E. (1984). Molecular orbital study of the dimers $(AH_n)_2$ formed from NH_3, OH_2, FH, PH_3, SH_2 and ClH. *Chemical Physics*, **122**, 413–30. [268]

Fu, T. Y., Scheffer, J. R. and Trotter, J. (1994). Crystal structure–reactivity correlations in the solid state photochemistry of *N*-(*tert*-butyl)succinimide. *Canadian Journal of Chemistry*, **72**, 1952–60. [95]

Fujii, S., Burley, S. K. and Wang, A. H.-J. (1987). Structure of an antigelling agent, L-phenylalanyl-glycyl-glycyl-D-phenylalanine trihydrate. *Acta Crystallographica, Section C*, **43**, 1008–10. [373, 375]

Fyfe, M. C. T. and Stoddart, J. F. (1997). Synthetic supramolecular chemistry. *Accounts of Chemical Research*, **30**, 393–401. [337]

Gallardo, O., Csöregh, I. and Weber, E. (1995). Crystal and molecular structure of the inclusion compound between 1,1′-binaphthyl-2,2′-dicarboxylic acid and acetylacetone (1:1). *Journal of Chemical Crystallography*, **25**, 769–76. [335]

Galoppini, E. and Gilardi, R. (1999). Weak hydrogen bonding between acetylene groups: the formation of diamondoid nets in the crystal structure of tetrakis(4-ethynylphenyl)methane. *Chemical Communications*, 173–4. [178]

Gardner, K. H. and Blackwell, J. (1974). The structure of native cellulose. *Biopolymers*, **13**, 1975–2001. [412, 413]

Garrell, R. L., Smyth, J. C., Fronczek, F. R. and Gandour, R. D. (1988). Crystal structure of the 2:1 acetonitrile complex of 18-crown-6. *Journal of Inclusion Phenomena*, **6**, 73–8. [307]

Gasco, A. M., Stilo, A. D., Fruttero, R., Sorba, G., Gasco, A. and Sabatino, P. (1993). Synthesis and structure of a trimer of the furoxan system with high vasodilator and platelet antiaggregatory activity. *Liebig's Annalen der Chemie*, 441–4. [107]

Gavezzotti, A. (1991a). Molecular packing and other structural properties of crystalline oxohydrocarbons. *Journal of Physical Chemistry*, **95**, 8948–55. [98]

Gavezzotti, A. (1991b). Generation of possible crystal structures from the molecular structure for low polarity organic compounds. *Journal of the American Chemical Society*, **113**, 4622–9. [321]

Gavezzotti, A. (1994). Are crystal structures predictable? *Accounts of Chemical Research*, **27**, 309–14. [321]

Gavezzotti, A. (1996). Organic crystals: engineering and design. *Current Opinion in Solid State and Materials Science*, **1**, 501–5. [315]

Gavezzotti, A. (ed.) (1997). *Theoretical aspects and computer modeling of the molecular solid state*. Wiley, Chichester. [321]

Gavezzotti, A. (1998). The crystal packing of organic molecules: challenge and fascination below 1000 Da. *Crystallography Reviews*, **7**, 5–121. [294]

Gerothanassis, I. P. and Vakka, C. (1994). ^{17}O NMR chemical shifts as a tool to study specific hydration sites of amides and peptides. Correlation with the IR amide I stretching vibration. *Journal of Organic Chemistry*, **59**, 2341–8. [116, 117]

Gessler, K., Krauss, N., Steiner, T., Betzel, C., Sarko, A. and Saenger, W. (1995). β-D-

Cellotetraose hemihydrate as a structural model for cellulose II. An X-ray diffraction study. *Journal of the American Chemical Society*, **117**, 11397–406. [411, 412]

Ghosh, P., Taube, H., Hasegawa, T. and Kuroda, R. (1995). Vanadium(II) salts in pyridine and acetonitrile solvents. *Inorganic Chemistry*, **34**, 5761–75. [95]

Gil, F. P. S. C., da Costa, A. M. A. and Teixeira-Dias, J. J. C. (1995). Conformational analysis of $C_mH_{2m+1}OCH_2OH$ (m = 1–4). The role of $C-H\cdots O$ intramolecular interactions. *Journal of Physical Chemistry*, **99**, 8066–70. [95]

Gillard, R. E., Raymo, F. M. and Stoddard, J. F. (1997). Controlling self-assembly. *Chemistry. A European Journal*, **3**, 1933–40. [338]

Gilli, G., Bellucci, F., Ferretti, F. and Bertolasi, V. (1989). Evidence for resonance-assisted hydrogen bonding from crystal-structure correlations on the enol form of the β-diketone fragment. *Journal of the American Chemical Society*, **111**, 1023–8. [81, 100]

Gilli, P., Ferretti, V., Bertolasi, V. and Gilli, G. (1994). Covalent nature of the strong homonuclear hydrogen bond. Study of the $O-H\cdots O$ system by crystal structure correlation methods. *Journal of the American Chemical Society*, **116**, 909–15. [14, 17]

Glasstone, S. (1937). The structure of some molecular complexes in the liquid phase. *Transactions of the Faraday Society*, 200–14. [29–30, 123]

Glidewell, C., Klar, R. B., Lightfoot, P., Zakaria, C. M. and Ferguson, G. (1996a). Hydrogen bonding in α-ferrocenyl alcohols. Structures of 1-ferrocenylethanol, 1-ferrocenyl-2-phenylethanol, 1-ferrocenyl-1-phenylpropan-1-ol, 1-ferrocenyl-1-phenyl-2-methylpropan-1-ol, 1-ferrocenyl-1-phenyl-2,2-dimethylpropan-1-ol, 1-ferrocenyl-1,2-diphenylethanol and diferrocenyl(phenyl)methanol. *Acta Crystallographica, Section B*, **52**, 110–21. [199–200]

Glidewell, C., Gottfried, M. J., Trotter, J. and Ferguson, G. (1996b). 1-Ferrocenyl-2-phenylethandione. *Acta Crystallographica, Section C*, **52**, 773–5. [200]

Glusker, J. P., Lewis, M. and Rossi, M. (1994). *Crystal structure analysis for chemists and biologists*. VCH, New York. [6, 21]

Goddard, R., Heinemann, O. and Krüger, C. (1997). Pyrrole and a co-crystal of 1H- and 2H-1,2,3-triazole. *Acta Crystallographica, Section C*, **53**, 1846–50. [190–1]

Goldberg, I. (1975). Structure and binding in molecular complexes of cyclic polyethers. I. 1,4,7,10,13,16-hexaoxacyclooctadecane (18-crown-6)-dimethyl acetylenedicarboxylate at 160°C. *Acta Crystallographica, Section B*, **31**, 754–62. [38, 305, 307]

Goldberg, I. (1978). A host–guest-type water adduct of 3,3′-(1,1′-bi-naphtol)-21-crown-5. *Acta Crystallographica, Section B*, **34**, 3387–90. [309–10]

Goldberg, I. (1984). Complexes of crown ethers with molecular guests. In *Inclusion compounds* (ed. J. L. Atwood, J. E. D. Davies and D. D. MacNicol) Vol. 2, pp. 261–335. Academic, London. [90, 91, 303]

Goldberg, I. and Kosower, E. M. (1982). Bimanes. 13. Crystal structure of 9,10-dioxa-syn-(hydro, chloro)bimane (3,7-dichloro-1,5-diazabicyclo[3.3.0]octa-3,6-diene-2,8-dione) and its relation to solid-state ultraviolet–visible spectroscopic shifts. *Journal of Physical Chemistry*, **86**, 332–5. [38]

Goldman, P. (1969). The carbon–fluorine bond in compounds of biological interest. *Science*, **164**, 1123–30. [342]

Görbitz, C. H. (1987). A redetermination of the crystal and molecular structure of glutathione (γ-L-glutamyl-L-cysteinylglycine) at 120 K. *Acta Chimica Scandinavica*, **41**, 362–6. [255, 256]

Görbitz, C. H. (1989). Hydrogen bonding of cysteine in the solid phase. Crystal and molecular structures of L-cysteine methyl ester·HCl and L-cysteine ethyl ester·HCl. *Acta Chimica Scandinavica*, **43**, 871–5. [263, 264]

Görbitz, C. H. and Dalhus, B. (1996). L-Cysteine, monoclinic form, redetermination at 120 K. *Acta Crystallographica, Section C*, **52**, 1756–9. [255, 256, 260]

Gordy, W. (1939). Spectroscopic evidence of hydrogen bonds: chloroform and bromoform in donor solvents. *Journal of Chemical Physics*, **7**, 163–6. [30]

Gough, K. M. and Millington, J. (1995). *Ab initio* interpretation of conformer stabilization through S⋯O and C−H⋯O bonding in the acetyl derivatives of two representative heterocyclic methine bases. *Canadian Journal of Chemistry*, **73**, 1287–93. [79]

Gould, R. O., Gray, A. M., Taylor, P. and Walkinshaw, M. D. (1985). Crystal environments and geometries of leucine, isoleucine, valine, and phenylalanine provide estimates of minimum nonbonded contact and preferred van der Waals interaction distances. *Journal of the American Chemical Society*, **107**, 5921–7. [360]

Grabowski, S. J. (1995). Π-electrons as an acceptor of proton in H-bonds in the structure of acetylene. *Polish Journal of Chemistry*, **69**, 223–8. [171]

Green, R. D. (1974). *Hydrogen bonding by C−H groups*. Macmillan, London. [4–5, 31, 120]

Grepioni, F., Cojazzi, G., Draper, S. M., Scully, N. and Braga, D. (1998). Crystal forms of hexafluorophosphate organometallic salts and the importance of charge-assisted C−H⋯F hydrogen bonds. *Organometallics*, **17**, 296–307. [214]

Gronert, S. (1993). Theoretical studies of proton transfers. 1. The potential energy surface of the identity reactions of the first- and second-row non-metal hydrides with their conjugate bases. *Journal of the American Chemical Society*, **115**, 10258–66. [12]

Grootenhuis, P. D. J., van Eerden, J., Dijkstra, P. J., Harkema, S. and Reinhoudt, D. N. (1987). Complexation of neutral molecules by pre-organized macrocyclic hosts. *Journal of the American Chemical Society*, **109**, 8044–51. [310, 311]

Gunter, M. J., Hockless, D. C. R., Johnston, M. R., Skelton, B. W. and White, A. H. (1994). Self-assembling porphyrin [2]-catenanes. *Journal of the American Chemical Society*, **116**, 4810–23. [338]

Hamilton, A. D. (ed.) (1996). *Supramolecular control of structure and reactivity*, Perspectives in supramolecular chemistry, Vol. 3. Wiley, Chichester. [293]

Hamilton, W. C. and Ibers, J. A. (1968). *Hydrogen bonding in solids*. W. A. Benjamin, New York. [7, 14, 22, 271]

Hancock, K. S. B. and Steed, J. W. (1998). C−H⋯π hydrogen bonding *vs.* anion complexation in rhodium(I) complexes of cyclotriveratrylene. *Chemical Communications*, 1409–10. [156]

Hanton, L. R., Hunter, C. A. and Purvis, D. H. (1992). Structural consequences of a molecular assembly that is deficient in hydrogen-bond acceptors. *Journal of the Chemical Society, Chemical Communications*, 1134–6. [161]

Hantzsch, A. (1910). Über die Isomeriegleichgewichte des Acetessigesters und die sogenannte Isorrhopesis seiner Salze. *Chemische Berichte*, **43**, 3049–76. [1]

Harakas, G., Vu, T., Knobler, C. B. and Hawthorne, M. F. (1998). Synthesis and crystal structure of 9,12-bis-(4-acetylphenyl)-1,2-dicarbadodecaborane(12): Self-assembly involving intermolecular carboranyl C−H hydrogen bonding. *Journal of the American Chemical Society*, **120**, 6405–6. [300]

Harata, K. (1998). Structural aspects of stereodifferentiation in the solid state. *Chemical Reviews*, **89**, 1803–27. [311]

Hardy, A. D. U. and MacNicol, D. D. (1976). Crystal and molecular structure of an O−H⋯π hydrogen-bonded system. 2,2-*bis*-(2-hydroxy-5-methyl-3-*t*-butylphenyl)propane. *Journal of the Chemical Society, Perkin Transactions 2*, 1140–2. [149, 266]

Harlow, R. L., Li, C. and Sammes, M. P. (1984*a*). Confirmation of a short intramolecular C—H···N hydrogen bond in 4,4-bisphenylsulphonyl-2,*N*,*N*-trimethylbutylamine by X-ray crystallography. *Journal of the Chemical Society, Chemical Communications*, 818–19. [38, 94]

Harlow, R. L., Li, C. and Sammes, M. P (1984*b*). Hydrogen bonds involving polar CH groups. Part 11. Further confirmation of intramolecular bonds to nitrogen 2-aminoalkyl-1,3-dithiane 1,1,3,3-tetraoxides by X-ray crystal structure analysis. *Journal of the Chemical Society, Perkin Transactions 1*, 547–9. [38, 94]

Hautzel, R., Anke, H. and Sheldrick, W. S. (1990). Mycenon, a new metabolite from a *mycena* species TA87202 (basidomycetes) as an inhibitor of isocitrate lyase. *The Journal of Antibiotics*, **43**, 1240–4. [166]

Heinekey, D. M., Millar, J. M., Koetzle, T. F., Payne, N. G., and Zilm, K. W. (1990). Structural and spectroscopic characterization of iridium trihydride complexes: evidence for proton–proton exchange coupling. *Journal of the American Chemical Society*. **112**, 909–19. [291]

Heinicke, J., Kadyrov, R., Kindermann, M. K., Koesling, M. and Jones, P. G. (1996). P/O ligand systems. Synthesis, reactivity, and structure of tertiary *o*-phosphanylphenol derivatives. *Chemische Berichte*, **129**, 1547–60. [239, 240]

Helmle, O., Csöregh, I., Weber, E. and Hens, T. (1997). Supramolecular crystalline complexes involving bulky hydroxy hosts: X-ray structure analysis of inclusion compounds with acetone and toluene. *Journal of Physical Organic Chemistry*, **10**, 76–84. [335]

Henschel, D., Hiemisch, L., Blaschette, A. and Jones, P. G. (1996). Occurrence of two supramolecular isomers in one crystal: Synthesis and structure of the monomeric 18-crown-6 complex $(CH_2CH_2O)_6.2$ MeN$(SO_2Me)SO_2Ph)$ (*in German*). *Zeitschrift für Naturforschung*, **51b**, 1313–19. [307]

Hensel, V., Lützow, K., Jacob, J., Gessler, K., Saenger, W. and Schlüter, A.-D. (1997). Repetitive construction of macrocyclic oligophenylenes. *Angewandte Chemie, International Edition in English*, **36**, 2654–6. [310, 311]

Hensen, K., Pullmann, P. and Bats, J. W. (1988). Crystal and molecular structures of chlorotrispyridinium-*bis*-(tetrachloroaluminate(III)) and a new modification of pyridinium chloride (*in German*). *Zeitschrift für Anorganische und Allgemeine Chemie*, **556**, 62–9. [251]

Hernàndez-Gasio, E., Mas, M., Molins, E., Rovira, C., Veciana, J., Almenar, J. J. B. and Coronado, E. (1994). Coexistence of alternative ferromagnetic and antiferromagnetic intermolecular interactions in organic compounds. Synthesis, structure, thermal stability, and magnetic properties of 2,4-hexadiynylenedioxybis[2-(*p*-phenylene)-4,4,5,5-tetramethyl-4,5-dihydro-1*H*-imidazol-1-oxyl] diradical. *Chemistry of Materials*, **6**, 2398–411. [336]

Hibbert, F. and Emsley, J. (1990). Hydrogen bonding and chemical reactivity. *Advances in Physical Organic Chemistry*, **26**, 255–379. [14]

Hiemisch, O., Henschel, D., Jones, P. G. and Blaschette, A. (1996). Polysulfonyl amines. LXXII. Triphenylcarbenium and triphenylphosphonium di(fluorosulfonyl)amines. Two crystal structures with ordered $(FSO_2)_2N^-$ sites (*in German*). *Zeitschrift für Anorganische und Allgemeine Chemie*, **622**, 829–36. [267, 269]

Hiller, W. and Rundel, W. (1993). Structure of 2,2-*bis*(3,5-di-*tert*-butylphenyl)propane-2′,2″-dithiol. *Acta Crystallographica, Section C*, **49**, 1127–30. [265, 266]

Hinchliffe, A. (1984). *Ab initio* study of the hydrogen bonded complexes $H_2X···HY$ (X=O, S, Se: Y=F, Cl, Br). *Journal of Molecular Structure (Theochem)*, **106**, 361–6. [12, 226]

Hinchliffe, A. (1985). *Ab initio* study of the hydrogen bonded complexes $NH_3\cdots HF$, $PH_3\cdots HF$, $AsH_3\cdots HF$, $AsH_3\cdots HCl$ and $AsH_3\cdots HBr$. *Journal of Molecular Structure (Theochem)*, **121**, 201–5. [226, 239]

Hinrichs, W., Kisker, C., Düvel, M., Müller, A., Tovar, K., Hillen, W. and Saenger, W. (1994). Structure of the tet repressor-tetracycline complex and regulation of antibiotic resistance. *Science*, **264**, 418–20. [377, 378]

Ho, T.-L. (1975). The hard soft acids bases (HSAB) principle and organic chemistry. *Chemical Reviews*, **75**, 1–20. [87]

Ho, T. -L. (1977). *Hard and soft acids and bases principle in organic chemistry*. Academic, New York. [87]

Hobza, P. and Zahradnik, R. (1988). Intermolecular interactions between medium-sized systems. Nonempirical and empirical calculations of interaction energies. Successes and failures. *Chemical Reviews*, **88**, 871–97. [127]

Hobza, P., Spirko, V., Selzle, H. L. and Schlag, E. W. (1998). Anti-hydrogen bond in the benzene dimer and other carbon proton donor complexes. *Journal of Physical Chemistry A*, **102**, 2501–4. [401]

Hofmann, D. W. M. and Lengauer, T. (1997). A discrete algorithm for crystal structure prediction of organic molecules. *Acta Crystallographica, Section A*, **53**, 225–35. [321]

Howard, J. A. K., Hoy, V. J., O'Hagan, D. and Smith, G. T. (1996). How good is fluorine as a hydrogen bond acceptor? *Tetrahedron*, **52**, 12613–22. [12, 202, 204, 205, 206]

Hubbard, S. J., Gross, K.-H. and Argos, P. (1994). Intramolecular cavities in globular proteins. *Protein Engineering*, **7**, 613–26. [430]

Huffman, J. C. and Haushalter, R. C. (1984). Preparation and crystal structure of $(Ph_4P)_2Te_4 \cdot 2\ CH_3OH$. *Zeitschrift für Anorganische und Allgemeine Chemie*, **518**, 203–9. [237]

Huggins, M. L. (1936). Hydrogen bridges in organic compounds. *Journal of Organic Chemistry*, **1**, 405–56. [2]

Hunter, C. A., Singh, J. and Thornton, J. (1991). π–π interactions. The geometry and energetics of phenylalanine–phenylalanine interactions in proteins. *Journal of Molecular Biology*, **218**, 837–46. [347]

Hunter, R., Haueisen, R. H. and Irving, A. (1994). The first water-dependent liquid clathrate. X-Ray evidence in the solid for a $C-H\cdots\pi$ (heteroarene)$\cdots H-C$ interaction. *Angewandte Chemie, International Edition in English*, **33**, 566–8. [192]

Ibberson, R. M. and Prager, M. (1996). The *ab initio* crystal structure determination of vapour-deposited methyl fluoride by high-resolution neutron powder diffraction. *Acta Crystallographica, Section B*, **52**, 892–5. [209, 212]

Iogansen, A. V. (1981). In *Hydrogen bonding* (ed. N. D. Sokolov), p. 112. Nauka, Moscow, (*in Russian*). [26]

Iogansen, A. V. and Rozenberg, M. Sh. (1971). Stretching and bending of hydrogen bonds as a function of their energy. *Doklady Akademii Nauk SSSR*, **197**, 117–20. English translation: pp. 204–7. [10]

Ishi-i, T., Sawada, T., Mataka, S. and Tashiro, M. (1996). Intramolecular $O-H\cdots\pi$ hydrogen bond found in [2.2]metacyclophane systems. Spectral properties and X-ray crystallographic analysis of 8-hydroxymethyl[2.2]metacyclophanes. *Journal of the Chemical Society, Perkin Transactions 1*, 1887–91. [151, 153]

Iwaoka, M. and Tomoda, S. (1994). First observation of a $C-H\cdots Se$ 'hydrogen bond'. *Journal of the American Chemical Society*, **116**, 4463–4. [237]

Iwaoka, M., Komatsu, H. and Tomoda, S. (1996). Deuterium-induced isotope effects of a $C-H\cdots Se$ 'hydrogen bond' on the IR and NMR Spectra of 6H,12H-dibenzo[*b,f*][1,5]diselenocin. *Bulletin of the Chemical Society of Japan*, **69**, 1825–8. [237]

Jacobs, H. and Kirchgässner, R. (1989). Chemical bonding in crystalline phases of caesium hydrogensulfides, CsHS and CsDS (*in German*). *Zeitschrift für Anorganische und Allgemeine Chemie*, **569**, 117–30. [262, 263]

James, S. L., Verspui, G., Spek, A. L. and van Koten, G. (1996). Organometallic polymers. An infinite organoplatinum chain in the solid state formed by (C≡CH⋯ClPt) hydrogen bonds. *Chemical Communications*, 1309–10. [219, 220]

Jaskólski, M. (1984). Crystal and molecular structure of 1,N^6-ethenoadenosine hydrochloride. *Journal of Crystallographic and Spectroscopic Research*, **14**, 45–57. [44–5, 104]

Jaulmes, S., Dugué, J., Agafonov, V., Céolin, R., Cense, J. M. and Lepage, F. (1993). Structure of N-(2,6-dimethylphenyl)-5-methylisoxazole-3-carboxamide and molecularorbital study of C—H⋯O bonded dimers. *Acta Crystallographica, Section C*, **49**, 1007–11. [107, 141]

Jedlovszky, P. and Turi, L. (1997). Role of C—H⋯O hydrogen bonds in ligands. A Monte Carlo simulation study of liquid formic acid using a newly developed pairpotential. *Journal of Physical Chemistry B*, **101**, 5429–36. [117]

Jeffrey, G. A. (1992). Accurate crystal structure analysis by neutron diffraction. In *Accurate molecular structures* (ed. A. Domenicano and I. Hargittai), pp. 270–98. Oxford University Press, Oxford. [7, 22]

Jeffrey, G. A. (1997). *An introduction to hydrogen bonding.* Oxford University Press, New York. [2, 4, 13, 20, 21, 26, 271, 400, 408, 414]

Jeffrey, G. A. and Maluszynska, H. (1982). A survey of hydrogen-bond geometries in the crystal structures of amino acids. *International Journal of Biological Macromolecules*, **4**, 173–85. [350]

Jeffrey, G. A. and Maluszynska, H. (1990). The stereochemistry of water molecules in the hydrates of small biological molecules. *Acta Crystallographica, Section B*, **48**, 546–9. [414]

Jeffrey, G. A. and Saenger, W. (1991). *Hydrogen bonding in biological structures.* Springer-Verlag, Berlin. [vii, 4, 6, 13, 17, 20, 21, 66, 247, 343, 347, 407, 408, 413, 414]

Jeng, M.-L. H. and Ault, B. S. (1990). Infrared matrix isolation study of hydrogen bonds involving C—H bonds: alkynes with bases containing second- and third-row donor atoms. *Journal of Physical Chemistry*, **94**, 1323–7. [239]

Jiang, J. C. and Tsai, M.-H. (1997). *Ab initio* study of the hydrogen bonding between pyrrole and hydrogen fluoride. A comparison of NH⋯F and FH⋯π interactions. *Journal of Physical Chemistry A*, **101**, 1982–8. [190]

Jitsukawa, K., Morioka, T., Masuda, H., Ogoshi, H. and Einaga, H. (1994). A novel amino acid-binding receptor based on weak non-covalent ligand–ligand interactions. *Inorganica Chimica Acta*, **216**, 249–51. [95]

Joesten, M. D. and Schaad, L. J. (1974). *Hydrogen bonding*, New York, Marcel Dekker. [125]

Jones, G. P., Cornell, B. A., Horn, E. and Tieklink, E. R. T. (1989). Crystal structures of dimethyl succinate and dimethyl oxalate: carbonyl group orientation for C^{13} chemical shielding tensor studies. *Journal of Crystallographic and Spectroscopic Research*, **19**, 715–23. [31, 32]

Jones, P. G., Hiemisch, O. and Blaschette, A. (1994). Formation and crystal structure of the complex ((18-Crown-6)(CH$_2$Cl$_2$)$_2$). *Zeitschrift für Naturforschung*, **49b**, 852–4. [304]

Jorgensen, W. L. and Severance, D. L. (1990). Aromatic–aromatic interactions. Free energy profiles for the benzene dimer in water, chloroform, and liquid benzene. *Journal of the American Chemical Society*, **112**, 4768–74. [134]

Joris, L., Schleyer, P. v. R. and Gleiter, R. (1968). Cyclopropane rings as proton-acceptor groups in hydrogen bonding. *Journal of the American Chemical Society*, **90**, 327–36. [195]

Joshi, B. S., Gawad, D. P., Pelletier, S. W., Kartha, G. and Bhandary, K. (1986). Isolation and structure (X-ray analysis) of ent-norsecurinine, an alkaloid from *Phyllanthus niruri*. *Journal of Natural Products*, **49**, 614–20. [422]

Kálmán, A. and Párkányi, L. (1997). Isostructurality of organic crystals: a tool to estimate the complementarity of homo- and heteromolecular associates. In *Advances in molecular structure research*, Vol. 3 (ed. I. Hargittai and M. Hargittai), 189–226, JAI Press, Greenwich, USA. [104]

Kapteijn, G. M., Grove, D. M., van Koten, G., Smeets, W. J. J. and Spek, A. L. (1993). A new mixed alkoxo aryloxo palladium complex with a bidentate nitrogen donor system. *Inorganica Chimica Acta*, **207**, 131–4. [45]

Kapteijn, G. M., Grove, D. M., Kooijman, H., Smeets, W. J. J., Spek, A. L. and van Koten, G. (1996). Chemistry of bis(aryloxo)palladium(II) complexes with N-donor ligands. Structural features of the palladium-to-oxygen bond and formation of O−H···O bonds. *Inorganic Chemistry*, **35**, 526–33. [117–18]

Karfunkel, H. R. and Gdanitz, R. J. (1992). *Ab initio* prediction of possible crystal structures for general organic molecules. *Journal of Computational Chemistry*, **13**, 1171–83. [322]

Karipides, A. and Miller, C. (1984). Crystal calcium 2-fluorobenzoate dihydrate. Indirect calcium···fluorine binding through a water-bridged outer-sphere intermolecular hydrogen bond. *Journal of the American Chemical Society*, **106**, 1494–5. [203, 204]

Kariuki, B. M., Harris, K. D. M., Philp, D. and Robinson, J. M. A. (1997). A triphenylphoshine oxide–water aggregate facilitates an exceptionally short C−H···O hydrogen bond. *Journal of the American Chemical Society*, **119**, 12 679–80. [41, 42, 81, 83]

Karle, I. L., Ranganathan, D. and Haridas, V. (1996). A persistent preference for layer motifs in self-assemblies of squarates and hydrogen squarates by hydrogen bonding [X−H···O; X=N, O, or C]. A crystallographic study of five organic salts. *Journal of the American Chemical Society*, **118**, 7128–33. [334]

Kaufmann, R., Knöchel, A., Kopf, J., Oehler, J. and Rudolph, G. (1977). Crystal and molecular structure of the malononitrile adduct of 18-crown-6 (1,4,7,10,13,16-hexaoxacyclooctadecane) (*in German*). *Chemische Berichte*, **110**, 2249–53. [304]

Kazarian, S. G., Hamley, P. A. and Poliakoff, M. (1993). Is intermolecular hydrogen-bonding to uncharged metal centers of organometallic compounds widespread in solution? A spectroscopic investigation in hydrocarbon, noble gas, and supercritical fluid solutions of the interaction between fluoro alcohols and $(\eta^5\text{-}C_5R_5)ML_2$ (R=H, Me; M=Co, Rh, Ir; L=CO, C_2H_4, N_2, PMe_3) and its relevance to protonation. *Journal of the American Chemical Society*, **115**, 9069–79. [276]

Keegstra, E. M. D., Spek, A. L., Zwikker, J. W. and Jenneskens, L. W. (1994). The crystal structure of 2-methoxy-1,4-benzoquinone: Molecular recognition involving intermolecular dipole–dipole and C−H···O hydrogen bond interactions. *Journal of the Chemical Society, Chemical Communications*, 1633–4. [98, 296, 297]

Keegstra, E. M. D., Meiden, V. v. d., Zwikker, J. W. and Jenneskens, L. W. (1996*a*). Self-organization of 2,5-di-*n*-alkoxy-1,4-benzoquinones in the solid state: Molecular recognition involving intermolecular dipole–dipole, weak C−H···O=C hydrogen bond and van der Waals interactions. *Chemistry of Materials*, **8**, 1092–105. [297]

Keegstra, E. M. D., Huisman, B-H., Paardekooper, E. M., Hoogesteger, F. J., Zwikker, J. W., Jenneskens, L. W. and Koojiman, H. (1996*b*). 2,3,5,6-Tetraalkoxy-1,4-benzoquinones and structurally related tetraalkoxy benzene derivatives: synthesis, properties and solid state packing motifs. *Journal of the Chemical Society, Perkin Transactions 2*, 229–40. [297]

Keller, M. B., Rzepa, H. S., White, A. J. P. and Williams, D. J. (1995). Refcode NAMLAO in the Cambridge Structural Database. [166]

Kemmink, J. and Creighton, T. E. (1993). Local conformations of peptides representing the entire sequence of bovine pancreatic trypsin inhibitor and their roles in folding. *Journal of Molecular Biology*, **234**, 861–78. [376]

Kiplinger, J. L., Arif, A. M., and Richmond, T. G. (1995). New bonding mode for cyanoacetylene. A tungsten(II) fluoride carbonyl complex in which cyanoacetylene serves as a four-electron donor alkyne ligand. *Inorganic Chemistry*, **34**, 399–401. [58, 211, 214]

Kitaigorodskii, A. I. (1973). *Molecular crystals and molecules*. Academic, New York. [294, 318, 327]

Klein, C. L., Cobbinah, I., Rouselle, D., Malmstrom, M. C. Sr. and Stevens, E. D. (1991). Structure of L-prolyl-L-tyrosine monohydrate. *Acta Crystallographica, Section C*, **47**, 2386–8. [424, 425]

Koch, U. and Popelier, P. L. A. (1995). Characterization of $C-H-O$ hydrogen bonds on the basis of charge density. *Journal of Physical Chemistry*, **99**, 9747–54. [80, 89]

Koellner, G., Kryger, G., Millard, C. M., Silman, I. and Sussman, J. L. and Steiner, T. (1999). The internal water molecules in acetylcholinesterase from *Torpedo californica*. In preparation. [362, 432, 433]

Koetzle, T. F. (1996). Personal communication. [279]

Koetzle, T. F. and Lehmann, M. S. (1976). Neutron diffraction studies of hydrogen bonding in α-amino acids. In *The hydrogen bond. recent developments in theory and experiments*, Vol. II, (ed. P. Schuster, G. Zundel and C. Sandorfy), pp. 457–69. North Holland, Amsterdam. [7]

Koetzle, T. F., Hamilton, W. C. and Parthasarathy, R. (1972). Precision neutron diffraction structure determination of protein and nucleic acid components. II. The crystal and molecular structure of the dipeptide glycylglycine monohydrochloride monohydrate. *Acta Crystallographica, Section B*, **28**, 2083–90. [34–5]

Kofranek, M., Lischka, H. and Karpfen, A. (1987). *Ab initio* stidies on hydrogen-bonded clusters. I. Linear and cyclic oligomers of hydrogen cyanide. *Chemical Physics*, **113**, 53–64. [83]

Kollman, P. A. and Allen, L. C. (1970). Theory of the hydrogen bond: *ab initio* calculations on hydrogen fluoride dimer and the mixed water–hydrogen fluoride dimer. *Journal of Chemical Physics*, **52**, 5085–93. [17]

Krebs., B. (1983). Thio- and seleno-compounds of main group elements – novel inorganic oligomers and polymers. *Angewandte Chemie, International Edition in English*, **22**, 113–34. [261]

Krebs, B. and Henkel, G. (1981). Investigations on compounds containing $S-H\cdots S$ hydrogen bonds. Crystal structure of diphenyldithiophosphinic acid at 140 and 293 K (*in German*). *Zeitschrift für Anorganische und Allgemeine Chemie*, **475**, 143–55. [261]

Krebs, B. and Henkel, G. (1987). $S-H\cdots S$ hydrogen bridges. The crystal structure of a new modification of trithiocarbonic acid. *Zeitschrift für Kristallographie*, **179**, 373–82. [261]

Krebs, B. and Jacobsen, H.-J. (1976). Preparation and structure of $Na_4GeSe_4 \cdot 14 H_2O$

(*in German*). *Zeitschrift für Anorganische und Allgemeine Chemie*, **421**, 97–104. [233]

Krebs, B., Henkel, G., Dinglinger, H. J. and Stehmeier, G. (1980). Neubestimmung der Kristallstruktur von Trithiokohlensäure α-H_2CS_3 bei 140 K (*in German*). *Zeitschrift für Kristallographie*, **153**, 285–96. [261]

Krebs, B., Henkel G. and Stücker, W. (1984). Ein neuer Wasserstoffbrücken-Typ. Intramolekulare S−H···S Brücken in Dithiotropolon (*in German*). *Zeitschrift für Naturforschung*, **39b**, 43–9. [259, 261]

Kroon, J. (1995). Personal communication. [60]

Kroon, J. and Kanters, J. A. (1974). Non-linearity of hydrogen bonds in molecular crystals. *Nature*, **248**, 667–9. [58, 59]

Kroon, J., Kanters, J. A., van Duijneveldt-van de Rijdt, J. C. G. M., van Duijneveldt, F. B. and Vliegenthart, J. A. (1975). O−H···O Hydrogen bonds in molecular crystals. A statistical and quantum chemical analysis. *Journal of Molecular Structure*, **24**, 109–29. [10, 110]

Kroon, J., Scherrenberg, R. L., Kooijman, H. and Kanters, J. A. (1990). The hydrogen-atom environment of the ether oxygen atom in crystal structures of some representative muscarinic agonists. *Journal of Molecular Structure*, **223**, 287–90. [34]

Kryger, G., Silman, I. and Sussman, J. L. (1998). Three-dimensional structure of a complex E2020 with acetylcholinesterase from *Torpedo californica*. *Journal of Physiology (Paris)*, **92**, 191–4. [379, 432, 433, 434]

Kubicki, M., Dutkiewicz, G. and Antkowiak, W. Z. (1995). 7,7-Dibromo-3-dibromomethylene-2,2-dimethylnorbornan-1-ol. *Acta Crystallographica, Section C*, **51**, 736–8. [223]

Kuchta, M. C. and Parkin, G. (1994). Multiple bonding between germanium and the chalcogens. The syntheses and structures of the terminal chalcogenido complexes (η^4-Me$_8$taa)GeE (E = S, Se, Te). *Journal of the Chemical Society, Chemical Communications*, 1351–2. [235, 236, 238]

Kuduva, S. S., Craig, D. C., Nangia, A. and Desiraju, G. R. (1999). Cubane carboxylic acids. Crystal engineering considerations and the role of C−H···O hydrogen bonds in determining O-H···O networks. *Journal of the American Chemical Society*, **121**, 1936–44. [316]

Kumar, S., Subramanian, K., Srinivasan, R., Rajagopalan, K., Schreurs, A. M. M., Kroon, J. *et al.* (1998*a*). N−H···C≡C−H hydrogen bonds as parts of cooperative networks: crystal structure of *N*-(*p*-methylphenyl)-*N*-prop-2-ynyl-urea. *Journal of Molecular Structure*, **488**, 51–5. [182, 185]

Kumar, S., Subramanian, K., Srinivasan, R., Rajagopalan, K. and Steiner, T. (1998*b*). Crystal structure of *N*-cyano-*N*-prop-2-ynyl-aniline and structural data on C≡C−H···N hydrogen bonds. *Journal of Molecular Structure*, **471**, 251–5. [194, 198–9]

Kumler, W. D. (1935). The effect of the hydrogen bond on the dielectric constants and boiling points of organic liquids. *Journal of the American Chemical Society*, **57**, 600–5. [29, 123, 209]

Kumpf, R. A. and Damewood Jr., J. R. (1988). Do nitromethane and malononitrile form C−H···O hydrogen bonds? Implications for molecular recognition by crown ethers. *Journal of the Chemical Society, Chemical Communications*, 621–2. [73]

Küppers, H., Takusagawa, F. and Koetzle, T. F. (1985). Neutron diffraction study of lithium hydrogen phthalate monohydrate. A material with two very short O−H···O hydrogen bonds. *Journal of Chemical Physics*, 82, 5636–47. [427]

Kvick, Å, Koetzle, T. F. and Thomas, R. (1974). Hydrogen bond studies. 89. A neutron diffraction study of hydrogen bonding in 1-methylthymine. *Journal of Chemical Physics*, **61**, 2711–19. [47]

Labar, D., Krief, A., Norberg, B., Evrard G. and Durant F. (1985). Stereochemical outcome of the reaction of some α-selenoalkyllithiums with 4-*t*-butyl cyclohexanone. *Bulletin de Societé Chimique de Belgie*, **94**, 1083–100. [232]

Lai, T. F. and Marsh, R. E. (1972). The crystal structure of adenosine. *Acta Crystallographica, Section B*, **28**, 1982–9. [392]

Lakshmi, S., Subramanian, K., Rajagopalan, K., Koellner, G. and Steiner, T. (1995*a*). 1β-Hydroxy-1α-propargyl-2β-methyl-2-(2-ethoxycarbonylvinyl) cycloheptane. *Acta Crystallographica, Section C*, **51**, 2327–9. [81, 82]

Lakshmi, S., Subramanian, K., Rajagopalan, K., Koellner, G. and Steiner, T. (1995*b*). Alkynyl contacts in 1β-hydroxy-1α-propargyl-2α-(2-ethoxycarbonylvinyl)-2,4,4-trimethylcyclopentane. *Acta Crystallographica, Section C*, **51**, 2325–7. [173]

Langley, P. J., Hulliger, J., Thaimattam, R. and Desiraju, G. R. (1998). Supramolecular synthons mediated by weak hydrogen bonding. Forming linear molecular arrays *via* $C \equiv C - H \cdots N \equiv C$ and $C \equiv C - H \cdots O_2N$ recognition. *New Journal of Chemistry*, 1307–9. [441, 442]

Latajka, Z. and Scheiner, S. (1987). Basis-sets for molecular interactions, 2. Application to H_3N–HF, H_3N–HOH, H_2O–HF, $(NH_3)_2$ and H_3CH–OH_2. *Journal of Computational Chemistry*, **8**, 674–81. [74]

Latimer, W. M. and Rodebush, W. H. (1920). Polarity and ionization from the standpoint of the Lewis theory of valence. *Journal of the American Chemical Society*, **42**, 1419–33. [1]

Law, R. V. and Sesamuna, Y. (1996). Nature of the non-bonded $(C - H) \cdots O$ interaction of ethers $CH_3O - (CH_2)_n - OCH_3$ ($n = 4$–8). *Journal of the Chemical Society, Faraday Transactions*, **92**, 4885–8. [96]

Ledoux, I., Zyss, J., Siegel, J. S., Brienne, J. and Lehn, J.-M. (1990). Second-harmonic generation from non-dipolar non-centrosymmetric aromatic charge-transfer molecules. *Chemical Physics Letters*, **94**, 440–4. [330]

Lee, J. Y., Lee, S. J., Choi, H. S., Cho, S. J., Kim, K. S. and Ha, T. K. (1995). *Ab initio* study of the complexation of benzene with ammonium cations. *Chemical Physics Letters*, **232**, 67–71. [160]

Le Fur, Y., Masse, R., Cherkaoui, M. Z. and Nicoud, J.-F. (1995). Crystal structure of 1-ethyl-2,6-dimethyl-4(1*H*)-pyridinone trihydrate: a potential nonlinear optical crystalline organic material transparent till the near ultraviolet range. *Zeitschrift für Kristallographie*, **210**, 856–60. [336]

Le Fur, Y., Beucher, M. B., Masse, R., Nicoud, J. F. and Levy, J. P. (1996). Crystal engineering of noncentrosymmetric structures based on 2-amino-5-nitropyridine and η-chloroacetic acid assemblies. *Chemistry of Materials*, **8**, 68–78. [336]

Legon, A. C. and Millen, D. J. (1987*a*). Directional character, strength, and nature of the hydrogen bond in gas-phase dimers. *Accounts of Chemical Research*, **20**, 39–46. [19, 27, 127]

Legon, A. C. and Millen, D. J. (1987*b*). Angular geometries and other properties of hydrogen-bonded dimers. A simple electrostatic interpretation of the success of the electron-pair model. *Chemical Society Reviews*, **16**, 467–98. [127, 414]

Legon, A. C. and Willoughby, L. C. (1984). The rotational spectrum and structure of the phosphine–hydrogen cyanide complex. *Chemical Physics*, **85**, 443–50. [239]

Legon, A. C., Aldrich, P. D. and Flygare, W. H. (1981). The rotational spectrum and molecular structure of the acetylene–HCl dimer. *Journal of Chemical Physics*, **75**, 625–30. [128, 129]

Legon, A. C., Aldrich, P. D. and Flygare, W. H. (1982). The rotational spectrum, chlorine nuclear quadrupole coupling constants, and molecular geometry of a hydrogen-bonded dimer of cyclopropane and hydrogen chloride. *Journal of the American Chemical Society*, **104**, 1486–90. [128, 129]

Lehn, J.-M. (1988). Supramolecular chemistry – scope and perspectives. Molecules, supermolecules and molecular devices. *Angewandte Chemie, International Edition in English*, **27**, 89–112. [293]

Lehn, J.-M. (1990). Perspectives in supramolecular chemistry – from molecular recognition towards molecular information processing and self-organization. *Angewandte Chemie, International Edition in English*, **29**, 1304–19. [293]

Lehn, J.-M. (1995). *Supramolecular Chemistry: Concepts and Perspectives*. VCH, Weinheim. [293, 294]

Leiserowitz, L. (1976). Molecular packing modes. Carboxylic acids. *Acta Crystallographica, Section B*, **32**, 775–802. [34, 37, 97, 98, 100, 120, 178]

Leiserowitz, L. and Hagler, A. T. (1983). The generation of possible crystal structures of primary amides. *Proceedings of the Royal Society of London, Series A*, **388**, 133–75. [25]

Leonard, G. A., McAuley-Hecht, K. E., Gibson, N. J., Brown, T., Watson, W. P. and Hunter, W. N. (1994). Guanine-1,N^6-ethenoadenine base pairs in the crystal structure of d(CGCGAATT(εdA)GCG). *Biochemistry*, **33**, 4755–61. [400]

Leonard, G. A., McAuley-Hecht, K., Brown, T. and Hunter, W. N. (1995). Do C−H···O hydrogen bonds contribute to the stability of nucleic acid–base pairs? *Acta Crystallographica, Section D*, **51**, 136–9. [400]

Leusen, F. J. J., Wilke, S., Verwer, P. and Engel, G. E. (1999) Computational approaches to crystal structure and polymorph prediction. In *Implications of molecular and materials structure for new technologies* (ed. J. A. K. Howard and F. H. Allen). pp. 303–14. Kluwer, Dortrecht. In press. [321, 322]

Levitt, M. and Perutz, M. F. (1988). Aromatic rings act as hydrogen bond acceptors. *Journal of Molecular Biology*, **201**, 751–4. [131–2, 365]

Lewis, F. D., Yang, J. S. and Stern, C. L. (1996). Crystal structures of secondary arenedicarboxamides. An investigation of arene–hydrogen bonding relationships in the solid state. *Journal of the American Chemical Society*, **118**, 12029–37. [300]

Lewis, T. J., Rettig, S. J., Sauers, R. R., Scheffer, J. R., Trotter, J. and Wu, C.-H. (1996). On the conformational analysis and photochemical reactivity of 1,6-cyclodecanedione. *Molecular Crystals Liquid Crystals*, **277**, 289–98. [95]

Li, Q. and Mak, T. C. W. (1997*a*). Hydrogen-bonded urea-anion host lattices. Part 4. Comparative study of inclusion compounds of urea with tetraethylammonium and tetraethylphosphonium chlorides. *Journal of Inclusion Phenomena and Molecular Recognition Chemistry*, **28**, 151–61. [224]

Li, Q. and Mak, T. C. W. (1997*b*). Inclusion compounds of thiourea and peralkylated ammonium salts. Part 5. Hydrogen-bonded host lattices built of thiourea and acetate ions. *Journal of Inclusion Phenomena and Molecular Recognition in Chemistry*, **28**, 183–204. [224]

Li, Q. and Mak, T. C. W. (1998). Hydrogen-bonded urea-anion host lattices. 6. New inclusion compounds of urea with tetra-*n*-propylammonium halides. *Acta Crystallographica, Section B*, **54**, 180–92. [224]

Li, C. and Sammes, M. P. (1983*a*). Hydrogen bonds involving polar CH groups. Part 8. The synthesis and X-ray crystal structure of 2-(2,2-dimethyl-3-piperidinopropyl)-1,3-dithian 1,1,3,3-tetraoxide. A sterically crowded cyclic disulphone. *Journal of the Chemical Society, Perkin Transactions 1*, 1299–302. [38]

Li, C. and Sammes, M. P. (1983*b*). Hydrogen bonds involving polar CH groups Part 9. Optimum structural parameters, and unequivocal demonstration of such intramolecular interactions in 2-substituted 1,3-dithian 1,1,3,3-tetraoxides. *Journal of the Chemical Society, Perkin Transactions 1*, 1303–9. [38]

Li, Q., Yip, W. H. and Mak, T. C. W. (1995). Hydrogen-bonded urea-anion host lattices. 2. Crystal structures of inclusion compounds of urea with tetraalkylammonium bicarbonates. *Journal of Inclusion Phenomena and Molecular Recognition in Chemistry*, **23**, 233–44. [225]

Liu, Q. and Hoffmann, R. (1995). Thoretical aspects of a novel mode of hydrogen–hydrogen bonding. *Journal of the American Chemical Society*, **117**, 10 108–12. [288]

Liu, S., Ji, X., Gilliland, G. L., Stevens, W. J. and Armstrong, R. N. (1993). Second-sphere electrostatic effects in the active site of glutathione S-transferase. Observation of an on-face hydrogen bond between the side chain of threonine 13 and the π-cloud of tyrosine 6 and its influence on catalysis. *Journal of the American Chemical Society*, **115**, 7910–11. [382]

Loewenthal, R., Sancho, J. and Fersht, A. R. (1992). Histidine–aromatic interactions in barnase. Elevation of histidine pK_a and contribution to protein stability. *Journal of Molecular Biology*, **224**, 759–70. [368]

Lommerse, J. P. M. and Cole, J. C. (1998). Hydrogen bonding to thiocyanate anions. Statistical and quantum-chemical analyses. *Acta Crystallographica, Section B*, **54**, 316–19. [196–8]

Lommerse, J. P. M., Stone, A. J., Taylor, R. and Allen, F. H. (1996). The nature and geometry of intermolecular interactions between halogens and oxygen or nitrogen. *Journal of the American Chemical Society*, **118**, 3108–16. [25]

Lowder, J. E., Kennedy, L. A., Sulzmann, K. G. P. and Penner, S. S. (1994). Spectroscopic studies of hydrogen bonding in hydrogen sulfide. *Journal of Quantum Spectroscopy and Radiation Transfer*, **10**, 17–23. [258]

Lu, K.-L., Su, C.-J., Lin, Y.-W., Gau, H.-M. and Wen, Y.-S. (1992). Synthesis of $Os_3(CO)_{10}(CNR)(NCMe)$ and its reaction with propyonic acid. *Organometallics*, **11**, 3832–7. [201]

Luisi, B., Orozco, M., Sponer, J., Luque, F. J. and Shakked, Z. (1998). On the potential of the amino nitrogen atom as a hydrogen bond acceptor in macromolecules. *Journal of Molecular Biology*, **279**, 1123–36. [393]

Lutz, B. T. G. and van der Maas, J. H. (1998). The sensorial potentials of the OH stretching mode. *Journal of Molecular Structure*, **436**, 213–31. [26]

Lutz, B. T. G., van der Maas, J. H. and Kanters, J. A. (1994). Spectroscopic evidence for $\equiv$C$-$H$\cdots$O interaction in crystalline steroids and reference compounds. *Journal of Molecular Structure*, **325**, 203–14. [43]

Lutz, B. T. G., Jacob, J. and van der Maas, J. H. (1996). Vibrational spectroscopic characteristics of $=$C$-$H$\cdots$O and N$-$H$\cdots\pi$ interaction in crystalline *N*-(2,6-dimethylphenyl)-5-methylisoxazole-3-carboxamide. *Vibrational Spectroscopy*, **12**, 197–206. [43, 141, 161]

Lutz, B., Kanters, J. A., van der Maas, J., Kroon, J. and Steiner, T. (1998). Spectroscopic and structural evidence for the hydrogen bond nature of C$\equiv$C$-$H$\cdots$C$=$C contacts in ethynyl steroids. *Journal of Molecular Structure*, **440**, 81–7. [186, 187–8]

Ma, J. C. and Dougherty, D. A. (1997). The cation–π interaction. *Chemical Reviews*, **97**, 1303–24. [159, 160]

MacGillivray, L. R. and Atwood, J. L. (1996). Structural reorganization of the doubly protonated [222] cryptand through cation–π and charge–charge interactions: synthesis and structure of its $[CoCl_4] \cdot 0.5\ C_6H_5CH_3$ salt. *Angewandte Chemie, International Edition in English*, **35**, 1828–30. [159]

Mackey, M. D. and Goodman, J. M. (1997). Conformational preferences of $R^1R^2C=O \cdot H_2BF$ complexes. *Chemical Communications*, 2383–4. [118]

Madhavi, N. N. L., Katz, A. K., Carrell, H. L., Nangia, A. and Desiraju, G. R. (1997). Evidence for the characterisation of the $C-H \cdots \pi$ interaction as a weak hydrogen bond. Toluene and chlorobenzene solvates of 2,3,7,8-tetraphenyl-1,9,10-anthyridine. *Chemical Communications*, 1953–4. [19, 156, 442, 443]

Magomedova, N. S., Sobolev, A. N., Belskii, V. K., Koroteev, M. P., Pugashova, N. M. and Nifantev E. E. (1991). Synthesis and structures of 1,2-*O*-isopropylidene-α-D-glucofuranose-3,6-*O*-cyclophosphate, thiono and selenonophosphates. *Phosphorus, Sulfur and Silicon*, **57**, 261–71. [233]

Mallinson, P. R., Wozniak, K., Smith, G. T. and McCormack, K. L. (1997*a*). A charge density analysis of cationic and anionic hydrogen bonds in a 'proton sponge' complex. *Journal of the American Chemical Society*, **119**, 11502–9. [79]

Mallinson, P. R., MacNicol, D. D., McCormack, K. L., Yufit, D. S., Gall, J. H. and Henderson, R. K. (1997*b*). Hexakis(mercaptomethyl)benzene. A structure possessing well ordered homodromic $[S-H \cdots S]_6$ interactions. *Acta Crystallographica, Section C*, **53**, 90–2. [260]

Malone, J. F., Murray, C. M., Charlton, M. H., Docherty, R. and Lavery, A. J. (1997). $X-H \cdots \pi$ (phenyl) interactions. Theoretical and crystallographic observations. *Journal of the Chemical Society, Faraday Transactions*, **93**, 3429–36. [12, 134, 135–6]

Mandel-Gutfreund, Y., Margalit, H., Jernigan, R. L. and Zhurkin, V. B. (1998). A role of $C-H \cdots O$ interactions in protein DNA recognition. *Journal of Molecular Biology*, **277**, 1129–40. [380, 402]

Marfurt, J. and Leumann, C. (1998). Evidence for $C-H \cdots O$ hydrogen bond assisted recognition of a pyrimidine base in the parallel DNA triple-helical motif. *Angewandte Chemie International Edition in English*, **37**, 175–7. [402–3]

Mascal, M. (1994). Noncovalent design principles and the new synthesis. *Contemporary Organic Synthesis*, 31–45. [325]

Mascal, M. (1998). Statistical analysis of $C-H \cdots N$ hydrogen bonds in the solid state. There *are* real precedents. *Chemical Communications*, 303–4. [49, 120]

Maverick, E., Seiler, P., Schweizer, W. B. and Dunitz, J. D. (1980). 1,4,7,10,13,16-Hexaoxacyclooctadecane: crystal structure at 100 K. *Acta Crystallographica, Section B*, **36**, 615–20. [90, 91]

Mayer-Jung, C., Moras, D. and Timsit, Y. (1998). Hydration and recognition of methylated CpG steps in DNA. *EMBO Journal*, **17**, 2709–18. [436]

Mazurek, J., Lis, T. and Jasztold-Howorko, R. (1995). Crystal and molecular structure of *N-p*-tolyl 5-benzoylamino-2-chloro-4-*p*-tolylamino-6-pyrimidine-carboxyamine. *Polish Journal of Chemistry*, **69**, 1679–86. [95]

Mazzanti, M., Veyrat, M., Ramasseul, R., Marchon, J. C., Turowska-Tyrk, I., Shang, M. and Scheidt, W. R. (1996). A new ruthenium(II) chiroporphyrin containing a multipoint recognition site. Enantioselective receptor of chiral aliphatic alcohols. *Inorganic Chemistry*, **35**, 3733–4. [95]

McDonald, I. K. and Thornton, J. M. (1994). Satisfying hydrogen bonding potentials in proteins. *Journal of Molecular Biology*, **238**, 777–92. [362]

McMullan, R. K., Kvick, Å. and Popelier, P. (1992). Structures of cubic and orthorhombic phases of acetylene by single-crystal neutron diffraction. *Acta Crystallographica, Section B*, **48**, 726–31. [171]

McPhail, A. T. and Sim, G. A. (1965). Hydroxyl–benzene hydrogen bonding. An X-ray study. *Chemical Communications*, 124–6. [130, 151, 343]

Meehan, P. R., Gregson, R. M., Glidewell, C. and Ferguson, G. (1997). Steric inhibition of molecular weaving: non-woven nets of $R_6^6(40)$ and $R_6^6(60)$ rings in 1,1,3-*tris*(2-methyl-4-hydroxy-5-*tert*-butylphenyl)butane–hexamethylenetetramine (1/1). *Acta Crystallographica, Section C*, **53**, 1637–40. [111]

Meot-Ner (Mautner), M. and Deakyne, C. A. (1985). Unconventional ionic hydrogen bonds. 1. $CH^{\delta+}\cdots X$. Complexes of quarternary ions with N- and π-donors. *Journal of the American Chemical Society*, **107**, 469–74. [160]

Metzger, S. and Lippert, B. (1996). A metalated guanine, cytosine base quartet with a novel GC pairing pattern involving H(5) of C. *Journal of the American Chemical Society*, **118**, 12467–8. [404]

Mikenda, W., Steinwender, E. and Mereiter, K. (1995). Hydrogen bonding in 2-hydroxybenzoic, 2-hydroxythiobenzoic, and 2-hydroxydithiobenzoic acid. A structural and spectroscopic study. *Monatshefte für Chemie*, **126**, 495–504. [255, 257]

Mills, W. J., Todd, L. J. and Huffman, J. C. (1989). Synthesis of quinuclidine-benzyl(ethylcarbamoyl)borane: the first boron analog of a phenylalanine derivative. *Journal of the Chemical Society, Chemical Communictions*, 900–2. [286]

Milstein, D., Calabrese, J. C. and William, I. D. (1986). Formation, structures and reactivity of *cis*-hydroxy, *cis*-methoxy, and *cis*-mercaptoiridium hydrides. Oxidative addition of water to Ir(I). *Journal of the American Chemical Society*, **108**, 6387–9. [289]

Mitchell, J. B. O., Nandi, C. L., McDonald, I. K., Thornton, J. M. and Price, S. L. (1994). Amino/aromatic interactions in proteins. Is the evidence stacked against hydrogen bonding? *Journal of Molecular Biology*, **239**, 315–31. [366, 367, 370, 373]

Moloney, M. J. and Foxman, B. M. (1995). Synthesis, structure and solid-state polymerization of dimethyl(propynoato)thallium. *Inorganica Chimica Acta*, **229**, 322–8. [180, 183]

Montelione, G. T., Arnold, E., Meinwald, Y. C., Stimson, E. R., Denton, J. B., Huang, S.-G. *et al.* (1984). Chain-folding initiation structures in ribonuclease A. Conformational analysis of *trans*-Ac-Asn-Pro-Tyr-NHMe and *trans*-Ac-Tyr-Pro-Asn-NHMe in water and in the solid state. *Journal of the American Chemical Society*, **106**, 7946–58. [432]

Moore, T. S. and Winmill, T. F. (1912). The state of amines in aqueous solution. *Journal of the Chemical Society*, **101**, 1635–76. [1]

Moorthy, J. N. and Venkatesan, K. (1994). Photobehaviour of crystalline 4-styrylcoumarin dimorphs: Structure–reactivity correlations. *Bulletin of the Chemical Society of Japan*, **67**, 1–6. [334]

Moorthy, J. N., Samant, S. D. and Venkatesan, K. (1994). Studies in crystal engineering: structure–reactivity correlations of substituted styrylcoumarins and related systems. *Journal of the Chemical Society, Perkin Transactions 2*, 1223–8. [334]

Mootz, D. and Deeg, A. (1992*a*). 2-Butyne and hydrogen chloride co-crystallized. Solid-state geometry of $Cl-H\cdots\pi$ hydrogen bonding to the carbon–carbon triple bond. *Journal of the American Chemical Society*, **114**, 5887–8. [139, 163]

Mootz, D. and Deeg, A. (1992*b*). Poly(hydrogen chlorides). Formation and crystal structure of the low-melting adducts $Me_2S\cdot4HCl$ and $Me_2S\cdot5HCl$ (*in German*). *Zeitschrift für Anorganische und Allgemeine Chemie*, **615**, 109–13. [263, 264]

Morokuma, K. (1971). Molecular orbital study of hydrogen bonds, III. Hydrogen bond in H_2CO-H_2O and H_2CO-2H_2O. *Journal of Physical Chemistry*, **55**, 1236–44. [17]

Morokuma, K. (1977). Why do molecules interact? The origin of electron donor–acceptor complexes, hydrogen bonding and proton affinity. *Accounts of Chemical Research*, **10**, 294–300. [17]

Muller, E. and Hyne, J. B. (1968). Hydrogen bonding in sulfanes. *Canadian Journal of Chemistry*, **46**, 3587–90. [259]

Müller, T. E., Mingos, M. P. and Williams, D. J. (1994). T-shaped intermolecular CH···π(C≡C) interactions in chloroform solvates of gold(I) ethyne complexes. *Journal of the Chemical Society, Chemical Communications*, 1787. [174–5]

Müller, G., Lutz, M. and Harder, S. (1996). Methyl group conformation-determining intermolecular C—H···O hydrogen bonds: structure of *N*-methyl-2-pyrrolidone. *Acta Crystallographica, Section B*, **52**, 1014–22. [97]

Murray-Rust, P. and Glusker, J. P. (1984). Directional hydrogen-bonding to sp^2 and sp^3 hybridized oxygen atoms and its relevance to ligand–macromolecule interactions. *Journal of the American Chemical Society*, **106**, 1018–25. [25, 63]

Murray-Rust, P., Stalling, W. C., Monti, C. T., Preston, R. K. and Glusker, J. P. (1983). Intermolecular interactions of the C—F bond. The crystallographic environment of the fluorinated carboxylic acids and related structures. *Journal of the American Chemical Society*, **105**, 3206–14. [202, 204]

Murugavel, R., Voigt, A., Chandrasekhar, V., Roesky, H. W., Schmidt, H.-G. and Noltemeyer, M. (1996). Silanediols derived from silanetriols. X-Ray crystal structures of (2,4,6-Me$_3$C$_6$H$_2$)N(SiMe$_3$)Si(OSiMe$_3$)(OH)$_2$ and (2,4,6-Me$_3$C$_6$H$_2$)N(SiMe$_3$)Si(OSiMe$_2$R)(OH)$_2$ [R = CH$_2$(2-NH$_2$-3,5-Me$_2$C$_6$H$_2$)]. *Chemische Berichte*, **129**, 391–5. [151]

Musah, R. A., Jensen, G. M., Rosenfeld, R. J., McRee, D. E. and Goodin, D. B. (1997). Variation in strength of an unconventional C—H to O hydrogen bond in a engineered protein cavity. *Journal of the American Chemical Society*, **119**, 9083–4. [380]

Nahringbauer, I. and Kvick, Å. (1977). 2-Amino-5-methylpyridine. *Acta Crystallographica, Section B*, **33**, 2902–5. [193]

Nakatsu, K., Yoshioka, H., Kunimoto, K., Kinugasa, T. and Ueji, S. (1978). 2,6-Diphenylphenol. A structure containing an intramolecular O—H···π hydrogen bond. *Acta Crystallographica, Section B*, **34**, 2357–9. [146]

Nangia, A. and Desiraju, G. R. (1998*a*). Supramolecular synthons and pattern recognition. In *Design of organic solids*, Topics in Current Chemistry. No. 198 (ed. E. Weber), pp. 57–95. Springer, Berlin. [25, 321, 324]

Nangia, A. and Desiraju, G. R. (1998*b*). Supramolecular structures – reason and imagination. *Acta Crystallographica, Section A*, **54**, 934–44. [25, 324]

Nangia, A. and Desiraju, G. R. (1998*c*). Axial and equatorial conformations of penicillins, their sulfoxides and sulfones: The role of N—H···S and C—H···O hydrogen bonds. *Journal of Molecular Structure*, **474**, 65–79. [113, 230, 340]

Narula, C. K., Janik, J. F., Duesler, E. N., Paine, R. T. and Scheffer, R. (1986). Convenient synthesis, separation and X-ray crystal structure determination of 1(*e*),3(*e*),5(*e*)-trimethylcycloborazane and 1(*e*),3(*e*),5(*a*)-trimethylcycloborazane *Inorganic Chemistry*, **25**, 3346–9. [284]

Nethaji, M., Pattabhi, V. and Desiraju, G. R. (1988). Structure of 3-(4-hydroxy-3-methoxyphenyl)-2-propenoic acid (ferulic acid). *Acta Crystallographica, Section C*, **44**, 275–7. [107]

Neuheuser, T., Hess, B. A., Reutel, C. and Weber, E. (1994). *Ab initio* calculations of supramolecular recogntion modes. Cyclic versus noncyclic hydrogen bonding in the formic acid/formamide system. *Journal of Physical Chemistry*, **98**, 6459–67. [12, 79–80]

Nijveldt, D. and Vos, A. (1988*a*). Single-crystal X-ray geometries and electron density distributions of cyclopropane, bicyclopropyl and vinylcyclopropane. I. Multipole refinements and dynamical electron density distributions. *Acta Crystallographica, Section B*, **44**, 289–96. [194]

Nijveldt, D. and Vos, A. (1988*b*). Single-crystal X-ray geometries and electron density distributions of cyclopropane, bicyclopropyl and vinylcyclopropane. II. Evidence for conjugation in vinylcyclopropane and bicyclopropyl. *Acta Crystallographica, Section B*, **44**, 296–307. [194]

Nishio, M. and Hirota, M. (1989). CH/π interaction. Implications in organic chemistry. *Tetrahedron*, **45**, 7201–45. [156]

Nishio, M., Umezawa, Y., Hirota, M. and Takeuchi, Y. (1995). The CH/π interaction. Significance in molecular recognition. *Tetrahedron*, **51**, 8665–701. [156]

Nishio, T., Mori, Y.-I., Iida, I. and Hosomi, A. (1996). Photoaddition of benzoxazole-2-thiones with alkenes. *Journal of the Chemical Society, Perkin Transactions 1*, 921–6. [264, 265]

Nishio, M., Hirota, M. and Umezawa, Y. (1998). *The CH/π interaction. Evidence, nature and consequences*. Wiley, New York. [4, 19, 156, 157, 160]

Nobeli, I., Price, S. L., Lommerse, J. P. M. and Taylor, R. (1997). Hydrogen bond properties of oxygen and nitrogen acceptors in aromatic heterocycles. *Journal of Computational Chemistry*, **18**, 2060–74. [27]

Noordik, J. H. and Jeffrey, G. A. (1977). The crystal structure of 3-amino-1,6-anhydro-3-deoxy-β-D-glucopyranose by neutron diffraction. *Acta Crystallographica, Section B*, **33**, 403–8. [45, 48]

Novoa, J. J. and Mota, F. (1997). Substituent effects in intermolecular $C(sp^3)-H\cdots O(sp^3)$ contacts: how strong can a $C(sp^3)-H\cdots O(sp^3)$ hydrogen bond be? *Chemical Physics Letters*, **266**, 23–30. [75]

Novoa, J. J., Tarron, B., Whangbo, M. H. and Williams, J. M. (1991). Interaction energies associated with short intermolecular contacts of $C-H$ bonds. *Ab-initio* computational study of the $C-H\cdots O$ contact interaction in CH_4-OH_2. *Journal of Chemical Physics*, **95**, 5179–86. [12, 73, 74]

Nowacki, W. (1942). Symmetrie und physikalisch-chemische Eigenschaften krystallisierter Verbindungen. I. Die Verteilung der Krystallstrukturen über die 219 Raumgruppen. *Helvetica Chimica Acta*, **25**, 863–78. [294]

Nowacki, W. (1943). Symmetrie und physikalisch-chemische Eigenschaften krystallisierter Verbindungen. II. Die allgemeinen Bauprinzipien organischer Verbindungen. *Helvetica Chimica Acta*, **26**, 459–62. [294]

Oi, T., Sekreta, E. and Ishida, T. (1983). *Ab initio* molecular orbital calculations on clusters of methyl fluoride. *Journal of Physical Chemistry*, **87**, 2323–9. [209]

Okabe, N. and Suga, T. (1995). 3-Iodo-L-tyrosine methanol solvate (1/1). *Acta Crystallographica, Section C*, **51**, 1700–1. [223, 224]

Oki, M. and Iwamura, H. (1967). Steric effects on the $O-H\cdots π$ interaction in 2-hydroxybiphenyl. *Journal of the American Chemical Society*, **89**, 576–9. [146]

Olovsson, I. and Jönsson, P.-G. (1976). X-Ray and neutron diffraction studies of hydrogen bonded systems. In *The hydrogen bond. Recent developments in theory and experiments*, Vol. II, (ed. P. Schuster, G. Zundel and C. Sandorfy), pp. 393–456. North Holland, Amsterdam. [7, 68]

Ornstein, R. L. (1988). Novel base pairing in nucleic acids. $C-H$ donor group mediated base pairing involving 3-methyluracil. *Journal of Molecular Structure (Theochem)*, **179**, 311–17. [400]

Ornstein, R. L. and Fresco, J. R. (1983). Correlation of crystallographically determined and computationally predicted hydrogen-bonded pairing configurations of nucleic acid bases. *Proceedings of the National Academy of Sciences of the USA*, **80**, 5171–5. [400]

Ornstein, R. L. and Zheng, Y-J. (1997). *Ab initio* quantum mechanics analysis of imidazole $C-H\cdots O$ water hydrogen bonding and a molecular mechanics forcefield correction. *Journal of Biomolecular Structure and Dynamics*, **14**, 657–65. [75, 429, 430]

Osterberg, C. E., King, M. A., Arif, A. M. and Richmond, T. G. (1990). Surprisingly basicity of low valent transition metal carbonyl fluorides. Crystallographic characterization of an sp^2-$CH\cdots F$ hydrogen bond. *Angewandte Chemie, International Edition in English*, **29**, 888–90. [213]

Paasch, K., Nieger, M. and Niecke, E. (1995). A simple route to the 1-aza-3-phosphaallyl system. Structure of a dimeric lithium complex and protonation to give the first NH_2-substituted phosphaalkene. *Angewandte Chemie, International Edition in English*, **34**, 2369–71. [241, 242]

Padiyar, G. S. (1998). Crystal and molecular structure of L-histidyl-L-serine trihydrate: occurrence of $C^\alpha-H\cdots O=C$ hydrogen bond motif similar to the motif in collagen triple helix and β-sheets. *Journal of Peptide Research*, **51**, 266–70. [353]

Park, S., Ramachandran, R., Lough, A. J. and Morris, R. H. (1994). A new type of intramolecular $H\cdots H\cdots H$ interaction involving $N-H\cdots H(Ir)\cdots H-N$ atoms. Crystal and molecular structure of $IrH(\eta_1\text{-}SC_5H_4NH)_2(\eta_2\text{-}SC_5H_4N)(PCy_3)]$ $BF_4\cdot 0.72CH_2Cl_2$. *Journal of the Chemical Society, Chemical Communications*, 2201–2. [289, 290]

Parkinson, G., Gunasekera, A., Vijtechovsky, J., Zhang, X., Kunkel, T. A., Berman, H. and Ebright, R. H. (1996). Aromatic hydrogen bond in sequence-specific protein DNA recognition. *Nature Structural Biology*, **3**, 837–41. [380, 381]

Parr, R. G. and Pearson, R. G. (1983). Absolute hardness. Companion parameter to absolute electronegativity. *Journal of the American Chemical Society*, **105**, 7512–16. [87]

Parry, G. S. (1954). The crystal structure of uracil. *Acta Crystallographica*, **7**, 313–30. [33]

Parthasarathy, R. (1969). Crystal structure of glycylglycine hydrochloride. *Acta Crystallographica, Section B*, **25**, 509–18. [66]

Parthasarathy, R., Ginell, S. L., De, N. C. and Chheda, G. B. (1974). Conformation of N_4-acetylcytidine, a modified nucleoside of tRNA, and stereochemistry of codon–anticodon interaction. *Biochemical and Biophysical Research Communications*, **83**, 657–63. [393, 394]

Parvez, M., Feldman, K. S. and Kosmider, B. J. (1987). Structure of 1-(1-hydroxy-1-methylethyl)-1-vinyl-3,5-dioxatricyclo[5.4.1.0²·⁶]dodeca-8,10-dien-12-ol. *Acta Crystallographica, Section C*, **43**, 1410–12. [189]

Patel, B. P., Wessel, J., Yao, W., Lee, J. C. Jr., Peris, E., Koetzle, T. F., *et al.* (1997). Intermolecular $N-H\cdots H-Re$ interactions involving rhenium polyhydrides. *New Journal of Chemistry*, **21**, 413–21. [288, 290]

Pauling, L. (1935). The structure and entropy of ice and of other crystals with some randomness of atomic arrangement. *Journal of the American Chemical Society*, **57**, 2680–4. [2]

Pauling, L. (1939). *The nature of the chemical bond*. Cornell University Press, Ithaca, New York. [2–3, 29]

Pauling, L. and Delbrück, M. (1940). Intermolecular forces in biological processes. *Science*, **92**, 77–9. [316]

Pearson, R. G. (1963). Hard and soft acids and bases. *Journal of the American Chemical Society*, **85**, 3533–9. [87]

Pearson, R. G. (1968). Hard and soft acids and bases, HSAB, Part II. *Journal of Chemical Education*, **45**, 643–8 [87]

Pearson, R. G. (1985). The transition-metal–hydrogen bond. *Chemical Reviews*, **85**, 41–9. [277, 287]

Pedersen, C. J. (1988). The discovery of crown ethers. *Angewandte Chemie, International Edition in English*, **27**, 1021–7. [293, 303]

Pedireddi, V. R. and Desiraju, G. R. (1992). A crystallographic scale of carbon acidity. *Journal of the Chemical Society, Chemical Communications*, 988–90. [52]

Pedireddi, V. R., Sarma, J. A. R. P. and Desiraju, G. R. (1992). Crystal engineering and solid state chemistry of some β-nitrostyrenes. *Journal of the Chemical Society, Perkin Transactions 2*, 311–20. [297–8]

Pedireddi, V. R., Jones, W., Chorlton, A. P. and Docherty, R. (1996). Creation of crystalline supramolecular assemblies using a C—H···O/O—H···N pair-wise hydrogen bond coupling. *Chemical Communications*, 997–8. [323]

Peng, J., Barr, M. E., Ashburn, D. A., Odom, J. D., Dunlap, R. B. and Silks, L. A. III (1994). Synthesis and characterization of chiral oxazolidine-2-selones. A general one-step procedure from readily available oxazolines. *Journal of Organic Chemistry*, **59**, 4977–87. [233, 234]

Pénicaud, A., Boubekeur, K., Batail, P., Canadell, E., Senzier, P. A. and Jerome, D. (1993). Hydrogen-bond tuning of macroscopic transport properties from the neutral molecular component site along the series of metallic organic–inorganic solvates. *Journal of the American Chemical Society*, **115**, 4101–12. [336]

Perlstein, J., Steppe, K., Vaday, S. and Ndip, E. M. N. (1996). Molecular self-assemblies. 5. Analysis of the vector properties of hydrogen bonding in crystal engineering. *Journal of the American Chemical Society*, **118**, 8433–43. [322]

Perutz, M. F. (1993). The role of aromatic rings as hydrogen-bond acceptors in molecular recognition. *Philosophical Transactions of the Royal Society of London, Series A*, **345**, 105–12. [376]

Perutz, M. F., Fermi, G., Abraham, D. J., Poyart, C. and Bursaux, E. (1986). Hemoglobin as a receptor of drugs and peptides. X-ray studies of the stereochemistry of binding. *Journal of the American Chemical Society*, **108**, 1064–78. [365, 376, 377]

Pfeiffer, P. (1913). Zur Theorie der Farblacke, II. *Liebig's Annalen der Chemie*, **398**, 137–96. [1]

Pfeiffer, P. (1927). *Organische Molekülverbindungen*, (2nd edn). Ferdinand Enke, Stuttgart. [294]

Philp, D. and Robinson, D. M. A. (1998). A computational investigation of co-operativity in weakly hydrogen-bonded assemblies. *Journal of the Chemical Society, Perkin Transactions 2*, 1643–9. [12, 88–9]

Pilkington, M., Wallis, J. D. and Larsen, S. (1995). An alkyne group with a pair of hydrogen bonds. The crystal structure of 2,2′-ethynylenedibenzeneboronic acid at 122 K. *Journal of the Chemical Society, Chemical Communications*, 1499–500. [168, 169]

Pilkington, M., Wallis, J., Smith, G. and Howard, J. A. K. (1996). Geometry distorting intramolecular interactions to an alkyne group in 1-(2-aminophenyl)-2-(2-nitrophenyl)ethyne. A joint experimental–theoretical study. *Journal of the Chemical Society, Perkin Transactions 2*, 1849–54. [168, 169]

Pimentel, G. C. and McClellan, A. L. (1960). *The hydrogen bond*. W. H. Freeman, San Francisco. [3, 253]

Pirkle, W. H. and Hauske, J. R. (1976). Conformational control by carbinyl hydrogens. Implications and applications. *Journal of Organic Chemistry*, **41**, 801–5. [158]

Platts, J. A. and Howard, S. T. (1997). C—H···C hydrogen bonding involving ylides. *Journal of the Chemical Society, Perkin Transactions 2*, 2241–8. [245]

Platts, J. A., Howard, S. T. and Bracke, B. R. F. (1996a). Directionality of hydrogen bonds to sulfur and oxygen. *Journal of the American Chemical Society*, **118**, 2726–33. [228]

Platts, J. A., Howard, S. T. and Wozniak, K. (1996*b*). Quantum chemical evidence for C−H···C hydrogen bonding. *Chemical Communications*, 63–4. [245]

Polse, J. L., Andersen, R. A. and Bergman, R. G. (1995). Cycloaddition and cycloreversion reactions of a monomeric Ti(IV) oxo complex with terminal and internal alkynes. A reversible oxametallacyclobutene/hydroxoacetylide interconversion. *Journal of the American Chemical Society*, **117**, 5393–4. [166]

Popelier, P. L. A. and Bader, R. F. W. (1992). The existence of an intramolecular C−H−O hydrogen bond in creatine and carbamoyl sarcosine. *Chemical Physics Letters*, **189**, 542–8. [80]

Potrzebowski, K. J., Michalska, M., Koziol, A. E., Kazmierski, S., Lis, T., Pluskowski, J. and Ciesielski, W. (1998). Structural implications of C-H···S contacts in organophosphorus compounds. Studies of 1,6-anhydro-2-*O*-tosyl-4-*S*-(5,5-dimethyl-2-thioxa-1,3,2-dioxaphosphorinan-2-yl)-β-D-glucopyranose by X-ray and solid-state NMR methods. *Journal of Organic Chemistry*, **63**, 4209–17. [231]

Power, L. F., Turner, K. E. and Moore, F. H. (1976). Crystal and molecular structure of 3-mercapto-1,3-diphenylprop-2-en-1-one by X-ray and neutron diffraction. *Journal of the Chemical Society, Perkin Transactions 2*, 249–52. [230]

Prey, V. and Berbalk, H. (1951). Zur Kenntnis des *o*-Oxyphenylacetylens (*in German*). *Monatshefte für Chemie*, **82**, 990–1007. [125]

Price, S. L. (1997). Intermolecular forces – from the molecular charge distribution to the molecular packing. In *Theoretical aspects and computer modeling of the molecular solid state* (ed. A. Gavezzotti), pp. 31–60. Wiley, Chichester. [17]

Price, S. L. and Stone, A. J. (1987). The electrostatic interactions in van der Waals complexes involving aromatic molecules. *Journal of Chemical Physics*, **86**, 2859–68. [17]

Prout, K., Burns, K., Watkin, D. J., Cooper, D. G., Durant, G. J., Ganellin, C. R. and Saeh, G. S. (1993). Structure of nine histamine H_2 antagonists related to *N*-cyano-*N'*-methyl-*N''*-[4-(2-methyl-2-pyridyl)butyl] guanidine. *Acta Crystallographica, Section B*, **49**, 547–59. [199]

Rablen, P. R., Lockman, J. W. and Jorgensen, W. L. (1998). *Ab initio* study of hydrogen-bonded complexes of small organic molecules with water. *Journal of Physical Chemistry A*, **102**, 3782–97. [134]

Ramachandran, G. N. and Sasisekharan, V. (1965). Refinement of the structure of collagen. *Biochimica et Biophysica Acta*, **109**, 314–16. [34, 343, 357]

Ramachandran, G. N., Ramakrishnan, C. and Sasisekharan, V. (1963). Stereochemistry of polypeptide chain configurations. *Journal of Molecular Biology*, **7**, 95–9. [34]

Ramachandran, G. N., Sasisekharan, V. and Ramakrishnan, C. (1966). Molecular structure of polyglycine II. *Biochimica et Biophysica Acta*, **112**, 168–70. [34, 35, 343, 357]

Ramakrishnan, C. (1988). Personal communication. [34]

Ramani, R., Venkatesan, K. and Marsh, R. E. (1978). Crystal structure and conformation of the cyclic dipeptide *cyclo*-(L-histidyl-L-aspartyl) trihydrate. *Journal of the American Chemical Society*, **100**, 949–53. [361]

Ratajczak-Sitarz, M. and Katrusiak, A. (1994*a*). Crystal and molecular structure of 3-chloro-5-tosylmethylpyridazine. *Polish Journal of Chemistry*, **68**, 255–60. [107]

Ratajczak-Sitarz, M. and Katrusiak, A. (1994*b*). Crystal and molecular structure of phenylsulphonylmethyl bromide. *Polish Journal of Chemistry*, **68**, 467–72. [107]

Read, W. G., Campbell, E. J. and Henderson, G. (1983). The rotational spectrum and molecular structure of the benzene—hydrogen chloride complex. *Journal of Chemical Physics*, **78**, 3501–8. [128, 129]

Reddy, P. J. and Chacko, K. K. (1993). Octa-coordinated environment for Ag⁺ involving an N_2O_6 donor set of two sandwiched pyridino crowns. X-Ray crystal structure of the complex between tribenzopyridino-15-crown-5 and silver nitrate (2:1). *Inorganica Chimica Acta*, **207**, 31–7. [107]

Reddy, D. S., Goud, B. S., Panneerselvam, K. and Desiraju, G. R. (1993). C—H···N mediated hexagonal network in the crystal structure of the 1:1 molecular complex 1,3,5-tricyanobenzene–hexamethylbenzene. *Journal of the Chemical Society, Chemical Communications*, 663–4. [327, 329]

Reddy, D. S., Craig, D. C., Desiraju, G. R. (1994). Organic alloys: Diamondoid networks in crystalline complexes of 1,3,5,7-tetrabromoadamantane, hexamethylenetetramine and carbon tetrabromide. *Journal of the Chemical Society, Chemical Communications*, 1457–8. [328, 332]

Reddy, D. S., Panneerselvam, K., Desiraju, G. R., Carrell, C. J. and Carrell, H. J. (1995). 1,3,5-Tricyanobenzene. *Acta Crystallographica, Section C*, **51**, 2352–4. [327]

Reddy, D. S., Craig, D. C. and Desiraju, G. R. (1996). Supramolecular synthons in crystal engineering. 4 – Structure simplification and synthon interchangeability in some organic diamondoid solids. *Journal of the American Chemical Society*, **118**, 4090–3. [178]

Reetz, M. T., Hütte, S. and Goddard, R. (1993). Tetrabutylammonium salts of CH-acidic carbonyl compounds. Real carbanions or supramolecules? *Journal of the American Chemical Society*, **115**, 9339–40. [119]

Rhine, W. E., Stucky, G. and Peterson, S. W. (1975). Stereochemistry of polynuclear compounds of the main group elements. A neutron and X-ray diffraction investigation of lithium–hydrogen–carbon interactions in $LiB(CH_3)_4$. *Journal of the American Chemical Society*, **97**, 6401–6. [281, 282]

Richardson, T. B., de Gala, S. and Crabtree, R. H. (1995). Unconventional hydrogen bonds. Intermolecular B—H···H—N interactions. *Journal of the American Chemical Society*, **117**, 12875–6. [283, 284]

Robertson, J. M. (1951). The measurement of bond lengths in conjugated molecules of carbon centres. *Proceedings of the Royal Society of London*, **A207**, 101–10. [318]

Robertson, K. N., Bakshi, P. K., Lantos, S, D., Cameron, T. S. and Knop, O. (1998). Crystal chemistry of tetraradial species. Part 9. The versatile BPh_4^- anion, or how organoammonium H(N) atoms compete for hydrogen bonding. *Canadian Journal of Chemistry*, **76**, 583–611. [142]

Rodham, D. A., Suzuki, S., Suenram, R. D., Lovas, F. J., Dasgupta, S., Goddard, W. A., III and Blake, G. A. (1993). Hydrogen bonding in the benzene–ammonia dimer. *Nature*, **362**, 735–7. [128, 129]

Rogers, R. D. and Green, L. M. (1986). A re-investigation of the crystal and molecular structure of (18-crown-6)·2 CH_3NO_2: D_{3d} stabilization via methyl hydrogen–crown oxygen 'hydrogen bonds'. *Journal of Inclusion Phenomena*, **4**, 77–84. [307]

Rogers, R. D. and Richards, P. D. (1987). Neutral solvent/crown ether interactions. 3. Re-orientation of the hydrogen bonds in the low-temperature (150°C). Structure of 18-crown-6·2 (CH_3NO_2). *Journal of Inclusion Phenomena*, **5**, 631–8. [307]

Rogers, R. D., Richards, P. D. and Voss, E. J. (1988). Neutral solvent/crown ether interactions. 4. Crystallization and low temperature (150°C). Structural characterization of 18-crown-6·2 (CH_3CN). *Journal of Inclusion Phenomena*, **6**, 65–71. [307]

Rohrer, D. C., Hazel, J. P., Duax, W. L. and Zeelen, F. J. (1976a) 11β-Methyl-19-nor-17α-pregn-4-en-20-yn-17β-ol $(C_{21}H_{30}O)$. *Crystal Structure Communications*, **5**, 543–46. [167]

Rohrer, D. C., Lauffenburger, J. C., Duax, W. L. and Zeelen, F. J. (1976*b*). Lynestrenol, 19-nor-17α-pregn-4-en-20-yn-17β-ol. *Crystal Structure Communications*, **5**, 539–42. [186]

Rohrer, D. C., Duax, W. L. and Zeelen, F. J. (1978). Conformational analysis of progestational agents. 11β-fluoro-19-nor-17α-pregn-4-en-20-yn-17β-ol. *Acta Crystallographica, Section B*, **34**, 3801–3. [186]

Romero, F. M., Ziessel, R., Cian, A. D., Fischer, J. and Turek, P. (1996). Synthesis, crystal structure and magnetic properties of novel stable nitronyl-nitroxide pyridine-based radicals (NIT)Py(X) C for Br, C≡CSiMe₃, C≡CH. *New Journal of Chemistry*, **20**, 919–24. [336]

Rosenberg, J. M., Seeman., N. C., Day, R. O. and Rich, A. (1976). RNA double helical fragment at atomic resolution. II. The crystal structure of sodium guanylyl-3′,5′-cytidine monohydrate. *Journal of Molecular Biology*, **104**, 145–67. [399]

Rosenstein, R. D., Oberding, M., Hyde, J. R., Zubieta, J., Karlin, K. D. and Seeman, N. C. (1982). The crystal structure of hypoxanthinium nitrate monohydrate, $C_5H_7N_5O_5$. *Crystal Structure Communications*, **11**, 1507–13. [47]

Rovira, C. and Novoa, J. J. (1998). Strength and directionality of the $C(sp^3)$—H···S(sp^3)interaction. An *ab initio* study using the H_2S···CH_4 model complex. *Chemical Physics Letters*, **279**, 140–50. [12, 75]

Rozas, I., Alkorta, I. and Elguero, J. (1997*a*). Unusual hydrogen bonds. H···π interactions. *Journal of Physical Chemistry A*, **101**, 9457–63. [134, 164, 185, 195]

Rozas, I., Alkorta, I. and Elguero, J. (1997*b*). Inverse hydrogen-bonded complexes. *Journal of Physical Chemistry A*, **101**, 4236–44. [291]

Rubin, J., Brennan, T. and Sundaralingam, M. (1972). Crystal and molecular structure of a naturally occurring dinucleoside monophosphate. Uridilyl-(3′,5′)-adenosine-hemihydrate. Conformational 'rigidity' of the nucleotide unit and models for polynucleotide chain folding. *Biochemistry*, **11**, 3121–8. [390]

Ruiz, E. and Alvarez, S. (1995). Host···guest interactions in the pyrrole and aniline Hofmann clathrates. A theoretical study. *Inorganic Chemistry*, **34**, 3260–9. [80]

Rutherford, J. S. and Calvo, C. (1969). The crystal structure of selenourea. *Zeitschrift für Kristallographie*, **128**, 229–58. [235]

Rzepa, H. S., Webb, M. L., Slawin, A. M. Z. and Williams, D. J. (1991). π Facial hydrogen-bonding in the chiral resolving agent (*S*)-2,2,2-trifluoro-1-(9-anthryl) ethanol and its racemic modification. *Journal of the Chemical Society, Chemical Communications*, 765–8. [140, 161]

Rzepa, H. S., Smith, M. H. and Webb, M. L. (1994). A crystallographic, AM1 and PM3 SCF-MO investigation of strong O—H···π alkene and alkyne hydrogen bonding interactions. *Journal of the Chemical Society, Perkin Transactions 2*, 703–7. [167, 185, 187, 189]

Saenger, W. (1984). *Principles of nucleic acid structure*. Springer, Berlin. [385, 388, 389, 394, 435]

Saenger, W. and Mikolajczyk, M. (1973). Crystal and molecular structure of 4-methyl-1,3,2-dioxaphosphorinane 2-oxide. A P—H···O hydrogen bond (*in German*). *Chemische Berichte*, **106**, 3519–23. [267, 268]

Saenger, W. and Steiner, T. (1998). Cyclodextrin inclusion complexes. Host–guest interactions and hydrogen-bonding networks. *Acta Crystallographica, Section A*, **54**, 798–805. [311]

Saenger, W., Jacob, J., Gessler, K., Steiner, T., Hoffmann, D., Sanbe, H. *et al.* (1998). Structures of the common cyclodextrins and their larger analogs – beyond the doughnut. *Chemical Reviews*, **98**, 1787–802. [311]

Salunke, D. M. and Vijayan, M. (1983). X-Ray studies on crystalline complexes involving amino acids and peptides. IX. Crystal structure of L-ornithine L-aspartate hemihydrate. *International Journal of Peptide and Protein Research*, **22**, 154–60. [422]

Sammes, M. P., Harlow, R. L. and Simonsen, S. H. (1976). Crystal structure, and infrared and proton magnetic resonance spectra of 3-cyanomethylsulphonyl-2-morpholinocyclohexene. Evidence for $C—H \cdots N$ intramolecular hydrogen bond. *Journal of the Chemical Society, Perkin Transactions 2*, 1126–30. [37–8]

Sarma, J. A. R. P. and Desiraju, G. R. (1986). The role of $Cl \cdots Cl$ and $C—H \cdots O$ interactions in the crystal engineering of 4-Å short-axis structures. *Accounts of Chemical Research*, **19**, 222–8. [25, 322]

Sarma, J. A. R. P. and Desiraju, G. R. (1987*a*). $C—H \cdots O$ interactions and the adoption of 4 Å short-axis crystal structures by oxygenated aromatic compounds. *Journal of the Chemical Society Perkin Transactions 2*, 1195–1202. [334]

Sarma, J. A. R. P. and Desiraju, G. R. (1987*b*). Mixed crystals of 6-chloro-3,4-methylenedioxycinnamic acid with 2,4- and 3,4-dichlorocinnamic acids. Structure, topochemistry and intermolecular interactions. *Journal of the Chemical Society, Perkin Transactions 2*, 1187–93. [334]

Sarma, J. A. R. P., Dhurjati, M. S. K., Ravikumar, K. and Bhanuprakash, K. (1994). Molecular and crystal engineering studies of two 2,4-dinitroalkoxystilbenes: An endeavor to generate efficient SHG crystal. *Chemistry of Materials*, **6**, 1369–77. [334]

Scheffer, J. R., Wong, Y.-F., Patil, A. O., Curtin, D. Y. and Paul, I. C. (1985). CPMAS ^{13}C NMR spectra of quinones, hydroquinones, and their complexes. Use of CMR to follow a reaction in the solid state. *Journal of the American Chemical Society*, **107**, 4898–904. [43–4]

Scheiner, S. (1997). *Hydrogen bonding. A theoretical perspective*. Oxford University Press, Oxford. [4, 21, 27, 80, 171]

Schick, G., Loew, A., Nieger, A., Airola, K. and Niecke, E. (1996). Syntheses and reactivity of aminobis(diorganylamino) phosphanes. *Chemische Berichte*, **129**, 911–17. [241]

Schleyer, P. v. R. and Allerhand, A. (1962). Strong hydrogen bonds to carbon in isocyanides. *Journal of the American Chemical Society*, **84**, 1322–3. [243]

Schleyer, P. v. R., Trifan, D. S. and Bacskai, R. (1958). Intramolecular hydrogen bonding involving double bonds, triple bonds and cyclopropane rings as proton acceptors. *Journal of the American Chemical Society*, **80**, 6691–2. [125]

Schmidbaur, H., Schier, A., Milewski-Marrla, B. and Schubert, U. (1982). Triphenylphosphonium-cyclopropylid. Röntgenstrukturanalyse eines pyramidalen Carbanions (*in German*). *Chemische Berichte*, **115**, 722–31. [246]

Schmutzler, R., Schomburg, D., Bartsch, R. and Stelzer, O. (1984). Reactions of (2-trimethylsiloxyphenyl)diphenylphosphane with phosphorous halides – Crystal and molecular structure of (2-hydroxyphenyl)-diphenylphosphonium bromide (*in German*). *Zeitschrift für Naturforschung* **39b**, 1177–84. [267, 269]

Schulz, G. E. and Schirmer, R. H. (1979). *Principles of protein structure*. Springer, Berlin. [348]

Seeman, N. C., Rosenberg, J. M., Suddath, F. L., Kim, J. J. P. and Rich, A. (1976). RNA double helical fragment at atomic resolution. I. The crystal and molecular structure of sodium adenylyl-3′,5′-uridine. *Journal of Molecular Biology*, **104**, 109–44. [395, 396, 399]

Seiler, P. and Dunitz, J. D. (1989). An eclipsed(sp^3)—CH_3 bond in a crystalline hydrated tricyclic orthoamide. Evidence for $C—H \cdots O$ hydrogen bonds. *Helvetica Chimica Acta*, **72**, 1125–34. [96]

Seiler, P., Isaacs, L. and Diederich, F. (1996). The X-ray crystal structure and packing of a hexakis-adduct of C_{60}. Temperature dependence of weak $C-H\cdots O$ interactions. *Helvetica Chimica Acta*, **79**, 1047–58. [110]

Sellmann, D., Lechner, P., Knoch, F. and Moll, M. (1992). Transition-metal complexes with sulfur ligands. 82. H_2S, S_2, and CS_2 molecules as ligands in sulfur-rich $[Ru(PPh_3)'S_4']$ complexes ('$S_4'^{2-}$ = 1,2-*bis*[(2-mercaptophenyl)thio]ethane(2^-)). *Journal of the American Chemical Society*, **114**, 922–30. [258]

Seminario, J. M., Concha, M. C. and Politzer, P. (1995). A density functional/molecular dynamics study of the structure of liquid nitromethane. *Journal of Chemical Physics*, **102**, 8281–2. [80]

Sennikov, P. G. (1994). Weak H-bonding by second-row (PH_3, H_2S) and third-row (AsH_3, H_2Se) hydrides. *Journal of Physical Chemistry*, **98**, 4973–81. [238, 239, 242, 270]

Senti, F. and Harker, D. (1940). The crystal structure of rhombohedral acetamide. *Journal of the American Chemical Society*, **62**, 2008–10. [2]

Seyeda, H., Armbruster, K. and Jansen, M. (1996). Synthesis and characterisation of ionic ozonides with bisquaternary ammonium counterions. *Chemische Berichte*, **129**, 997–1001. [108]

Shalaby, M. A., Fronczek, F. R. and Younathan, E. S. (1994). Conformations and structure studies of sugar lactones in the solid state. Part I. The molecular structure of L-rhamnono- and L-mannono-1,4-lactones. *Carbohydrate Research*, **264**, 181–90. [107]

Shalamov, A. E., Agashkin, O. V., Yanovskii, A. I., Struchkov, Y. T., Logunov, A. P., Revenko, G. P. and Bosyakov, Yu. G. (1990). X-Ray diffraction study of 2,3-diphenyl-4-morpholinomethyl-5-ethynyl-2-thioxo-2-phosphabicyclo[4.4.0]decan-5-ol (*in Russian*). *Zhurnal Strukturnoi Khimii*, **31**, 153–6. [231]

Shallcross, F. V. and Carpenter, G. B. (1958). The crystal structure of cyanoacetylene. *Acta Crystallographica*, **11**, 490–6. [30, 31, 441]

Shallenberger, R. S. (1982). *Advanced sugar chemistry*. Ellis Horwood, Chichester. [341]

Shallenberger, R. S. and Acree, T. E. (1967). Molecular theory of sweet taste. *Nature*, **216**, 480–2. [341]

Shallenberger, R. S., Acree, T. E. and Lee, C. Y. (1969). Sweet taste of D- and L-sugars and amino-acids and the steric nature of their chemo-receptor site. *Nature*, **221**, 555–6. [341]

Shannon, R. D. (1976). Revised effective radii and systematic studies of interatomic distances in halides and chalcogenides. *Acta Crystallographica, Section A*, **32**, 751–67. [247]

Sharma, C. V. K. and Desiraju, G. R. (1994). $C-H\cdots O$ hydrogen bond patterns in crystalline nitro compounds: studies in solid state molecular recognition. *Journal of the Chemical Society, Perkin Transactions 2*, 2345–52. [83, 100, 299]

Sharma, C. V. K., Panneerselvam, K., Pilati, T. and Desiraju, G. R. (1992). Molecular recognition via $C-H\cdots O$ hydrogen bonding. Crystal structure of the 1:1 complex 4-nitrobenzoic acid-4-(*N,N*-dimethylamino)benzoic acid. *Journal of the Chemical Society, Chemical Communications*, 832–3. [329, 332]

Sharma, C. V. K., Panneerselvam, K., Pilati, T. and Desiraju, G. R. (1993). Molecular recognition involving an interplay of $O-H\cdots O$, $C-H\cdots O$ and $\pi\cdots\pi$ interactions. The anomalous crystal structure of the 1:1 complex 3,5-dinitrobenzoic acid–4-(*N,N*-dimethylamino)benzoic acid. *Journal of the Chemical Society, Perkin Transactions 2*, 2209–16. [65, 100]

Sharma, C. V. K., Panneerselvam, K. and Desiraju, G. R. (1995). Methyl 3,5-dinitro-*trans*-cinnamate. *Acta Crystallographica, Section C*, **51**, 1364–6. [108]

Shea, J. A. and Kukolich, S. G. (1983). The rotational spectrum and molecular structure of the furan–HCl complex. *Journal of Chemical Physics*, **78**, 3545–51. [128, 130]

Shea, J. A., Bumganer, R. E. and Henderson, G. (1984). The microwave spectrum and properties of the propyne–HF complex. *Journal of Chemical Physics*, **80**, 4605–9. [127, 128, 129]

Shefter, E. and Trueblood, K. N. (1965). The crystal and molecular structure of D(+)-barium uridine-5′-phosphate. *Acta Crystallographica*, **18**, 1067–77. [38, 343, 389–90]

Sheldrick, W. S. and Morr, M. (1980). The stereochemistry of 3′-N-substituted 3′-deoxyadenosines. *Acta Crystallographica, Section B*, **36**, 2328–38. [393]

Shimon, L. J. W., Vaida, M., Addadi, L., Lahav, M. and Leiserowitz, L. (1990). Molecular recognition at the solid-solution interface: a 'relay' mechanism for the effect of solvent on crystal growth and dissolution. *Journal of the American Chemical Society*, **112**, 6215–20. [301]

Shimoni, L. and Glusker, J. P. (1994). The geometry of intermolecular interactions in some crystalline fluorine-containing organic compounds. *Structural Chemistry*, **5**, 383–97. [88, 202, 206]

Shimoni, L., Carrell, H. L., Glusker, J. P. and Coombs, M. M. (1994). Intermolecular effects in crystals of 11-(trifluoromethyl)-15,16-dihydrocyclopenta[*a*]phenanthren-17-one. *Journal of the American Chemical Society*, **116**, 8162–8. [209]

Shoemaker, K. R., Fairman, R., Schultz, D. A., Robertson, A. D., York, E. J., Stewart, J. M. and Baldwin, R. L. (1990). Side-chain interactions in the C-peptide helix. Phe8···His12$^+$. *Biopolymers*, **29**, 1–11. [369]

Shoham, G., Burley, S. K. and Lipscomb, W. N. (1989). Structure of *cyclo*-(-L-prolylglycyl-)$_2$ trihydrate. *Acta Crystallographica, Section C*, **45**, 1944–8. [422]

Shubina, E. S., Belkova, N. V., Krylov, A. N., Vorontsov, E. V., Epstein, L. M., Gusev, D. G. *et al.* (1996). Spectroscopic evidence for intermolecular M—H···H—OR hydrogen bonding. Interaction Of WH(CO)$_2$(NO)L$_2$ hydrides with acidic alcohols. *Journal of the American Chemical Society*, **118**, 1105–12. [288]

Singh, J. and Thornton, J. M. (1990). SIRIUS. An automated method for the analysis of the preferred packing arrangements between protein groups. *Journal of Molecular Biology*, **211**, 595–615. [360, 366]

Sklar, N., Senko, M. E. and Post, B. (1961). Thermal effects in urea. The crystal structure at 140 °C and at room temperature. *Acta Crystallographica*, **14**, 716–20. [224]

Smith, B. D., Haller, K. J. and Shang, M. (1993). Evidence for intramolecular C—H···O hydrogen bonds determining N,N′-diacylindigo crystal structure conformations. *Journal of Organic Chemistry*, **58**, 6905–7. [94]

Souhassou, M., Aubry, A., Boussard, G. and Marraud, M. (1986). C$_{phenyl}$—H···O and C$_{phenyl}$—H···N interactions. Crystal structure of (Z)-4-benzylidene-2-methyl-5(4H)-oxazolone and intermediate for dehydropeptide synthesis. *Angewandte Chemie, International Edition in English*, **25**, 1753–5. [107]

Spaniel, T., Görls, H. and Scholz, J. (1998). (1,4-Diaza-1,3-diene)titanium and niobium halides: unusual structures with intramolecular C—H···halogen hydrogen bonds. *Angewandte Chemie, International Edition in English*, **37**, 1862–5. [219, 221]

Sponer, J., Leszczynski, J. and Hobza, P. (1997). Thioguanine and thiouracil: hydrogen-bonding and stacking properties. *Journal of Physical Chemistry A*, **101**, 9489–95. [400]

Spurr, R. A. and Byers, F. (1958). The intensity of the S—H stretching fundamental. Dimerization of mercaptans. *Journal of Physical Chemistry*, **62**, 425–8. [259]

Starikov, E. B. and Steiner, T. (1997). Computational support for the suggested contribution of C—H···O=C interactions to the stability of nucleic acid base pairs *Acta Crystallographica, Section D*, **53**, 345–7. [399, 400, 401]

Starikov, E. B. and Steiner, T. (1998). Structural and computational analysis of published neutron diffraction data shows that crystalline vitamin B_{12} coenzyme contains a strong intramolecular $N-H \cdots Ph$ hydrogen bond. *Acta Crystallographica, Section B*, **54**, 94–6. [152, 154]

Starikov, E. B., Saenger, W. and Steiner, T. (1998). Quantum chemical calculations on the weak polar host–guest interactions in crystalline cyclomaltoheptaose (β-cyclodextrin)-but-2-yne-1,4-diol heptahydrate. *Carbohydrate Research*, **307**, 343–6. [314]

Stec, B., Yamano, A., Whitlow, M. and Teeter, M. (1997). Structure of human plasminogen kringle 4 at 1.68 Å and 277 K. A possible structural role of disordered residues. *Acta Crystallographica, Section D*, **53**, 169–78. [361]

Steiner, T. (1994a). Effect of acceptor strength on $C-H \cdots O$ hydrogen bond lengths as revealed by and quantified from crystallographic data. *Journal of the Chemical Society, Chemical Communications*, 2341–2. [53, 54, 235]

Steiner, T. (1994b). Reduction of thermal vibrations by $C-H \cdots X$ hydrogen bonding. Crystallographic evidence from terminal alkynes. *Journal of the Chemical Society, Chemical Communications*, 101–2. [70]

Steiner, T. (1995a). Weak hydrogen bonding. Part 1. Neutron diffraction data of amino acid $C_\alpha - H$ suggest lengthening of the covalent $C-H$ bond in $C-H \cdots O$ interactions. *Journal of the Chemical Society, Perkin Transactions 2*, 1315–19. [60, 68, 69]

Steiner, T. (1995b). Lengthening of the $N-H$ bond in $N-H \cdots N$ hydrogen bonds. Preliminary structural data and implications of the bond valence concept. *Journal of the Chemical Society, Chemical Communications*, 1331–2. [68]

Steiner, T. (1995c). Co-operative $C \equiv C - H \cdots C \equiv C - H$ interactions. Crystal structure of DL-prop-2-ynylglycine and database study of terminal alkynes. *Journal of the Chemical Society, Chemical Communications*, 95–6. [183]

Steiner, T. (1995d). Water molecules which apparently accept no hydrogen bonds are systematically involved in $C-H \cdots O$ interactions. *Acta Crystallographica, Section D*, **51**, 93–7. [361, 421, 422, 426]

Steiner, T. (1996a). $C-H \cdots O$ hydrogen bonds in crystals. *Crystallographic Reviews*, **6**, 1–57. [30, 48, 120, 304, 311, 410]

Steiner, T. (1996b). Triphenylpropargylphosphonium bromide. *Acta Crystallographica, Section C*, **52**, 2263–6. [247, 250]

Steiner, T. (1996c). Glycyl-L-histidinium chloride dihydrate. An unusual histidine conformation. *Acta Crystallographica, Section C*, **52**, 2266–9. [361]

Steiner, T. (1997a). Unrolling the hydrogen bond properties of $C-H \cdots O$ interactions. *Chemical Communications*, 727–34. [20, 59, 60, 63, 120]

Steiner, T. (1997b). L-Histidylglycinium dichloride. *Acta Crystallographica, Section C*, **53**, 255–7. [253]

Steiner, T. (1997c). L-Histidylglycinium chloride. *Acta Crystallographica, Section C*, **53**, 730–2. [353, 356]

Steiner, T. (1998a). Lengthening of the covalent $X-H$ bond in heteronuclear hydrogen bonds quantified from neutron diffraction data. *Journal of Physical Chemistry A*, **102**, 7041–52. [8, 68, 69, 230]

Steiner, T. (1998b). The terminal alkynes: a versatile model for weak directional interactions in crystals. In *Advances in molecular structure research*, Vol. 4 (ed. I. Hargittai and M. Hargittai) pp. 43–77. JAI Press, Stamford, CT, USA. [31, 171]

Steiner, T. (1998c). Donor and acceptor strengths in $C-H \cdots O$ hydrogen bonds quantified from crystallographic data of small solvent molecules. *New Journal of Chemistry*, 1099–103. [51, 54, 55, 56]

Steiner, T. (1998*d*). Structural evidence for the aromatic–(i + 1)-amine hydrogen bond in peptides: L-Tyr-L-Tyr-L-Leu monohydrate. *Acta Crystallographica, Section D*, **54**, 584–8. [149, 373, 374, 375]

Steiner, T. (1998*e*). N−H⋯S and N−H⋯Ph hydrogen bonding in 1-(2-fluorophenyl) thiourea. *Acta Crystallographica, Section C*, **54**, 1121–3. [162]

Steiner, T. (1998*f*). Structural evidence for resonance-assisted O−H⋯S hydrogen bonding. *Chemical Communications*, 411–12. [230]

Steiner, T. (1998*g*). Chloroform molecules donate hydrogen bonds to S, Se and Te acceptors. Evidence from a published series of terminal chalcogenido complexes. *Journal of Molecular Structure*, **447**, 39–42. [235, 236, 238]

Steiner, T. (1998*h*). Hydrogen bond distances to halide ions in organic and organometallic crystal structures: up-to-date database study. *Acta Crystallographica, Section B*, **54**, 456–63. [246, 247, 248, 249]

Steiner, T. (1998*i*). The propargylammonium halide salts. Acetylenic hydrogen bond donors in a competitive situation. *Journal of Molecular Structure*, **443**, 149–53. [249, 251]

Steiner, T. (1999*a*). Weak hydrogen bonds. In *Implications of molecular and materials structure for new technologies* (ed. J. A. K. Howard and F. H. Allen), pp. 185–96. Kluwer, Dordrecht. [113, 114]

Steiner, T. (1999*b*). Not all short C−H⋯O contacts are hydrogen bonds: the prototypical example of contacts to C=O⁺−H. *Chemical Communications*, 313–14. [412, 413]

Steiner, T. and Desiraju, G. R. (1998). Distinction between the weak hydrogen bond and the van der Waals interaction. *Chemical Communications*, 891–2. [61, 62, 120]

Steiner, T. and Mason, S. A. (2000). Short N⁺−H⋯Ph hydrogen bonds in ammonium tetraphenylborate characterized by neutron diffraction. *Acta Crystallographica, Section B*, **56**, 254–60.

Steiner, T. and Saenger, W. (1992*a*). Geometry of C−H⋯O hydrogen bonds in carbohydrate crystal structures. Analysis of neutron diffraction data. *Journal of the American Chemical Society*, **114**, 10146–54. [45, 48, 60, 72, 312, 408, 409, 410]

Steiner, T. and Saenger, W. (1992*b*). Geometric analysis of non-ionic O−H⋯O hydrogen bonds and non-bonding arrangements in neutron diffraction studies of carbohydrates. *Acta Crystallographica, Section B*, **48**, 819–27. [48]

Steiner, T. and Saenger, W. (1993*a*). Role of C−H⋯O hydrogen bonds in the coordination of water molecules. Analysis of neutron diffraction data. *Journal of the American Chemical Society*, **115**, 4540–7. [4, 115, 415, 416, 417, 418, 419, 430, 431]

Steiner, T. and Saenger, W. (1993*b*). C−H⋯O hydrogen bonds as integral part of water coordination. The example of the ordered water cluster in vitamin B₁₂ coenzyme at 15 K. *Acta Crystallographica, Section D*, **49**, 592–3. [418, 420]

Steiner, T. and Saenger, W. (1994). Lengthening of the covalent O−H bond in O−H⋯O hydrogen bonds re-examined from low temperature neutron diffraction data of organic compounds. *Acta Crystallographica, Section B*, **50**, 348–57. [68]

Steiner, T. and Saenger, W. (1995*a*). Weak polar host–guest interactions stabilizing a molecular cluster in a cyclodextrin cavity C−H⋯O and C−H⋯π contacts in β-cyclodextrin but-2-yne-1,4-diol heptahydrate. *Journal of the Chemical Society, Chemical Communications*, 2087–8. [314]

Steiner, T. and Saenger, W. (1995*b*). Crystal structure of anhydrous heptakis-(2,6-di-*O*-methyl)-cyclomaltoheptaose (dimethyl-β-cyclodextrin). *Carbohydrate Research*, **275**, 73–82. [410]

Steiner, T. and Saenger, W. (1996). Crystal structure of anhydrous hexakis-(2,3,6-tri-*O*-

methyl)-cyclomaltohexaose (permethyl-α-cyclodextrin) grown from hot water and from cold NaCl solution. *Carbohydrate Research*, **282**, 53–63. [410]

Steiner, T. and Saenger, W. (1998*a*). Relief of steric strain by intramolecular C—H···O interactions. Structural evidence for the 1,4-substituted cyclohexanes. *Journal of the Chemical Society, Perkin Transactions*, 371–8. [113, 409]

Steiner, T. and Saenger, W. (1998*b*). Closure of cavity in permethylated cyclodextrins assisted by glucose inversion, flipping and kinking. *Angewandte Chemie, International Edition in English*, **37**, 3404–7. [410, 411]

Steiner, T. and Tamm, M. (1998). Weak hydrogen bonds from Cp donors to C≡C acceptors. *Journal of Organometallic Chemistry*, **570**, 235–9. [175, 176]

Steiner, T., Starikov, E. B., Amado, A. M. and Teixeira-Dias, J. J. C. (1995*a*). Weak hydrogen bonding. Part 2. The hydrogen bonding nature of short C—H···π contacts. Crystallographic, spectroscopic and quantum mechanic studies of some terminal alkynes. *Journal of the Chemical Society, Perkin Transactions 2*, 1321–26. [84, 153, 172, 173, 176, 178, 181]

Steiner, T., Koellner, G., Gessler, K. and Saenger, W. (1995*b*). Isostructural replacement of an N—H···O by a C—H···O hydrogen bond in complex stabilisation. Crystal structures of β-cyclodextrin complexed with diethanolamine and with pentane-1,5-diol. *Journal of the Chemical Society, Chemical Communications*, 511–12. [104, 313–14, 445]

Steiner, T., Kanters, J. A. and Kroon, J. (1996*a*). Acceptor directionality of sterically unhindered C—H···O=C hydrogen bonds donated by acidic C—H groups. *Chemical Communications*, 1277–8. [63]

Steiner, T., Starikov, E. B. and Tamm, M. (1996*b*). Weak hydrogen bonding. Part 3. A benzyl group accepting equally strong hydrogen bonds from O—H and C—H donors. 5-Ethynyl-5*H*-dibenzo[*a,d*]cyclohepten-5-ol. *Journal of the Chemical Society, Perkin Transactions 2*, 67–71. [141, 152, 161]

Steiner, T., Tamm, M., Lutz, B. and van der Maas, J. (1996*c*). First example of cooperative O—H···C≡C—H···Ph hydrogen bonding. Crystalline 7-ethynyl-6,8-diphenyl-7*H*-benzocyclohepten-7-ol. *Chemical Communications*, 1127–8. [166, 177]

Steiner, T., Tamm, M., Grzegorzewski, A., Schulte, N., Veldman, N., Schreurs, A. M. M. *et al.* (1996*d*). Weak hydrogen bonding. Part 5. Experimental evidence for the long-range nature of C≡C—H···π interactions. Crystallographic and spectroscopic studies of three terminal alkynes. *Journal of the Chemical Society, Perkin Transactions 2*, 2441–6. [172, 180]

Steiner, T., Hirayama, F. and Saenger, W. (1996*e*). Crystal structures of hexakis-(2,6-di-*O*-methyl)cyclomaltohexaose (dimethyl-α-cyclodextrin) crystallized from acetone and crystallized from hot water. *Carbohydrate Research*, **296**, 83–96. [315]

Steiner, T., Lutz, B. T. G., van der Maas, J., Veldman, N., Schreurs, A. M. M., Kroon, J. and Kanters, J. A. (1997*a*). Spectroscopic evidence for cooperativity effects involving C—H···O hydrogen bonds. Crystalline mestranol. *Chemical Communications*, 191–2. [43, 86, 87, 166]

Steiner, T., van der Maas, J. and Lutz, B. T. G. (1997*b*). Constructing a short C—H···O hydrogen bond. The crystalline complex of triphenylsilyl-acetylene and triphenylphosphinoxide. A very unusual crystal structure with $Z = 7$ and $Z' = 3.5$. *Journal of the Chemical Society, Perkin Transactions 2*, 1287–91. [42, 44, 47]

Steiner, T., Mason, S. A. and Tamm, M. (1997*c*). Neutron diffraction study of aromatic hydrogen bonds — 5-ethynyl-5*H*-dibenzo[*a,d*]-cyclohepten-5-ol at 20 K. *Acta Crystallographica, Section B*, **53**, 843–8. [140]

Steiner, T., Lutz, B. T. G., van der Maas, J. H., Schreurs, A. M. M., Kroon, J. and Tamm,

M. (1998a). Very long C—H···O contacts can be weak hydrogen bonds. Experimental evidence from crystalline [Cr(CO)$_3${η^6-[7-exo-C≡CH)C$_7$H$_7$]}]. *Chemical Communications*, 171–2. [41, 42]

Steiner, T., Schreurs, A. M. M., Kanters, J. A. and Kroon, J. (1998b). Water molecules hydrogen bonding to aromatic acceptors of amino acids. The structure of Tyr-Tyr-Phe dihydrate and a crystallographic database study on peptides. *Acta Crystallographica, Section D*, **54**, 25–31. [373, 423, 424, 425]

Steinwender, E. and Mikenda, W. (1990). O—H···O(S) hydrogen bonds in 2-hydroxy(thio)benzamides. Survey of spectroscopic and structural data. *Monatshefte für Chemie*, **121**, 809–20. [230]

Steinwender, E., Lutz, E. T. G., Van der Maas, J. H. and Kanters, J. A. (1993). 2-Ethymyladamantan-2-ol: a model compound with distinct OH—π and CH—O hydrogen bonds. *Vibrational Spectroscopy*, **4**, 217–29. [43]

Stevens, R. C., Bau, R., Milstein, D., Blum, O. and Koetzle, T. F. (1990). Concept of the H(δ$^+$)···H(δ$^-$) interaction. A low temperature neutron diffraction study of cis-[IrH(OH)(PMe$_3$)$_4$]PF$_6$. *Journal of the Chemical Society, Dalton Transactions*, 1429–32. [288, 289]

Stewart, R. F. and Jensen, L. H. (1967). Redetermination of the crystal structure of uracil. *Acta Crystallographica*, **23**, 1102–5. [33]

Suárez, D., González, J., Sordo, T. L. and Sordo, J. A. (1994). *Ab initio* study of the thermal and Lewis acid-catalyzed hetero Diels–Alder reactions of 1,3-butadiene and isoprene with sulfur dioxide. *Journal of Organic Chemistry*, **59**, 8058–64. [79]

Subramanian, S. and Zaworotko, M. J. (1994). Exploitation of the hydrogen bond: recent developments in the context of crystal engineering. *Coordination Chemistry Reviews*, **137**, 357–401. [325]

Subramanian, K., Lakshmi, S., Rajagopalan, K., Koellner, G. and Steiner, T. (1996). Cooperative hydrogen bond cycles involving O—H···π and C—H···O hydrogen bonds as found in a hydrated di-alkyne. *Journal of Molecular Structure*, **384**, 121–6.

Subramanian, K., Kumar, S. and Steiner, T. (1999). Unpublished results. [162, 180, 184]

Sudha, L., Selvan, J. S., Subramanian, K., Steiner, T., Koellner, G., Srinivasan, N. and Ramdas, K. (1996). 1,3-bis(2-Methylphenyl)-2-(4-morpholino)-isothiourea. *Acta Crystallographica, Section C*, **52**, 2047–9. [92, 93]

Sukhodub, L. F. (1987). Interactions and hydration of nucleic acid bases in vacuum. Experimental study. *Chemical Reviews*, **87**, 589–606. [401]

Sundaralingam, M. (1966). Stereochemistry of nucleic acid constituents. III. Crystal and molecular structure of adenosine 3'-phosphate dihydrate (adenylic acid b). *Acta Crystallographica*, **21**, 495–505. [38, 343, 392]

Sundaralingam, M. (1974). Principles governing nucleic acid and polynucleotide conformations. In *Structures and conformation of nucleic acids and protein–nucleic acid interactions* (ed. M. Sundaralingam and S. T. Rao), pp. 487–524. University Park Press, Baltimore. [391]

Sundaralingam, M. (1998). Personal communication. [35]

Sundaralingam, M. and Jensen, L. H. (1965). Refinement of the structure of salicylic acid. *Acta Crystallographica*, **18**, 1053–8. [257]

Surange, S. S., Kumaran, G., Rajappa, S., Pal., D. and Chakrabarti, P. (1997). Push–pull butadienes. Evidence for a possible C—H···S hydrogen bond in 4-(methylthio)-4-nitro-1-(pyrrolidin-1-yl)buta-1,3-diene. *Helvetica Chimica Acta*, **80**, 2329–36. [231]

Sussman, J. L., Seeman, N. C., Kim, S.-H. and Berman, H. (1972). Crystal structure of a naturally occurring dinucleotide phosphate. Uridylyl 3',5'-adenosine phosphate. Model for RNA chain folding. *Journal of Molecular Biology*, **66**, 403–21. [38, 390]

Sussman, J. L., Harel, M., Frolow, F., Oefner, C., Goldman, A., Toker, L. and Silman, I.

(1991). Atomic structure of acetylcholinesterase from *Torpedo californica*. A prototypic acetylcholine-binding protein. *Science*, **253**, 872–9. [159, 379]

Sutor, D. J. (1958). The structures of pyrimidines and purines. VII. The crystal structure of caffeine. *Acta Crystallographica*, **11**, 453–8. [33]

Sutor, D. J. (1962). The C—H···O hydrogen bond in crystals. *Nature*, **195**, 68–9. [32, 123, 389]

Sutor, D. J. (1963). Evidence for the existence of C—H···O hydrogen bonds in crystals. *Journal of the Chemical Society*, 1105–10. [32–3, 343, 389]

Suzuki, S., Green, P. G., Bumgarner, R. E., Dasgupta, S., Goddard, W. A., III and Blake, G. A. (1992). Benzene forms hydrogen bonds with water. *Science*, **257**, 942–5. [128, 129, 161]

Suzuki, T., Fujii, H. and Miyashi, T. (1992). Molecular recognition through C—H···O hydrogen bonding in charge-transfer crystals: Highly selective complexation of 2,4,7-trinitrofluoroenone with 2,6-dimethylnaphthalene. *Journal of Organic Chemistry*, **57**, 6744–8. [299]

Svinning, T. and Soerum, H. (1975). Crystal structure of acetylcholine bromide. *Acta Crystallographica, Section B*, **31**, 1581–6. [252]

Szczesniak, M. M., Chalasinski, G., Cybulski, S. M. and Cieplak, P. (1993). *Ab-initio* study of the potential energy surface of CH_4–H_2O. *Journal of Chemical Physics*, **98**, 3078–85. [74]

Takahara, P. M., Frederick, C. A. and Lippard, S. J. (1996). Crystal structure of the anti-cancer drug cisplatin bound to duplex DNA. *Journal of the American Chemical Society*, **118**, 12309–21. [405, 406]

Takusagawa, F., Higuchi, T. and Shimada, A. (1974). The crystal structure of pyrazinic acid. *Bulletin of the Chemical Society of Japan*, **47**, 1409–13. [316]

Takusagawa, F., Koetzle, T. F., Srikrishnan, T. and Parthasarathy, R. (1979). C—H···O interactions and stacking of water molecules between pyrimidine bases in 5-nitro-1-(β-D-ribosyluronic acid)-uracil monohydrate [1-(5-nitro-2,4-dioxopyrimidinyl-β-ribofuranoic acid monohydrate]: a neutron diffraction study at 80 K. *Acta Crystallographica, Section B*, **35**, 1388–94. [38, 90, 91, 390]

Takusagawa, F., Koetzle, T. F., Kou, W. W. H. and Parthasarathy, R. (1981). Structure of *N*-acetyl-L-cysteine: X-ray ($T = 295$ K) and neutron ($T = 16$ K) diffraction studies. *Acta Crystallographica, Section B*, **37**, 1591–6. [254, 255, 256]

Tang, T.-H., Hu, W.-J. and Cui, Y.-P. (1990). A quantum chemical study on selected π-type hydrogen-bonded systems. *Journal of Molecular Structure (Theochem)*, **207**, 319–26. [132–4, 164, 185]

Taylor, R. and Kennard, O. (1982). Crystallographic evidence for the existence of C—H···O, C—H···N and C—H···Cl hydrogen bonds. *Journal of the American Chemical Society*, **104**, 5063–70. [38–40, 49, 53, 59, 60, 120, 123, 231]

Tchertanov, L. and Pascard, C. (1996). Statistical analysis of noncovalent interactions of anion groups in crystal structures. II. Hydrogen bonding of thiocyanate anions. *Acta Crystallographica, Section B*, **52**, 685–90. [196, 197]

Teff, D. J., Huffman, J. C. and Caulton, K. G. (1997). μ-Benzene in a heterobimetallic fluoroalkoxide. CH···FC hydrogen bonding? *Inorganic Chemistry*, **36**, 4372–80. [208, 212]

Teller, R. G. and Bau, R. (1981). Crystallographic studies of transition metal hydride complexes. In: *Structure and bonding*, Vol. 44, pp. 1–82. Springer, Berlin. [277]

Tezuka, T., Nakagawa, M., Yokoi, K., Nagawa, Y., Yamagaki, T. and Nakanishi, H. (1997). Studies of steric effects. Spectroscopic evidence for through-space interaction in CH_3···O in crowded alcohols. *Tetrahedron Letters*, 4223–6. [117]

Thaimattam, R., Reddy, D. S., Xue, F., Mak, T. C. W., Nangia, A. and Desiraju, G. R.

(1998). Interplay of strong and weak hydrogen bonding in molecular complexes of some 4,4-disubstituted biphenyls with urea, thiourea and water. *Journal of the Chemical Society, Perkin Transactions. 2*, 1783–9. [224, 225, 302]

Thalladi, V. R., Panneerselvam, K., Carrell, C. J., Carrell, H. L. and Desiraju, G. R. (1995). Hexagonal supramolecular networks in the crystal structure of the 1:1 molecular complex trimethylisocyanurate–1,3,5-trinitrobenzene. *Journal of the Chemical Society, Chemical Communications*, 341–2. [327, 330, 331]

Thalladi, V. R., Brasselet, S., Bläser, D., Boese, R., Zyss, J., Nangia, A. and Desiraju, G. R. (1997). Engineering of an octupolar non-linear optical crystal: tribenzyl isocyanurate. *Chemical Communications*, 1841–2. [331, 333]

Thalladi, V. R., Weiss, H.-C., Bläser, D., Boese, R., Nangia, A. and Desiraju, G. R. (1998a). C—H⋯F Interactions in the crystal structures of some fluorobenzenes. *Journal of the American Chemical Society*, **120**, 8702–10. [206–7, 208, 209, 210, 211]

Thalladi, V. R., Brasselet, S., Weiss, H. C., Bläser, D., Katz, A. K., Carrell, H. L. *et al.* (1998b). Crystal engineering of some 2,4,6-triaryloxy-1,3,5-triazines: Octupolar non-linear materials. *Journal of the American Chemical Society*, **120**, 2563–77. [335]

Thanki, N., Thornton, J. M. and Goodfellow, J. M. (1988). Distribution of water around amino acids in proteins. *Journal of Molecular Biology*, **202**, 637–57. [428–9]

Thomas, K. A., Smith, G. M., Thomas, T. B. and Feldman, R. J. (1982). Electronic distributions within protein phenylalanine aromatic rings are reflected by the three-dimensional oxygen atom environments. *Proceedings of the National Academy of Sciences of the USA*, **79**, 4843–7. [360]

Toda, F., Tanaka, K., Stein, Z. and Goldberg, I. (1995). Structure, solid-state photochemistry and reactivity in asymmetric synthesis of 3,4-bis(diphenylmethylene)-*N*-methylsuccinimide. *Acta Crystallographica, Section B*, **51**, 856–63. [108]

Toda, F., Liu, H., Miyahara, I. and Hirotsu, K. (1997). Potential host compounds: 1,1′-bis(diarylhydroxymethyl)ferrocenes. *Journal of the Chemical Society, Perkin Transactions 2*, 85–8. [218]

Tomic, S., van Eijck, B. P., Kojic-Prodic, B., Kroon, J., Magnus, V., Nigovic, B. *et al.* (1995). Synthesis and conformational analysis of the plant hormone (auxin) related 2-(indol-3-yl)ethyl and 2-phenylethyl β-D-xylopyranosides and their 2,3,4-tri-*O*-acetyl derivatives. *Carbohydrate Research*, **270**, 11–32. [107]

Tomori, H., Yoshihara, H. and Ogura, K. (1996). Facile optical resolution of a dibenzopyrazinoazepine derivative and the nature of molecular recognition of amines by chiral 2,3-di-*O*-(arylcarbonyl)tartaric acids. *Bulletin of the Chemical Society of Japan*, **69**, 3581–90. [108]

Trotter, J. (1960). A three-dimensional analysis of the crystal structure of *p*-benzoquinone. *Acta Crystallographica*, **13**, 86–95. [37]

Truter, M. R. (1967). Comparison of photographic and counter observations for the X-ray crystal structure analysis of thiourea. *Acta Crystallographica*, **22**, 556–9. [225]

Tüchsen, E. and Woodward, C. (1987). Assignment of asparagine-44 side-chain primary amide ¹H HMR resonances and the peptide amide N¹H resonance of glycine-37 in basic pancreatic trypsin inhibitor. *Biochemistry*, **26**, 1918–25. [364, 365]

Turi, L. and Dannenberg, J. J. (1993). Molecular orbital studies of C—H⋯O H-bonded complexes. *Journal of Physical Chemistry*, **97**, 7899–909. [12, 73]

Turi, L. and Dannenberg, J. J. (1994). Molecular orbital study of crystalline 1,3-cyclohexanedione. 2. Aggregates in two and three dimensions. *Chemistry of Materials*, **6**, 1313–16. [73]

Turi, L. and Dannenberg, J. J. (1995). Molecular orbital studies of the nitromethane–ammonia complex. An unusually strong $C-H\cdots N$ hydrogen bond. *Journal of Physical Chemistry*, **99**, 639–41. [73]

Tyce, P. A. and Powell, D. B. (1965). Hydrogen bonding in trithiocarbonic acid and related thio acids. *Spectrochimica Acta*, **21**, 835–9. [259]

Ueji, S., Nakatsu, K., Yoshioka, H. and Kinoshita, K. (1982). X-Ray and IR studies on crystal and molecular structure of 4-nitro-2,6-diphenylphenol. Stereochemistry of bifurcated $O-H\cdots\pi$ hydrogen bonds. *Tetrahedron Letters*, 1173–6. [147, 148]

Umeyama, H. and Morokuma, K. (1977). The origin of hydrogen bonding: an energy decomposition study. *Journal of the American Chemical Society*, **99**, 1316–32. [17]

Umezawa, Y. and Nishio, M. (1998). CH/π interactions as demonstrated in the crystal structure of guanine-nucleotide binding proteins, src homology-2 domains and human growth hormone in complex with their specific ligands. *Bioorganic and Medicinal Chemistry*, **6**, 493–504. [157]

Umezawa, Y., Tsuboyama, S., Honda, K., Uzawa, J. and Nishio, M. (1998). CH/π interaction in the crystal structure of organic compounds. A database study. *Bulletin of the Chemical Society of Japan*, **71**, 1207–13. [137, 139, 152]

Ung, A. T., Gizachew, D., Bishop, R., Scudder, M. L., Dance, I. G. and Craig, D. C. (1995). Structure and analysis of helical tubulate inclusion compounds formed by 2,6-dimethylbicyclo[3.3.1]nonane-*exo*-2-*exo*-6-diol. *Journal of the American Chemical Society*, **117**, 8745–56. [334]

Valle, G., Crisma, M., Toniolo, C., Sen, N., Sukumar, M. and Balaram, P. (1988). Crystallographic characterization of the conformation of the 1-amino-cyclohexane-1-carboxylic acid residue in simple derivatives and peptides. *Journal of the Chemical Society, Perkin Transactions 2*, 393–8. [152, 154]

Vampa, G., Benvenuti, S., Severi, F., Malmusi, L. and Antolini, L. (1995). The methylation, oxidation and crystallographic characterisation of imidazole derivatives. *Journal of Heterocyclic Chemistry*, **32**, 227–34. [111]

Van Brocklin, H. F., Liu, A., Welch, M. J., O'Neil, J. P. and Katzenellenbogen, J. A. (1994). The synthesis of 7a-methyl-substituted estrogens labeled with fluorine-18. Potential breast tumor imaging agents. *Steroids*, **59**, 34–45. [342]

Van de Bovenkamp, J., Matxain, J. M., van Duijneveldt, F. B. and Steiner, T. (1999). Combined *ab-initio* computational and statistical investigation of a model $C-H\cdots O$ hydrogen bonded dimer as occurring in 1,4-benzoquinone. *Journal of Physical Chemistry A*, **103**, 2784–92. [75–9, 110]

Van der Sluys, L. S., Eckert, J., Eisenstein, O., Hall, J. H., Huffman, J. C., Jackson, S. A. *et al.* (1990). An attractive '*cis*-effect' of hydride on neighbor ligands: experimental and theoretical studies on the structure and intramolecular rearrangement of $Fe(H)_2(\eta^2\text{-}H_2)(PEtPh_2)_3$. *Journal of the American Chemical Society*, **112**, 4831–41. [291]

Van Geerestein, V. J. (1987). Structure of the methanol solvate of 11β-chloro-13-ethyl-18-norlynestrenol. *Acta Crystallographica, Section C*, **43**, 2232–5. [186]

Van Geerestein, V. J. and Leeflang, V. G. (1988). Structure of 19-nor-17α-pregna-4,15dien-20-yn-17β-ol. *Acta Crystallographica, Section C*, **44**, 378–9. [186]

Van Meerssche, M., Declerq, J. P., Germain, G., Soubrier-Payen, B., Lindner, H. J. and Kitschke, B. (1984). Etudes structurales en serie helicenique. II. Structure du (hydroxy-1-ethyl)-1-hexahelicene (*in French*). *Bulletin de Societé Chimique de Belgie*, **93**, 445–8. [151, 153]

Van Mourik, T. and van Duijneveldt, F. B. (1995). *Ab initio* calculations on the $C-H\cdots O$ hydrogen-bonded systems CH_4–H_2O, CH_3NH_2–H_2O and $CH_3NH_3^+$–H_2O. *Journal of Molecular Structure (Theochem)*, **341**, 63–73. [17, 74, 75]

Van Staveren, C. J., Aarts, V. M. L. J., Grootenhuis, P. D. J., van Eerden, J., Harkema, S. and Reinhoudt, D. N. (1986). Complexation of crown ethers with neutral molecules. 2. Comparison of free macrocycles and their complexes with malononitrile in solution. *Journal of the American Chemical Society*, **108**, 5271–6. [304]

Vasco-Mendez, N. L., Panneerselvam, K., Pinera, E. D. and Soriano-Garcia, M. (1996). Crystal structure of 2-amino-5-chlorobenzophenone. *Analytical Sciences*, **12**, 677–8. [108]

Vásquez, G. B., Ji, X., Fronticelli, C. and Gilliland, G. L. (1998). Human carboxyhemoglobin at 2.2 Å resolution: structure and solvent comparisons of R-state, R2-state and T-state hemoglobins. *Acta Crystallographica, Section D*, **54**, 355–66. [192, 366]

Verdonk, M. L., Boks, G. J., Kooijman, H., Kanters, J. A. and Kroon, J. (1993). Stereochemistry of charged nitrogen–aromatic interactions and its involvment in ligand–receptor binding. *Journal of Computer-Aided Molecular Design*, **7**, 173–82. [160]

Vishveshwara, S. (1978). *Ab-initio* molecular studies on C—H···X hydrogen bonded systems. *Chemical Physics Letters*, **59**, 26–9. [38]

Viswamitra, M. A., Radhakrishnan, R., Bandekar, J. and Desiraju, G. R. (1993). Evidence for O—H···C and N—H···C hydrogen bonding. *Journal of the American Chemical Society*, **115**, 4868–9. [85, 86, 166, 185]

Voet, D. and Voet, J. G. (1995). *Biochemistry* (2nd edn.). Wiley, New York. [347, 360]

Vögtle, F. (1991). *Supramolecular chemistry*. Wiley, Chichester. [303]

Wagner, G., Braun, W., Havel, T. F., Schaumann, T., Go, N. and Wüthrich, K. (1987). Protein structures in solution by nuclear magnetic resonance and distance geometry. The polypeptide fold of the bovine pancreatic trypsin inhibitor determined using two different algorithms, DISGEO and DISMAN. *Journal of Molecular Biology*, **196**, 611–39. [365]

Wahl, M. and Sundaralingam, M. (1997). C—H···O hydrogen bonding in biology. *Trends in Biochemical Sciences*, **22**, 97–102. [387, 400, 404]

Wahl, M. C., Rao, S. T. and Sundaralingam, M. (1996). The structure of r(UUCGCG) has a 5′-UU-overhang exhibiting Hoogsteen-like *trans* U.U base pairs. *Nature Structural Biology*, **3**, 24–31. [403, 404]

Waksman, G., Kominos, D., Robertson, S. C., Pant, N., Baltimore, D., Birge, R. B. *et al.* (1992). Crystal structure of the phosphotyrosine recognition domain SH2 of v-*src* complexed with tyrosine-phosphorylated peptides. *Nature*, **358**, 646–53. [377, 378]

Warren, C. J., Dhingra, S. S., Ho, D. M., Haushalter, R. C. and Bocarsly, A. B. (1994). Electrochemical synthesis of new Sb—Te zintl anions by cathodic dissolution of Sb_2Te_3 electrodes. Structures of $Sb_2Te_5^{4-}$ and $Sb_6Te_9^{4-}$. *Inorganic Chemistry*, **33**, 2709–10. [238]

Weber, G. (1983). The structure of a 1:1 host guest complex between dimethyl sulphate and 18-crown-6. *Journal of Molecular Structure*, **98**, 333–6. [307]

Weber, E. (1987). Solid-state inclusion compounds. Site-specific binding of organic molecules to a macrocylic host. *Molecular Crystals and Liquid Crystals*, **156**, 371–81. [335]

Weber, E. (1989). New developments in crown ether chemistry: Lariat spherand and second sphere complexes. In *Crown compounds and analogs* (ed. S. Patai and Z. Rappoport), pp. 305–57. Wiley, Chichester. [303]

Weber, E., Franken, S., Puff, H. and Ahrendt, H. (1986). Enclave inclusion of nitromethane by a new crown host. X-Ray crystal structure of the inclusion complex and host selectivity properties. *Journal of the Chemical Society, Chemical Communications*, 467–9. [335]

Weber, E., Franken, S., Ahrendt, J. and Puff, H. (1987). Chameleon behaviour of a crown host at molecular complexation. *Journal of Organic Chemistry*, **52**, 5291–2. [335]

Weber, E., Hecker, M., Koepp, E., Orlia, W., Czugler, M. and Csöregh, I. (1988). New trigonal lattice hosts: Stoichiometric crystal inclusions of laterally trisubstituted benzenes – X-ray crystal structure of 1,3,5-tris-(4-carboxyphenyl)benzene·dimethylformamide. *Journal of the Chemical Society, Perkin Transactions 2*, 1251–7. [335]

Weber, E., Köhler, H.-J. and Reuter, H. (1991). Macroring-neutral molecular complexation. Synthesis of biconcave pyridino hosts, complex formation and X-ray crystal structures of two inclusion compounds. *Journal of Organic Chemistry*, **56**, 1236–42. [335]

Weber, E., Hens, T., Gallardo, O. and Csöregh, I. (1996). Roof-shaped hydroxy hosts. Synthesis, complex formation and X-ray structures of inclusion compounds with EtOH, nitromethane and benzene. *Journal of the Chemical Society, Perkin Transactions 2*, 737–45. [149]

Wehman-Ooyevaar, I. C. M., Grove, D. M., Kooijman, H., van der Sluis, P., Spek, A. L. and van Koten, G. (1992*a*). A hydrogen atom in an organoplatinum–amine system. Synthesis and spectroscopic and crystallographic characterization of novel zwitterionic complexes with a $Pt(II) \cdots H - N^+$ unit. *Journal of the American Chemical Society*, **114**, 9916–24. [276]

Wehman-Ooyevaar, I. C. M., Grove, D. M., de Vaal, P., Dedieu, A. and van Koten, G. (1992*b*). A hydrogen atom in an organoplatinum–amine system. 2. The first isolated arylplatinum(IV) hydrides. Synthetic and theoretical study (SCF and CAS-SCF calculations) on model complexes with a $N - M(IV) - H/M(II)^- \cdots H - N^+$ unit (M = Pd, Pt). *Inorganic Chemistry*, **31**, 5484–93. [276]

Wei, K.-T. and Ward, D. L. (1976). Crystal and molecular structures of ammonium fluoroacetate, $C_2H_6FNO_2$ and ammonium difluoroacetate, $C_5H_5F_2NO_2$. *Acta Crystallographica, Section B*, **32**, 2768–73. [203]

Weisinger-Lewin, Y., Frolow, F., McMullan, R. K., Koetzle, T. F., Lahav, M. and Leiserowitz, L. (1989). Reduction in crystal symmetry of a solid solution: a neutron diffraction study at 15 K of the host/guest system asparagine/aspartic acid. *Journal of the American Chemical Society*, **111**, 1035–40. [418]

Weiss H.-C., Bläser, D., Boese, R., Doughan, B. M. and Haley, M. M. (1997*a*). C — H ··· π interactions in ethynylbenzenes. The crystal structures of ethynylbenzene and 1,3,5-triethynylbenzene and a redetermination of the structure of 1,4-diethynylbenzene. *Chemical Communications*, 1703–4. [176, 178]

Weiss, H.-C., Boese, R., Smith, H. L. and Haley, M. M. (1997*b*). ≡C — H ··· π *versus* ≡C — H ··· Halogen interactions. The crystal structures of the 4-halogenoethynylbenzenes. *Chemical Communications*, 2403–4. [207–8, 210, 211]

Weller, F., Borgholte, H., Stenger, H., Vogler, S. and Dehnicke, K. (1989). Synthesen und Kristallstrukturen der Kronenether-Komplexe (18-Krone-6)·2CH₃CN, [Na-15-Krone-5][ReO₄]·CH₃CN und [Na-15-Krone-5]PF₆ (*in German*). *Zeitschrift für Naturforschung*, **44b**, 1524–30. [307]

Werner, A. (1902). Über Haupt- und Nebenvalenzen und die Constitution der Ammoniumverbindungen. *Liebig's Annalen der Chemie*, **322**, 261–97. [1]

Wéry, A. S. J., Gutiérrez-Zorrilla, J. M., Luque, A., Ugalde, M. and Román, P. (1996). Phase transitions in metavanadates. Polymerization of tetrakis (*tert*-butylammonium)-*cyclo*-tetrametavanadate. *Chemistry of Materials*, **8**, 408–13. [108]

Wessel, J., Lee, J. C. Jr., Peris, E., Yap, G. P. A., Fortin, J. B., Ricci, J. S. *et al.* (1995). An unconventional intermolecular three-center $N - H \cdots H_2Re$ hydrogen bond in crystalline [ReH$_5$(PPh$_3$)$_3$]•indole•C$_6$H$_6$. *Angewandte Chemie, International Edition in English*, **34**, 2507–9. [288]

West, R., Powell, D. L., Whatley, L. S., Lee, M. K. T. and Schleyer, P. v. R. (1962). The relative strengths of alkyl halides as proton acceptor groups in hydrogen bonding. *Journal of the American Chemical Society*, **84**, 3221–2. [202]

Westerhaus, W. J., Knop, O. and Falk, M. (1980). Infrared spectra of the ammonium ion in crystals. Part IX. Ammonium tetraphenylborate, NH$_4$B(C$_6$H$_5$)$_4$. Crystal structure at room temperature and at 120 K and evidence of hydrogen bonding. *Canadian Journal of Chemistry*, **58**, 1355–64. [142, 159]

Westhof, E. (ed.) (1993). *Water and biological macromolecules*. CRC Press, Boca Raton, USA. [413, 435]

Westhof, E., Dumas, P. and Moras, D. (1985). Crystallographic refinement of yeast aspartic acid transfer RNA. *Journal of Molecular Biology*, **184**, 119–45. [395, 397]

Wiberg, K. B., Waldron, R. F., Schulte, G. and Saunders, M. (1991). Lactones. 1. X-Ray crystallographic studies of nonanolactone and tridecanolactone. Nature of C—H···O nonbonded interactions. *Journal of the American Chemical Society*, **113**, 971–7. [95]

Wiechert, D., Mootz, D. and Dahlems, T. (1997). The formic acid 1D array with H bonds all reversed. Structure of a co-crystal with hydrogen fluoride. *Journal of the American Chemical Society*, **119**, 12665–6. [209, 213]

Wiewiórowska, M. D. B., Alejska, M., Sarzynska, J., Surma, K., Figlerowics, M. and Wiewiórowski, M. (1992). A new outlook on the nature of short intramolecular non-bonded contacts between 6C—H···O5' in the crystals of pyrimidine nucleosides and their salts. *Journal of Molecular Structure*, **275**, 167–81. [111]

Williams, D. E. (1974). Coulombic interactions in crystalline hydrocarbons. *Acta Crystallographica, Section A*, **30**, 71–7. [318]

Williams, M. A., Goodfellow, J. M. and Thornton, J. M. (1994). Buried waters and internal cavities in monomeric proteins. *Protein Science*, **3**, 1224–35. [430]

Winter, H., van de Gampel, J. C., de Boer, J. L., Meetsma, A. and Spek, A. L. (1987). Crystal structure of the novel compound (1α,3β,5α)-1,3-diphenyl-5-hydrido-5-isopropoxy-1λ^6,3λ^6,2,4,6,5λ^5-dithiatriazaphosphorine-1,3-dioxide. A comparison with related structures. *Phosphorus and Sulfur*, **32**, 145–51. [267, 268]

Wlodawer, A., Walter, J., Huber, R. and Sjölin, L. (1984). Structure of bovine pancreatic trypsin inhibitor. Results of joint neutron and X-ray refinement of crystal form II. *Journal of Molecular Biology*, **180**, 301–29. [364]

Woodbridge, E. L., Tso, T.-L., McGrath, M. P., Hehre, W. J. and Lee, E. K. C. (1986). Infrared spectra of matrix-isolated monomeric and dimeric hydrogen sulfide in solid O$_2$. *Journal of Chemical Physics*, **85**, 6991–4. [12]

Worth, G. A. and Wade, R. C. (1995). The aromatic (i + 2) amine interaction in peptides. *Journal of Physical Chemistry*, **99**, 17473–82. [132, 133, 370–3]

Wozniak, K., Wilson, C. C., Knight, K. S., Jones, W. and Grech, E. (1996). Neutron diffraction of a complex of 1,8-bis(dimethylamino)naphthalene with 1,2-dichloromaleic acid. *Acta Crystallographica, Section B*, **52**, 691–6. [72]

Wright, W. B. and Meyers, E. A. (1980). Tris(acetylacetonato)cobalt(III)-2-selenourea, C$_{17}$H$_{29}$CoN$_4$O$_6$Se$_2$. *Crystal Structure Communications*, **9**, 1173–80. [225, 233]

Wulf, O. R., Liddel, U. and Hendricks, S. B. (1936). The effect of *ortho* substitution on the absorption of the OH group of phenol in the infrared. *Journal of the American Chemical Society*, **58**, 2287–93. [123, 124, 125, 146, 218]

Xiao, G., Liu, S., Ji, X., Johnson, W. W., Chen, J., Parsons, J. F. *et al.* (1996). First-sphere and second-sphere electrostatic effects in the active site of a class mu glutathione transferase. *Biochemistry*, **35**, 4753–65. [382]

Xu, C., Anderson, G. K., Brammer, L., Braddock-Wilking, J. and Rath, N. P. (1996). Synthesis, molecular structures and fluxional behavior of dppm-bridged complexes of platinum(II) with linear gold(I), trigonal silver(I), or tetrahedral mercury(II) centers. *Organometallics*, **15**, 3972–9. [214, 215, 219, 220]

Xu, W., Lough, A. J. and Morris, R. H. (1997). Weak M—H···H—C interactions observed in ruthenium and iridium complexes containing hydride, amine and bulky phosphine ligands. *Canadian Journal of Chemistry*, **75**, 475–82. [288]

Yang, J., Yin, J., Abboud, K. A. and Jones, W. M. (1994). Metalloindenes of molybdenum, tungsten and ruthenium. *Organometallics*, **13**, 971–8. [201]

Yanson, I. K., Teplisky, A. B. and Sukhodub, L. F. (1979). Experimental studies of molecular interactions between nitrogen bases of nucleic acids. *Biopolymers*, **18**, 1149–70. [401]

Yap, G., Rheingold, A. L., Das, P. and Crabtree, R. H. (1995). A three-center hydrogen bond in 2,6-diphenylpyridinium tetrachloroaurate. *Inorganic Chemistry*, **34**, 3474–6. [217, 219]

Yathindra, N. and Sundaralingam, M. (1973). Correlation between the backbone and side chain conformations in 5′-nucleotides. The concept of a 'rigid' nucleotide conformation. *Biopolymers*, **12**, 297–314. [391]

Yip, B.-C., Yaw, O.-L., Ong, L.-H., Fun, H.-K. and Sivakumar, K. (1995). Coumarin 311. *Acta Crystallographica, Section C*, **51**, 2087–9. [107]

Yoshida, Z. and Osawa, E. (1966). Hydrogen bonding of phenol to π electrons of aromatics, polyolefins, heteroaromatics, fulvenes and azulenes. *Journal of the American Chemical Society*, **88**, 4019–26. [126, 199]

Yoshida, Z., Ishibe, N. and Kusumoto, H. (1969). Hydrogen bonding between phenol and the cyclopropane ring. *Journal of the American Chemical Society*, **91**, 2279–83. [126]

Yoshida, Z., Ishibe, N. and Ozoe, H. (1972). Hydrogen bonding of phenol with acetylenes and allenes. *Journal of the American Chemical Society*, **94**, 4948–52. [126]

Youngs, W. J., Kinder, J. D., Bradshaw, J. D. and Tessier, C. A. (1993). Synthesis of a nickel(0) complex of a methoxy-substituted tribenzocyclotriyne. X-Ray crystallographic evidence for an intermolecular C—H—Ni agostic interaction. *Organometallics*, **12**, 2406–7. [280]

Zabel, V., Saenger, W. and Mason, S. A. (1986). Neutron diffraction study of the hydrogen bonding in β-cyclodextrin undecahydrate at 120 K. From dynamic flip-flops to static homodromic chains. *Journal of the American Chemical Society*, **108**, 3664–73. [419]

Zaccaro, J., Beucher, M. B., Ibanez, A. and Masse, R. (1996). Crystal structure of 2-amino-5-nitropyridinium dihydrogenphosphate monophosphoric acid: Influence of the polyanion charge on the formation of centrosymmetric structure. *Journal of Solid State Chemistry*, **124**, 8–16. [336]

Zimmerman, S. C. (1997). Putting molecules behind bars. *Science*, **276**, 543–4. [315]

Zimmermann, H. E. and Zuraw, M. J. (1989). Photochemistry in a box. Photochemical reactions of molecules entrapped in crystal lattices. Mechanistic and exploratory organic photochemistry. *Journal of the American Chemical Society*, **111**, 7974–89. [189]

Zyss, J. and Ledoux, I. (1994). Nonlinear optics in multipolar media: theory and experiments. *Chemical Reviews*, **94**, 77–105. [330]

Index